Integrated Membrane Systems and Processes

Edited by

ANGELO BASILE

Institute on Membrane Technology – Italian National Research Council (ITM-CNR), University of Calabria, Cosenza, Italy

and

CATHERINE CHARCOSSET

Laboratoire d'Automatique et de Génie des Procédés, CNRS, Université Lyon 1, Villeurbanne Cedex, France

WILEY

This edition first published 2016

Registered office
John Wiley & Sons Ltd, The Atrium, Southern Gate, Chichester, West Sussex, PO19 8SQ, United Kingdom

For details of our global editorial offices, for customer services and for information about how to apply for permission to reuse the copyright material in this book please see our website at www.wiley.com.

Library of Congress Cataloging-in-Publication Data

Integrated membrane systems and processes / edited by Angelo Basile and Catherine Charcosset.
pages cm
Includes bibliographical references and index.
ISBN 978-1-118-73908-2 (cloth)
1. Membrane filters. 2. Filters and filtration. I. Basile, Angelo, editor. II. Charcosset, Catherine, editor.
TP156.F5157 2016
660′.284245–dc23 2015024794

A catalogue record for this book is available from the British Library.

ISBN: 9781118739082

Set in 10/12pt Times by Aptara Inc., New Delhi, India.

1 2016

Contents

List of Contributors

Abbas Aghaeinejad-Meybodi, Chemical Engineering Department, Sahand University of Technology, Tabriz, Iran

Angelo Basile, Institute on Membrane Technology – Italian National Research Council (ITM-CNR), University of Calabria, Rende (Cs), Italy

Merit Bodner, Institute of Chemical Engineering and Environmental Technology, Graz University of Technology, Graz, Austria

Alfredo Cassano, Institute on Membrane Technology – Italian National Research Council (ITM-CNR), University of Calabria, Cosenza, Italy

Catherine Charcosset, Laboratoire d'Automatique et de Génie des Procédés, CNRS, Université Lyon 1, Villeurbanne Cedex, France

Anthony G. Fane, Singapore Membrane Technology Centre, Nanyang Technological University, Singapore

Kamran Ghasemzadeh, Chemical Engineering Department, Urmia University of Technology, Urmia, Iran

Raja Ghosh, Department of Chemical Engineering, McMaster University, Ontario, Canada

Viktor Hacker, Institute of Chemical Engineering and Environmental Technology, Graz University of Technology, Graz, Austria

Adolfo Iulianelli, Institute on Membrane Technology – Italian National Research Council (ITM-CNR), University of Calabria, Cosenza, Italy

Simona Liguori, Institute on Membrane Technology – Italian National Research Council (ITM-CNR), University of Calabria, Cosenza, Italy

Sayed S. Madaeni, Membrane Research Center, Chemical Engineering Department, Faculty of Engineering, Razi University, Tagh Bostan, Kermanshah, Iran

Mohammadmahdi Malmali, Ralph E. Martin Department of Chemical Engineering, University of Arkansas, Fayetteville, AR, USA

Maxime Pontié, L'UNAM, Laboratory GEPEA UMR CNRS 6144, Nantes university, Nantes, France

Ehsan Salehi, Department of Chemical Engineering, Faculty of Engineering, Arak University, Arak, Iran

Alexander Schenk, Institute of Chemical Engineering and Environmental Technology, Graz University of Technology, Graz, Austria

Bart Van der Bruggen, Department of Chemical Engineering, ProcESS – Process Engineering for Sustainable Systems, KU Leuven, Belgium; Faculty of Engineering and the Built Environment, Tshwane University of Technology, Private Bag X680, Pretoria 0001, South Africa

Goran T. Vladisavljević, Chemical Engineering Department, Loughborough University, Leicestershire, UK

S. Ranil Wickramasinghe, Ralph E. Martin Department of Chemical Engineering, University of Arkansas, Fayetteville, AR, USA

Jinsong Zhang, Singapore Membrane Technology Centre, Nanyang Technological University, Singapore

Preface

Membrane science and technology have shown an impressive growth since the early 1960s with the discovery of an effective method for the preparation of asymmetric cellulose acetate membranes. Membrane technology is now recognized for a number of advantages such as operational simplicity, low energetic requirements, good stability under a wide range of operative conditions, easy control and scale-up, and large flexibility. With the increasing understanding and development of membrane techniques, it became possible to integrate various operations with the purpose to improve performance in terms of product quality, plant compactness, environmental impact, and energy use. Hybrid or integrated membrane processes can be classified into several categories. In some processes, adsorption or reaction may be included in the membrane itself, like in membrane reactors, ion-exchange membranes, adsorptive membranes, and others. Other hybrid or integrated membrane processes combine several membrane separation steps, one step being dependent on the former one, in a multistage configuration. Finally, membrane filtration may be associated with other unit operations like adsorption on activated carbon or ion-exchange resins.

This book is split into two parts. The first part covers several examples of integrated membrane systems and processes in the water, food, biotechnology, and pharmaceutical fields. Chapter 1 (Charcosset) reviews and discusses processes based on the integration of ultrafiltration, microfiltration, nanofiltration, reverse osmosis, and other membrane processes such as membrane distillation and membrane bioreactors. Examples of seawater desalination, wastewater treatment, and separation in biotechnology and food industries are also given. Chapter 2 (Ghosh) reviews developments in the area of integrated membrane process for bioseparations. By using integrated membrane processes, the number of separation steps required for purification of biological macromolecules such as proteins and nucleic acids could be significantly reduced. Most integrated membrane processes for biological applications have been developed based on microfiltration and ultrafiltration. In Chapter 3 (Cassano), the most relevant applications of integrated membrane systems in specific areas of the agro-food production, such as fruit juice, milk, and whey processing, are reviewed and discussed, highlighting technological advances and improvements over conventional methodologies. In Chapter 4 (Malmali and Wickramasinghe), it is shown that membrane-based separation processes provide the opportunity for significant process intensification through development of a continuous hydrolysis process. However, design of a continuous hydrolysis process is likely to involve the use of more than one membrane unit in series. Chapter 5 (Vladisavljević) is an overview of membrane processes used for the preparation of liquid–liquid and gas–liquid micro- and nano-dispersions, including direct and premix membrane emulsification, membrane micromixing, and membrane micro- and nano-bubbling. Integration of membrane emulsification and downstream processing for the preparation of structured microparticles is also covered. In Chapter 6 (Van der Bruggen), integration of nanofiltration with other processes is described on the basis of different levels with increasing interdependency: pretreatment prior to nanofiltration and nanofiltration as a pretreatment for further processing, as part of processes in series, in the integration of processes and hybridization, and in membrane cascades. It is shown that by integrating processes, the performance of the overall system can be substantially enhanced, far beyond the limitations of a single membrane. Chapter 7 (Pontié and Charcosset) highlights recent developments in hybrid membrane processes for desalination, such as membrane pre-treatment before reverse osmosis, and natural water treatment, such as activated carbon adsorption or oxidation associated with ultrafiltration or microfiltration. In Chapter 8 (Zhang and Fane), integrated membrane systems for industry wastewater treatment are presented, providing more sustainable solutions for energy and cost-saving, that is, smaller footprint, approaching to zero discharge, etc. The concept of integrated membrane system as well as some case studies for wastewater treatment is introduced and discussed. The second part considers various examples of integrated membrane systems and processes in inorganic applications, IGCC power plants, solar reformers, fuel cells, pervaporation systems, as well as adsorption-membrane integrated processes. Chapter 9 (Basile, Iulianelli, and Liguori) points out the relevance of membrane reactor technology. In particular, once applied for hydrogen generation, special attention is payed to the concept of a membrane reactor used for reforming processes. Furthermore, an overview of different kinds of inorganic membranes useful in membrane reactor applications is given, highlighting the role of the Pd-based membranes and their benefits and drawbacks. Chapter 10 (Ghasemzadeh, Basile, and Tilebon) reviews the application of membranes (microporous ceramic, palladium based, polymeric, and mixed ionic and electronic membranes) for hydrogen, carbon dioxide, and oxygen separation in the integrated gasification combined cycle (IGCC) power plant. Moreover, the role of membrane technology on the efficiency of IGCC is also discussed. Chapter 11 (Hacker, Bodner, and Schenk) investigates the integration of a membrane reactor with a fuel cell. Polymer electrolyte fuel cells (PEFCs) convert the chemical energy into electrical energy. The influences of impurities of fuel and air on the performance and the lifetime of PEFCs are discussed, and the advantages of the combination of a reformer unit and a membrane reactor to circumvent these lifetime limitations for a sustainable power production out of hydrocarbon fuels are shown. In Chapter 12 (Ghasemzadeh, Basile, and Aghaeinejad-Meybodi), an overview of the main issues dealing with the coupling between solar plant and membrane reactor is presented as a novel technology, in which different processes such as water electrolysis, water decomposition, and steam reforming are investigated.

Chapter 13 (Madaeni and Salehi) deals with adsorption-membrane integrated systems/processes. Integration of adsorption and membrane systems is a promising solution for the problems of individual processes. Membrane adsorption bioreactors and membrane adsorbents are the most attempted adsorption/membrane integrations with an improved separation performance.

The editors would take this opportunity to particularly thank all the authors for their excellent work and availability in preparing and reviewing the chapters of this book. Special thanks also to the various experts of Wiley: their specialistic help highly contributed to improve the final presentation of this book.

Angelo Basile
Catherine Charcosset

1

Ultrafiltration, Microfiltration, Nanofiltration and Reverse Osmosis in Integrated Membrane Processes

Catherine Charcosset
Laboratoire d'Automatique et de Génie des Procédés, CNRS, Université Lyon 1, Villeurbanne Cedex, France

1.1 Introduction

Membrane science and technology have known an impressive growth since the early 1960s when Loeb and Sourirajan discovered an effective method for the preparation of asymmetric cellulose acetate membranes with increased permeation flux without significant changes in selectivity. Pressure-driven separation techniques such as microfiltration (MF), ultrafiltration (UF), nanofiltration (NF) and reverse osmosis (RO) have then been extensively studied and developed in industries including desalination and wastewater treatment, biotechnology and pharmaceutics, chemical and food industries. Other membrane processes have been developed and found industrial applications such as gas separation and pervaporation, membrane distillation (MD), electrodialysis (ED), membrane bioreactor (MBRs), and membrane contactors. Membrane technology is usually recognized for the following advantages: operational simplicity, low energetic requirements, good stability under a wide range of operative conditions, high eco-compatibility, easy control and scale-up, large flexibility [1].

With the increasing understanding and development of membrane techniques, it became possible to integrate various membrane operations in the same process with the purpose to improve performance in terms of product quality, plant compactness, environmental impact, and energy use. The concept of integrated membrane processes appears clearly at the end

Integrated Membrane Systems and Processes, First Edition. Edited by Angelo Basile and Catherine Charcosset.

of the 1990s [1] when several applications were reported such as hybrid process NF–ED for treatment of pulp bleaching effluents [2], multistages UF, NF and RO for removal of contaminants from wastewater effluents [3] and RO–MD for seawater desalination [4]. In the following years, it became more and more obvious that other combinations could have significant impact [5], such as MBR–RO for wastewater treatment [6], pressure-driven membrane processes–MD for the treatment of wastewaters [7], and multistages pressure-driven membrane processes for high-resolution separations of biomolecules from food and biotechnology feeds [8].

In this chapter, some general backgrounds on membrane processes are first recalled including pressure-driven processes (MF, UF, NF, RO), and MD, ED and MBRs. Examples of membrane integrated processes are then given such as multistages pressure driven membrane processes and pressure-driven membrane processes associated to MD, ED or MBRs. Applications concern seawater desalination, wastewater treatment, separation in biotechnology and food industries and chemical production. These hybrid membrane techniques are further detailed in the following chapters of the book as well as other integrated membrane processes. Integrated membrane processes including gas and vapour separation and catalytic membrane reactors are considered in the second part of this book. Another important aspect of integrated membrane processes concern their association with processes other than membranes. This is also considered in the following chapters.

1.2 Membrane Processes

Various membrane operations are available for a wide range of industrial applications. Pressure-driven membrane processes include MF, UF, NF and RO. Other membrane unit operations include MD, ED and MBRs.

1.2.1 Ultrafiltration, Microfiltration and Nanofiltration

UF is a size exclusion pressure-driven separation process which came into use in the 1960s when Loeb and Sourirajan discovered the preparation of asymmetric cellulose acetate membranes [9]. UF membranes typically have pore sizes in the range of 10–1000 Å and are capable of retaining species in the molecular weight range of 300–1,000,000 Da. Operating pressures are usually in the range of 0.2–4 bar. Typical rejected species include biomolecules, polymers and colloidal particles, as well as emulsions and micelles. UF is found in a very large range of industries such as food, biotechnology and pharmaceutics, chemicals and water production.

MF is a pressure-driven separation process similar to UF with membranes typically having nominal pore sizes on the order of 0.1–1.0 μm [9]. MF applications include concentrating, purifying or separating macromolecules, colloids and suspended particles from solution. MF processing is widely used, for example, in the food industry for applications such as wine, juice and beer clarification, for wastewater treatment, and plasma separation from blood for therapeutic and commercial uses.

NF dates back to the 1970s when RO membranes with a relatively high water flux operating at relatively low pressures were developed [10, 11]. Such low-pressure RO membranes were termed NF membranes. NF is a pressure-driven membrane process, involving pressures between 5 and 20 bar, used to separate ions and molecules in the molecular weight range of 200–2000 g mol^{-1}. NF membranes have relatively high charge and are typically characterized by lower rejection of monovalent ions than that of RO membranes, but maintaining high rejection of divalent ions. Applications include pretreatment before desalination, water treatment, food industry, chemical processing industry, pulp and paper industry, metal and acid recovery, etc.

1.2.2 Reverse Osmosis

RO became commercially viable in the 1960s when Loeb and Sourirajan discovered asymmetric membranes. RO is a pressure-driven process that separated two solutions with different concentrations across a semi-permeable membrane [12]. In RO, the pressure difference Δp between the concentrated side and the dilute side is larger than a certain value that depends upon the difference of the respective concentrations and is called the osmotic pressure difference $\Delta\pi$. The direction of flow is reversed as observed in osmosis and water flows from the concentrate to the dilute side. The rate at which water crosses the membrane is then proportional to the pressure differential that exceeds $\Delta\pi$. In order to overcome the feed side osmotic pressure, fairly high feed pressure is required. In seawater desalination it commonly ranges from 55 to 70 bar. Operating pressures for the purification of brackish water are lower due to the lower osmotic pressure caused by lower feed water salinity. The most commonly used applications of RO are desalination, brackish water and wastewater treatment and concentrating food and biotechnological preparations.

1.2.3 Membrane Distillation

MD is a thermally driven membrane process in which a hydrophobic microporous membrane separates a hot and cold stream of water [13]. The hydrophobic nature of the membrane prevents the passage of liquid water through the pores while allowing the passage of water vapour (Figure 1.1). The temperature difference produces a vapour pressure gradient

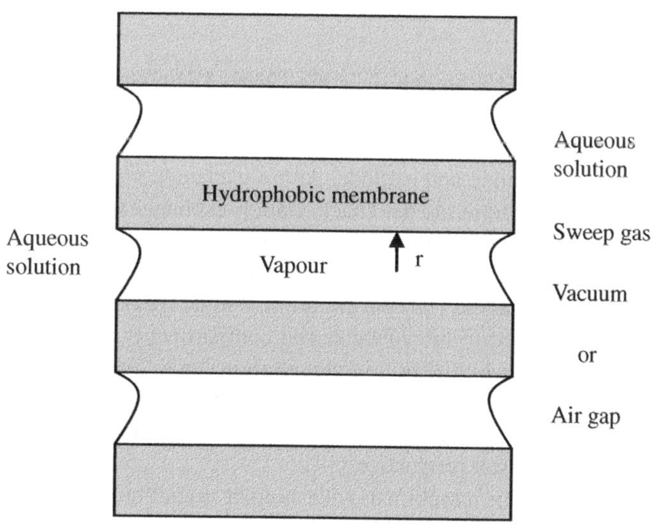

Figure 1.1 *Schematic diagram illustrating the principle of membrane distillation.*

which causes water vapour to pass through the membrane and condense on the colder surface. The result is a distillate of very high purity. MD has been developed into four different configurations, differing by the method employed to impose the vapour pressure difference across the membrane. The permeate side of the membrane may consist of a condensing fluid in direct contact with the membrane, a condensing surface separated from the membrane by an air gap, a sweeping gas, or a vacuum. MD has been applied for water desalination, waste treatment, and food processing like milk and juice concentration, biomedical applications such as water removal from blood and treatment of protein solutions [14]. In desalination by MD, the heated seawater is in direct contact with one side of the membrane. Salts and organic matter stay in the feed while pure water diffuses through the membrane.

Osmotic distillation (OD) is a variant of MD for which the driving force is a difference in concentration. OD uses the hydrophobic microporous membrane to separate two aqueous solutions having different solute concentrations: a dilute solution on one side and a hypertonic salt solution (concentrated brine stripper) on the opposite side [15]. The hydrophobic nature of the membrane prevents penetration of the pores by aqueous solutions, creating air gaps within the membrane. The water vapour pressure gradient across the membrane determines a transfer of vapour across the pores from the high vapour pressure phase to the low one. This migration of water vapour results in the concentration of the feed and dilution of the osmotic agent solution. OD can proceed at ambient temperature and is an attractive process for the concentration of solutions containing thermo-sensitive compounds such as fruit juices and pharmaceuticals.

Membrane crystallization (MCr) [16] has been proposed as an extension of MD: solutions, concentrated above their saturation limit by solvent evaporation through microporous hydrophobic membranes, reach a supersaturated state in which crystals nucleate and grow. The crystallizing solution flows along the membrane fibres. The driving force of the process is a vapour pressure gradient between both sides of the membrane which may be activated by heating the feed solution. MCr is mainly applied at laboratory scale for the formation of crystals with well-controlled properties and the treatment of brine disposal from RO plants.

1.2.4 Electrodialysis

The general principle of ED is known since the 1940s. The process is based on the movement of charged species in an electrical field: anions move towards the anode, while cations are attracted by the cathode [17]. The movement of the ions is controlled by ion-selective membranes between the anode and cathode. Anion-exchange membranes (AEM) are permeable for anions, while cations are held back. Cation-exchange membranes (CEMs) show the opposite behaviour. The ED stack is divided into several cells by AEM and CEM in an alternating sequence (Figure 1.2). The basic unit of an ED stack consists of a pair of diluted and concentrated compartments. The concentration of ionic species is reduced in the diluted compartments and increased in the concentrated compartments. One major advantage of ED compared to RO is that a higher brine concentration can be achieved because there is no osmotic pressure limitation. Some of the more important large scale industrial applications of conventional ED include brackish water desalination, waste treatment, demineralization of food products and table salt production [17].

Conventional ED can be combined with bipolar membranes in a process termed bipolar membrane electrodialysis (BMED) [17]. Bipolar membranes are composed of

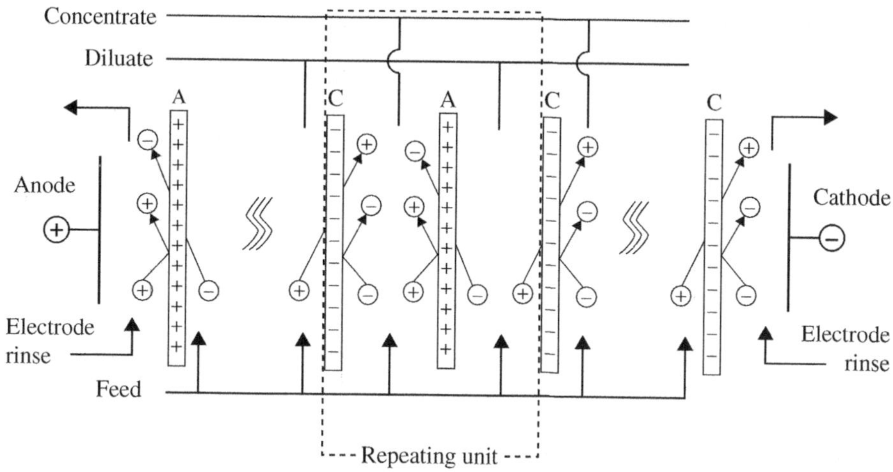

Figure 1.2 *Schematic diagram illustrating the principle of electrodialysis.* Source: *Reproduced from Reference 17 with permission from Elsevier.*

cation- and anion-exchange layers with a 4–5 nm thick transition layer arranged between two electrodes; they are installed in alternating series in an electrodialysis stack. Commercial plants of BMED are utilized to produce acids and bases from the corresponding salts.

1.2.5 Membrane Bioreactors

MBRs have been studied from the 1980s as alternative approaches to classical methods of immobilizing microorganisms, such as enzymes, antibodies and activated sludge. The microorganisms are suspended in solution and compartmentalized by a membrane in a reactor or immobilized within the membrane matrix itself. In the first method, the system consists of a traditional stirred tank reactor combined with a membrane separation unit, such as UF and MF. In the second method, the membrane acts both as a support for the microorganisms and as a separation unit.

Today, membrane bioreactor systems are applied at industrial scale for water treatment such as industrial wastewater, domestic wastewater and specific municipal wastewater [18, 19]. Conventional treatment of wastewater usually consists of a three-stage process: sedimentation of solids in the feed water followed by aerobic degradation of the organic matter using activated sludge and then a second sedimentation process to remove the biomass. An MBR can displace the two physical separation processes by filtering the biomass through an MF or UF membrane. MBRs present several advantages compared to activated sludge plants including their compactness (up to five times more compact than conventional plants), reduced sludge production, and higher product water quality [20]. The two main MBR configurations are immersed and external configurations which are characterized by different operating conditions (membrane material, filtration mode, shear stress, etc.). Membranes are usually flat sheet or hollow fibres (immersed configuration) or multitube (external configuration).

1.3 Combination of Various Membrane Processes

Membrane processes can be associated in the overall purification or production schemes in industrial applications. They often include pressure-driven membrane processes as pre-treatment before other purification steps or as a final treatment. Pressure-driven membrane processes may be associated with other processes such as MD, MBRs and ED. Some examples in the field of wastewater treatment, desalination, food industry and biotechnology and pharmaceutics are given in section 1.3.1.

1.3.1 Pressure-Driven Separation Processes

Due to their different retention properties, pressure-driven membrane processes can be associated in a cascade configuration in a purification scheme. At a first step, setting, flocculation, sand or cartridge filtrations serve to remove very large particles. Then, MF can be used to remove large compounds, such as suspended particles, colloidal materials and bacteria. The obtained suspension can be treated by UF to separate or remove macromolecules, colloids, solutes in the molecular weight range of 300–1,000,000 Da. NF or RO can then be used to remove very small molecules and salts. In a general way, it can be seen that different membrane processes in a cascade are closely linked. The membrane at step n reduces fouling of the membrane at step $n + 1$, thus increasing the permeation flux and reducing the flux decay. At step $n + 1$, the membrane performance depends on steps $<n$. As these integrated processes are used to treat industrial suspensions such as wastewater effluent and food and biological suspensions, it can be foreseen that their behaviour will be very difficult to predict.

1.3.1.1 Pre-treatment in Seawater Desalination

The applicability of membranes in seawater desalination has increased in the last 30 years with more than 2000 RO installations operating worldwide. Many desalination plants have been changed to use membrane processes because they are much more energetically efficient than thermal techniques. In seawater RO desalination, pre-treatment serves to reduce fouling potential, increase membrane life, maintain performance level and minimize scaling on the RO membrane surface [12]. Most RO plants used chemical and physical pre-treatment without membrane technologies. Physical pre-treatment generally uses flocculation, settling, sand filtration and cartridge filtration to obtain feed water with a low silt density index (SDI), where SDI describes the fouling potential of the feed water and is determined in filtration tests with MF membranes. Chemical pre-treatment includes chlorination to disinfect the water and prevent biological growth, coagulation and flocculation agents, pH adjustment, antiscaling agents to reduce precipitation of salts on the RO membrane surface and dechlorination prior to the RO stage to avoid damage of the membrane by oxidation.

The interest in membrane pre-treatment prior to RO has been recognized for many years [21] but has been limited by high cost compared to conventional pre-treatment. With advances in membrane technology and increasing requirements on water quality, the use of membrane pre-treatment prior to RO is now a suitable alternative to conventional pre-treatment [12]. It is generally estimated that membrane pre-treatment will rapidly grow in the coming years. Both UF and MF membranes, used as pre-treatment units, are able to remove suspended particles, colloidal materials, bacteria, virus and pathogenic microorganisms from raw water. They guarantee an SDI of the RO feed water generally below 2.5 even with strong fluctuation of raw water quality, enabling operation with a high and stable permeate flux even in long-term operation. They usually require less chemical addition than conventional pre-treatment, which is characterized by a rather high consumption of chemicals. In addition, membrane systems require significantly less space than conventional pre-treatment.

Membrane pre-treatment prior to RO has been successfully applied at laboratory scale and pilot scale. At laboratory scale, Kumar et al. [22] compared MF and UF membranes using dead-end filtration before RO experiments. Natural seawater was first filtered through a 1 μm prefilter. MF pretreatment was found more effective in reducing RO fouling than conventional filtration. UF pretreatment with the 100 kDa membrane did not decrease fouling compared to the MF membrane. The 20 kDa UF membranes was the most effective in reducing fouling but operated at higher pressures for the same flux as the MF and 100 kDa UF membranes. At three different locations, Vial et al. [23] operated a long-term pilot equipped with 0.1 μm hollow fibre Microza RO membranes for pre-treatment of Mediterranean seawater. Depending on the location, the pilots operated with or without pre-treatment using ferric chloride and with or without daily sodium hypochlorite backwash. The system optimization yielded stable, reproducible permeate flow and permeate SDI below 1.8. Water quality allows RO operation at high recovery, enhancing the total system running cost.

UF was also demonstrated to provide excellent pre-treatment to RO at various desalination sites, for example in Saudi Arabia, Gulf of Mexico, the Red Sea and the Mediterranean [24], and at Qingdao Jiaozhou Bay, the Yellow Sea in China [25]. For example, Pearce et al. [24] operated UF pre treatment to RO desalination for a 6-month period at Jeddah Port, Saudi Arabia, as an alternative to its conventional pre-treatment facility, which could not meet targeted feed water quality during periods of algal bloom and storms. An average filtrate SDI of 2.2 was obtained, approximately two units better than the existing conventional pre-treatment. Xu et al. [25] utilized UF hollow fibre membranes as pretreatment prior to RO desalination at Qingdao Jiaozhou Bay, the Yellow Sea in China (Figure 1.3).

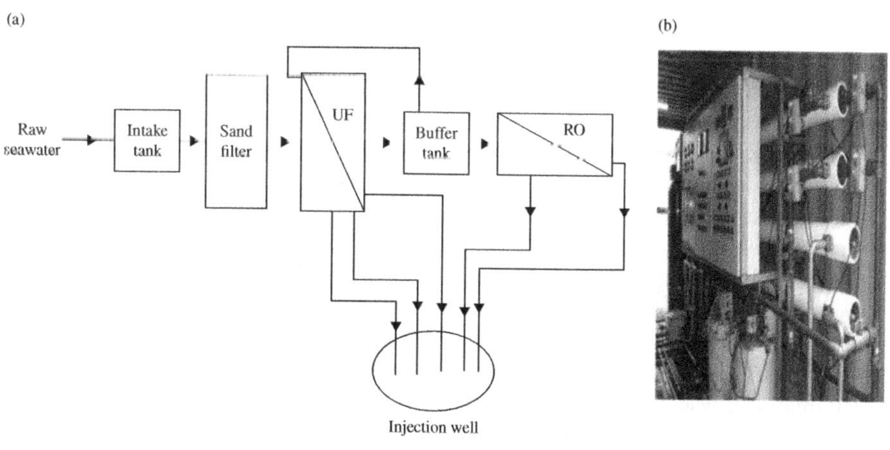

Figure 1.3 *The UF–RO pilot system for seawater desalination. (a) Schematic diagram of the pilot system. (b) Photograph of the pilot system.* Source: *Reproduced from Reference 25 with permission from Elsevier.*

During the experimental periods, the UF–RO system was run stably and successfully, without any chemicals required for disinfection, flocculation, enhanced chemical backwash and cleaning.

The use of NF upstream RO is also a possible alternative as water hardness can thus be strongly reduced, and the most part of multivalent ions can be rejected [26,27]. In addition, monovalent species are retained by 10–50%, depending on their molecular weight and electrostatic interactions with the membrane. As a consequence, the osmotic pressure of RO feed and retentate streams is decreased, thus allowing the system to operate at high water recovery factors with a considerable decrease in scale forming components. For application at industrial scale, the Saline Water Conversion Corporation (Saudi Arabia) has developed NF pretreatment prior to RO. Initial work on a pilot plant scale, resulted in its application in one of the commercial RO plants at Ummlujj in operation since September 2000. Al-Hajouri et al. [28] detailed the long-term operation of the NF–RO plant as well as different research programs which were undertaken and the results obtained.

The comparison of overall costs of membrane pre-treatment and conventional pre-treatment may depend on site-specific factors [12]. Fluctuations in the feed water quality in terms of turbidity and total dissolved solids (TDS) as well as algae bloom can cause problems for conventional pre-treatment, which might result in additional cost. Membrane pre-treatments are less sensitive to fluctuations of feed water quality and supply the RO stage with superior water quality for long-term operation. Membrane pre-treatment therefore might be able to increase RO membrane life and ensure stable operation even under adverse conditions and could thus lead to overall cost reductions.

1.3.1.2 Treatment of Waste Effluents

As industrial effluents are very complex mixtures of contaminants such as suspended solids, molecules and salts, several pressure-driven membrane processes with different properties may be needed to obtain a complete treatment. MF, UF, NF and/or RO in a cascade purification scheme can be adequate alternatives to other processes such as coagulation/flocculation, sand or cartridge filtration, adsorption, etc. The application of membrane processes in the treatment of industrial wastewaters can give a reduction of the environmental impact, a simplification of cleaning procedures of aqueous effluents, an easy re-use of sludge, a decrease of disposal costs and a saving of chemicals, water and energy [5]. These processes have been reported for treatment of several waste effluents from the leather, coke, dairy olive and oil industries.

In the leather industry, traditionally considered as one of the most polluting industries, conventional chrome tanning produces spent liquors containing significant amounts of chromium and other polluting substances, both organic and inorganic. Cassano et al. [29] recovered and concentrated chromium salts through an integrated membrane process at laboratory scale (Figure 1.4). The spent tanning liquors were subjected to a preliminary UF step to remove most suspended solids and fat substances. The permeate obtained from the UF treatment was then subjected to NF in which chromium salts were concentrated to a final value of about 10 g/L. The UF pretreatment reduced fouling of the NF membrane, thus improving the permeation flux and reducing the flux decay.

In coking industries, the treatment of desulphurization wastewater and the recovery of usable substances such as suspended sulphur (SS) and ammonium salts, for example,

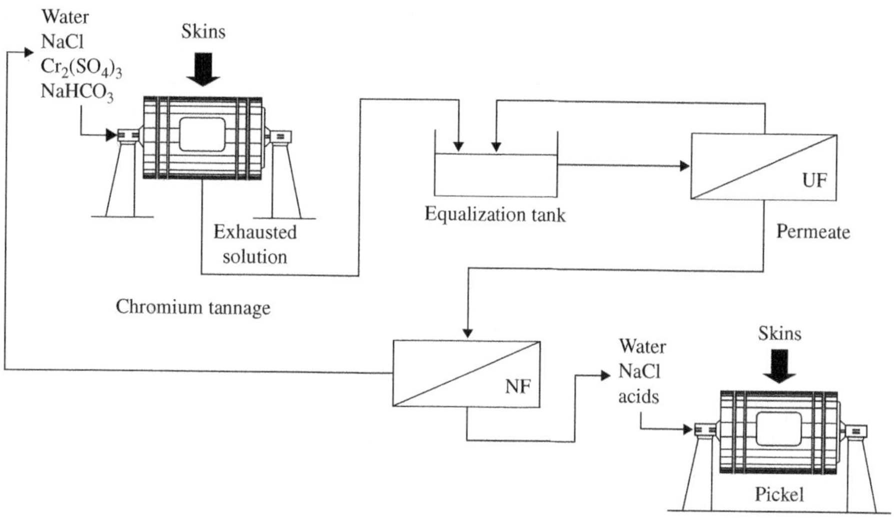

Figure 1.4 *Scheme of the UF/NF process for the recovery of chromium from spent tanning effluents.* Source: *Reproduced with permission from Reference 29. Copyright 2007 American Chemical Society.*

$(NH_4)^2S_2O_3$ and NH_4SCN, is of major concern and can avoid severe environmental problems in the case of improper disposal. Yin et al. [30] proposed an integrated membrane process consisting of UF, NF and RO to treat the wastewater. The permeate resulting from the UF treatment was then introduced to an NF process in which bivalent ammonium salts were separated from monovalent ammonium salts. The final RO process was repeatedly applied to the treatment of the NF permeate to separate monovalent salts and water. Both the NF and RO retentates were circulated in the system for further dialysis to obtain pure salt products.

The dairy industry generates a large amount of wastewater, which contains high levels of suspended solids, ammonia, protein and other nutrients. UF, NF and RO have been proposed for dairy wastewater treatment to produce purified water for water reuse or recover nutrients. A two-stage membrane process with UF + NF was investigated to produce water for discharge and recover the nutrient in wastewater [31]. The recovered nutrient could be used for feed production. UF operation could remove the protein in raw wastewater and decrease the membrane fouling in the NF process. Compared with RO operation, transmembrane pressure was lower and the membrane flux was higher than in NF operation.

A simplified integrated membrane system was also reported to be successful in the treatment of olive mill wastewaters [32]. The integrated membrane process included an initial UF step with 0.02 μm nominal pore size hollow fibre membranes for the removal of suspended solids from olive mill wastewaters. The first UF permeate was treated by a second UF process by using a flat-sheet membrane having a molecular weight cut-off of 1000 Da. Finally, the resulting permeate stream was submitted to an NF process to obtain a concentrated phenolic solution. With this integrated process, different fractions were produced: (1) a concentrated solution containing organic substances at high molecular weight (retentate of both UF processes), (2) a concentrated solution (NF retentate) enriched in polyphenolic compounds suitable for cosmetic, food and pharmaceutical industries and (3) a water stream (NF permeate) which can be reused in the olive oil extraction process as process water or in the integrated membrane system as membrane cleaning solution or in the diafiltration step to increase the yield of polyphenols in UF permeates.

1.3.1.3 Food and Biotechnology Industries

In food and biotechnology industries, feeds are also complex and multicomponent. Membrane process performances for biomolecule fractionation, concentration and purification from these feeds are then much more limited than from simple ones. These last years, a number of solutions have been proposed to achieve high-resolution separations. For example, Zydney and van Reis [33] have exploited a number of different strategies to obtain high performance tangential flow filtration (HPTFF), a complete procedure which includes: (1) proper choice of pH and ionic strength to maximize differences in the hydrodynamic volume of the product and impurity, (2) use of electrically charged membranes to enhance the retention of like charged proteins, (3) operation in the pressure-dependent regime to maximize the selectivity and (4) use of a diafiltration mode to wash impurities through the membrane. Cascade or multistage membrane systems have been proposed from the mid-2000s as an alternative to achieve these high-resolution separations. MF, UF and even NF membranes, with different or similar properties, can be considered at the different stages of the cascade [34]. In such systems, the manner in which the different flows stream within the cascade (e.g. recycling of retentates) contributes towards efficiency of separation and not merely the number of stages [35].

In the biotechnology industry, separation of biomolecules from fermentation broths is a common operation. Zhou et al. [36] investigated the use of a two-stage tangential flow filtration process for the separation of hyaluronic acid from fermentation broth. MF membranes (0.45 and 0.20 μm size) and UF membranes (300 and 100 kDa) were used to achieve the separation in series. The two-stage membrane process was undertaken with two separating schemes: the first using MF followed by UF with pure water as diafiltrate, the second similar to the first except permeate from previous UF stage as diafiltrate for MF stage. The two schemes could effectively separate and purify hyaluronic acid with above 77% overall yield and about 1000 purification factor. The second scheme seemed to be more effective for its higher overall yield (89%) and saving water. A three-stage process (MF, UF, NF) was also designed for the purification of sweeteners from *Stevia rebaudiana* Bertoni [37]. Retentions of the sweeteners for a synthetic mixture and plant extract were measured in combination with flux decline. Starting from an extract purity of 11% with the overall process, a purity of 37% and a yield of 30% could be reached. It was concluded that this process should be seen as a pre-treatment prior to other purification steps, for instance crystallization.

In the dairy industry, there is commercial interest in the production of individual whey proteins with well-characterized functional and biological properties such as α-lactalbumin (α-LA) and β-lactoglobulin (β-LG). A two-stage tangential flow filtration system has been proposed for the purification of both α-LA and β-LG from whey protein isolate [8]. Separation was achieved using 100 and 30 kDa membranes in series. Two purification strategies were examined (Figure 1.5). In Strategy I, the 100 kDa membrane was used in the first stage

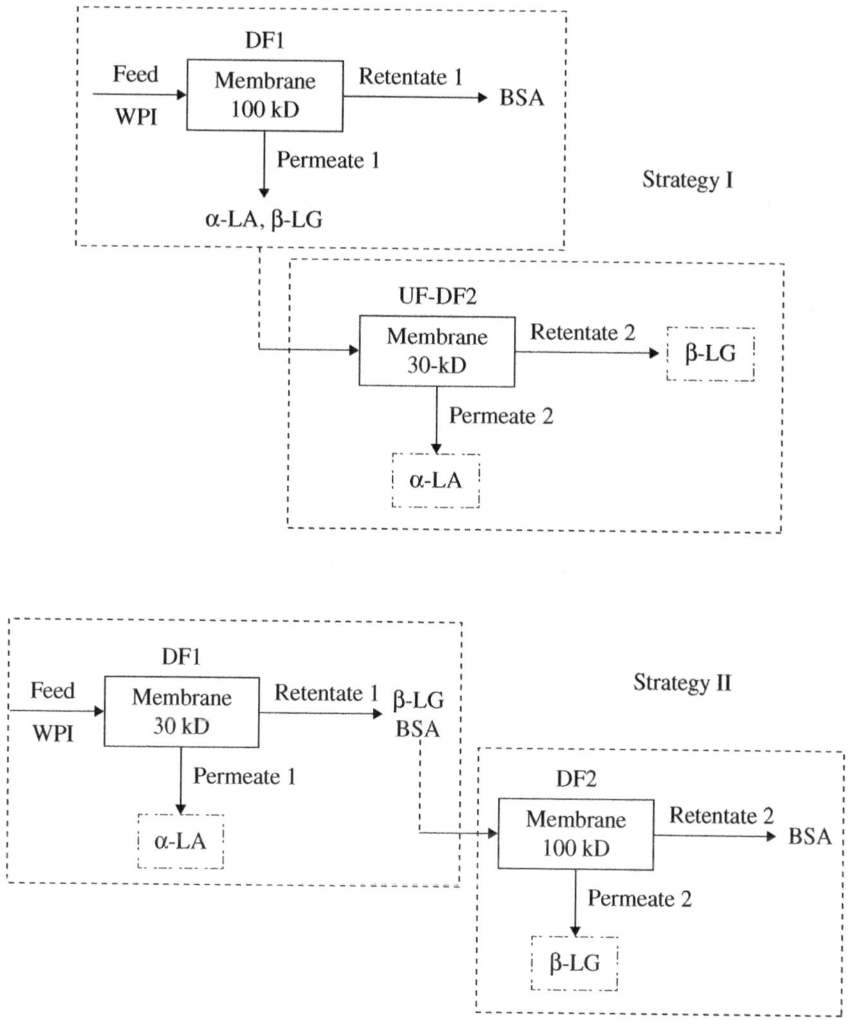

Figure 1.5 *Schematic diagram showing two separation strategies for the purification of α-LA and β-LG from whey protein isolate. Each block represents a separate diafiltration process.* Source: *Reproduced from Reference 8 with permission from Elsevier.*

to remove bovine serum albumin (BSA) while collecting α-LA and β-LG in the permeate solution. The collected permeate was then used as a feed in the second stage where the α-LA and β-LG were separated using a 30 kDa membrane. The membrane combination was reversed in Strategy II, with the 30 kDa membrane used in stage I to obtain purified α-LA in the permeate solution while retaining the β-LG and BSA. The collected retentate was then separated in stage II using a 100 kDa membrane. In order to achieve high degrees of purification and yield during the protein separation, the filtrations in stages I and II were both performed using a diafiltration process to effectively wash the more permeable protein(s) through the membrane. In both cases, the α-LA purification was greater than

10-fold at 90% yield. The recovery of β-LG was more challenging since it was obtained in the permeate from one stage and the retentate from the other. The authors concluded that staged membrane processes for high-resolution separations require further experimental and theoretical investigations to reach optimal performances [8].

1.3.2 Membrane Distillation and Pressure-Driven Membrane Processes

MD can make a significant contribution in increasing the efficiency of the RO process [4]. In RO, the osmotic effect often does not permit reaching values of interest. In MD, concentration polarization phenomena do not have the same limiting effect. Therefore, pure water can be obtained by MD from highly concentrated feeds, with which RO cannot operate. Hybrid pressure-driven membrane processes and MD have been implemented mainly in desalination units, in production of high concentrate solutions in the beverage industry and in treatment of wastewaters.

Seawater desalination by RO is limited by the osmotic pressure; therefore, a high recovery factor is not attainable. Consequently, large volumes of brine are discharged into the sea and the permeate flow rate is limited. From the end of the 1990s [4], MD has been investigated to concentrate RO brine and increase the global recovery of the process. In the following years, a specific attention was paid to optimize operating conditions such as a highly permeable membrane, high feed temperature, low permeate pressure and a turbulent fluid regime. At these operating conditions, Mericq et al. [38] obtained high permeate flow rates even at a very high salt concentration (300 g L^{-1}). At high salt concentrations, scaling occurred (mainly due to calcium precipitation) but had only a limited impact on the permeate flux (24% decrease for a permeate-specific volume of 43 L m^{-2} for the highest concentration of salt). A global recovery factor of 89% was obtained by coupling RO and MD. When MCr unit follows NF and/or RO, the highly concentrated brine does not represent waste but rather the mother liquor in which crystals may nucleate and grow. Like MD, MCr leads to a further increase of the overall water recovery factor. Before the MCr unit, Ca^{2+} ions are precipitated as carbonates by adding Na_2CO_3. This is necessary in order to avoid calcium sulphate precipitation, which causes scale and drastically limits the recovery of magnesium sulphate [26].

The possibility of integrating MD after pressure-driven membrane processes has also been reported in the beverage industry for producing various high concentrated fruit juices (orange, apple, kiwi fruit, passion fruit, etc.) [39,40]. Separating the suspended solids and pectins from juices by UF or MF decreases viscosity and increases flux of RO and/or OD, maximizing yield and minimizing nutrient and flavour losses. For example, Galaverna et al. [41] reported the production of concentrated blood orange juice according to the following scheme: an initial clarification of freshly squeezed juice by UF; the clarified juice was successively concentrated by RO, used as a pre-concentration technique (up to 25–30 °Brix), then OD, up to a final concentration of about 60 °Brix (Figure 1.6). The integrated membrane process was presented as a valuable alternative to obtain high quality concentrated juice, as the final product showed a very high antioxidant activity and a very high amount of natural bioactive components. An integrated membrane process, which involved UF, RO and OD was also reported to concentrate anthocyanin, a natural red colorant from red radish [42]. The integrated membrane process had the advantages of achieving higher concentration of anthocyanin compared to that of the individual membrane

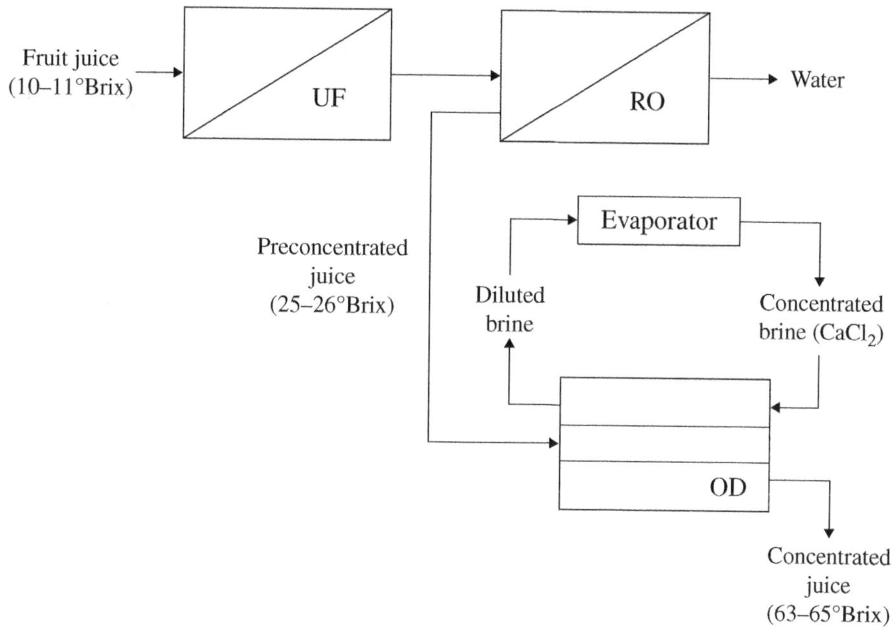

Figure 1.6 *Scheme of the integrated membrane process for the production of concentrated orange juice.* Source: *Reproduced from Reference 41 with permission from Elsevier.*

processes. Final concentration of 26 °Brix (from 1 °Brix) was achieved, with an increase in the concentration of anthocyanin from 40 to 980 mg/100 mL.

Hybrid membrane processes including MF, UF and NF and MD have also been used for treatment of wastewater such as purification of oily wastewater [7], drained wastewater [43], and textile dye bath wastewater [44]. For example, Gryta et al. [7] performed the treatment of oily wastewater collected from a harbour without pre-treatment by a combination of tubular UF and capillary MD as a final purification method. The permeate obtained from the UF process generally contains less than 5 ppm of oil. A further purification of the UF permeate by MD resulted in a complete removal of oil from wastewater and a very high reduction of the total organic carbon (TOC) (99.5%) and TDS (99.9%).

1.3.3 Electrodialysis and Pressure-Driven Membrane Processes

From the 1990s, pressure-driven membrane processes have been proposed as a pre-treatment step prior to ED for treatment of different types of wastewaters. A hybrid process involving NF–ED was designed to recover water from alkaline pulp bleaching effluent having a high content of organic and organochlorinated compounds and salts [2]. NF was selected to remove organic compounds and undertake partial desalination. A higher degree of water purity was further achieved using ED. In another application, ceramic MF membranes were reported as a pre-treatment step prior to ED for removal of colour and contaminants from paper industry wastewaters [45]. The hybrid pilot plant was found more efficient than the single ED process since the ceramic MF membrane eliminated the suspended colloids.

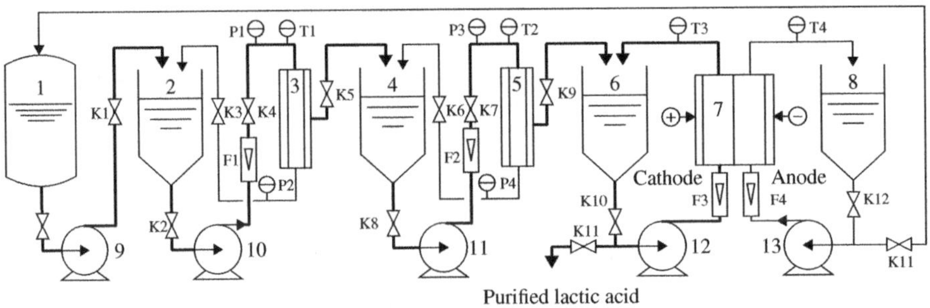

Figure 1.7 *Scheme of the integrated membrane process for the purification of lactic acid from a fermentation broth neutralized with sodium hydroxide: (1) fermentor, (2) UF feed tank, (3) ceramic membrane module, (4) NF feed tank, (5) NF membrane module, (6) acid tank, (7) BMED stack, (8) caustic tank, (9–13) pump, (P1–P4) pressure gauges, (T1–T4) temperature gauges, (F1–F4) flow meters.* Source: *Reproduced with permission from Reference 47. Copyright 2013 American Chemical Society.*

In addition, hybrid pressure-driven processes-ED is a common technique used for the conversion of organic acids such as lactic acid. Lactic acid is a common additive for flavour and preservation in a large number of industries including food and pharmaceuticals [46]. Production is based on an initial biological fermentation step. In order to increase the fermentation yield, the pH is adjusted by addition of a base. The resulting fermentation broth contains the lactic acid as a calcium, ammonium or sodium salt and several organic and inorganic fermentation residues. The largest impurities (i.e. bacterial cells, high molecular weight residues) are eliminated in a first step of clarification that can be done by MF. The fluid is then concentrated by ED. The acid salt is then converted into its free acid form by bipolar electrodialysis (BED) without addition or production of any by-product. In this process, hardness caused by calcium and magnesium cations in the fermentation broth may affect the lifetime of CEMs in the BMED stack. Therefore, Wang et al. [47] proposed a hybrid process using NF as a pre-treatment to remove these ions before BMED (Figure 1.7).

1.3.4 Membrane Bioreactors and Pressure-Driven Separation Processes

MBRs have been extensively implemented these last years for the treatment of wastewaters. With the increasing requirements for water quality, the MBR technique can be complemented by RO or NF. From the early 2000s, MBR–RO has been developed and implemented at municipal wastewater plants. The MBR process can remove above 95% organic carbon and completely remove suspended solids from wastewaters by biodegradation and membrane retention [48]. Subsequently, the RO membrane eliminates dissolved solids, organic compounds, nutrients and pathogens in MBR effluent to produce high quality reclaimed water.

For example, a landfill leachate treatment plant located at Chung-Nam Province in Korea with a capacity of 50 $m^3\ d^{-1}$ consisted of an aeration basin equipped with MBR submerged hollow fibre membranes; the effluent was further treated with spiral wound RO elements (Filmtec, USA) [6]. Another MBR–RO system with the capacity of 20 $m^3\ d^{-1}$ was reported for the treatment of domestic sewage at Bedok Water Reclamation Plant in Singapore

[49]. The MBR–RO process produced the same or higher water quality (in terms of TOC, NH_4 and NO_3) compared to the conventional activated sludge process, the ASP–MF–RO process. RO membranes in the MBR–RO process could be operated at 22 Lm^{-2} h^{-1} during 5 months, which was 30% higher than that in the ASP–UF–RO process (17 Lm^{-2} h^{-1}).

The MBR–RO technology is limited by membrane fouling, which reduces productivity and increases energy costs, in particular due to membrane fouling in RO systems [48]. The colloidal, organic and inorganic substances in MBR effluents promote organic fouling and scaling of RO membranes. In addition, biofilm development on RO membranes becomes biofouling that decreases RO performance. The soluble polysaccharides and soluble transparent exopolymer particles accumulated on the RO membrane are major factors related to RO fouling. However, the MBR–RO technology was found to reduce fouling compared to filtration–RO [50]. In addition, inline MF pretreatment of the RO feed water can significantly reduce fouling of the RO membranes [48].

1.3.5 Other Processes and Pressure-Driven Separation Processes

A large range of other combinations of membrane processes are possible. Some of them are briefly discussed in sections 1.3.5.1 to 1.3.5.3. Pressure-driven membrane processes may be associated to membrane chromatography for purification of biomolecules, membrane emulsification for beverage and dairy production and membrane extraction for detoxification of biomass hydrolysates.

1.3.5.1 Membrane chromatography and pressure-driven membrane processes

Membrane chromatography is an alternative to bead chromatography for the purification of biological products [51]. Membrane chromatography was developed from the end of the 1990s when Brandt et al. [52] reported Protein A hollow fibre devices for purification of fibronectin from blood plasma and purification of IgG. The benefit of these adsorptive membranes is to maintain high efficiencies at high flow-rates, using large biomolecules with small diffusivities, reducing biomolecule degradation and denaturation. The available interaction modes include affinity interaction, ion exchange, hydrophobic interaction, reversed-phase and multistage chromatography. Membrane chromatography is currently being employed for the purification and polishing of different biomolecular species, including purification of monoclonal antibodies and DNA, and virus capture in biotechnology and food industries.

UF or MF can be implemented before one or several chromatographic steps as a pretreatment of the feed solution. Moreover, UF can be used as a final step for polishing, desalting and buffer exchange. Yu et al. [53] reported the purification of humanized monoclonal antibody expressed in tobacco juice by ion-exchange followed by hydrophobic interaction membrane chromatography. Ion exchange was used for capture and purification, hydrophobic interaction membrane chromatography for high-resolution purification, followed by UF for polishing, desalting and buffer exchange. Using this scheme, both high monoclonal antibody purity and high recovery (77% of monoclonal antibody spiked into the tobacco extract) were achieved. In the dairy industry, membrane chromatography can also be incorporated in integrated purification schemes. Bhattacharjee et al. [54] obtained a relative separation of β-lactoglobulin (β-LG) from whey protein concentrate by fractionation of protein using two-stage UF with 30 and 10 Da membranes followed by ion-exchange membrane chromatography (Figure 1.8). Prior to UF, centrifugation, MF and a four-stage

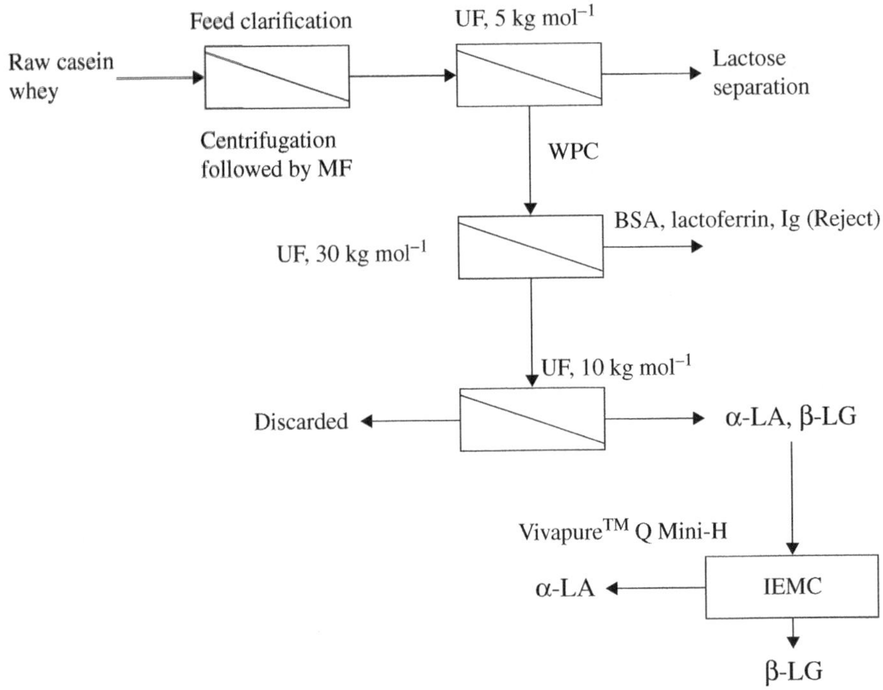

Figure 1.8 Schematic diagram for the fractionation of β-lactoglobulin from casein whey using UF and ion-exchange membrane chromatography (IEMC). Source: *Reproduced from Reference 54 with permission from Elsevier.*

discontinuous diafiltration were carried out to obtain whey protein concentrate from raw casein whey. At the end of the overall purification scheme, 87.6% purity of β-LG (on total protein basis) was obtained.

1.3.5.2 Membrane Emulsification

Membrane emulsification has been developed from the end of the 1980s as an alternative to other methods of emulsification [55]. The dispersed phase is pressed through the membrane pores, while the continuous phase flows along the membrane surface. Droplets grow at pore openings until they detach. The resulting droplet size is controlled primarily by the choice of the membrane and not by the generation of turbulent droplet break-up. Besides the possibility of using shear-sensitive ingredients, emulsions with narrow droplet size distributions can be produced. A large range of parameters, including flow configurations, membranes, surfactants, pressures, and cross-flow rates, can be varied to modify the dispersed flux and the distribution size. A very large range of colloids can be produced such as simple oil/water and water/oil emulsions to multiple emulsions of different types, solid/oil/water dispersions, coherent solids (silica particles, solid lipid microspheres, solder metal powder), and structured solids (solid lipid microcarriers, gel microbeads, polymeric microspheres, core–shell microcapsules and hollow polymeric microparticles) [56]. Although membrane emulsification has not been integrated so far to other membrane processes, it can be foreseen that such hybrid processes could be attractive in particular in the food and biotechnology industries [57]. For the preparation of beverage emulsions, membrane emulsification can be a suitable alternative to classical homogenization. The flavour products (e.g. orange oils, lemon oils, etc.) can be converted into a water-dispersible emulsion using membrane emulsification with appropriate production conditions. In the dairy industry, the preparation of emulsions with reconstituted milk is usually done by high-pressure homogenization. In such cases, the milk components (casein micelles, whey proteins, and free milk fat globule membranes in buttermilk) may be deteriorated. Membrane emulsification is therefore a suitable alternative to other emulsification processes. It could then be associated to other common membrane operations such as bacteria removal and milk globular fat fractionation using cross-flow MF for the production of drinking milk and cheese milk [58].

1.3.5.3 Membrane Extraction in Biorefinery

The first generation of biofuels, biodiesel, was obtained from plant oil and bioethanol obtained from sugarcane or corn [59]. The second generation is not based on plant products of the food chain but on lignocellulosic biomass. These last 10 years, membrane processes have known an increasing use in biorefinery applications including pervaporation for alcohol recovery and UF of canola oil, as well as new developments such as the UF/NF of lignin in a solvent-based lignocellulose conversion process or the recovery of amino acids via ED. Membrane extraction has also been reported for detoxification of biomass hydrolysates and it can be foreseen that such techniques can be included in integrated membrane processes. A typical biochemical process for biomass conversion includes a thermochemical pretreatment step to improve enzymatic cellulose hydrolysis and to release hemicellulosic sugars from the polymer matrix. Compounds that are toxic to microorganisms in subsequent fermentation steps have to be eliminated and membrane extraction is a possible alternative. For example, Grzenia et al. [60] investigated the use of membrane extraction to detoxify or remove these toxic compounds from corn stover hydrolysates pretreated using dilute sulphuric acid. The organic octanol phase is pumped outside polypropylene hollow fibres while the aqueous hydrolysate is pumped inside the fibres. The organic phase/aqueous phase interface located at the inside surface of the fibres (as the fibres are hydrophobic) is stabilized by maintaining the pressure of the aqueous phase at a value equal to or greater than the organic phase pressure. The hollow fibre membrane immobilizes the organic phase/aqueous phase interface. Octanol and oleyl alcohol were used as organic phase solvents and Alamine 336 as the aliphatic amine extractant. Reactive extraction of sulphuric, acetic, formic and levulinic acid was observed while 5-hydroxymethylfurfural and furfural were extracted due to their distribution in the organic solvent.

1.4 Conclusion

In this chapter, several examples of hybrid membrane processes have been presented, mostly pressure-driven membrane processes in a cascade configuration or associated to other membrane techniques such as MD, MBR and membrane chromatography. A number of these hybrid processes are applied at industrial plants for applications such as reducing fouling of RO for desalination, elimination of contaminants in industrial or municipal wastewater and production of biomolecules with high purities from feeds in food or biotechnology industries. A number of other hybrid systems are expected to be used in industries such as membrane emulsification in the overall production of food, beverage, cosmetic and pharmaceutical preparations, or membrane extraction in biofuels production.

The feeds to be treated are complex and multicomponent suspensions; therefore, the optimization and prediction of these processes is a real challenge. However, very few articles report theoretical modelling with experimental validation. In the following chapters, several examples of hybrid membrane processes will be presented. This will provide a possible basis to better understand and optimize these techniques.

List of Abbreviations

AEM	Anion exchange membrane
ASP	Activated sludge process
BED	Bipolar electrodialysis
BMED	Bipolar membrane electrodialysis
CEM	Cation exchange membrane
ED	Electrodialysis
HPTFF	High performance tangential flow filtration
MBR	Membrane bioreactor
MCr	Membrane crystallization
MD	Membrane distillation
MF	Microfiltration
NF	Nanofiltration
OD	Osmotic distillation
RO	Reverse osmosis
SDI	Silt density index
SS	Suspended sulphur
TDS	Total dissolved solids
TOC	Total organic carbon
UF	Ultrafiltration

References

1. Drioli, E. and Romano, M. (2001) Progress and new perspectives on integrated membrane operations for sustainable industrial growth. *Industrial and Engineering Chemistry Research*, **40**, 1277–1300.
2. Geraldes, V. and de Pinho, M.N. (1995) Process water recovery from pulp bleaching effluents by an NF/ED hybrid process. *Journal of Membrane Science*, **102**, 209–221.
3. Cassano, A., Drioli, E. and Molinari, R. (1997) Recovery and reuse of chemicals in unhairing, degreasing and chromium tanning processes by membranes. *Desalination*, **113**, 251–261.

4. Drioli, E., Laganà, F., Criscuoli, A. and Barbieri, G. (1999) Integrated membrane operations in desalination processes. *Desalination*, **122**, 141–145.
5. Drioli, E. and Fontananova, E. (2009) Integrated membrane processes. In *Membrane Operations. Innovative Separations and Transformations* (eds E. Drioli and L. Giorno), Wiley-VCH, Weinheim, pp. 265–283.
6. Ahn, W.Y., Kang, M.S., Yim, S.K. and Choi, K.H. (2002) Advanced landfill leachate treatment using an integrated membrane process. *Desalination*, **149**, 109–114.
7. Gryta, M., Karakulski, K. and Morawski, A.W. (2001) Purification of oily wastewater by hybrid UF/MD. *Water Research*, **35**, 3665–3669.
8. Cheang, B. and Zydney, A.L. (2004) A two-stage ultrafiltration process for fractionation of whey protein isolate. *Journal of Membrane Science*, **231**, 159–167.
9. Zeman, L.J. and Zydney, A.L. (1996) *Microfiltration and Ultrafiltration, Principles and Applications*, Marcel Dekker, New York.
10. Schäfer, A.I., Fane, A.G. and Waite, T.D. (eds) (2005) *Nanofiltration: Principles and Applications*, Elsevier.
11. Hilal, N., Al-Zoubi, H., Darwish, N.A., Mohammad, A.W. and Abu Arabi, M. (2004) A comprehensive review of nanofiltration membranes: Treatment, pretreatment, modelling, and atomic force microscopy. *Desalination*, **170**, 281–308.
12. Fritzmann, C., Löwenberg, J., Wintgens, T. and Melin, T. (2007) State-of-the-art of reverse osmosis desalination. *Desalination*, **216**, 1–76.
13. Lawson, K. W. and Lloyd, D.R. (1997) Membrane distillation. *Journal of Membrane Science*, **124**, 1–25.
14. El Bourawi, M.S., Ding, Z., Ma, R. and Khayet, M. (2006) A framework for better understanding membrane distillation separation process. *Journal of Membrane Science*, **285**, 4–29.
15. Hogan, P.A., Canning, R.P., Peterson, P.A., Johnson, R.A. and Michaels, A.S. (1998) A new option: Osmotic distillation. *Chemical Engineering Progress*, **94**, 49–61.
16. Curcio, E., Criscuoli, A. and Drioli, E. (2001) Membrane crystallizers. *Industrial and Engineering Chemistry Research*, **40**, 2679–2684.
17. Strathmann, H. (2010) Electrodialysis, a mature technology with a multitude of new applications. *Desalination*, **264**, 268–288.
18. Judd, S. (2007) The status of membrane bioreactor technology. *TRENDS in Biotechnology*, **26**, 109–116.
19. Wisniewski, C. (2007) Membrane bioreactor for water reuse. *Desalination*, **203**, 15–19.
20. Drioli, E. and Macedonio, F. (2012) Membrane engineering for water engineering. *Industrial and Engineering Chemistry Research*, **51**, 10051–10056.
21. Okabe, T., Hirai, M., Ohta, K., Ishii, T. and Goto, T. (1993) Preliminary experiment on seawater pretreatment with UF membranes for RO desalination. *Desalination*, **93**, 235–244.
22. Kumar, M., Adham, S. and Pearce, W. (2006) Investigation of seawater reverse osmosis fouling and its relationship to pre-treatment type. *Environmental Science and Technology*, **40**, 2037–2044.
23. Vial, D., Doussau, G. and Galindo, R. (2003) Comparison of three pilot studies using Microza® membranes for Mediterranean seawater pre-treatment. *Desalination*, **156**, 43–50.

24. Pearce, G.K., Talo, S., Chida, K., Basha, A. and Gulamhusein A. (2004) Pre-treatment options for large scale SWRO plants: Case study of UF trials at Kindasa, Saudi Arabia, and conventional pre-treatment in Spain. *Desalination*, **167**, 175–189.
25. Xu, J., Ruan, G., Chu, X., Yao, Y., Su, B. and Gao, C. (2007) A pilot study of UF pretreatment without any chemicals for SWRO desalination in China. *Desalination,* **207**, 216–226
26. Drioli, E., Curcio, E., Di Profio, G., Macedonio, F. and Criscuoli, A. (2006) Integrating membrane contactors technology and pressure-driven membrane operations for seawater desalination, energy, exergy and costs analysis. *Chemical Engineering Research and Design*, **84**, 209–220.
27. El-Zanati, E. and El-Khatib, K.M. (2007) Integrated membrane-based desalination system. *Desalination*, **205**, 15–25
28. Al-Hajouri, A.A., Al-Amoudi, A.S. and Farooque, A.M. (2013) Long term experience in the operation of nanofiltration pretreatment unit for seawater desalination at SWCC SWRO plant, *Desalination and water treatment*, **51**, 1861–1873.
29. Cassano, A., Pietra, L.D. and Drioli, E. (2007) Integrated membrane process for the recovery of chromium salts from tannery effluents. *Industrial and Engineering Chemistry Research*, **46**, 6825–6830
30. Yin, N., Liu, F., Zhong, Z. and Xing, W. (2010) Integrated membrane process for the treatment of desulfurization wastewater. *Industrial and Engineering Chemistry Research*, **49**, 3337–3341.
31. Gong, Y.-W., Zhang, H.-X. and Cheng, X.-N. (2012) Treatment of dairy wastewater by two-stage membrane operation with ultrafiltration and nanofiltration. *Water Science and Technology,* **655**, 915–919.
32. Cassano, A., Conidi, C., Giorno, L. and Drioli, E. (2013) Fractionation of olive mill wastewaters by membrane separation techniques. *Journal of Hazardous Materials*, **248–249**, 185–193.
33. Zydney, A.L. and van Reis, R. (2001) High performance tangential flow filtration. In *Membrane Separations in Biotechnology* (ed. W.K. Wang), Marcel Dekker, New York, pp. 277–298.
34. Caus, A., Braeken, L., Boussu, K. and Van der Bruggen, B. (2009) The use of integrated countercurrent nanofiltration cascades for advanced separations. *Journal of Chemical Technology and Biotechnology*, **84**, 391–398.
35. Mayani, M., Filipe, C.D.M. and Ghosh, R. (2010) Cascade ultrafiltration systems—Integrated processes for purification and concentration of lysozyme. *Journal of Membrane Science*, **347**, 150–158.
36. Zhou, H., Ni, J., Huang, W. and Zhang, J. (2006) Separation of hyaluronic acid from fermentation broth by tangential flow microfiltration and ultrafiltration. *Separation and Purification Technology,* **52**, 29–38
37. Vanneste, J., Sotto, A., Courtin, C.M., Van Craeyveld, V., Bernaerts, K., Van Impe, J., Vandeur, J., Taes, S. and Van der Bruggen, B. (2011) Application of tailor-made membranes in a multi-stage process for the purification of sweeteners from *Stevia rebaudiana. Journal of Food Engineering,* **103**, 285–293.
38. Mericq, J.-P., Laborie, S. and Cabassud, C. (2010) Vacuum membrane distillation of seawater reverse osmosis brines. *Water Research*, **44**, 5260–5273.

39. Bailey, A.F.G., Barbe, A.M., Hogan, P.A., Johnson, R.A. and Sheng, J. (2000) The effect of ultrafiltration on the subsequent concentration of grape juice by osmotic distillation. *Journal of Membrane Science*, **164**, 195–204.
40. Cassano, A., Drioli, E., Galaverna, G., Marchelli, R., Di Silvestro, G. and Cagnasso, P. (2003) Clarification and concentration of citrus and carrot juices by integrated membrane processes. *Journal of Food Engineering*, **57**, 153–163.
41. Galaverna, G. Di Silvestro, G. Cassano, A., Sforza, S., Dossena, A., Drioli, E. and Marchelli, R. (2008) A new integrated membrane process for the production of concentrated blood orange juice: Effect on bioactive compounds and antioxidant activity. *Food Chemistry*, **106**, 1021–1030.
42. Patil, G. and Raghavarao, K.S.M.S. (2007) Integrated membrane process for the concentration of anthocyanin. *Journal of Food Engineering*, **78**, 1233–1239.
43. Chunrui, W., Yue, J., Huayan, C., Xuan, W. and Xiaolong, L. (2010) Membrane distillation and novel integrated membrane process for reverse osmosis drained wastewater treatment. *Desalination and Water Treatment*, **18**, 286–291.
44. Vergili, I., Kaya, Y., Sen, U., Gönder, Z.B. and Aydiner, C. (2012) Techno-economic analysis of textile dye bath wastewater treatment by integrated membrane processes under the zero liquid discharge approach. *Resources, Conservation and Recycling*, **58**, 25–35.
45. Nataraj, S.K., Sridhar, S., Shaikha, I.N., Reddy, D.S. and Aminabhavi, T.M. (2007) Membrane-based microfiltration/electrodialysis hybrid process for the treatment of paper industry wastewater. *Separation and Purification Technology*, **57**, 185–192.
46. Bouchoux, A., Roux-de Balmann, H. and Lutin, F. (2006) Investigation of nanofiltration as a purification step for lactic acid production processes based on conventional and bipolar electrodialysis operations. *Separation and Purification Technology*, **52**, 266–273.
47. Wang, K., Li, W., Fan, Y. and Xing, W. (2013) Integrated membrane process for the purification of lactic acid from a fermentation broth neutralized with sodium hydroxide. *Industrial and Engineering Chemistry Research*, **52**, 2412–2417.
48. Wu, B., Kitade, T., Chong, T.H., Uemura, T. and Fane, A.G. (2013) Impact of membrane bioreactor operating conditions on fouling behavior of reverse osmosis membranes in MBR–RO processes. *Desalination*, **311**, 37–45.
49. Qin, J.J., Kekre, K.A., Tao, G.H., Oo, M.H., Wai, M.N., Lee, T.C., Viswanath, B. and Seah, H. (2006) New option of MBR-RO process for production of NEWater from domestic sewage. *Journal of Membrane Science*, **272**, 70–77.
50. Kent, F.C., Farahbakhsh, K., Mahendran, B., Jaklewicz, M., Liss, S.N., and Zhou, H. (2011) Water reclamation using reverse osmosis: Analysis of fouling propagation given tertiary membrane filtration and MBR pretreatments. *Journal of Membrane Science*, **382**, 328–338.
51. Charcosset, C. (2012) *Membrane Processes in Biotechnology and Pharmaceutics*, Elsevier, ISBN: 978-0-444-56334-7.
52. Brandt, S., Goffe, R.A., Kessler, S.B., O'Connor, J.L. and Zale, S.E. (1988) Membrane-based affinity technology for commercial scale purifications. *Biotechnology*, **6**, 779–782.

53. Yu, D.Q., McLean, M.D., Hall, J.C. and Ghosh, R. (2008) Purification of monoclonal antibody from tobacco extract using membrane-based bioseparation techniques. *Journal of Membrane Science*, **323** 159–166.
54. Bhattacharjee, S., Bhattacharjee, C. and Datta, S. (2006) Studies on the fractionation of β-lactoglobulin from casein whey using ultrafiltration and ion-exchange membrane chromatography. *Journal of Membrane Science*, **275**, 141–150.
55. Nakashima, T., Shimizu, M. and Kukizaki, M. (2000) Particle control of emulsion by membrane emulsification and its application. *Advanced Drug Delivery Reviews*, **45**, 47–56.
56. Vladisavljević, G.T. and Williams, R.A. (2005) Recent developments in manufacturing emulsions and particulate products using membranes. *Advances in Colloid and Interface Science*, **113**, 1–20.
57. Charcosset, C. Membrane emulsification in integrated processes for innovative food formulation. In *Integrated Membrane Operations in the Food Production* (eds. A. Cassano and E. Drioli), Verlag Walter de Gruyter & Co., Berlin, 2013.
58. Daufin, G., Escudier, J.P., Carrère, H., Berot, S., Fillaudeau, L. and Decloux, M. (2001) Recent and emerging applications of membrane processes in the food and dairy industry. *Chemical Engineering Research and Design*, **79**, 89–102.
59. Abels, C., Carstensen, F. and Wessling, M. (2013) Membrane processes in biorefinery applications. *Journal of Membrane Science*, **444**, 285–317. http://dx.doi.org/10.1016/j.memsci
60. Grzenia, D.L., Schell, D.J., and Wickramsinghe, S.R. (2010) Detoxification of biomass hydrolysates by reactive membrane extraction. *Journal of Membrane Science*, **348**, 6–12.

2

Bioseparations Using Integrated Membrane Processes

Raja Ghosh

Department of Chemical Engineering, McMaster University, Ontario, Canada

2.1 Introduction

Integrated membrane processes are those that combine multiple separation sub-processes into a single-unit operation, with one or more of these sub-processes involving membrane(s) as the means for separation. Such integrated processes are operated as regular single-step separation processes. Early research and development activity in this area of separation technology took place in the 1980s (Drioli, 1987; Daugulis, 1988; Belfort, 1989). Things have come a long way since then, both in terms of technology development and the breadth of use (Drioli et al., 2011; Charcosset, 2012; Cassano and Drioli, 2013). Integrated membrane separation is utilized for a diverse range of applications such as membrane reactors and bioreactors (Gan et al., 2002), gas separations (Bernardo et al., 2009), food and beverage processing (Cassano et al., 2011), water and waste water treatment (Glucina et al., 2000), crystallization (Zarkadas and Sirkar, 2006), desalination (Charcosset, 2009) and pharmaceutical processing (Shang et al., 2011). In this chapter, application of integrated membrane processes in the area of bioseparations is reviewed.

The increasing use of membranes in bioseparations in general is driven by several factors, key amongst these being simplicity, scalability and productivity (van Reis and Zydney, 2007). In addition, membrane-based processes are known to be benign to sensitive biological products and are generally more amenable to sterile processing of such products due to ease of material containment and minimal use of additives and reagents. The increasing availability of inexpensive, pre-sterilized and disposable membrane-based separation

Integrated Membrane Systems and Processes, First Edition. Edited by Angelo Basile and Catherine Charcosset.

devices, which eliminate cleaning and validation, is likely to make membrane separation processes more widely accepted in biopharmaceutical manufacturing. A wide range of membranes is now available, and this has opened up options with regards to membrane selection for a myriad of applications.

The purification of biological macromolecules such as proteins and nucleic acids typically involves multiple steps. Therefore, integrated membrane processes, which cut down the number of steps in a separation process, and thereby increase process efficiency and product recovery, are likely to be very attractive in the area of protein purification. The application of membranes in bioseparation has largely been focused around the purification of macromolecular biopharmaceuticals (Charcosset, 2006). Two types of membrane processes have dominated biological separations: microfiltration and ultrafiltration. Most integrated membrane processes have been developed based on these.

2.2 Integrated Bioseparation Processes Involving Microfiltration

Microfiltration, which is based on the use of membranes with micron to sub-micron-sized pores, is primarily used for retaining fine particles such as bacteria and viruses. It is widely used for sterilizing pharmaceutical products such as injectables and ophthalmic preparations, and for concentrating particulate biological products such as viral and bacterial vaccines. Proteins are too small to be directly retained or effectively fractionated by microfiltration membranes. However, as will be discussed in this chapter, researchers have examined different indirect strategies by which proteins could be selectively retained, and thereby fractionated using microfiltration membranes. In one such approach, the target protein is reversibly transformed into a particulate form by precipitation, and the particles thus formed are retained by microfiltration (Ghosh, 2004). After removal of soluble impurities, the solution condition within the membrane device is changed such that the protein particles are converted back into soluble protein molecules which pass through the membrane, and are thereby recovered in the permeate stream.

An integrated membrane process for separation of human plasma proteins human serum albumin and human immunoglobulin G was reported by the author (Ghosh, 2004). This process, which was operated in the pulse-mode, and combined *in-situ* salt-induced protein precipitation, microfiltration and membrane adsorption as basis for separation, was carried out within the same membrane filtration device. Human immunoglobulin G was retained by two mechanisms, that is, retention of its precipitated form by microfiltration and retention of its dissolved form by membrane adsorption, while human serum albumin, which did not either precipitate or bind under the operating condition, was removed in the permeate stream. Retained human immunoglobulin G was then recovered by lowering the salt concentration within the membrane device which resulted in its simultaneous desorption and dissolution. Using this technique, nearly complete separation of the two proteins could be achieved in a single step, the purity of each being in excess of 96%, and their recoveries being above 95%. The above study demonstrated the potential for application of precipitation–microfiltration-type integrated separations in the area of protein separation. Several follow-on studies have examined the use of such integrated membrane separations for protein purification.

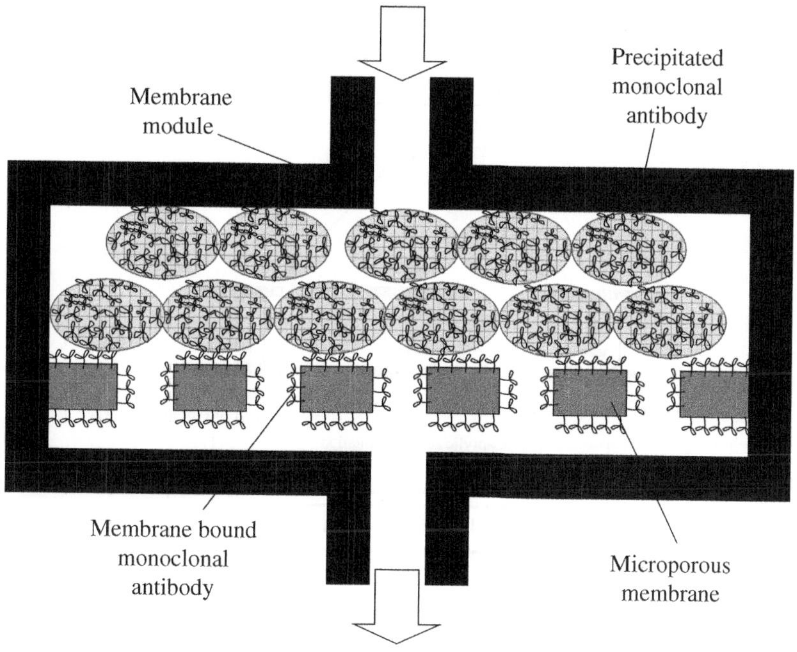

Figure 2.1 *Purification of monoclonal antibody using salt-induced precipitation-based integrated membrane separation process.* Source: *Reproduced from Wang et al. (2006) with permission from Elsevier.*

Wang et al. (2006) described a rapid and scalable integrated-precipitation–microfiltration technique for single-step purification of humanized monoclonal antibodies from mammalian cell culture supernatant. As shown in Figure 2.1, immunoglobulin G class monoclonal antibody was selectively retained within a membrane module using a combination of microfiltration of antibody precipitate and hydrophobic interaction-based membrane-adsorption antibody of dissolved antibody. The soluble impurities were first washed out of the system after which the initially retained monoclonal antibody was recovered in a highly pure form in the permeate by decreasing the salt concentration within the membrane device. In another paper, Wang et al. (2009) compared three different integrated membrane separation techniques for immunoglobulin G purification from human serum (see Figure 2.2). A combination of salt-induced precipitation, microfiltration and membrane adsorption using the set-up shown in Figure 2.3 resulted in 93% recovery of immunoglobulin G from human serum with the purity being 85%. When caprylic acid was used as the precipitating agent, in place of salt, the recovery was reduced to 55% while the purity was boosted to almost 95%. A two-step process combining salt-based integrated membrane separation and caprylic-induced acid precipitation gave the best combination of recovery and purity, that is, 73% and 93%, respectively. The factors affecting purity, recovery and membrane fouling are also discussed in this paper.

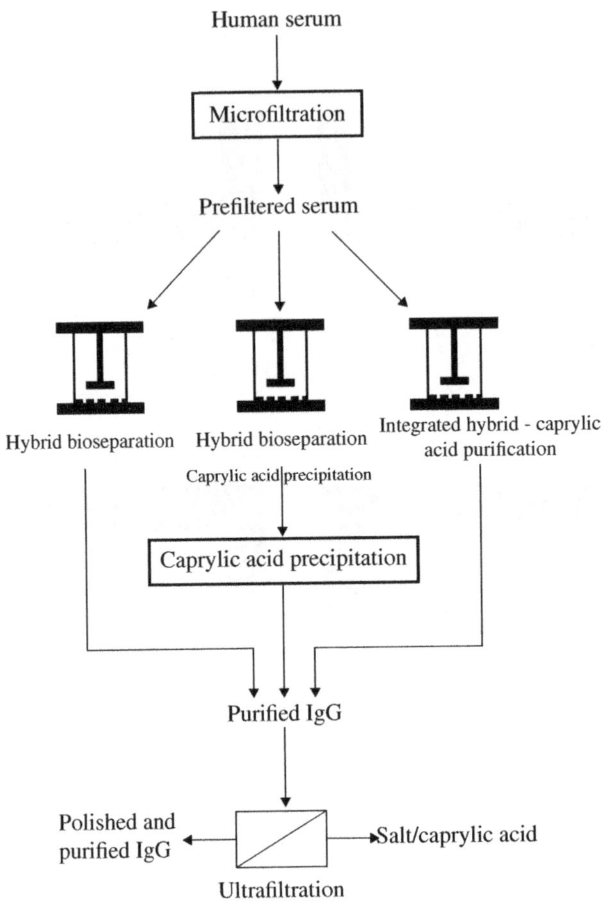

Figure 2.2 *Integrated membrane processes for purification of IgG from human serum.* Source: *Reproduced from Wang et al. (2009) with permission from Elsevier.*

Venkiteshwaran et al. (2008) reported efficient recovery of immunoglobulin G from its mixture with bovine serum albumin using an integrated process combining salt-induced precipitation and constant-permeate flux membrane microfiltration under low fouling conditions. Ammonium sulphate-induced precipitation was used to selectively precipitate the antibody. Cross-flow microfiltration was then carried out in the diafiltration mode for efficient removal of bovine serum albumin. The selectivity of immunoglobulin G/bovine serum albumin separation was about 36 while the immunoglobulin G purity was about 93%. Immunoglobulin G could be completely resolubilized without any aggregation by lowering of salt concentration. Factors such as membrane pore-size, axial Reynolds number and recirculation rate were shown to influence both the extent of membrane fouling and recovery of protein.

Wang et al. (2008) have discussed an integrated-precipitation–microfiltration technique for efficient and scalable purification of equine immunoglobulin G from horse serum. The

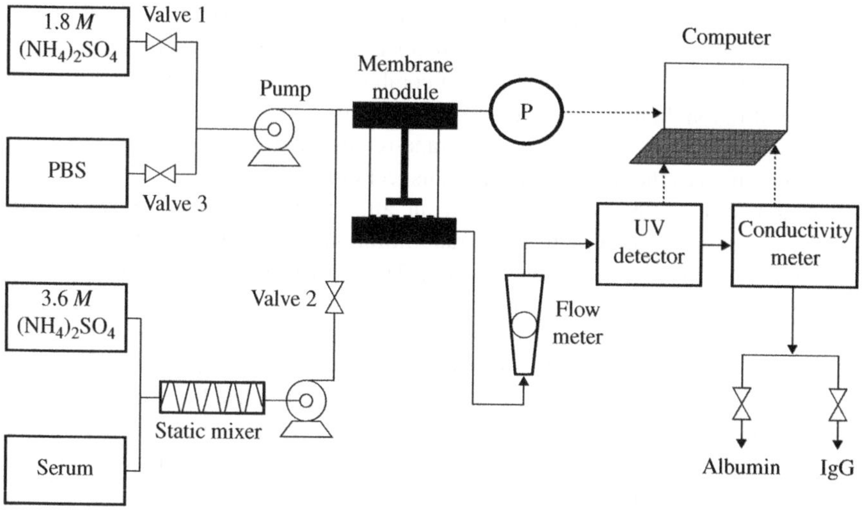

Figure 2.3 *Experimental set-up used for purification of immunoglobulin G from human serum.* Source: *Reproduced from Wang et al. (2009)) with permission from Elsevier.*

recovery of immunoglobulin G from the membrane module was enhanced by manipulating the hydrodynamic conditions within the membrane device. Also, membrane fouling was reduced by pH adjustment during the process. An ultrafiltration step was used in conjunction with the above-integrated technique to remove ammonium sulphate and low molecular weight impurities from the purified equine antibody. The purity and recovery were 88% and 90%, respectively, both being significantly higher than corresponding values reported in similar studies.

An alternative approach for protein retention by microfiltration relies on the adsorption of the target protein on adsorbent particles, large enough to be retained by the microfiltration membrane (see Figure 2.4). After impurities are removed from the system, the solution

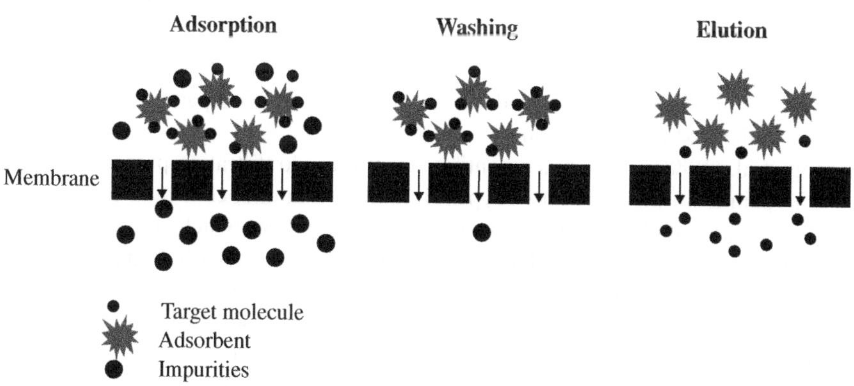

Figure 2.4 *Integrated membrane separation process based on adsorption and microfiltration.*

condition within the membrane device is changed such that desorption takes place and the target protein is recovered in the permeate stream. Ling and Mattiasson (1989) have reported a method for biomolecule purification by a combination of affinity-based adsorption and membrane filtration. In this study, alcohol dehydrogenase and lactate dehydrogenases were purified using silica nanoparticles as membrane-retained adsorbent. The low filtration flow rate was identified as the limiting factor in this technique.

Chen and Wang (1991) have reported an affinity microfiltration process, combining affinity chromatography with membrane filtration for purification of lactoferrin and immunoglobulin G from Cheddar cheese whey. The adsorbents used in this process included heparin-Sepharose, protein G-Sepharose, and protein-G-bearing streptococcal cells, while 0.2 μm microfiltration membranes were used for retailing these macro-ligands. Very high purity and recoveries were reported. Herak and Merrill (1990) have examined a variety of affinity cross-flow filtration processes for purifying biomolecules. In these studies, specific affinity ligand such as Cibacron blue, protein-A and concanavalin A tethered to agarose particles were used for targeted binding of specific proteins such as human serum albumin, immunoglobulin G and horseradish peroxidase, respectively. Agarose particles of two different size ranges, that is, 40–150 μm and 11–15 μm were used in batch affinity cross-flow filtration processes conducted in both constant-volume and volume-reduction modes and compared with model predictions.

2.3 Integrated Bioseparation Processes Involving Ultrafiltration

Ultrafiltration, which utilizes membranes with pores in the tens of nanometre size range, has been widely used for purifying proteins. The separation capability of ultrafiltration could be augmented by combining it with adsorption, in a manner similar to microfiltration as described in the latter part of the previous section. Such integrated membrane processes are generally described as affinity ultrafiltration even for non-affinity adsorbent-based separations. Affinity separations are traditionally carried out using packed columns where the predominance of diffusive mass transport in the adsorption and elution steps, along with other issues such as steric hindrance and high-pressure drop could lead to challenges in scaling-up. Some of these problems are addressed by affinity ultrafiltration, where the affinity macro-ligands are free to move around and the adsorption if therefore unhindered. The sequence of steps in affinity ultrafiltration is similar to that in affinity microfiltration (see Figure 2.4), the only difference being with regards to the type of membrane used. Affinity ultrafiltration has been used for a wide range of bioseparation applications as reviewed in the next few paragraphs.

Luong et al. (1988a) described a continuous affinity ultrafiltration process for purification of trypsin from pig pancreas. Trypsin inhibitor m-aminobenzamidine, attached to a water-soluble high molecular weight (>100,000 Da) polymer which could be retained by an ultrafiltration membrane served as the macro-ligand. While trypsin bound to the macro-ligand, impurities could pass through the membrane. Trypsin was eventually eluted using either arginine or benzamidine. The process was made continuous using a unique arrangement for recirculation of the macro-ligand as well as the eluant. Trypsin obtained by this process was pure (~97%), while its yield was about 77%. In another paper from the same research group, Male et al. (1990) discussed the purification of urokinase using an affinity polymer, formed by copolymerizing *N*-acryloyl-m-aminobenzamidine and acrylamide in the absence of oxygen. The recovered urokinase had purity and activity comparable to commercial grade product.

Powers et al. (1990) have described how affinity-ligand-modified liposomes could be used as macro-ligands in affinity ultrafiltration processes. The macro-ligand was prepared by covalently attaching p-aminobenzamidine to the surface of small unilamellar liposomes consisting of dimyristoyl phosphatidylcholine, cholesterol, and dimyristoyl phosphatidyl ethanolamine, via the hydrophilic spacer arm diglycolic acid. The feasibility of using the above for trypsin purification has been discussed in this paper. Trypsin was eluted from the macro-ligand using benzamidine, which is a trypsin inhibitor. The recovery and purity of trypsin were 68% and 98%, respectively, while the affinity-modified liposomes remained stable for 30 days.

Luong et al. (1988b) described an approach for modelling batch affinity ultrafiltration operated in the cross-flow filtration mode. The mathematical model could describe the dynamic behaviour of the process and could predict the product profile in the effluent stream with high accuracy. Ghosh et al. (1996a) described a simple mathematical model for the washing phase of an affinity ultrafiltration process, suitable for predicting the concentration of different components, including impurities. In a follow-on paper, Ghosh et al. (1996b) described a model for predicting the of concentration profile during the elution phase of an affinity ultrafiltration process. The scope of the above models has been extended in a subsequent publication from the same research group (Swami et al., 2001).

The use of affinity ultrafiltration for low molecular weight solute fractionation, particularly chiral separation, has been widely reported. Romero and Zydney (2001) used the protein bovine serum albumin as a stereospecific binding agent for the amino acid L-tryptophan. A selectivity factor of 11 was obtained in the separation of L-tryptophan and D-tryptophan in experiments carried out at low amino acid concentrations. The purity and recovery of L-tryptophan were 90% and 60%, respectively. Model calculations were found to be in good agreement with experimental data. Romero and Zydney (2002a) examined the performance of a multi-stage diafiltration process operated in the tangential flow filtration mode for the above chiral separation. A two-stage system gave a purification factor of more than 20, and 90% recovery of L-tryptophan. In another paper Romero and Zydney (2002b) examined the effects of pH and salt concentration on the separation of D- and L-tryptophan using bovine serum albumin as macro-ligand. Maximum binding of L-tryptophan to bovine serum albumin was observed in the pH 7 and 10 range and the highest purification factors were also obtained within this pH range. Further, Romero and Zydney (2002c) used purification – yield diagrams to examine the effects of affinity-ligand concentration and binding constants on the efficiency of separation, and the trade-offs between purification factor and yield. Optimum product yield and purification factor was obtained at an intermediate ligand concentration due to the competing effects of the intrinsic selectivity and the rate of impurity removal.

Shinkazh et al. (2011) have described a modified form of affinity membrane separation termed countercurrent tangential chromatography, and which is shown to be suitable for large-scale protein, which they purification. This technology, which is based on chromatographic resin flow (in the form of a slurry) through a series of static mixers and hollow fibre membrane modules overcomes some of the limitations of conventional column chromatography. It enables high-resolution protein separation at significantly reduced buffer consumption rates. A two-stage batch countercurrent tangential chromatography process for the separation of model proteins myoglobin and bovine serum albumin yielded the purified target protein at greater than 99% purity and 94% recovery. In a follow-on paper, Napadensky et al. (2013) examined the same technology for purification of an immunoglobulin G-4 sub-class monoclonal antibody from a simulated feed using protein-A as affinity resin. The antibody recovery and purity obtained were 94% and 97%, respectively, while the antibody productivity was more than 40 g protein per litre of resin per hour, which was significantly greater than can be obtained by packed-bed columns.

The efficiency of ultrafiltration-based separation processes can be increased by using cascade or multi-stage systems, frequently integrating different types of membranes (Ghosh, 2003). Cascade ultrafiltration systems are particularly attractive when multiple product streams need to be obtained from a given separation process. The use of multiple stages in a cascade system by itself is not enough to ensure good separation. The manner in which the different types of membranes are integrated within a cascade is of critical importance. Also, how the different internal flow streams within the system are directed as well as the respective flow rates are also just as important. Overall, cascade systems typically have a narrow operating window within which they operate efficiently and these processes need to be very precisely optimized.

Mohanty and Ghosh (2008) have described a three-stage, countercurrent cascade ultrafiltration configuration involving different types of tangential flow membrane (see Figure 2.5) for continuous purification of monoclonal antibody Campath-1H, directly from simulated mammalian cell-culture supernatant. Using this system, one of the major limitations of

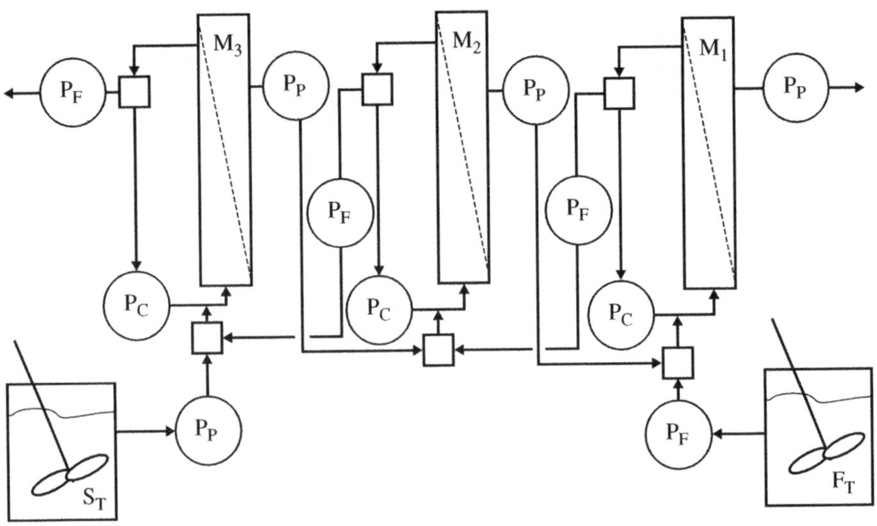

Figure 2.5 *Experimental set-up for continuous three-stage, countercurrent, cascade ultrafiltration. F_T, feed tank; M_1, stage 1 membrane module; M_2, stage 2 membrane module; M_3, stage 1 membrane module; P_C, cross flow pump; P_F, feed pump; P_P, permeate pump; S_T, sweep tank.* Source: *Reproduced from Mohanty and Ghosh (2008) with permission from Elsevier.*

conventional ultrafiltration, that is, the trade-off between recovery and purity was overcome. The system was optimized to improve both purity and recovery of the antibody. The performance of the three-stage cascade system was compared with a single-stage and a two-stage system and was found to be significantly superior, both in terms of product recovery and purification factor. The highest purity obtained using the three-stage cascade system was 85%, which was excellent considering that the cell culture supernatant was used as the feed without any pre-purification. The experimental value for purity and recovery were found to be in reasonably good agreement with results obtained by simulation.

Mayani et al. (2010) demonstrates the feasibility of using cascade ultrafiltration systems for efficient purification of the enzyme lysozyme from chicken egg white. A three-stage continuous cascade ultrafiltration system was able to generate a pure but dilute lysozyme product stream (96% purity) with the recovery being over 75%. The efficiency of separation was found to depend on process variables such as the retentate to feed ratio as well as the internal recycle rates within the cascade. A material-balance-based mathematical model was developed for the system and the experimental and simulation results were found to be in good agreement. A four-stage cascade system which focused on both purification and concentration was able to generate a continuous 97% pure lysozyme product stream which was almost twice as concentrated as that obtained from the three-stage system, with the recovery being over 71%. This four-stage system could be operated with a high degree of buffer recycling and is therefore likely to be attractive for economic large-scale protein purifications in general.

More recently, Mayani et al. (2013) have reported the purification of recombinant monoclonal antibody from tissue extract obtained from transgenic tobacco plant. This type of separation is technically challenging in several ways. To begin with, very large volumes of materials containing very low antibody titre need to be processed. Also, the antibody is present along with large quantities of native tobacco proteins and other impurities. In such a situation, the conventional approach of capturing antibody using protein-A-based packed-bed chromatography is not particularly suitable. A two-stage cascade ultrafiltration process was used to remove about 97% impurities while at the same time recover the antibody as completely as possible. Using the two-stage cascade system, 32-fold antibody enrichment was obtained. Once the impurity load was reduced using ultrafiltration, the partially purified antibody could be purified to the final pure form using chromatography. Overall, the recombinant human antibody expressed in tobacco (about 12.5 mg antibody/kg tobacco tissues) could be purified to greater than 95% purity with the overall recovery being over 50%.

2.4 Conclusion

Multiple steps are typically used for purification of biological macromolecules such as proteins. This increases the processing cost and reduces product recovery. Therefore, integrated membrane processes, which reduce the number of separation steps by combining them into single-unit operations, are likely to be very attractive in the area of protein bioseparation. The integration of membrane separation with either precipitation or adsorption has been successfully utilized in a range of bioseparation processes. Depending on the requirements of the specific application, either a microfiltration or an ultrafiltration membrane could be used. While precipitation-membrane separations have mainly been used for purifying proteins, adsorption-membrane separations have been used for both protein purification of small molecules such as amino acids. In precipitation-based processes, a microfiltration membrane serves to retain the precipitated protein particles which are several microns or bigger in size. In adsorption-based processes, both micro- and ultrafiltration membranes could be used. The efficiency of ultrafiltration-based purification processes could be significantly enhanced using cascade or multi-stage systems. Integrated membrane separation techniques, in addition to being convenient from an operational perspective, have been shown to be highly efficient, both in terms of high purity and high product recovery.

References

Belfort, G. (1989) Membranes and bioreactors: A technical challenge in biotechnology, *Biotechnology and Bioengineering*, **33**, 1047–1066.

Bernardo, P., Drioli, E., Golemme, G. (2009) Membrane gas separation: A review/state of the art, *Industrial and Engineering Chemistry Research*, **48**, 4638–4663.

Cassano, A., Conidi, C., Drioli, E. (2011) Clarification and concentration of pomegranate juice (*Punica granatum L.*) using membrane processes, *Journal of Food Engineering*, **107**, 366–373.

Cassano, A., Drioli, E. (eds.) (2013) *Integrated Membrane Operations*, Walter de Gruyter, Berlin.

Charcosset, C. (2006) Membrane processes in biotechnology: an overview, *Biotechnology Advances*, **24**, 482–492.

Charcosset, C. (2009) A review of membrane processes and renewable energies for desalination, *Desalination*, **245**, 214–231.

Charcosset, C. (2012) *Membrane Processes in Biotechnology and Pharmaceutics*, Elsevier, Oxford.

Chen, J-.P., Wang, C-.H. (1991) Microfiltration affinity purification of lactoferrin and immunoglobulin G from cheese whey, *Journal of Food Science*, **56**, 701–706.

Daugulis, A.J. (1988) Integrated reaction and product recovery in bioreactor systems, *Biotechnology Progress*, **4**, 113–122.

Drioli, E. (1987) Membrane development for seawater desalination, *Desalination*, **63**, 57–69.

Drioli, E., Stankiewicz, A.I., Macedonio, F. (2011) Membrane engineering in process intensification – An overview, *Journal of Membrane Science*, **380**, 1–8.

Gan, Q., Allen, S.J., Taylor, G. (2002) Design and operation of an integrated membrane reactor for enzymatic cellulose hydrolysis, *Biochemical Engineering Journal*, **12**, 223–229.

Ghosh, R (2003) Novel cascade ultrafiltration configuration for continuous, high-resolution protein–protein fractionation: a simulation study, *Journal of Membrane Science*, **226**, 85–99.

Ghosh, R. (2004) Separation of human albumin and IgG by a membrane-based integrated bioseparation technique involving simultaneous precipitation, microfiltration and membrane adsorption, *Journal of Membrane Science*, **237**, 109–117.

Ghosh, R., Sanyal, S.K., Mukherjea, R.N., Bhattacharya, P. (1996a) Modeling and simulation of the washing phase of an affinity ultrafiltration system, *Separation Science and Technology*, **31**, 125–131.

Ghosh, R., Sanyal, S.K., Mukherjea, R.N., Bhattacharya, P. (1996b) Modeling and simulation of the elution phase of an affinity ultrafiltration system, *Separation Science and Technology*, **31**, 679–685.

Glucina, K., Alvarez, A., Laîné, J.M. (2000) Assessment of an integrated membrane system for surface water treatment, *Desalination*, **132**, 73–82.

Herak, D.C., Merrill, E.W. (1990) Affinity cross-flow filtration: some new aspects, *Biotechnology Progress*, **6**, 33–40.

Ling, T.G.I., Mattiasson, B. (1989) Membrane filtration affinity purification (MFAP) of dehydrogenases using cibacron blue, *Biotechnology and Bioengineering*, **34**, 1321–1325.

Luong, J.H.T., Male, K.B., Nguyen, A.L. (1988a) A continuous affinity ultrafiltration process for trypsin purification, *Biotechnology and Bioengineering*, **31**, 516–520.

Luong, J. H. T., Male, K.B., Nguyen, A.L., Mulchandani, A. (1988b) Mathematical modeling of affinity ultrafiltration process, *Biotechnology and Bioengineering*, **32**, 451–459.

Male, K. B., Nguyen, A.L., Luong, J.H.T. (1990) Isolation of urokinase by affinity ultrafiltration, *Biotechnology and Bioengineering*, **35**, 87–93.

Mayani, M., Filipe, C.D.M., Ghosh, R. (2010) Cascade ultrafiltration systems—Integrated processes for purification and concentration of lysozyme, *Journal of Membrane Science*, **347**, 150–158.

Mayani, M., Filipe, C.D.M., McLean, M.D., Hall, J.C., Ghosh, R. (2013) Purification of transgenic tobacco-derived recombinant human monoclonal antibody, *Biochemical Engineering Journal*, **72**, 33–41.

Mohanty, K., Ghosh, R. (2008) Novel tangential-flow countercurrent cascade ultrafiltration configuration for continuous purification of humanized monoclonal antibody, *Journal of Membrane Science*, **307**, 117–125.

Napadensky, B., Shinkazh, O., Teella, A., Zydney, A.L. (2013) Continuous countercurrent tangential chromatography for monoclonal antibody purification, *Separation Science and Technology*, **48**, 1289–1297.

Powers, J.D., Kilpatrick, P.K., Carbonell, R.G. (1990) Trypsin purification by affinity binding to small unilamellar liposomes, *Biotechnology and Bioengineering*, **36**, 506–519.

Romero, J., Zydney, A.L. (2001) Chiral separations using ultrafiltration with a stereoselective binding agent, *Separation Science and Technology*, **36**, 1575–1594

Romero, J., Zydney, A.L. (2002a) Staging of affinity ultrafiltration processes for chiral separations, *Journal of Membrane Science*, **209**, 107–119.

Romero, J., Zydney, A.L. (2002b) pH and salt effects on chiral separations using affinity ultrafiltration, *Desalination*, **148**, 159–164.

Romero, J., Zydney, A.L. (2002c) Affinity ultrafiltration: effects of ligand binding on selectivity and process optimization, *Biotechnology and Bioengineering*, **77**, 256–265.

Shang, X., Yu, D., Ghosh, R. (2011) Integrated solid-phase synthesis and purification of PEGylated protein, *Biomacromolecules*, **12**, 2772–2779.

Shinkazh, O., Kanani, D., Barth, M., Long, M., Hussain, D., Zydney, A.L. (2011) Countercurrent tangential chromatography for large-scale protein purification, *Biotechnology and Bioengineering*, **108**, 582–591.

Swami, V., Chowdhury, R., Bhattacharya, P. (2001) Mathematical modelling of the washing phase of a semi-batch affinity ultrafiltration system for enzyme separation, *Biochemical Engineering Journal*, **9**, 41–48.

van Reis, R., Zydney, A. (2007) Bioprocess membrane technology, *Journal of Membrane Science*, **297**, 16–50.

Venkiteshwaran, A., Heider, P., Teysseyre, L., Belfort, G. (2008) Selective precipitation-assisted recovery of immunoglobulins from bovine serum using controlled-fouling cross-flow membrane microfiltration, *Biotechnology and Bioengineering*, **101**, 957–966.

Wang, L., Kanani, D.M., Ghosh, R. (2006) Purification of humanized monoclonal antibodies by membrane-based hybrid bioseparation technique, *Journal of Immunological Methods*, **314**, 1–8.

Wang, L., Sun, X., Raja Ghosh, R. (2008) Purification of equine IgG using membrane based enhanced hybrid bioseparation technique: A potential method for manufacturing hyperimmune antibody, *Biotechnology and Bioengineering*, **99**, 625–633.

Wang, L., Mah, K.Z., Ghosh, R. (2009) Purification of human IgG using membrane based hybrid bioseparation technique and its variants: A comparative study, *Separation and Purification Technology*, **66**, 242–247.

Zarkadas, D.M., Sirkar, K.K. (2006) Antisolvent crystallization in porous hollow fiber devices, *Chemical Engineering Science*, **61**, 5030–5048.

3

Integrated Membrane Processes in the Food Industry

Alfredo Cassano
Institute on Membrane Technology – Italian National Research Council (ITM-CNR), University of Calabria, Cosenza, Italy

3.1 Introduction

With a market volume of 800–850 million of Euro/year, the food industry represents the second biggest worldwide industrial market for membranes after water and wastewater treatment with a growth of around 7.5% per year [1]. Main applications concern the dairy industry (i.e. whey protein concentration, milk protein standardization, etc.) followed by beverages (wine, beer, fruit juices, etc.) and egg products [2].

Pressure-driven membrane operations such as microfiltration (MF), ultrafiltration (UF), nanofiltration (NF) and reverse osmosis (RO) are key technologies in the food industry. The basic properties of these processes make them ideal for the treatment of both food products and by-products; high selectivity, minimization of thermal damage, low energy consumption, modularity and easy scale-up, no phase change and use of chemical additives are typical advantages over conventional separation technologies. Other membrane processes, such as pervaporation (PV), electrodialysis (ED), membrane bioreactors (MBRs) and membrane contactors (MCs), have been also investigated in recent years for selected applications in the food industry.

Potentialities of membrane unit operations in different areas of the agro-food production are well known. New processes are based on the combination of different membrane unit operations or between membrane operations and conventional separation technologies (i.e. adsorption, centrifugation, evaporation). These combinations give interesting benefits that

Integrated Membrane Systems and Processes, First Edition. Edited by Angelo Basile and Catherine Charcosset.

cannot be achieved when developed as one concept and offer interesting perspectives in the revision of traditional flow sheets in food processing. In addition, the development of hybrid processes, within the logic of process intensification strategy, offers new and much more opportunities in terms of competitiveness, improvement of quality, process or product novelty and environmental friendliness [3].

This chapter gives an outlook on the most relevant applications of integrated membrane operations in specific areas of the agro-food production such as fruit juice, milk and whey processing where the combination of different membrane technologies has been largely explored on both laboratory and industrial scale. Technological advances and improvements over conventional methodologies will be analysed and discussed.

3.2 Fruit Juice Processing

3.2.1 Fruit Juice Clarification

The global market for fruit and vegetable juices reached a worth of almost $84 billion in 2010 and is forecasted to reach 72.29 billion litres by the year 2017 [4]. Demand for fruit juices continues to rise mainly due to public perception of juices as a healthy natural source of nutrients and increased public interest in health issues. Indeed, fruit juice intake has been consistently associated with reduced risk of many cancer types and stroke [5,6].

Clarification, stabilization, fractionation and concentration of fruit juices are typical operations in which membrane processes such as MF, UF, NF, RO and MCs compete with traditional operations. The introduction of these technologies in the manufacture of fruit juices represents a key point for the production of additive-free juices with high quality and natural fresh taste.

Fruit juices are colloidal mixtures of both soluble and suspended solids containing low molecular weight solutes such as sugars, organic acids, vitamins, pigments and high molecular weight solutes such as proteins, polysaccharides (pectin, cellulose, hemicelluloses, lignin and starch), microorganisms, etc. Natural pectin (in fruit juice typical concentration is up to 1.0%) and proteins are responsible for cloudiness and post bottling haze formation [7].

Juice clarification aims at removing suspended solids in order to produce a clear juice and to avoid haze formation after bottling the juice [8]. At the same time, the removal of suspended solids improves the efficiency of some post-treatments such as bitterness, tartness and acids removal with adsorbent resins, deacidification by ED, concentration by membrane technologies [9].

Conventional methods of juice clarification are based on the use of fining agents (gelatin, diatomaceous earth, silica sol, bentonite, etc.) and different techniques including centrifugation, decantation, depectinization and filtration. These processes are generally labour-intensive and time-consuming. In addition, fining agents impart a typical bitter taste to the juice and are characterized by risks of dust inhalation with consequent health problems due to handling and disposal; they cannot be reused after juice clarification creating environmental problems and significant disposal costs.

MF and UF processes provide cost-effective alternatives to traditional fining and clarification methodologies. Advantages of MF and UF over conventional fruit juice clarification methodologies are in terms of increased juice yield, possibility of operating in a single step, reduction in enzyme utilization and working times, easy cleaning and maintenance of the equipment, elimination of needs for pasteurization, no requirement of filter aids and production with a continuous simplified process [10]. Overall extraction efficiency of UF processes are of about 95–97% if compared to 90–93% of traditional processing methodologies [11].

MF and UF membranes typically separate the raw juice into a fibrous concentrated pulp (retentate) and a clarified fraction free of spoilage microorganisms (permeate). These membranes retain large species such as microorganisms, lipids, pectin or proteins and allow low molecular weight solutes (sucrose, acids, salts, aroma and flavour compounds) to permeate through the membrane together with water.

Concentration polarization and fouling phenomena represent important limitations in the performance of these processes. Concentration polarization is defined as a localized increase in the concentration of rejected solutes at the membrane surface due to the convective transport of solutes leading to a rapid permeate flux decay during early period of filtration followed by a more or less gradual flux decline towards a steady-state value. Fouling mechanisms are related to physicochemical interactions between the membrane and accumulated materials, such adsorption on the membrane pore walls and pore blocking by solute penetration. Membrane fouling manifests itself as a decay in permeate flux and an alteration of membrane selectivity [7].

The degree of membrane fouling determines the frequency of cleaning, membrane lifetime and the membrane area needed for a specific application; therefore it has a significant effect on the cost, design and operation of membrane plants.

Membrane characteristics, juice properties and operating conditions such as recirculating flow, transmembrane pressure (TMP) and temperature (T) are important factors influencing membrane fouling, and consequently, the membrane productivity in the clarification of fruit juices.

The adsorption of colloids is strongly affected by the membrane material. In addition, the surface structure and flux resistance of the fouling layer is affected by the membrane morphology. Smooth-surfaced membranes produce dense fouling layers with a greater flow resistance per unit thickness of foulant when compared with rough membranes [12].

In general, the permeate flux increases with membrane pore size or molecular weight cut-off (MWCO) while the retention of sugars, pigments and potential haze precursors (phenolics and proteins) decreases as membrane pore size is increased [13].

The influence of operating conditions on membrane fouling in cross-flow MF and UF of different fruit juices has been widely investigated in literature [14–21]. It is well known that the TMP–permeate flux curve shows a linear dependence of permeate flux with pressure at lower values of the applied TMP while at higher pressures the permeate flux approaches a limiting value independent of further increase in pressure. Higher flow rates enhance the wall shear stress on the membrane surface reducing the concentration polarization layer with a consequent improvement of permeation flux. According to the film model, permeation rates are also increased by increasing the juice temperature due to a reduction of juice viscosity and an increase of diffusion coefficients of macromolecules [14].

A decrease in permeate flux is observed when the juice concentration is increased [20].

Fouling mechanisms involved in the MF and UF of different fruit juices have been also widely investigated [22–26] and different mathematical models have been proposed in order to describe the flux decline based on equations applied to dead-end filtration mechanisms at constant pressure [27] and opportunely modified [28, 29].

The selection of MF or UF membranes for fruit juice clarification should take in consideration the viscosity of the raw juice, the retention of specific compounds, the membrane performance and, in addition, the membrane cost. Pulpy juices, with high solids content and viscosity require the use of large-bore tubular modules or plate and frame modules with large spacers. On the other hand, particle clogging can occur in the narrow flow path of hollow fibre or spiral-wound membrane modules reducing permeate flux and interfering with subsequent chemical and mechanical cleaning. Consequently, the use of thin channel devices requires additional pre-treatments of the raw juice to reduce the overall content of solids.

Commercially available membranes used in fruit juice clarification are usually made with organic polymers such as polysulphone (PS), polyethersulphone (PES), polypropylene (PP), polyamide (PA), cellulose acetate (CA), polytetrafluoroethylene (PTFE), poly(vinylidene fluoride) (PVDF). Ceramic membranes are also of interest for their greater resistance to chemical degradation and much longer life even if they are more expensive if compared with polymeric ones.

The most used configurations for the clarification of fruit juices at industrial level are tubular (inner diameter 5–10 mm), capillary (1–1.5 mm) and plate-and-frame membrane modules. The tubular configuration is associated with low packing density and high membrane replacement costs. Vibrating membrane systems able to develop high shear rate at the membrane surface are particularly appropriate for juice with high pulp content.

3.2.2 Fruit Juice Concentration

Fruit juices are marketed in three main forms: fresh juices, obtained by simple squeezing and mild pasteurization (fresh squeezed); not from concentrate (NFC) juices, obtained by freezing after squeezing and juices reconstituted from concentrate (RFC). A large part of the market is based on the latter products, as the concentration process (up to 60 °Brix of total dissolved solids) allows to reduce storage volumes (thus reducing transport and storage costs) and to improve their microbiological stability.

Fruit juices are usually concentrated by multi-stage vacuum evaporation. This process results in a severe loss of fresh juice flavours as well as a partial degradation of ascorbic acid and natural antioxidants. The process is frequently accompanied by a certain discoloration and the appearance of a typical cooked taste due to thermal effects. All these phenomena lead to a remarkable qualitative degradation of the juice [30].

Freeze concentration has long been considered a valid alternative to the use of multi-stage vacuum evaporators. In this process, water is removed as ice so that thermo-sensitive compounds are well preserved. However, some drawbacks have made the process not practical at the industrial level. Some problems are related to the production of fine ice crystals produced by rapid freezing which cannot be separated from the juice. Moreover, the achievable concentration (about 40 °Brix) is lower than the values obtained by evaporation (in average 60–65 °Brix). High investment costs and remarkable energy consumption are additional limitations [31].

Membrane concentration is a promising alternative for improving product quality in fruit juice processing. Advances and developments of membrane technology in this field have been reviewed by Jiao et al. [32].

The use of RO as a concentration technique to remove water from fruit juices has been of interest to the fruit juice industry for about 40 years. Some advantages over traditional evaporation are in terms of preservation of thermo-sensitive compounds (as the process can be performed at low temperature), simple system design, increase in aroma retention, less energy consumption and lower equipment costs [33].

RO is based on the use of dense membranes operating at a TMP higher than the osmotic pressure of the solution in order to allow the permeation of water from the side at high solute concentration to the low one. Typically, high retention of nutritional, aroma and flavour compounds can be achieved with the attainment of high-quality products. Energy savings of 1/30 to 1/10 have been estimated in comparison with single or triple effect evaporators [13].

Several research works using RO have been published over the past years for the concentration of a variety of fruit juices such as apple [34–36], citrus [37–39], blackcurrant [40], grape [41], pineapple [42], watermelon [43]. These studies pay special attention to the retention of juice constituents (especially aroma compounds and sugars) and the permeate flux which are strongly affected by the type of membranes and the operating conditions used during the process. The TMP also affects the maximum concentration that can be obtained.

In general, PA membranes exhibit greater retention of flavours and higher fluxes when compared with CA membranes [36]. Increased pressures can result in higher flux, reduced process time and higher rejections of aroma compounds. At higher temperature, both permeate fluxes and permeation rate of volatiles are increased [44, 45].

Fouling and concentration polarization phenomena, as in other pressure-driven membrane operations, are still important drawbacks of RO. However, the main disadvantage of RO in fruit juice concentration is its inability to reach the concentration of standard products due to high osmotic pressure limitations. Different studies have shown that the final concentration of fruit juices is limited by membranes and equipment to about 25–30 °Brix with the most efficient flux and solute recovery, suggesting the use of RO as preconcentration technique instead of a single-stage concentration unit.

Osmotic distillation (OD) and membrane distillation (MD) represent interesting and innovative systems to overcome osmotic pressure limits of RO operations. In these processes, the driving force for mass transfer is a vapour pressure difference across the membrane generated by either a temperature gradient (in MD) or a water activity difference (in OD). Both processes are based on the use of macroporous hydrophobic membranes acting as a barrier between two phases permitting gas/liquid mass transfer of the components without dispersion of one phase into the other. In OD, the membrane separates a dilute solution (the juice to be concentrated) from a salt solution (hypertonic solution) preventing penetration of the pores by aqueous solutions due to its hydrophobic nature. The difference in solute concentration generates a vapour pressure gradient across the membrane causing a vapour transfer across the pores from the dilute solution to the concentrated one [46, 47]. The process can be carried out at room temperature and low pressures (typical pressures encountered in OD are lower than 140 kPa) so avoiding thermal and mechanical damage of juice components.

Salt solutions (mainly $CaCl_2$, $MgSO_4$, K_2HPO_4 and NaCl) are typically used as stripping solution as the downstream side of the OD membranes. Macroporous membranes manufactured with hydrophobic polymers such as polyethylene (PE), PP, PTFE and PVDF are available with a pore size of defined ranges for OD applications. Membrane modules with a shell-and-tube configuration designed for both laboratory and industrial applications are manufactured by Membrana (Charlotte, United States) and named as Liqui-Cel Extra-Flow membrane contactors. They contain microporous PP hollow fibres with external diameter of 300 micron, a mean pore diameter of 0.03 micron and a porosity of about 40%. Total membrane surface areas range from 1.4 to 135 m^2 [48].

The removal of water by OD has been investigated for a variety of fruit juices including pineapple [49, 50], camu camu [51], passion fruit [9,52], noni [53], kiwi [54], orange [55] and apple [56].

The water flux in OD is affected by different parameters such as flow rate, temperature and concentration of both feed and brine solutions. Flow rates affect the thickness of the boundary layer at the membrane surface and, consequently, the resistance to the mass transfer. Brine and feed concentrations affect the vapour pressure gradient through the membrane which is related to the magnitude of the driving force [57].

In MD, a net water flux occurs from the warm side to the cold side due to the existing temperature gradient between two aqueous solutions separated by a macroporous hydrophobic membrane. Different configurations can be realized depending on the mechanism to obtain the required driving force. In direct contact membrane distillation (DCMD) water at lower temperature than the juice is used as a condensing fluid in the permeate side of the membrane [58]. This configuration offers some key advantages due to the possibility to operate at low temperatures (24–48°C) and to achieve high-quality concentrates with high content of soluble solids (60–70 °Brix). Some detailed studies concern the concentration of orange [59], apple [60, 61] and blackcurrant juices [62].

In vacuum membrane distillation (VMD), the membrane separates the liquid feed (the juice) from a downstream gaseous phase kept under vacuum. This configuration has been studied for the recovery of volatile aroma compounds from different juices such as blackcurrant [63] and pear [64] juices.

3.2.3 Integrated Systems in Fruit Juice Processing

Different combinations based on the use of membrane operations have been suggested for replacement of conventional juice processing unit operations. Most developed applications are related to the clarification and concentration of juices as well as to the recovery of aroma compounds.

A summary of integrated processes in fruit juice concentration including membrane operations is reported in Table 3.1.

Apple juice is one of the most common fruit juices in the world, with world production led by China, Poland, the United States and Germany. The juice is traditionally produced by maceration and pressing of apples. The resulting expelled juice may be further treated by enzymatic and centrifugal clarification to remove the starch and pectin and then pasteurized for packaging in glass, metal or aseptic processing system containers, or further treated by dehydration processes to a concentrate.

An integrated membrane process for the production of apple juice concentrate was investigated by Aguiar et al. [65]. The raw juice obtained from apples cv. Fuji was enzymatically treated with commercial enzymes (mainly cellulase and xylanase) and then clarified by MF.

Table 3.1 Integrated operations in fruit juice processing.

Juice type	Processes	Membrane types	Final concentration (°Brix)	Reference
Blackcurrant	RO, NF, DCMD, VMD	VMD, CMD: tubular PP; NF, RO: tubular PA	66	79
Apple	EMR, RO, PV, TE	EMR: tubular PVDF; RO: spiral-wound PA; PV: plate-and-frame (different materials)	70–72	67
Melon	MF, OD	MF: multichannel ceramic; OD: hollow fibre PP	55	82
Apple	MF, RO, OD	MF: polymeric membrane; RO: plate-and-frame thin film composite; OD: flat sheet PTFE	51.2	65
Kiwi	UF, OD	UF: tubular PVDF; OD: hollow fibre PP	>60	70
Black currant	UF, RO	UF: tubular PES; RO: flat sheet PA	22–25	81
Apple	UF, OD, MD, coupled MD/OD, TE	UF: PES membranes; MD, OD: hollow fibre PP	65	66
Cactus pear	UF, OD	UF: tubular PVDF; OD: hollow fibre PP	61	73
Blood orange	UF, RO, OD	UF: tubular PVDF; RO: spiral-wound PA; OD: hollow fibre PP	60	69
Citrus and carrot	UF, RO, OD	UF: tubular PVDF; RO: spiral-wound PA; OD: hollow fibre PP	60–63	76
Kiwi	UF, OD, PV	UF: tubular PVDF; OD: hollow fibre PP; PV: flat sheet PDMS	>60	71
Black currant	MF, RO, OD	MF: tubular ceramic; RO: flat sheet; OD: hollow fibre PP	63–72	80
Orange	MF, OD	UF: tubular ceramic; OD: hollow fibre PP	62	68
Bergamot	UF, OD	UF: hollow fibre PS; OD: hollow fibre PP	54	75
Acerola	MF, RO	MF: tubular PES; RO: plate-and-frame thin film composite	29.2	85
Pomegranate	UF, OD	UF: hollow fibre PEEK; OD: hollow fibre PP	52	74
Camu camu	MF, RO, OD	MF: tubular α-alumina; RO: plate-and-frame thin film composite; OD: flat sheet PTFE/PP	53	83
Acerola	MF, RO, OD	MF and RO: data not reported; OD: flat sheet PTFE/PP	55	84

MF, microfiltration; UF, ultrafiltration; RO, reverse osmosis; OD, osmotic distillation; NF, nanofiltration; DCMD, direct contact membrane distillation; VMD, vacuum membrane distillation; EMR, enzymatic membrane reactor; PS, polysulphone; PES, polyethersulphone; PP, polypropylene; PTFE, polytetrafluoroethylene; PEEK, polyetheretherketone; PVDF, polyvinylidene fluoride; PA, polyamide; PDMS, polydimethylsiloxane.

The clarified juice was preconcentrated by RO (up to 29 °Brix) and a final concentration (up to 53 °Brix) was performed by using a flat sheet PTFE membrane and 5.5*M* $CaCl_2$ as stripping solution. The evaluation of physicochemical and sensorial properties of the processed juice revealed an overall loss of phenolic compounds of about 18% and a loss of the more volatile compounds (ethanol and 1-butanol). Some less volatile compounds such 1-hexanol and butyl acetate were well preserved along the RO and OD processes.

The impact of different concentration methodologies (including thermal evaporation (TE), MD, OD and coupled operation of OD and MD) on the quality of apple juice after a preliminary clarification by fining agents (gelatine and bentonite) and UF was investigated by Onsekizoglu et al. [66]. The clarified juice with an initial content of total soluble solids (TSS) of 12 °Brix was concentrated up to 65 °Brix. The membrane-based concentration process was very efficient in preserving the nutritional and sensorial quality of the fresh juice especially regarding the retention of the bright natural colour and pleasant aroma which are considerably lost during TE. In particular, bioactive compounds such as phenolic compounds, organic acids and sugars were very stable against all the concentration processes including TE. The formation of 5-hydroxymethylfurfural (HMF), the potential indicator of Maillard reactions, was resultant in the juice concentrated by TE, while the aroma profile, valuated on the basis of the trans-2-hexenal content (the most abundant aroma compound in the apple juice), was remarkably lost. Adversely, the coupled operation of OD and MD led to a high retention of trans-2-hexenal because of the low operating temperature and short processing time.

An integrated process for producing apple juice concentrates was proposed by Álvarez et al. [67] in order to overcome some of the problems of conventional apple juice processing methods. The process involved an initial clarification step of the raw apple juice by using an enzymatic membrane reactor (EMR) followed by a preconcentration of the clarified juice by RO, the recovery of aroma compounds from the preconcentrated juice by RO and a final concentration step up to 72 °Brix by using TE. The flow sheet of the process is depicted in Figure 3.1.

The clarified juice showed a turbidity much lower than that of conventionally clarified juices (7.8 NTU) and almost all sugars, acids and phenolic compounds was retained in the preconcentrated juice. In the PV step, the enrichment factor and the overall mass transfer coefficient of aroma compounds was in the range 100–1000 and 5–500 kg/m^2, respectively, indicating that the process is well suited for aroma recovery from apple juice.

An economic evaluation of the investigated process indicated a reduction of the total capital investment of 14% and an increasing in process yield of 5% when compared with the conventional process. Total manufacturing costs decreased by 8% due to less energy requirement to concentrate the juice. Membrane replacement accounted only for 2% of operating costs and membrane life was estimated to be 2, 3 and 2 years for UF, RO and PV membranes, respectively.

Integrated membrane processes for the clarification and concentration of orange juice have been implemented on pilot and semi-industrial scales. In the process proposed by Cissé et al. [68], the single-strength juice was clarified by using a MF unit equipped with a ceramic membrane (Membralox IP19-40, 0.2 μm, 0.22 m^2) and the clarified juice was then concentrated up to 62 °Brix by using a pilot plant equipped with an OD membrane module (PP hollow fibres with average pore diameter of 0.2 μm) with a total area of 10.2 m^2. Calcium chloride was used as the stripping solution. In selected operating conditions, the

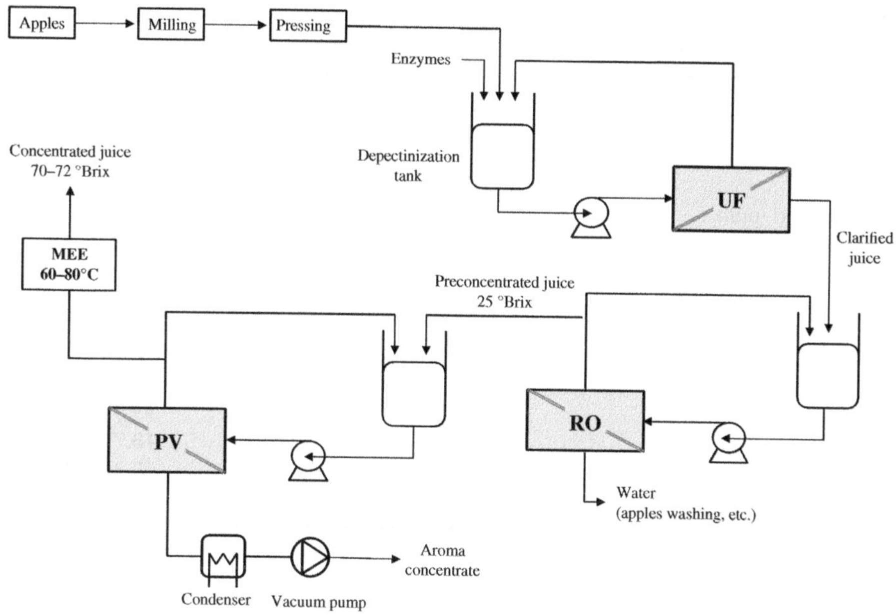

Figure 3.1 *Integrated membrane process for the production of apple juice concentrate and apple juice aroma.* Source: *Adapted from Reference 67 with permission from Elsevier.*

evaporation fluxes decreased from 0.7 to 0.67 L/m^2h when TSS reached 45 °Brix and to 0.59 L/m^2h when TSS reached 62 °Brix.

Most aroma compounds and vitamin C were recovered in the MF permeate while terpenic hydrocarbons and carotenoids were retained due to their apolar properties and association with the insoluble solids of the retentate fraction. The quality of concentrate was very similar to that of the clarified juice in terms of organic acids and sugar content.

The effect of an integrated membrane process on the preservation of bioactive antioxidant components of blood orange juice was investigated by Galaverna et al. [69]. The freshly squeezed juice was clarified by UF with PVDF tubular membranes and then preconcentrated with spiral-wound RO membrane up to 25–30 °Brix. The final concentration up to 60 °Brix was made by using an OD laboratory plant equipped with a Liqui-Cel Extra-Flow 2.5 × 8 inch MC. An integrated membrane process in which the clarified juice was directly concentrated by OD was also investigated. The results related to the quantification of antioxidant compounds and antioxidant activity in the UF–RO–OD sequence are reported in Table 3.2. In the concentrated juice, a decrease of 15% of the total antioxidant activity (TAA) was detected due to the partial degradation of vitamin C and anthocyanins. Adversely, in the juice concentrated by TE, a greater degradation of such compounds was observed with a 26% reduction of TAA. The juice concentrated by OD presented a brilliant red colour and a pleasant aroma, characteristics that were significantly lost during the thermal treatment.

An integrated membrane process for the concentration of kiwifruit juice was investigated by Cassano et al. [70]. The depectinized juice was clarified by a tubular UF membrane (Koch Series-Cor HFM 251, PVDF, 0.23 m^2) in optimized operating conditions and then concentrated by OD with the Liqui-Cel Extra-Flow 2.5 × 8 inch MC up to a TSS content higher

Table 3.2 *Analyses of antioxidant compounds in blood orange juice samples concentrated by integrated membrane process and thermal evaporation.*

Parameter	Fresh juice	UF permeate	RO retentate*	OD retentate*	TE retentate*
TSS (°Brix)	12.6	12.4	21.4	60.6	56.3
Ascorbic acid (ppm)	701.0	636.5	610.6	594.2	486.3
Cyanidin-3-glucoside (ppm)	22.7	22.2	17.9	17.2	–
Cyanidin-3-glucoside-6″-malonyl (ppm)	24.8	24.1	19.5	18.8	–
Total anthocyanins (ppm)	56.3	55.0	44.7	43.3	35.8
Sinapic acid (ppm)	6.6	6.6	6.0	5.6	3.7
Caffeic acid (ppm)	6.8	6.8	7.4	7.6	14.6
Ferulic acid (ppm)	51.3	51.7	51.1	51.0	11.8
p-Coumaric acid (ppm)	33.5	34.9	34.3	33.5	13.9
Narirutin (ppm)	50.7	50.8	50.2	48.7	38.2
Hesperidin (ppm)	45.1	45.5	46.6	45.3	35.2
TAA (mM trolox)	8.65	8.21	7.47	7.33	6.40

Source: Reproduced from Reference 69 with permission from Elsevier.
*Values referred at 12.6 °Brix except for TSS.
UF, ultrafiltration; RO, reverse osmosis; OD, osmotic distillation; TE, thermal evaporation; TSS, total soluble solids; TAA, total antioxidant activity.

than 60 °Brix. Calcium chloride dihydrate at 60% w/w was used as brine solution. The performance of the OD process is illustrated in Figure 3.2. The evaporation flux is strictly correlated to the driving force of the process represented by the water activity difference (Δa_w) between the juice and the brine (Figure 3.2a). When the original concentration of the stripping solution was restored (Figure 3.2b), an increase in the evaporation flux value was observed. The TSS content of the juice increased exponentially (Figure 3.2b) leading to an increase in the juice viscosity (Figure 3.2c). At low TSS content, the evaporation flux decay was mainly attributed to the dilution of the stripping solution; at TSS content higher than 38 °Brix, the flux decay was mainly affected by the juice viscosity and, consequently, by the juice concentration.

A little decrease of the TAA (about 4.3%) was observed in the clarified juice (UF permeate) in comparison with the original juice; in the concentrated juice at 65.8 °Brix, an 8% reduction of the TAA was detected with respect to the clarified juice. A PV step can be integrated in the proposed process in order to recover aroma compounds from the juice. PV studies performed on fresh, clarified and concentrated juices revealed higher enrichment factors for most part of aroma compounds in the fresh juice rather than in clarified and concentrated juices suggesting the use of PV before any concentration step [71]. In order to reduce concentration polarization phenomena, PV can be replaced by VMD as proposed in the combined process for must concentration by Bandini and Sarti [72].

Integrated membrane processes based on the use of UF and OD were also proposed for the clarification and concentration of cactus pear [73], pomegranate [74] bergamot [75] and carrot [76] juices. As reported in Table 3.3, concentrated juices retained antioxidant compounds of the fresh juices suggesting their potential use in food and nutraceutical applications.

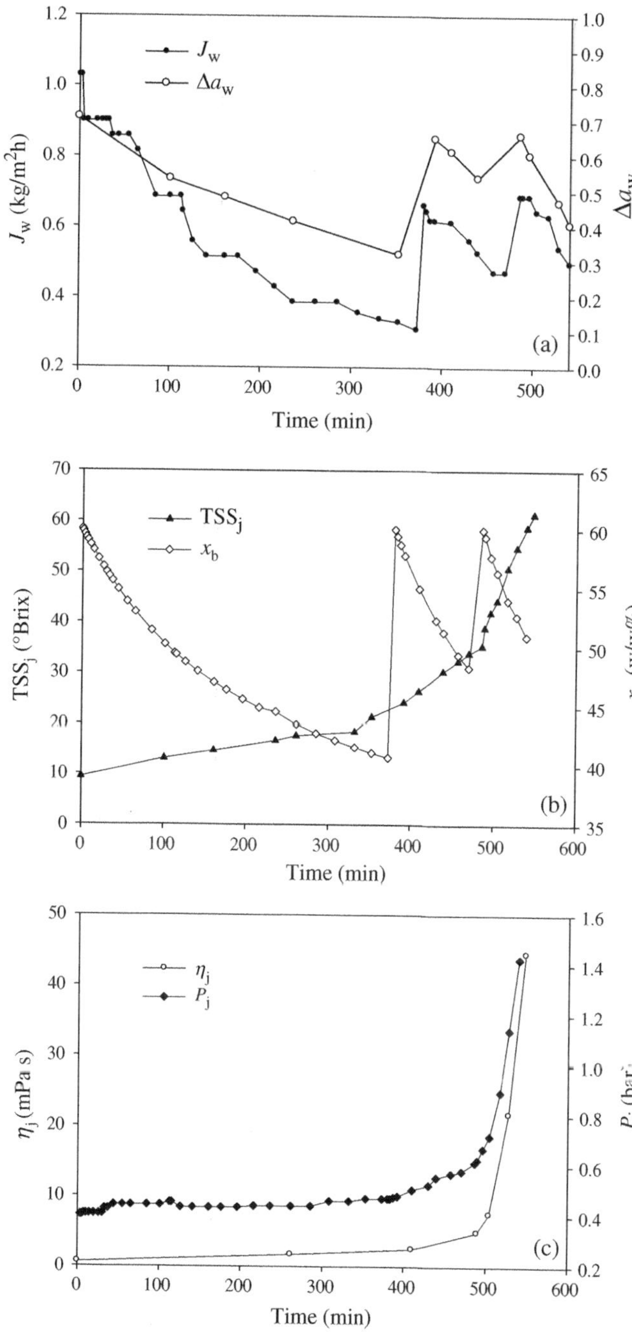

Figure 3.2 *Concentration of clarified kiwifruit juice by osmotic distillation. Time course of: (a) evaporation flux and water activity difference between juice and brine; (b) TSS and brine concentration; (c) kiwifruit juice viscosity and shell side pressure.* Source: *Adapted from Reference 70 with permission from Elsevier.*

Table 3.3 *Analytical measurements of cactus pear juice clarified and concentrated by integrated membrane process.*

Parameter	Fresh juice	UF permeate	OD retentate*
TSS (°Brix)	13.4	13.0	58.0
Ascorbic acid (mg/L)	5.0	693.0	696.0
Citric acid (mg/L)	416.0	395.0	375
Glutamic acid (g/L)	2.06	2.05	2.05
Betaxanthins (mg/L)	57.1	53.4	52.5
Betacyanins (mg/L)	19.9	12.7	12.4
TAA (mM Trolox)	5.0	4.8	4.6

Source: Reproduced from Reference 73 with permission from Elsevier.
*Values referred at 13.4 °Brix except for TSS.
UF, ultrafiltration; OD, osmotic distillation; TSS, total soluble solids; TAA, total antioxidant activity.

Pilot-plant facilities based for the concentration of fruit and vegetable juices were implemented in Mildura and Melbourne (Australia). The Melbourne facility, designed by Zenon Environmental (Burlington, Ontario), was a hybrid plant consisting of UF and RO pretreatment stages and an OD section containing two 19.2 m^2 Liqui-Cel membrane modules. Fresh fruit juices were concentrated up to 65–70 °Brix at an average throughput of 50 L/h [46]. The Mildura plant, designed by Vineland Concentrates and Celgard LLC, contained 22 Liqui-Cel membrane modules (4 × 28 inches type) for a total interfacial area of 425 m^2. It was used for the concentration of grape juices to make wine from reconstituted concentrate. The installation produced approximately 20–25 L/h of 68 °Brix concentrate [77]. A multi-step membrane process for water removal from grape juice based on the use of MD and OD after a MF clarification step was also investigated by Rektor et al. [78].

A conceptual process design for the concentration of blackcurrant juice was proposed by Sotoft et al. [79] to replace traditional TE and aroma recovery on the basis of laboratory and pilot scale tests. The process layout is depicted in Figure 3.3. The plant consists of a VMD unit for aroma recovery, a combined RO/NF unit for juice preconcentration and a DCMD unit for final concentration. The VMD unit produces a permeate stream enriched in aroma compounds and a retentate of 12.6 °Brix. The permeate stream is then forwarded to a traditional distillation unit where 66% of aroma compounds can be recovered. The combination of NF with RO in the preconcentration step allows to overcome limitations of RO due to high osmotic pressure of concentrated liquids. In particular, the retentate stream coming from the VMD unit is preconcentrated by using RO dense membranes with a retention of sugar compounds higher than 99.5%. The RO retentate (25.3 °Brix) is forwarded to the NF unit equipped with low rejection (<70% for sugars) membranes. The NF permeate is recirculated back to the RO unit while the retentate (31.5 °Brix) is submitted to a final DCMD step to produce a concentrated juice up to 65–70 °Brix.

A production of 17,283 tons of 66 °Brix concentrated juice/year was estimated assuming a specific membrane surface area for each membrane unit (810 m^2 for VMD, 630 m^2 for RO, 170 m^2 for NF and 4250 m^2 for DCMD). An operating cost of 0.40 €/kg of concentrated juice was calculated assuming a membrane lifetime of 1 year. Production costs of 0.30 and 0.27 €/kg were estimated for membrane lifetimes of 2 and 3 years, respectively. These predictions are highly competitive if compared with the production costs of conventional evaporators.

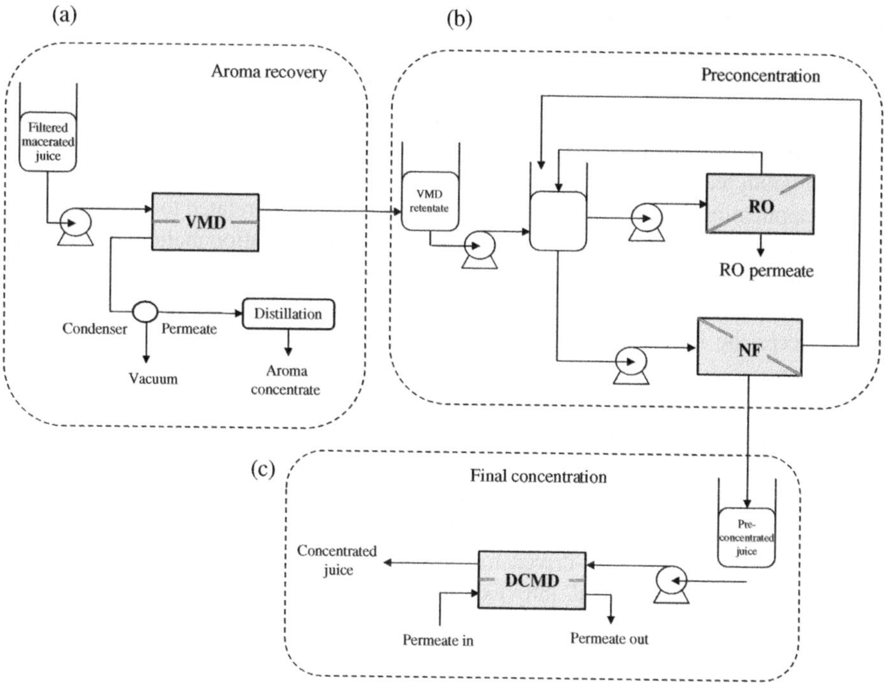

Figure 3.3 *Process layout for aroma recovery and juice concentration based on membrane processes. (a) Aroma recovery by VMD followed by distillation. (b) Preconcentration by RO combined with preconcentration with NF. (c) Final concentration by DCMD.* Source: *Reproduced from Reference 79 with permission from Elsevier.*

Concentrated blackcurrant juices up to 72 °Brix were also produced on laboratory and large scale membrane processes based on the use of MF (clarification), RO (preconcentration) and OD (concentration) [80, 81]. The anthocyanin content of the concentrated juice was more than three times higher than that measured in the raw juice.

The effects of MF and OD on the physico-chemical, nutritional and microbiological qualities of melon juice were studied by Vaillant et al. [82]. The enzymatically treated juice was clarified by MF with a ceramic multichannel membrane (Membralox 1P19-40, Pall-Exekia, France) with a filtration area of 0.24 m^2 and an average pore size of 0.2 μm. The clarified juice was concentrated by a PP hollow fibre membrane with a total area of 10.2 m^2. The clarified juice presented physico-chemical and nutritional properties similar to those of the fresh juice except for the absence of suspended solids and carotenoids which were retained by the MF membrane. Therefore, the UF retentate was proposed as raw material for the extraction of β-carotene or for the production of functional drinks. The composition of the clarified juice was well preserved in the OD retentate at 55 °Brix.

Integrated membrane processes for the clarification and concentration of tropical juices such as camu camu [83] and acerola [84, 85] juices were also investigated on a laboratory scale. High-quality concentrated juices with a TSS content of 53–55 °Brix were produced in the final OD step with no losses of vitamin C content and antioxidant activity.

3.3 Milk and Whey Processing

The dairy industry is the area of the agro-food sector in which the membrane technology is widely developed. In particular, pressure-driven membrane operations find large application for both milk and whey processing with improved separation capabilities when compared with traditional technologies.

Main applications of cross-flow MF in the dairy industry are related to: (i) the separation of bacteria and somatic cells from milk and whey; (ii) fat separation including the defatting of whey and globular milk fat removal; (iii) fractionation of milk proteins.

UF is the most widely used process in the world dairy industry: the largest applications concern the fractionation of cheese whey and preconcentration of milk for cheese manufacture.

NF membranes due to their intermediate selectivity between UF and RO (200–1000 Da) are useful for whey demineralization and valorization of whey proteins.

RO is actually used in dairies to concentrate whey and to recover water from waste. Although some applications for concentration of sheep milk have been reported [86] multiple-effect evaporators are preferred due to their greater efficiencies of operation.

The integration of pressure-driven membrane processes in milk and dairy processing chains has been implemented into existing installations leading to improvements in the quality of existing products, development of new products such as functional and health-promoting foods, enhanced process efficiency, better control of the environmental impact [87].

3.3.1 Integrated Systems in Milk Processing

MF is able to retain, partially or totally, microorganisms, casein micelles and fat globules competing with centrifugal separation (e.g. bactofugation) and pasteurization to extend milk shelf life. In addition, being a non-thermal process, MF preserves thermosensitive milk compounds avoiding losses of nutritional value and undesired milk changes.

In the bactocatch system developed by Alfa-Laval MF is combined with a high-temperature treatment (130°C for 4s) for the removal of bacteria from skim milk [88]. In this approach high cross-flow velocities (6–8 m/s) and recirculation of permeate are used in order to create a uniform transmembrane pressure (UTMP) along the MF membrane. MF membranes with pore size of 1.4 micron are able to retain up to 99.95% of total bacterial cells with decimal reduction factor above 3.5 [89]. The use of organic and ceramic reversed asymmetric membranes with pore size of 0.87 μm combined with an optimized permeate back-pulsing system allows a better control of concentration polarization and fouling phenomena leading to high fluxes (500 L/m^{2}h) and high retention of spores even at low cross-flow velocities (0.5–1 m/s) [90]. The shelf life of milk is increased from 12 to 18 days at 8°C without affecting its sensory quality.

UF membranes retain proteins, fats and insoluble and bound salts from milk allowing the permeation of lactose and soluble salts. They can be used to standardize the content of protein and total solids in milk, thus overcoming variations in milk composition.

A combination of MF and UF processes with diafiltration (DF) and a correct adjustment of pH and temperature enables the production of milk protein concentrates (MPCs) containing 50–85% of proteins. MPCs are ideal for a large range of applications, including

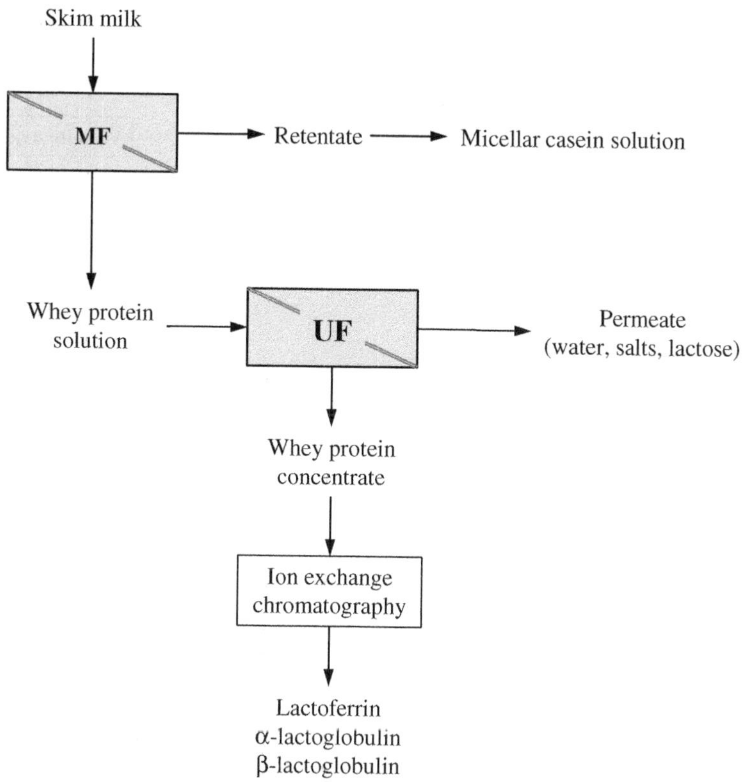

Figure 3.4 *Fractionation of caseins with a combination of membrane processes and ion exchange chromatography.*

cheese products, cultured products, dairy-based beverages, weight management products, powdered dietary supplements and sports nutrition products.

The fractionation of caseins with a combination of MF and UF membranes (Figure 3.4) is another application of interest in the dairy industry. MF membranes with pore size of 0.2 μm separate skim milk in a permeate fraction consisting of the major whey proteins and a retentate with high concentration of native caseins that can be used for cheese making. A desired ratio of casein to total protein can be achieved by manipulating the MF process design. The MF permeate can be further treated by UF to produce high-quality whey protein concentrates (WPCs). A further separation of WPCs into lactoferrin, β-lactoglobulin and α-lactoglobulin can be achieved by using ion-exchange chromatography. PVDF membranes with pore size of 0.2 and 0.3 μm (PTI Advanced Filtration, United States) and ceramic membranes with pore size of 0.2 μm (TAMI Industries, France) provided comparable performance in the treatment of skim milk for casein concentrate manufacture when operated at comparable values of TMP, cross-flow velocity and temperature [91].

The partial removal of β-lactoglobulin from skim milk through an integrated process MF/UF was proposed by Quiblier et al. [92] and then patented by Inra (Rennes Cedex, France). The process allows the production of milk powder with improved cheese-making

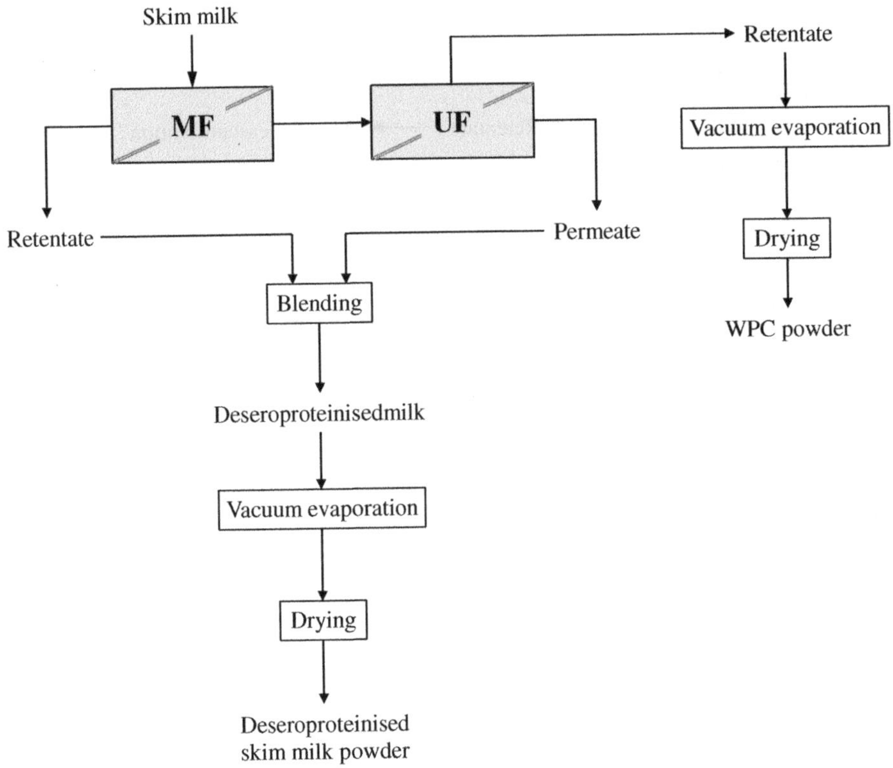

Figure 3.5 *Integrated membrane system for the production of whey protein depleted skim milk powder.* Source: *Adapted from Reference 92.*

ability in comparison with low heat milk powders. A first MF step with membranes of pore size in the range 0.1–0.2 mm allows a partial or total removing of whey proteins from skim milk. Then the MF permeate is treated with UF membranes with a MWCO of 20 kDa and the UF permeate is blended with the MF retentate. The blend is finally treated by vacuum evaporation and spray drying (Figure 3.5).

Mozzarella produced from recombination of milk powder showed similar composition and properties of the cheese produced with raw fresh milk. In addition, the cheesemaking yields were 7.3 higher when compared with those of the conventional process [93].

An integrated membrane process for obtaining a lactose-free milk product was patented by Tossavainen and Sahlstein in 2010 [94]. The process is based on an initial UF treatment of standardized milk in terms of fat content in order to produce a permeate containing about 4.3% lactose and about 0.4% ash with a dry matter content in the range 5–5.5%. The UF permeate is then submitted to a NF treatment preferably carried out up to a volume concentration factor (VCF) 5, generating a retentate containing lactose (about 90% of the dry matter) and a dry matter content of 20–24%. The NF permeate contains mainly univalent salts and urea. The NF permeate is finally concentrated by RO, preferably up to a VCF 12. The RO retentate containing 0.5–2% ash and 1–3% dry matter is used to increase the salt concentration of the UF retentate. The obtained low-lactose milk product can be

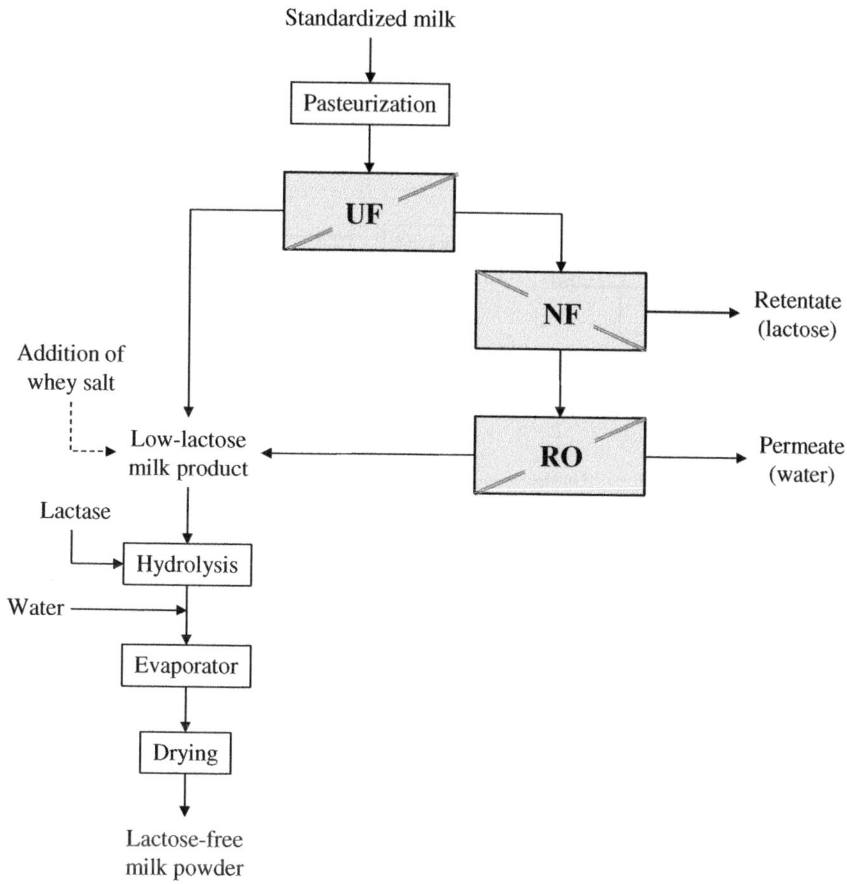

Figure 3.6 *Integrated membrane process for the production of a lactose-free milk products.* Source: *Adapted from Reference 94.*

subjected to hydrolysis where the residual lactose is converted into monosaccharides by means of lactase resulting in a substantially lactose-free milk product (Figure 3.6). The obtained RO permeate, containing mainly water, can be reused in the equipment as rinsing water. A significant advantage of the process is the possibility to eliminate lactose from milk without affecting the organoleptic characteristics of the milk product since the salts removed can be restored or replaced.

3.3.2 Integrated Systems in Cheesemaking

The UF of milk for the production of cheese is one of the most consolidated membrane-based processes in the dairy industry. The advantages of milk concentration by UF are mainly represented by better equipment operations, higher cheese yields, saving of rennet, improvement of quality and nutritional value due to the incorporation of whey proteins in the cheese. Cheese milk can be ultrafiltered at three different concentration levels depending on the cheese varieties: a low-concentrated retentate (LCR) is obtained at low concentration

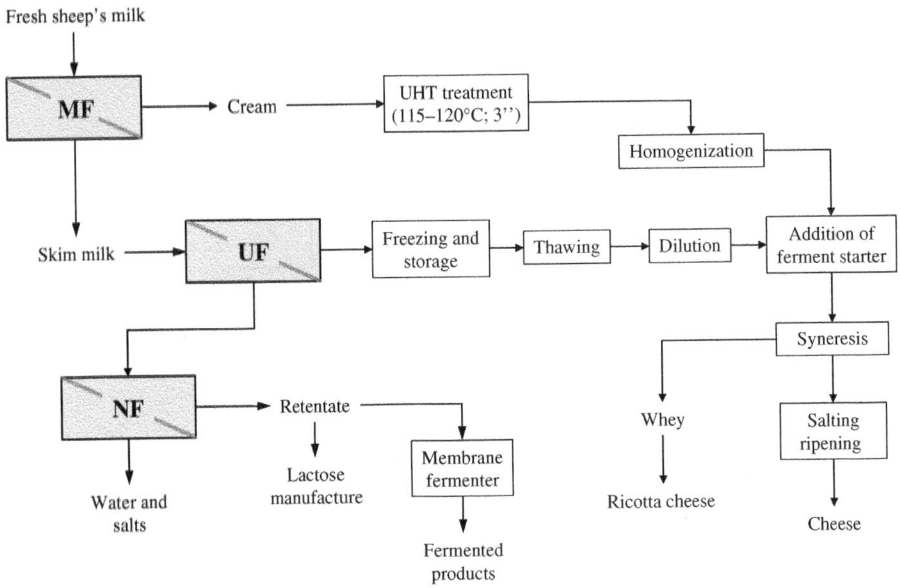

Figure 3.7 Integrated membrane process in cheese manufacture.

factors (1.2–2) in the production of Mozzarella, Cheddar, Cottage cheeses, etc. Although at these concentration factors the protein content is slightly increased (up at 4.5–5%), a complete milk standardization can be obtained to overcome variations in milk composition. A medium-concentrated retentate (MCR) can be obtained at concentration factors in the range 2–6 in the preparation of Feta, Gouda and Havarti cheese [87]. Higher concentrations lead to a liquid pre-cheese (LPC) with a composition similar to that of the final cheese preventing whey syneresis and reducing whey drainage [95]. Camembert, Quarg and Mascarpone are typically manufactured starting from LPC.

A typical integrated process including pressure-driven membrane operations for the production of cheese is depicted in Figure 3.7. MF membranes produce a low-fat milk (free of lipid-like substances and caseins) and a retentate (cream) containing bacteria and fat. Concentration by UF of skim milk produces a WPC allowing the permeation of lactose and soluble salts. Concentrated casein micelles can be recombined with cream for the production of cheese.

The UF permeate contains the same lactose concentration of the original milk. It can be concentrated by NF and the lactose concentrate (with 20–25% of lactose) can be used in the sweet industry or as a raw material for fermentation processes by using different membrane filtration processes [96].

3.3.3 Integrated Systems in Whey Processing

Whey is a by-product of the cheese industry with a low content of solids (up to 5–6%) and high biological oxygen demand (BOD_5 = 30–50 g/L for 1000 L of whey) which make difficult and costly its disposal. Whey processing is one of the most important applications of the membrane technology in the dairy industry. Whey concentration, fractionation, demineralization and purification can be achieved by using a combination of different membrane processes. At the same time, the development of high value-added products from whey compounds is of great interest.

MF membranes can be used to produce defatted whey in order to reduce membrane fouling in the following UF step during the manufacture of WPCs. Lipoproteins tend to form aggregates through calcium bridging when subjected to moderate heat treatment and then aggregates can be removed with MF membranes having a pore size of 0.14 micron. The MF process exhibits a higher defatting efficiency when compared with centrifugation and allows also a significant removal of bacteria.

The production of WPCs by UF is today a well-established application in the dairy industry. The protein content in the WPC can be modified in the range 35–85% through a proper combination of UF and DF. The removal of fat and bacteria by a MF pretreatment allows the production of whey protein isolates (WPIs) with a protein content of 90%. WPCs can be further fractionated in β- and α-lactoglobulins by properly selecting pH, ionic strength and temperature [97] or used for the manufacture of caseinomacropeptides with pharmotherapeutic values.

The concentration of whey and/or its partial demineralization can be achieved by NF. NF membranes allow univalent ions (Na^+, K^+, Cl^-) to pass while larger ions such as Ca^{2+} or PO_4^{3-} are retained. Desalination degrees up to 40% can be achieved through a combination of NF with DF.

The integration of pressure-driven membrane operations with evaporation and crystallization techniques in whey processing is depicted in Figure 3.8. WPCs after whey defatting are produced by UF. The NF treatment of the UF permeate allows to recover lactose in

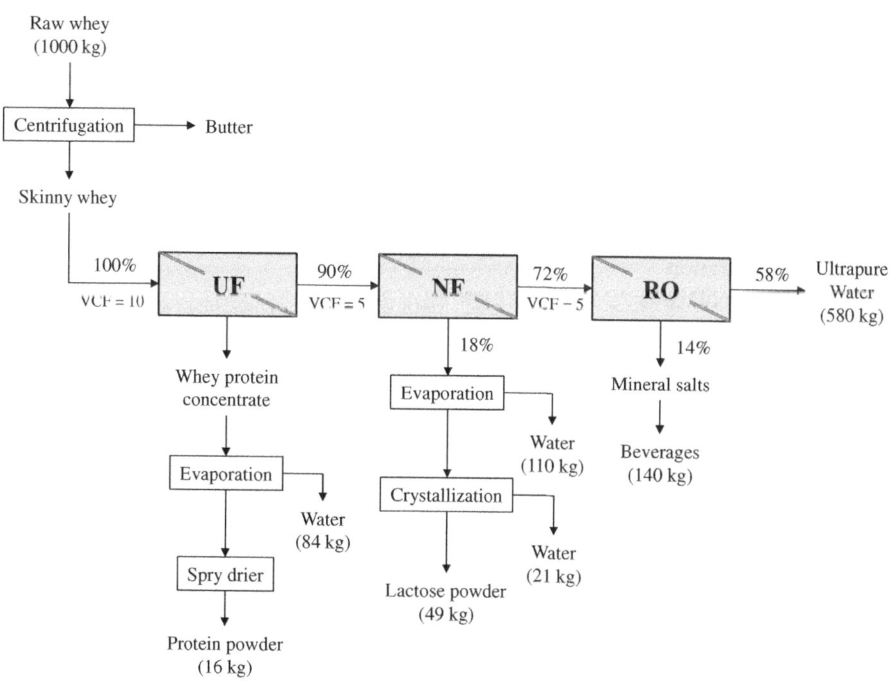

Figure 3.8 Integrated membrane process in whey processing.

the retentate stream together with a partial demineralization. Finally, mineral salts can be concentrated by RO membranes with a simultaneous production of purified water.

According to the mass balance of the process, 1.6% of protein powder (from WPC) and 4.9% of lactose powder together with mineral salts are recovered from the initial raw whey. In addition, more than 80% of water, as permeate from the RO process and after condensation from evaporative processes, can be recovered from the raw whey and reused in the dairy factory. The target quality of the treated water depends on the requirements for the type of reuse (cleaning, cooling, boiler feed, etc.); it must be similar to the characteristics of drinking water but for some reuses, the requirements can be more drastic.

3.4 Conclusions

The combination of conventional and membrane separation technologies offer a wide range of advantages in the agro-food productions in terms of improvement of the product quality, energy saving and reduction of water consumption and environment impact. In this chapter, hybrid combinations in the field of fruit juice and milk processing have been presented illustrating some interesting applications developed on both laboratory and industrial scales. The implementation of integrated processes in these areas, and more in general in different agro-food productions, allows to redesign traditional flow sheets of the industrial production providing an essential contribution to the concept of process intensification.

The development of new membranes and module designs for better control of membrane fouling as well as of membranes with high selectivity and stability for long-term applications are expected innovations for the expansion of membrane technology in food-related applications.

List of Abbreviations

BOD	Biochemical oxygen demand
CA	Cellulose acetate
DCMD	Direct contact membrane distillation
DF	Diafiltration
ED	Electrodialysis
EMR	Enzymatic membrane reactor
LCR	Low-concentrated retentate
LPC	Liquid pre-cheese
MBRs	Membrane bioreactors
MCR	Medium-concentrated retentate
MCs	Membrane contactors
MPCs	Milk protein concentrates
MF	Microfiltration
MD	Membrane distillation
MWCO	Molecular weight cut-off
NF	Nanofiltration
NFC	Not from concentrate

OD	Osmotic distillation
PA	Polyamide
PE	Polyethylene
PES	Polyethersulphone
PP	Polypropylene
PS	Polysulphone
PTFE	Polytetrafluoroethylene
PV	Pervaporation
PVDF	Poly(vinylidene fluoride)
RCF	Reconstituted from concentrate
RO	Reverse osmosis
T	Temperature
TAA	Total antioxidant activity
TE	Thermal evaporation
TSS	Total soluble solids
TMP	Transmembrane pressure
UF	Ultrafiltration
VCF	Volume concentration factor
VMD	Vacuum membrane distillation
WPCs	Whey protein concentrates

References

1. F. Lipnizki, Cross-flow membrane applications in the food industry, *in Membrane Technology, Volume 3: Membranes for Food Applications*, K.V. Peinemann, S.P. Nunes, and L. Giorno (Eds), Wiley-VCH Verlag GmbH & Co., Weinheim, 2010.
2. G. Daufin, J.P. Escudier, H. Carrère, S. Bérot, L. Fillaudeau and M. Decloux, Recent and emerging applications of membrane processes in the food and dairy industry, *Food Bioprod. Process.*, **79**, 89–102 (2001).
3. E. Drioli and M. Romano, Progress and new perspectives on integrated membrane operations for sustainable industrial growth, *Ind. Eng. Chem. Res.*, **40**, 1277–1300 (2001).
4. N.A. Memon, Fruit and vegetable juices global market, *Pak. Food J.*, 20–23 (2012).
5. E.B. Feldman, Fruits and vegetables and the risk of stroke, *Nutr. Rev.*, **59**, 24–27 (2001).
6. H. Wu, Q. Dai, M.J. Shrubsole, R.M. Ness, D. Schlundt, W.E. Smalley, H. Chen, M. Li, Y. Shyr and W. Zheng, Fruit and vegetable intakes are associated with lower risk of colorectal adenomas. *J. Nutr.*, **139**, 340–344 (2009).
7. M. Cheryan, *Ultrafiltration and Microfiltration Handbook*, Technomic Publishing Company, Lancaster, 1998.
8. A. Kilara and J.P. Van Buren, Clarification of apple juice, in *Processed Apple Products*, D.L. Downing (Ed), Van Nostrand Reinhold, New York, 1989.
9. F. Vaillant, S. Jeanton, E.M. Dornier, G. O'Brien, M. Reynes and M. Decloux, Concentration of passion fruit juice on an industrial pilot scale using osmotic evaporation, *J. Food Eng.*, **47**, 195–202 (2001).

10. S.S. Koseoglu, J.T. Lawhon and E.W. Lusas, Use of membranes in citrus juice processing, *Food Technol.*, **44**, 90–97 (1990).
11. R.J. Swientek, Ultrafiltration expanding role in food and beverage processing, *Food Process.*, **47**, 71–83 (1986).
12. K. Riedl, B. Girard and R.W. Lencki, Influence of membrane structure on fouling layer morphology during apple juice clarification, *J. Membr. Sci.*, **139**, 155–166 (1998).
13. B. Girard and L. Fukumoto, Membrane processing of fruit juices and beverages: a review, *Crit. Rev. Food Sci. Nutr.*, **40**, 91–157 (2000).
14. A. Cassano, M. Marchio and E. Drioli, Clarification of blood orange juice by ultrafiltration: analyses of operating parameters, membrane fouling and juice quality, *Desalination*, **212**, 15–27 (2007).
15. A. Cassano, L. Donato and E. Drioli, Ultrafiltration of kiwifruit juice: operating parameters, juice quality and membrane fouling, *J. Food Eng.*, **79**, 613–621 (2007).
16. J.P.F. de Bruijn, A. Venegas, J.A. Martinez and R. Bórquez, Ultrafiltration performance of Carbosep membranes for the clarification of apple juice, *LWT-Food Sci. Technol.*, **36**, 397–406 (2003).
17. R.C. de Oliveira, R. Caleffi Docê and S.T.D de Barros, Clarification of passion fruit juice by microfiltration: analyses of operating parameters, study of membrane fouling and juice quality, *J Food Eng.*, **111**, 432–439 (2012).
18. L.R. Fukumoto, P. Delaquis and B. Girard, Microfiltration and ultrafiltration ceramic membranes for apple juice clarification, *J. Food Sci.*, **63**, 845–850 (1998).
19. A. Laorko, L. Zhenyu, S. Tongchitpakdee, S. Chantachum and W. Youravong, Effect of membrane property and operating conditions on phytochemical properties and permeate flux during clarification of pineapple juice, *J. Food Eng.*, **100**, 514–521 (2010).
20. F. Tasselli, A. Cassano and E. Drioli, Ultrafiltration of kiwifruit juice using modified poly(ether ether ketone) hollow fibre membranes, *Sep. Purif. Technol.*, **57**, 94–102 (2007).
21. G.T. Vladisavljević, P. Vukosavljević and B. Bukvić, Permeate flux and fouling resistance in ultrafiltration of depectinized apple juice using ceramic membranes, *J. Food Eng.*, **60**, 241–247 (2003).
22. S.T.D. de Barros, C.M.G. Andrade, E.S. Mendes and L. Peres, Study of fouling mechanism in pineapple juice clarification by ultrafiltration, *J. Membr. Sci.*, **215**, 213–224 (2003).
23. R. Jiraratananon and A. Chanachai, A study of fouling in the ultrafiltration of passion fruit juice, *J. Membr. Sci.*, **111**, 39–48 (1996).
24. P. Rai, G.C. Majumdar, S. Sharma and S. De, Understanding ultrafiltration performance with mosambi juice in an unstirred batch cell, *J. Food Process Eng.*, **28**, 166–180 (2005).
25. P. Rai and S. De, Clarification of pectin-containing juice using ultrafiltration, *Curr. Sci.*, **96**, 1361–1371 (2009).
26. S. Todisco, L. Peña, E. Drioli and P. Tallarico, Analysis of the fouling mechanism in microfiltration of orange juice, *J. Food Process. Preserv.*, **20**, 453–466 (1996).
27. J. Hérmia, Constant pressure blocking filtration laws. Application to power-law non-Newtonian fluids, *Trans. IChemE*, **60**, 183–187 (1982).
28. R.W. Field, D. Wu, J.A. Howell and B.B. Gupta, Critical flux concept for microfiltration fouling, *J. Membr. Sci.*, **100**, 259–272 (1995).

29. R.W. Field, D. Wu, J.A. Howell and B.B. Gupta, Mass transport and design of membrane system, *in Industrial Membrane Separation Technology*, K. Scott and R. Hughes (Eds), Chapman and Hall, London, 1996.
30. E. Arena, B. Fallico and E. Maccarone, Evaluation of antioxidant capacity of blood orange juices as influenced by constituents, concentration process and storage, *Food Chem.*, **74**, 423–427 (2001).
31. O. Jariel, M. Reynes, M. Courel, N. Durand, M. Dornier and P. Deblay, Comparison of some fruit juice concentration techniques, *Fruits*, **51**, 437–450 (1996).
32. B. Jiao, A. Cassano and E. Drioli, Recent advances on membrane processes for the concentration of fruit juices: a review, *J. Food Eng.*, **63**, 303–324 (2004).
33. N.K. Rastogi and K.S.M.S. Raghavarao, Advances in reverse osmosis technology for the processing of fruit juices, *in Membrane Processes for Sustainable Growth*, A. Basile and A. Cassano (Eds), Nova Science Publishers, Inc., New York, 2013.
34. V. Alvarez, S. Alvarez, F.A. Riera and R. Alvarez, Permeate flux prediction in apple juice concentration by reverse osmosis, *J. Membr. Sci.*, **127**, 25–34 (1997).
35. S. Alvarez, F.A. Riera, R. Alvarez and J. Coca, Permeation of apple aroma compounds in reverse osmosis, *Sep. Purif. Technol.*, **14**, 209–220 (1998).
36. M.J. Sheu and R.C. Wiley, Preconcentration of apple juice by reverse osmosis, *J. Food Sci.*, **48**, 422–429 (1983).
37. G.M. Benjamin and G.I. Albert, Concentration of orange juice by reverse osmosis, *J. Food Process Eng.*, **10**, 217–230 (1988).
38. R.J. Braddock, S. Nikdel and S. Nagy, Composition of some organic and inorganic compounds in reverse osmosis-concentrated citrus juices, *J. Food Sci.*, **53**, 508–512 (1988).
39. B.G. Medina and A. Garcia, Concentration of orange juice by reverse osmosis, *J. Food Process Eng.*, **10**, 217–230 (1998).
40. N. Pap, Sz. Kertész, E. Pongrácz, L. Myllykoski, R.L. Keiski, Gy. Vatai, Zs. László, S. Beszédes and C. Hodúr, Concentration of blackcurrant juice by reverse osmosis, *Desalination*, **241**, 256–264 (2009).
41. P.D. Gurak, L.M.C. Cabral, M.H.M. Rocha-Leão, V.M. Matta and S.P. Freitas, Quality evaluation of grape juice concentrated by reverse osmosis, *J. Food Eng.*, **96**, 421–426 (2010).
42. R.P. Bowden and A.R. Isacs, Concentration of pineapple juice by reverse osmosis, *Food Aust.*, **41**, 850–851 (1989).
43. D.K. Das Gupta and K.S. Jayaramam, Studies on the membrane concentration of watermelon juice, *J. Sci. Ind. Res.*, **55**, 966–970 (1996).
44. F. Chou, R.C. Wiley and D.V. Schlimme, Reverse osmosis and flavor retention in apple juice concentration. *J. Food Sci.*, **56**, 484–487 (1991).
45. A. Pozderović and T. Moslavac, Apple juice aroma concentration from evaporator condensate by reverse osmosis: I. Influence of process parameters on retention and final aroma concentration, *Acta Aliment.*, **28**, 71–83 (1999).
46. P.A. Hogan, R.P. Canning, P.A. Peterson, R.A. Johnson and A.S. Michaels, A new option: osmotic distillation, *Chem. Eng. Prog.*, **94**, 49–61 (1998).
47. M.S.M. Lefebvre, Method of performing osmotic distillation, US Patent 4,781,837, 1988.

48. Liqui-Cel Extra-Flow products, http://www.liqui-cel.com/product-information/data-sheets.cfm (accessed on May 27, 2013).
49. C. Hongvaleerat, L.M.C. Cabral, M. Dornier, M. Reynes and S. Nigsanond, Concentration of pineapple juice by osmotic evaporation, *J. Food Eng.*, **88**, 548–552 (2008).
50. P.E. Shaw, M. Lebrun, M.N. Ducamp, M.J. Jordan and K.L. Goodner, Pineapple juice concentrated by osmotic evaporation, *J. Food Qual.*, **25**, 39–49 (2002).
51. R.B. Rodrigues, H.C. Menez, L.M.C. Cabral, M. Dornier, G.M. Rios and M. Reynes, Evaluation of reverse osmosis and osmotic evaporation to concentrate camu-camu juice (*Myciaria dubia*), *J. Food Eng.*, **63**, 97–102 (2004).
52. P.E. Shaw, M. Lebrun, M. Dornier, M.N. Ducamp, M. Courel and M. Reynes, Evaluation of concentrated orange and passion fruit juices prepared by osmotic evaporation, *LWT-Food Sci. Technol.*, **34**, 60–65 (2001).
53. H. Valdés, J. Romero, A. Saavedra, A. Plaza and V. Bubnovich, Concentration of noni juice by means of osmotic distillation, *J. Membr. Sci.*, **330**, 205–213 (2009).
54. A. Cassano and E. Drioli, Concentration of clarified kiwifruit juice by osmotic distillation, *J. Food Eng.*, **79**, 1397–1404 (2007).
55. V.D. Alves and I.M. Coelhoso, Orange juice concentration by osmotic evaporation and membrane distillation: a comparative study, *J. Food Eng.*, **74**, 125–133 (2006).
56. M. Cissé, F. Vaillant, S. Bouquet, D. Pallet, F. Lutin, M. Reynes and M. Dornier, Athermal concentration by osmotic evaporation of roselle extract, apple and grape juices and impact on quality, *Innov. Food Sci. Emerg. Technol.*, **12**, 352–360 (2011).
57. M. Courel, M. Dornier, J.M. Herry, G. M. Rios and M. Reynes, Effect of operating conditions on water transport during the concentration of sucrose solutions by osmotic distillation, *J. Membr. Sci.*, **170**, 281–289 (2000).
58. K.W. Lawson and D.R. Loyd, Membrane distillation II. Direct contact MD, *J. Membr. Sci.*, **120**, 123–133 (1996).
59. V. Calabrò, B. Jiao and E. Drioli, Theoretical and experimental study on membrane distillation in the concentration of orange juice, *Ind. Eng. Chem. Res.*, **33**, 1803–1808 (1994).
60. S. Gunko, S. Verbych, M. Bryka and N. Hilal, Concentration of apple juice using direct contact membrane distillation, *Desalination*, **190**, 117–124 (2006).
61. F. Laganà, G. Barbieri and E. Drioli, Direct contact membrane distillation: modelling and concentration experiments, *J. Membr. Sci.*, **166**, 1–11 (2000).
62. A. Kozák, E. Békássy-Molnár and G. Vatai, Production of black-currant juice concentrate by using membrane distillation, *Desalination*, **241**, 309–314 (2009).
63. R. Bagger-Jørgensen, A.S. Meyer, C. Varming and G. Jonsson, Recovery of volatile aroma compounds from black currant juice by vacuum membrane distillation, *J. Food Eng.*, **64**, 23–31 (2004).
64. N. Diban, O.C. Voinea, A. Urtiaga and I. Ortiz, Vacuum membrane distillation of the main pear aroma compound: experimental study and mass transfer modelling, *J. Membr. Sci.*, **326**, 64–75 (2009).
65. I.B. Aguiar, N.G.M. Miranda, F.S. Gomes, M.C.S. Santos, D.G.C. Freitas, R.V. Tonon and L.M.C. Cabral, Physicochemical and sensory properties of apple juice concentrated by reverse osmosis and osmotic evaporation, *Innov. Food Sci. Emerg. Technol.*, **16**, 137–142 (2012).

66. P. Onsekizoglu, K. Savas Bahceci and M. Jale Acar, Clarification and concentration of apple juice using membrane processes: A comparative quality assessment, *J. Membr. Sci.*, **352**, 160–165 (2010).
67. S. Álvarez, F.A. Riera, R. Álvarez, J. Coca, F.P. Cuperus, S. Bouwer, G. Boswinkel, R.W. van Gemert, J.W. Veldsink, L. Giorno, L. Donato, S. Todisco, E. Drioli, J. Olsson, G. Trägårdh, S.N. Gaeta and L. Panyor, A new integrated membrane process for producing clarified apple juice and apple juice aroma concentrate, *J. Food Eng.*, **46**, 109–125 (2000).
68. M. Cissé, F. Vaillant, A. Perez, M. Dornier and M. Reynes, The quality of orange juice processed by coupling crossflow microfiltration and osmotic evaporation, *Int. J. Food Sci. Technol.*, **40**, 105–116 (2005).
69. G. Galaverna, G. Di Silvestro, A. Cassano, S. Sforza, A. Dossena, E. Drioli and R. Marchelli, A new integrated membrane process for the production of concentrated blood orange juice: effect on bioactive compounds and antioxidant activity, *Food Chem.*, **106**, 1021–1030 (2008).
70. A. Cassano, B. Jiao and E. Drioli, Production of concentrated kiwifruit juice by integrated membrane processes, *Food Res. Int.*, **37**, 139–148 (2004).
71. A. Cassano, A. Figoli, A. Tagarelli, G. Sindona and E. Drioli, Integrated membrane process for the production of highly nutritional kiwifruit juice, *Desalination*, **189**, 21–30 (2006).
72. S. Bandini and G.C. Sarti, Concentration of must through vacuum membrane distillation, *Desalination*, **149**, 253–259 (2002).
73. A. Cassano, C. Conidi, R. Timpone, M. D'Avella and E. Drioli, A membrane-based process for the clarification and the concentration of the cactus pear juice, *J. Food Eng.*, **80**, 914–921 (2007).
74. A. Cassano, C. Conidi and E. Drioli, Clarification and concentration of pomegranate juice (Punica granatum L.) using membrane processes, *J. Food Eng.*, **107**, 366–373 (2011).
75. A. Cassano, C. Conidi, E. Drioli, A membrane-based process for the valorization of the bergamot juice, *Sep. Sci. Technol.*, **48**, 537–546 (2013).
76. A. Cassano, E. Drioli, G. Galaverna, R. Marchelli, G. Di Silvestro, P. Cagnasso, Clarification and concentration of citrus and carrot juices by integrated membrane processes, *J. Food Eng.*, **57**, 153–163 (2003).
77. D. Thompson, The application of osmotic distillation for the wine industry, *Aust. N. Z. Grapegrow. Winemak.*, **11**, 11–14, 1991.
78. A. Rektor, G. Vatai and E. Békássy-Molnár, Multi-step membrane processes for concentration of grape juice, *Desalination*, **191**, 446–453 (2006).
79. L.F. Sotoft, K.V. Christensen, R. Andrénsen and B. Norddahl, Full scale plant with membrane based concentration of blackcurrant juice on the basis of laboratory and pilot scale tests, *Chem. Eng. Proc.*, **54**, 12–21 (2012).
80. A. Kozák, S. Bánvölgyi, I. Vincze, I. Kiss, E. Békássy Molnár and G. Vatai, Comparison of integrated large-scale and laboratory scale membrane processes for the production of black currant juice concentrate, *Chem. Eng. Proc.*, **47**, 1171–1177 (2008).
81. S. Bánvölgyi, S. Horváth, E. Stefanovits-Bányai, E. Békássy-Molnár and G. Vatai, Integrated membrane process for blackcurrant (*Ribes nigrum L.*) juice concentration, *Desalination*, **241**, 281–287 (2009).

82. F. Vaillant, M. Cisse, M. Chaverri, A. Perez, M. Dornier, F. Viquez and C. Dhuique-Mayer, Clarification and concentration of melon juice using membrane processes, *Innov. Food Sci. Emerg. Technol.*, **6**, 213–220 (2005).
83. A.L.R. Souza, M.M. Pagani, M Dornier, F.S. Gomes, R.V. Tonon and L.M.C. Cabral, Concentration of camu-camu juice by the coupling of reverse osmosis and osmotic evaporation processes, *J. Food Eng.*, **119**, 7–12 (2013).
84. M.M. Pagani, M.H. Rocha-Leão, A.B. Barbosa Couto, J.P. Pinto, A.O. Riberio, F. dos Santos Gomes and L.M.C. Cabral, Concentration of aceroal (*Malpighia emrginata* DC.) juice by integrated membrane separation process, *Desalin. Water Treat.*, **27**, 130–134 (2011).
85. V.M. Matta, R.H. Moretti and L.M.C. Cabral, Microfiltration and reverse osmosis for clarification and concentration of acerola juice, *J. Food Eng.*, **61**, 477–482 (2004).
86. P.L. Voutsinas, M.C. Katsiari, C.P. Pappas and H. Mallatou, Production of yoghurt from sheep's milk which had been concentrated by reverse osmosis and stored frozen. 1. Physicochemical, microbiological and physical stability characteristics of concentrates, *Food Res. Int.*, **29**, 403–409 (1996).
87. M. Rosemberg, Current and future applications for membrane processes in the dairy industry, *Trends Food Sci. Technol.*, **6**, 12–19 (1995).
88. S. Holm, R. Malmberg and K. Svenssom, Method and plant producing milk powder with low bacterial content, World Patent WO 86/01687 (1986).
89. L.V. Saboya and J.L. Maubois, Current developments of microfiltration technology in the dairy industry, *Lait*, **80**, 541–553 (2000).
90. A. Guerra, G. Jonsson, A. Rasmussen, E. Waagner Nielsen and D. Edelsten, Low cross-flow velocity microfiltration of skim milk for removal of bacterial spores, *Int. Dairy J.*, **7**, 849–861 (1997).
91. K. Karasau, N. Glennon, N.D. Laurence, G.W. Stevens, A.J. O'Connor, A.R. Barber, S. Yoshikawa and S. Kentish, A comparison between ceramic and polymeric membrane systems for casein concentrate manufacture, *Int. J. Dairy Technol.*, **63**, 284–289 (2010).
92. J.P. Quiblier, C. Ferron-Baumy, G. Garric G. and J.L. Maubois, Procédé de traitement des laits permettant au moins de conserver leur aptitude fromagère, French Patent No. 0.542.583 B1 (1992).
93. A. Garem, P. Schuck and J.L. Maubois, Cheesemaking properties of a new dairy-based powder made by a combination of microfiltration and ultrafiltration, *Lait*, **80**, 25–32 (2000).
94. O. Tossavainen and J. Sahlstein, Process for producing a lactose-free milk product, US Patent 7,829,130 B2 (2010).
95. V.V. Mistry and J.L. Maubois, Application of membrane separation technology to cheese production, *in Cheese: Chemistry, Physics and Microbiology*, P.F. Fox (Ed), Chapman & Hall, London, 1992.
96. R. Atra, G. Vatai, E. Bekassy-Molnar and A. Balint, Investigation of ultra and nanofiltration for utilization of whey protein and lactose, *J. Food Eng.*, **67**, 325–332 (2005).
97. A.L. Zydney, Protein separations using membrane filtration: new opportunities for whey fractionation, *Int. Dairy J.*, **8**, 243–250 (1998).

4

Continuous Hydrolysis of Lignocellulosic Biomass via Integrated Membrane Processes

Mohammadmahdi Malmali and S. Ranil Wickramasinghe
Ralph E. Martin Department of Chemical Engineering, University of Arkansas, Fayetteville, AR, USA

4.1 Introduction

Prior to the discovery of crude oil, society met all its energy needs from plant or lignocellulosic biomass. Today increasing world energy demand coupled with decreasing fossil fuel reserves, continues to drive the quest for new energy sources. In addition, political and environmental concerns are driving the development of sustainable energy sources. Biofuels derived from lignocellulosic biomass are one of the leading renewable energy candidates to replace fossil-based transportation fuels [1–3]. Unlike first generation biofuels, such as bioethanol from sugarcane and corn stover, production of second generation biofuels from lignocellulosic biomass is far more complex. Development of efficient separation and purification operations are essential for the production of competitive second generation drop-in biofuels. Membrane-based separation processes are attractive as they could lead to significant process intensification and hence reduced operating costs [4].

Three main strategies exist for conversion of lignocellulosic biomass into liquid fuels and chemical intermediates: gasification, pyrolysis, and hydrolysis [1]. Here, we focus on hydrolysis of lignocellulosic biomass followed by fermentation. Figure 4.1 is a schematic representation of a lignocellulose to biofuel biorefinery. The overall process consists of the

Integrated Membrane Systems and Processes, First Edition. Edited by Angelo Basile and Catherine Charcosset.

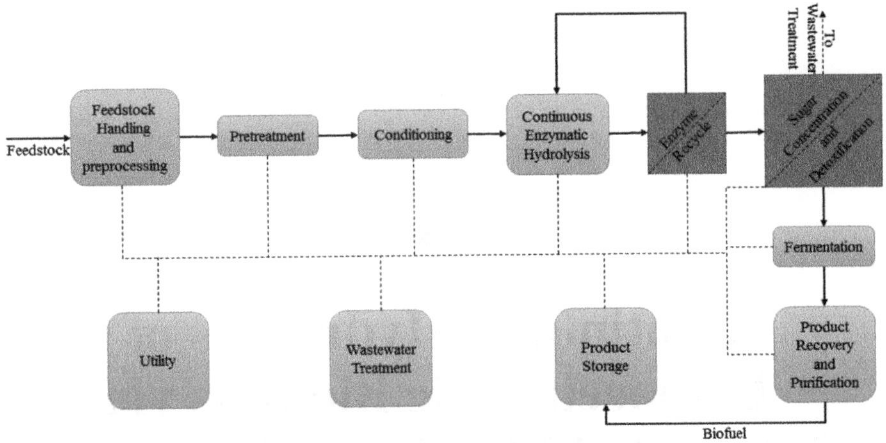

Figure 4.1 Schematic process diagram for bioconversion of lignocellulosic biomass integrated with membrane separation/concentration steps; solid lines show the process flow streams and dashed lines show the utility and water stream connections between different areas.

following main steps: feed stock handing and preprocessing; pretreatment and conditioning; saccharification and fermentation; product recovery and purification; wastewater treatment and recovery; and product storage. Feed stock storage and preprocessing includes receiving the lignocellulosic biomass and typically processing the biomass into a slurry that can be pumped into the pretreatment reactor. Lignocellulosic biomass feed stocks can include waste materials such as agricultural wastes, crop residues, forestry wastes, municipal wastes, and aquatic biomass. Handling and preprocessing depends on the type of lignocellulosic biomass. Typically large amounts of water will be needed in order to create a slurry that can be pumped into the pretreatment reactor. Thus recovery and recycle of the wastewater will be essential.

Many different mechanical and thermochemical methods have been proposed to pretreat lignocellulosic biomass [5]. Frequently, sulfuric acid has been used to pretreat the lignocellulosic biomass. Dilute sulfuric acid (typically 0.1–1.0 wt%) effectively hydrolyses the hemicellulose present to its monomeric sugars (mainly xylose) and enhances the enzymatic digestibility of the cellulose present [6, 7]. After dilute acid pretreatment, the pH of the hydrolysate is between 1 and 2. Further, compounds that inhibit subsequent bioconversion of the solubilized sugars to the desired biofuel are also produced. These compounds are often referred to as inhibitory compounds. Acetic acid is produced due to hydrolysis of acetyl groups present in the hemicellulose. Pentose sugar degradation products such as furfural are also produced [8]. In addition, other degradation compounds such as 5-hydroxymethylfurfural, levulinic acid, and formic acid may also be formed [9]. Consequently, hydrolysate detoxification or conditioning is essential prior to fermentation. In this step, the pH of the hydrolysate is adjusted and these inhibitory compounds are removed. Saccharification typically refers to the enzymatic hydrolysis of cellulose to glucose using a cocktail of cellulase enzymes [2]. Some processes make use of simultaneous saccharification and fermentation. Depending on the biofuel being produced, various purification schemes have been developed for product recovery and storage.

Numerous possibilities exist for integration of membrane-based separation processes in a lignocellulose to biofuel biorefinery. Huang et al. [10] and more recently, Abels et al. [11] have reviewed the use of membrane technologies for future biorefineries. Here, we focus on enzymatic hydrolysis of cellulose as the cost of the enzymes has been a major inhibitory factor for the commercialization of biomass conversion technology [12]. Development of a continuous enzymatic hydrolysis process, where the enzymes may be reused is of considerable interest. A further disadvantage of the commonly used batch hydrolysis of cellulose, is that the conversion rate is often limited by inhibition due to glucose and cellobiose.

In this chapter, we begin by reviewing previous studies that have focused on the development of a membrane reactor for continuous enzymatic hydrolysis of lignocellulosic biomass. Next, we present results for an integrated submerged membrane reactor. The focus is on the hydrolysis of cellulose present in the biomass. As the product sugar concentration from a membrane rector will be low, concentration of the product stream will be essential prior to fermentation. Thus the second part of this chapter focuses on the use of nanofiltration membranes to concentrate sugars present in the hydrolysate. Concentration of sugars and not inhibitory compounds present in the hydrolysate will be challenging and will require careful selection of the nanofiltration membrane and operating conditions. We include recent results on the use of Design of Experiments software to help identify appropriate nanofiltration membranes and optimize the operating conditions. Finally, we close this chapter by highlighting some of the challenges involved in developing an integrated submerged membrane reactor and nanofiltration step for continuous enzymatic hydrolysis.

4.2 Continuous Enzymatic Hydrolysis

Several studies [13–18] have focused on the use of external ultrafiltration membranes for continuous removal of glucose in the permeate. The retentate containing residual cellulose and cellulase enzyme is recycled to the enzyme reactor as shown in Figure 4.2a. In addition,

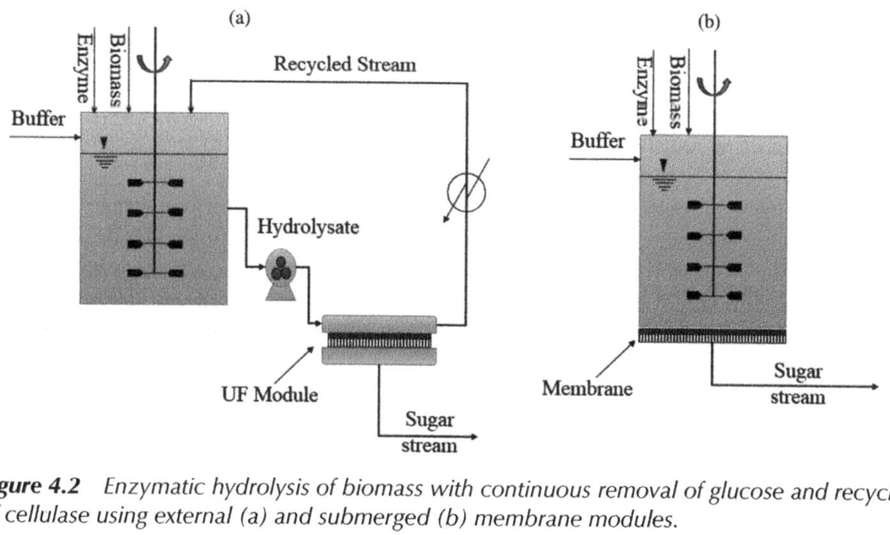

Figure 4.2 Enzymatic hydrolysis of biomass with continuous removal of glucose and recycle of cellulase using external (a) and submerged (b) membrane modules.

Andrić et al. [19] have considered the use of dialysis for removal of glucose. All of these studies indicate the importance of glucose removal in order to avoid product inhibition. Andrić et al. [20] have reviewed the design of reactors for minimum inhibition during enzymatic hydrolysis of lignocellulosic biomass. They showed that membrane reactors that permit continuous removal of glucose and recycle of cellulase enzyme can lead to higher rates of hydrolysis of cellulose as well as higher usage of the enzyme.

Glucose fluxes obtained using ultrafiltration membranes are generally higher than dialysis membranes. Use of an external membrane as shown in Figure 4.2a, requires pumping the feed through the membrane module. Given the high viscosity and non-Newtonian behavior of a lignocellulosic biomass slurry, this can be difficult and costly [21–23]. The maximum solids concentration in the slurry is limited and membrane fouling is a concern. In addition, using an external ultrafiltration membrane requires feed pressures of up to 700 kPa. Obtaining these feed pressures will require additional capital investments and increased operating costs which will adversely impact the economic viability of the process. These concerns have limited the development of membrane reactors containing external dialysis or ultrafiltration membranes for hydrolysis of lignocellulosic biomass.

Knutsen and Davis [24] have investigated the use of combined sedimentation followed by ultrafiltration for recovery of cellulase enzyme. They noted that the inclusion of a sedimentation step prior to ultrafiltration leads to much higher permeate fluxes during ultrafiltration. Sedimentation enabled the removal of large insoluble biomass particles significantly reducing membrane fouling. Mores et al. [25] have considered the use of microfiltration and ultrafiltration for recovery of cellulase enzyme. They indicate that using 0.22 μm pore size membranes, 75% of the cellulase enzyme (46–100 kDa MW) could be recycled in active form due to the fact that the majority of the enzyme is bound to the biomass. The much lower pressures required for microfiltration (up to 100 kPa) is a significant benefit resulting in lower operating costs. While addition of a sedimentation step improves membrane performance and reduces the rate of membrane fouling, since the rate of sedimentation depends on the gravitational force acting on the particles, it is not easily controlled and is often slow. Further additional equipment is required which adds to the overall cost of the process.

Carstensen et al. [26] have reviewed the use of membranes for *in-situ* product recovery. Two modes of operation exist: external membranes and submerged membranes. As described above, most previous studies have made use of external membranes. Various configurations have been described for external membranes [27–33]. However, all configurations require pressurizing and pumping the feed through the membrane module. The permeate is generally removed at close to atmospheric pressure.

Submerged membranes on the other hand involve placing the membrane inside the enzyme reactor (Figure 4.2b). Thus, there is no need to pump the contents of the enzyme reactor through an external loop. Further, pressurization of the feed is not required as the permeate is withdrawn at sub-atmospheric pressure. Difficulties involved with pumping high viscosity non-Newtonian biomass slurries are avoided. In addition, by ensuring rapid mixing membrane fouling due to deposition of cellulose and cellulase enzyme on the membrane surface is minimized. Carstensen et al. [26] describe nine different submerged membrane reactor configurations. Commercial use of submerged membrane reactors has been limited due to scale-up difficulties and a lack of commercial units [34].

Several investigators have studied the use of submerged membrane reactors for enzyme recovery and recycle. Greiner et al. [35] used an ultrafiltration membrane for retention of hydrogenase enzyme, while Isono et al. [36] used a submerged membrane module for recycle of lipase. Shi et al. [37] investigated the recovery and recycle of free nuclease. Wang et al. [38] developed a cascade consisting of two submerged membrane reactors. Importantly, all of these studies were conducted at laboratory scale using reactor volumes between 10 and 380 mL.

Development of an integrated membrane reactor for enzymatic hydrolysis of lignocellulosic biomass may be an enabling technology for the development of a continuous hydrolysis process. In the next section, we describe some of our recent results. Since ultrafiltration membranes with molecular weight cutoffs of around 40 kDa are required for rejection of cellulase (46–100 kDa), the pressure driving force across the membrane will be around 700 kPa. Generation of an appropriate suction pressure is unlikely. Consequently, we have focused on the development of a submerged membrane rector containing a microfiltration membrane. The membrane is designed to retain insoluble biomass particles and adsorbed cellulase enzyme.

4.3 Integrated Submerged Membrane System

Continuous enzymatic hydrolysis experiments were conducted using a submerged membrane reactor consisting of a flat sheet microfiltration membrane placed at the bottom of the reactor vessel. The stirring speed was 100 rpm. Figure 4.2b is a schematic representation of the submerged membrane reactor. Hydrolysis experiments were conducted using α-cellulose (Sigma-Aldrich, St. Louis, MO). The cellulase enzyme consisted of Cellic CTec2 (Novozymes North America, Franklinton, NC). The membrane disc consisted of a 0.65-μm polyethersulfone membrane (Pall Corporation, Port Washington, NY). Table 4.1 summarizes the experimental conditions used. In experiments 1 and 3, the reactor was charged and held for 5 and 0.5 hours, respectively, before permeate withdrawal while for experiments 2 and 4, permeate was withdrawn immediately. Fresh feed was added at the same rate at which permeate was withdrawn ensuring the feed volume remained constant. In addition, reactor residence times of 1 hour and 10 hours were investigated.

The variation of cellulose and glucose concentration is shown in Figure 4.3. As can be seen in the absence of preholding time, the glucose concentration in the permeate is lower while the cellulose concentration is higher. Since a microfiltration membrane has been used only enzyme bound to the biomass will be retained. Free enzyme will not be retained. Thus

Table 4.1 *Experimental conditions used for continuous enzymatic hydrolysis.*

Test	Residence time (h)	Pre-holding time (h)	Permeate flux (mL min^{-1})	Reactor volume (mL)	Permeate conc. ($g_{glucose}$ L^{-1})	Production ($mg_{glucose}$ min^{-1})
1	10	5	2.5	1200	15.00	37.5
2	10	0	2.5	1200	10.41	26.25
3	1	0.5	5	300	10.95	54.75
4	1	0	5	300	5.04	25.2

Initial cellulose loading was 100 g L^{-1}. In all experiments the initial cellulase enzyme loading was 50 mg enzyme/g cellulose. Fresh feed consisting of 25 g L^{-1} cellulose and 4 and 10 mg enzyme/g cellulose, 1 and 10 hours residence time, respectively, was added at the same rate at which permeate was removed such that the reactor volume remained constant.

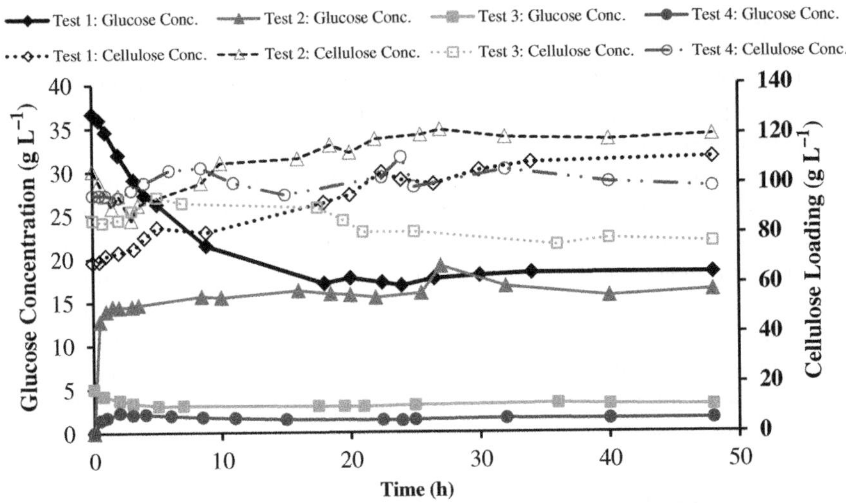

Figure 4.3 Variation of cellulose loading and glucose concentration during continuous enzymatic hydrolysis. Solid lines with filled markers show the glucose concentrations in the permeate; dashed lines with open markers show the cellulose concentrations in the reactor (readable from secondary axis).

the preholding time is essential for the enzyme to bind to the biomass present. Comparing results for 10 and 1 hour residence times indicates the effect of glucose inhibition on the rate of hydrolysis. For a preholding time of 5 hours and residence time of 10 hours the initial glucose concentration is above 30 g L^{-1} but falls to a steady state value of about 15 g L^{-1}. For a preholding time of 0.5 hour and residence time of 1 hour, a steady state glucose concentration of 10.95 g L^{-1} is attained almost immediately. These results highlight the importance of optimizing both the preholding time and residence time in order to maximize the rate of cellulose hydrolysis.

The glucose concentration in the permeate from the submerged membrane reactor is in the range of 5–15 g L^{-1} depending on operating conditions. However, after conventional batch enzymatic hydrolysis, glucose concentration is over 50 g L^{-1} [39]. Consequently, concentration of glucose and other sugars present will be essential as the viability of a membrane reactor for continuous enzymatic hydrolysis will depend on obtaining an aqueous sugar stream where the sugar concentration is optimized for the following fermentation step. In the next section, the use of membrane filtration for concentration of aqueous sugar streams is discussed.

4.4 Sugar Concentration

Ultrafiltration membranes typically have pore sizes from 1–100 nm and are capable of rejecting species from 1 to 500 kDa [40]. The molecular weights of common sugars in biomass hydrolysates are given in Table 4.2. Consequently, ultrafiltration membranes are unlikely to efficiently concentrate aqueous sugar streams.

Table 4.2 *Common sugars in biomass hydrolysates.*

	Glucose	Xylose	Arabinose	Galactose	Mannose	Sucrose	Cellubiose
MW	180	150	150	180	180	342	342
Type	Glucan	Xylan	Arabinan	Galactan	Mannan	Glucose and fructose monomers	Two glucose monomers
Source	Cellulose	Hemicellulose	Hemicellulose	Hemicellulose	Hemicellulose	Biomass	Biomass

Reverse osmosis membranes find their greatest application in the desalination of sea water to produce drinking water. While reverse osmosis membranes would effectively concentrate an aqueous sugar stream, there are a number of disadvantages. Feed pressure can range from 6–8 MPa for sea water desalination and 1–3 MPa for brackish water desalination. The high energy costs required to pressurize the feed will most likely adversely impact the commercial viability of the process. Further, fouling of the membrane, as a high solid concentration feed is pumped through the module, is a major concern. Finally, besides dissolved sugars, other dissolved inhibitory compounds such as 5-hydroxymethylfurfural, furfural, and acetic acid will also be concentrated in the permeate. These compounds should not be concentrated in the aqueous sugar stream as they could reduce the efficiency of the subsequent fermentation step [8].

Nanofiltration or low-pressure reverse osmosis membranes were initially developed for water softening applications [41]. These membranes were originally designed for rejection of multivalent ions where rejection can be as high as 99%. In addition, they exhibit over 90% rejection of dissolved organic compounds with molecular weights over 150–300 g mol^{-1}. A few recent studies have investigated the possibility of using nanofiltration membranes to concentrate sugars but not inhibitory compounds.

Weng et al. [42] investigated separation of acetic acid, during dilute acid hydrolysis of rice straw from xylose. They indicated a separation factor of acetic acid over xylose and arabinose of 49 and 52, respectively. Their work indicates that the actual separation factor depends on many variables such as pH and the presence of other species in the feed. As nanofiltration membranes contain few real pores, transport is via diffusion through the polymeric barrier layer [43]. Our recent work indicates the importance of pH on the flux, rejection, and selectivity of nanofiltration membranes [44]. Qi et al. [45] also indicate the importance of feed pH and the presence of other species on glucose and xylose rejection.

Maiti et al. [46] indicate that removal of inhibitory compounds and concentration of monomeric sugars is possible using nanofiltration membranes. However, as nanofiltration membrane performance depends on size exclusion as well as surface interactions between the membrane and dissolved species, feed pH, ionic strength, and the concentration of the various dissolved solutes will affect membrane performance. Consequently, membrane performance is difficult to predict as it will depend on the interplay between all of these parameters. Thus selection of an optimized membrane that maximizes concentration of sugars in the retentate but also maximizes removal of inhibitory compounds in the permeate will be complex. In the next section, we propose the use of a design of experiments software in order to rapidly screen commercially available nanofiltration membranes.

4.5 Sugar Concentration and Hydrolysate Detoxification by Nanofiltration

All experiments were performed in dead-end flow mode and conducted using a stirred cell HP4750 (Strelitech Corporation, Kent, WA). The cell is designed to operate at a maximum feed pressure of 6.9 MPa using 49 mm membrane discs with 14.6 cm^2 active membrane area. We used high-pressure nitrogen cylinder to pressurize the cell. Three commercially available Alpha Laval (Wood Dale, IL) membranes (RO90, RO98, and RO99) and a Dow FilmTec (Edina, MN) membrane (NF90) were tested with model sugar feed solutions. Based on our results from model sugar feed solution tests, membranes that displayed higher permeabilities, while showing low rejections for inhibitory compounds, were further tested with corn stover hydrolysate. The Alpha Laval membranes are marketed as low-pressure (brackish water) reverse osmosis membranes while the Dow FilmTec membrane is marketed as a nanofiltration membrane. Table 4.3 gives further information on the four membranes tested here. As can be seen, the Alpha Laval membranes are generally tighter (lower NMWCO).

Table 4.4 gives the composition of the various model sugar feed solutions and the real corn stover hydrolysate tested. The model sugar feed solutions were prepared by adding the various constituents to DI water. The corn stover hydrolysate was obtained from the National Renewable Energy Laboratory (Golden, CO). The hydrolysate was centrifuged to remove particulate matter, and the pH of the hydrolysate was 4.75. The hydrolysate was diluted 5- to 10-fold so that the total sugars concentration was in the range 14–27 g L^{-1}, and the xylose to glucose ratio was 1:1.3. The total concentration of inhibitory compounds was 0.8–1.6 g L^{-1}.

Table 4.3 Membranes tested together with manufacturer's specifications.

Membrane	NF90 [47]	RO90 [48]	RO98 [49]	RO99 [50]
Manufacturer	Dow FilmTec, Edina, MN	Alpha Laval, Lund, Sweden	Alpha Laval, Lund, Sweden	Alpha Laval, Lund, Sweden
Structure	Thin film composite polyamide, polysulfone, and polyester support	Thin film composite polyamide, polyester support	Thin film composite polyamide, polypropylene support	Thin film composite polyamide, polyester support
Molecular weight cut off and rejection	200 Da 90–96% NaCl >98% $MgSO_4$	>90% NaCl	>97% NaCl	>98% NaCl
Charge at pH 7.0	Negative	Negative	Negative	Negative
Operating range	5–45°C 3–10 pH upto 4.1 MPa	5–50°C 3–10 pH upto 5.5 MPa	5–60°C 2–11 pH upto 5.5 MPa	5–50°C 3–10 pH upto 5.5 MPa

Table 4.4 *Aqueous sugar feed streams tested.*

Compound	Model feed solution concentration (g L^{-1})	Actual hydrolysate concentration (g L^{-1})
Glucose	3.75, 7.5, 15	77.872
Xylose	1.25, 2.5, 5	58.791
Acetic acid	0, 0.6, 1.2	3.445
5-Hydroxymethylfurfural	0, 0.2, 0.4	0.093
Furfural	0, 0.1, 0.2	4.508

The middle column gives the concentrations of the various components present in the model feed streams. The right-hand column gives the concentration of the various components present in actual hydrolysate.

4.6 Statistical Design of Experiments

Design Expert 8.0.0 (Stat-Ease Inc., Minneapolis, MN) trial version was used to design a set of experiments to determine and optimize the experimental conditions for concentration of model feed solutions. Response surface methodology was used to find the smallest number of experiments to detect the most influential parameters. It is an empirical statistical technique applied to situations where multiple factors influence different responses. One categoric factor with four levels (different membranes), and four continuous numeric factors with three levels were investigated. Numeric factors were total sugars concentrations with xylose to glucose ratio of 1:3, total inhibitory compounds concentration with acetic acid to 5-hydroxymethylfurfural to furfural ratio of 6:2:1 and pH.

We used a D-optimal model for response surface methodology. The goal was to identify and assess the most important operating parameter and minimize the number of tests while working with both continuous and categoric factors. Our design resulted in 30 required models, 5 lack-of-fit and 5 replicate points. A total of 35 experiments were designed with five replications (Table 4.5 [51]). Experiments marked "*" were run in duplicate as suggested by the design of experiments software. A total of seven responsive variables namely xylose rejection (R1), glucose rejection (R2), acetic acid rejection (R3), furfural rejection (R4), 5-hydroxymethylfurfural rejection (R5), permeability (R6), and permeability decline (R7) were chosen as responses to analyze the results. Analysis of variance (ANOVA) enabled us to identify the significant terms in polynomial approximation for our response variables. Numerical optimization was used to search the design space and find the combination of conditions that satisfy the defined goals of sugar concentration and detoxification. A graph of desirability for each membrane was generated using the numerical optimization of obtained solutions.

4.7 Analysis of Variance using Response Surface Methodology

Adjusted and predicted R^2 values derived from fitting results suggest that the linear model best fits for all responses. R^2 value is a measure of the amount of variation around the mean obtained with the model. F-value is a tool to compare the variance with residual variance. F-values showed that all models used for each response are significant and the

Table 4.5 *Experimental conditions for model hydrolysate solutions.*

Run #	Total sugars concentration ($g\ L^{-1}$)	Total inhibitory compounds concentration ($g\ L^{-1}$)	Transmembrane pressure (bar)	pH	Membrane
1	5.0	0	40	5.75	NF90
2	5.0	0.90	40	3.00	NF90
3	5.0	0.90	20	8.50	NF90
4	5.0	1.80	30	3.00	NF90
5	12.5	0	20	3.00	NF90
6	12.5	1.80	40	8.50	NF90
7	20.0	0	30	8.50	NF90
8	20.0	0	40	3.00	NF90
9	20.0	1.80	20	5.75	NF90
10	20.0	1.80	40	3.00	NF90
11	5.0	0	20	8.50	RO90
12	5.0	0	40	3.00	RO90
13	5.0	1.80	20	3.00	RO90
14	5.0	1.80	40	8.50	RO90
15	20.0	0	40	8.50	RO90
16	20.0	0	20	3.00	RO90
17	20.0	1.80	20	8.50	RO90
18	20.0	1.80	40	3.00	RO90
19*	5.0	0	20	3.00	RO98
20*	5.0	0	40	8.50	RO98
21	5.0	1.80	40	3.00	RO98
22	5.0	1.80	20	8.50	RO98
23*	12.5	0.90	30	5.75	RO98
24	20.0	0	20	8.50	RO98
25	20.0	0	40	3.00	RO98
26*	20.0	1.80	20	3.00	RO98
27*	20.0	1.80	40	8.50	RO98
28	5.0	0	30	8.50	RO99
29	5.0	0.90	20	3.00	RO99
30	5.0	1.80	40	5.75	RO99
31	12.5	0	40	3.00	RO99
32	12.5	1.80	20	8.50	RO99
33	20.0	0	20	5.75	RO99
34	20.0	0.90	40	8.50	RO99
35	20.0	1.80	30	3.00	RO99

Source: Reproduced from Reference 51 with permission from Elsevier.
*Experiments that were run in duplicate.

highest probability that these F-values could occur due to noise was lower than ~2.58% for glucose rejection. The significance of all five process variables (total sugars concentration, total inhibitory compounds concentration, feed pressure, pH, and type of membrane) on responsive variables have been correlated with their respective p-values. The p-values of the response variables are summarized in Table 4.6.

Table 4.6 Analysis of variance for response variables.

Source	R1: *p*-value	R2: *p*-value	R3: *p*-value	R4: *p*-value	R5: *p*-value	R6: *p*-value	R7: *p*-value
Model	0.0152	0.0258	0.0003	<0.0001	0.0356	0.0213	<0.0001
A-Total sugars concentration	0.1061	0.1401	0.4425	0.7584	0.5177	0.1588	<0.0001
B-Pressure	0.8008	0.8367	0.6480	0.0715	0.2314	0.6161	0.0140
C-pH	0.2128	0.2073	<0.0001	<0.0001	0.0013	0.6745	0.8441
D-Total compounds concentration	0.0094	0.0121	0.9150	0.1651	0.5936	0.0511	0.3552
E-Membrane type	0.045	0.049	0.5717	0.0007	0.1798	0.0152	0.0222
Lack of fit	0.0127	0.0063	0.9727	0.2930	0.7987	<0.0001	0.2012
Adequate precision	7.087	6.681	7.622	14.511	6.765	5.677	9.765

The *p*-value is a measure of the probability of seeing the observed *F*-value if the null hypothesis is true. A model or term is significant if the *p*-value is less 0.05 (significant at 5% level). Analysis of variance of the linear model for each response shows that the model is statistically significant for each response. Comparing the *p*-values show that total inhibitory compounds concentration had maximum impact on xylose ($p > 0.0094$) and glucose ($p > 0.0121$) rejections. Acetic acid ($p > 0.0001$), furfural ($p > 0.0001$), and 5-hydroxymethylfurfural ($p > 0.0013$) rejections were highly influenced by the pH of feed solution. Total sugars concentration ($p > 0.0001$) and pressure ($p > 0.0140$) as well as membrane type ($p > 0.0222$) have the maximum influence on permeability decline while permeability was only influenced by membrane type ($p > 0.0152$). Lack-of-fit for each response is also summarized in Table 4.6. In addition, adequate precision reported shows the signal-to-noise ratio. A ratio greater than four for adequate precision is desirable, and larger numbers indicated an adequate signal. Table 4.6 shows that all response signals are adequate.

We used the numerical optimization mode to search the design space and find factor settings that provide us with our desired goals. Desirability 3D plots reflect the desirable range of each response. The range of desirability is from zero to one. The desirability function is defined as following:

$$D = (d_1 \times d_2 \times \cdots \times d_n)^{\frac{1}{n}} = \left(\prod_{i=1}^{n} d_i\right)^{\frac{1}{n}}$$

where n is the number of responses, and d_i is the response desirability value. Maximizing the permeability and sugar rejections while minimizing the inhibitory compound rejection and permeability decline were set as our goal with different priority to analyze the desirability 3D plots. Figure 4.4 shows the response surface methodology desirability plots depicting the effect of total sugars concentration (mg/L) and feed pressure (bar) on desirability for the four different membranes (NF90, RO90, RO98, and RO99). NF90 and RO99 membranes showed the highest desirability for sugar concentration and inhibitory compound detoxification (Figures 4.5a and 4.5d). There is an optimum for total sugar concentration, total inhibitory compound concentration, pressure, and pH for each membrane. Lower total sugar concentration and lower operating pressure are required for higher desirability. NF90

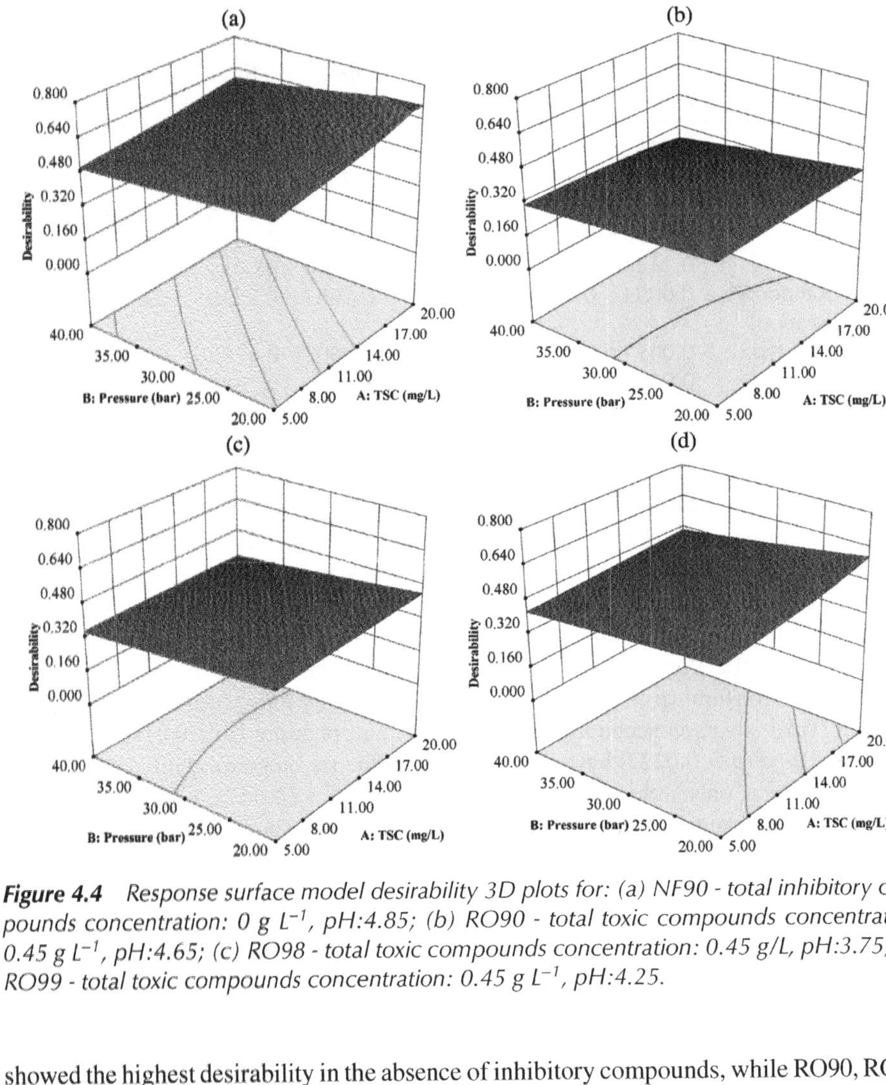

Figure 4.4 *Response surface model desirability 3D plots for: (a) NF90 - total inhibitory compounds concentration: 0 g L^{-1}, pH:4.85; (b) RO90 - total toxic compounds concentration: 0.45 g L^{-1}, pH:4.65; (c) RO98 - total toxic compounds concentration: 0.45 g/L, pH:3.75; (d) RO99 - total toxic compounds concentration: 0.45 g L^{-1}, pH:4.25.*

showed the highest desirability in the absence of inhibitory compounds, while RO90, RO98, and RO99 showed the highest desirability at total inhibitory compounds concentration of ~0.45 g/L. In addition, desirability of NF90, RO90, RO98, and RO99 were maximum at pH values of 4.85, 4.65, 3.75, and 4.25, respectively. Analysis of data indicated that total inhibitory compounds concentration is the most influential parameter on desirability of each membrane. Total sugar concentration and pressure are less influential on desirability of the membranes. In general, model sugar feed solution concentration tests highlighted that:

- NF90 displayed the highest desirability of the four membranes. This is not surprising as it has a large nominal molecular weight cut off and relatively low NaCl rejection. RO90 and RO98 displayed desirabilities that were similar over the range of operating conditions tested. However, according to the manufacturer, RO98 has much higher NaCl rejection rating. Interestingly, RO99 which has the highest rated NaCl rejection, does display

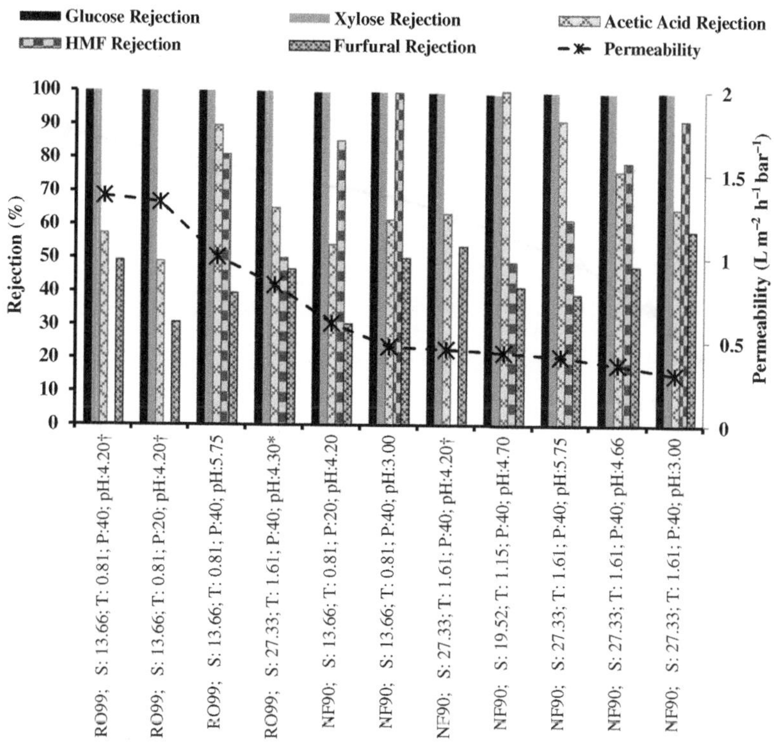

Figure 4.5 *Rejection of glucose, xylose, acetic acid, 5-hydroxymethylfurfural and furfural by RO99 and NF90 for feed streams consisting of real hydrolsyates. (∗) These experiments were conducted in duplicate. Average results are shown. (†) In these experiments, 5-hydroxymethylfurfural concentration was below the limit of detection. S: total sugar concentration (g L^{-1}) (glucose and xylose), T: total inhibitory compound concentration (g L^{-1}) (acetic acid, 5-hydroxymethylfurfural and furfural); P; pressure (bar). Dashed line shows the variation of permeability read using the secondary y-axis.*

desirabilities higher than RO90 and RO98, which could be due to the characteristics of the RO99 functional layer.

- The feed pH can affect both the membrane and dissolved solutes present. The inhibitory compounds investigated here interact with the membrane. The highest desirability occurs in the absence of the inhibitory compounds. In addition, higher pressures reduce the membrane permeability.
- Based on the results obtained from our model sugar feed tests, RO90 and RO98, which showed the lowest desirabilities, were discarded from our categoric parameters for real hydrolysate concentration tests. We conducted the real corn stover hydrolysate concentration tests with NF90 and RO99. Rejection and permeability results for real hydrolysate tests are presented in Figure 4.5. As was the case for model sugar feed solutions, it was observed that for both membranes, an approximately steady flux was obtained between 15 and 25 minutes of operation. Consequently, membrane permeability and rejection results are given after 20 minutes of operation. In addition, results are given in the order of decreasing permeability. Permeability is defined as the permeate flux divided by the feed pressure.

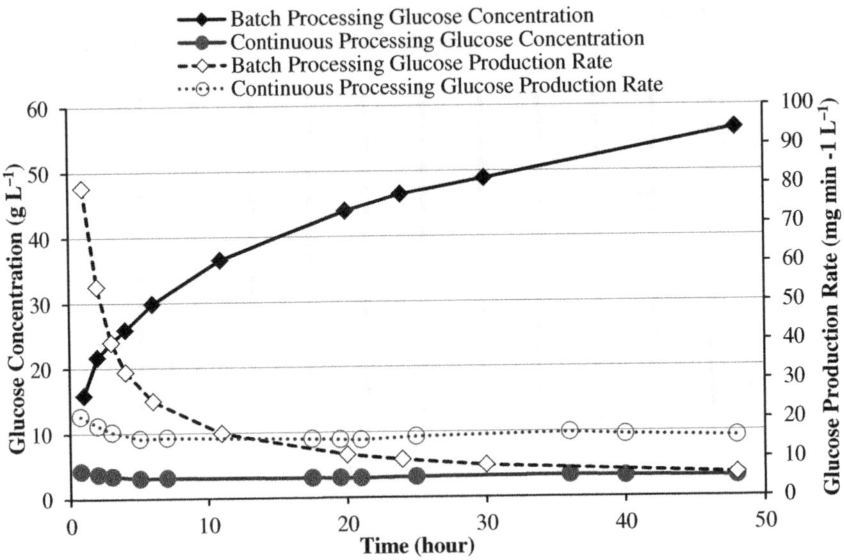

Figure 4.6 Comparison of glucose concentration and production rates for a batch and continuous (Test 3 of Table 1) hydrolysis process. Solid line with filled markers show the glucose concentration in the permeate; dashed line with open markers show the normalized rate of glucose production per unit volume of the reactor (readable from secondary axis).

The observed permeabilities for NF90 and RO99 for real hydrolysates are less than for model sugar feed solutions. Interestingly, the permeability of NF90 is always less than RO99 under the experimental condition tested with actual hydrolysate. For model sugar feed solution tests, NF90 displayed slightly higher desirabilities than RO99 in our design space. This fact shows that: (1) the presence of dissolved and suspended solutes in the real hydrolysate may lead to more severe interactions between membranes and dissolved/suspended solids; (2) higher permeability does not mean higher desirability, necessarily. There are other parameters such as sugars and inhibitory compound rejection, as well as permeability decline that must also be considered. More detailed discussions on simultaneous concentration and detoxification of biomass slurry is reported elsewhere [51].

Our results indicate that given the difficulty in selecting an optimized nanofiltration membrane for concentration of an aqueous sugar stream after enzymatic hydrolysis, a design of experiments software can be used to help identify promising membranes using model sugar feed solutions. Dead end filtration experiments permit rapid screening of membranes under a range of conditions. After identifying promising membranes further testing can be conducted using real hydrolysates. Again dead end filtration can be used to quickly assess membrane performance before conducting longer tangential flow testing.

4.8 Future Challenges

Development of an integrated membrane processes for continuous enzymatic hydrolysis could lead to significant process intensification which will be essential when developing economically viable biorefineries. Figure 4.6 compares glucose concentration and production rates for a batch and continuous hydrolysis process. As can be seen the rate of glucose production is about three times higher for the continuous process. However, the glucose concentration in the permeate is significantly lower since we are continuously removing glucose while in batch experiments the glucose continues to accumulate in the reactor. Thus inclusion of a nanofiltration step is essential in order to concentrate the glucose to optimum levels for subsequent fermentation [52,53]. In a recent publication Qian et al. [54] indicate that development of a catalytic membrane that catalyzes hydrolysis of hemicellulose and cellulose could lead to the combining of pretreatment, conditioning and enzymatic hydrolysis into one step (Figure 4.1). Such a process could lead to further process intensification.

4.9 Conclusion

Implementation of integrated membrane processes will require selection of appropriate membranes and module configurations. Development of new advanced membrane materials such as catalytic membranes that can be produced cost effectively could lead to more cost effective biorefineries. However, it will be important to consider integration of new membrane processes into the overall biorefinery. The required feed conditions for the membrane process and the conditions of the product streams from the membrane process will affect the upstream and downstream unit operations.

Acknowledgements

Experimental work was funded by the US Department of Energy (Contract No. DE-AC36-08-GO28308) and the National Renewable Energy Laboratory via subcontract AFA-1-11658-01.

List of Abbreviations

ANOVA	Analysis of variance
DI	Deionized water
MW	Molecular weight
NMWCO	Normal molecular weight cut-off

List of Symbols

D	Desirability function
d_i	Response desirability value
h	Hour
n	Number of responses
P	Pressure
p	p-value
S	Total sugar concentration
T	Total inhibitory compounds concentration

References

1. G.W. Huber, S. Iborra, A. Corma, Synthesis of transportation fuels from biomass: chemistry, catalysts, and engineering, *Chem. Rev.*, **106**, 4044–4098, 2006.
2. M.E. Himmel, S.-Y. Ding, D.K. Johnson, W.S. Adney, M.R. Nimlos, J.W. Brady, T.D Foust, Biomass recalcitrance: engineering plants and enzymes for biofuels production, *Science*, **315**, 804–807, 2007.
3. A.E. Farrell, R.J. Plevin, B.T. Turner, A.D. Jones, M. O'Hare, D.M. Kammen, Ethanol can contribute to energy and environmental goals, *Science*, **311**, 506–508, 2006.
4. E. Drioli, A. Brunetti, G. Di Profio, G. Barbieri, Process intensification strategies and membrane engineering, *Green Chem.*, **14**, 1561–1572, 2012.
5. P. Kumar, D.M. Barrett, M.J. Delwiche, P. Stroeve, Methods for pretreatment of lignocellulosic biomass for efficient hydrolysis and biofuel production, *Ind. Eng. Chem. Res.*, **48**, 3713–3729, 2009.
6. J.C. Parajó, H. Domínguez, J. Domínguez, Biotechnological production of xylitol. Part 3: Operation in culture media made from lignocellulose hydrolysates, *Bioresour. Technol.*, **66**, 25–40, 1998.
7. D.L. Grzenia, D.J. Schell, S.R. Wickramsinghe, Detoxification of biomass hydrolysates by reactive membrane extraction, *J. Memb. Sci.*, **348**, 6–12, 2010.
8. D.L. Grzenia, D.J. Schell, S.R. Wickramasinghe, Membrane extraction for removal of acetic acid from biomass hydrolysates, *J. Memb. Sci.*, **322**, 189–195, 2008.
9. S. Bower, R. Wickramasinghe, N.J. Nagle, D.J. Schell, Modeling sucrose hydrolysis in dilute sulfuric acid solutions at pretreatment conditions for lignocellulosic biomass, *Bioresour. Technol.*, **99**, 7354–7362, 2008.
10. H.-J. Huang, S. Ramaswamy, U.W. Tschirner, B.V. Ramarao, A review of separation technologies in current and future biorefineries, *Sep. Purif. Technol.*, **62**, 1–21, 2008.
11. C. Abels, F. Carstensen, M. Wessling, Membrane processes in biorefinery applications, *J. Memb. Sci.*, **444**, 285–317, 2013.
12. C.E. Wyman, What is (and is not) vital to advancing cellulosic ethanol, *Trends Biotechnol.*, **25**, 153–157, 2007.
13. R.G. Henley, R.Y.K. Yang, P.F. Greenfield, Enzymatic saccharification of cellulose in membrane reactors, *Enzyme Microb. Technol.*, **2**, 206–208, 1980.
14. F. Alfani, D. Albanesi, M. Cantarella, V. Scardi, A. Vetromile, Kinetics of enzymatic saccharification of cellulose in a flat-membrane reactor, *Biomass*, **2**, 245–253, 1982.
15. I. Ohlson, G. Trägårdh, B. Hahn-Hägerdal, Enzymatic hydrolysis of sodium-hydroxide-pretreated sallow in an ultrafiltration membrane reactor, *Biotechnol. Bioeng.*, **26**, 647–653, 1984.
16. S. Kinoshita, J.W. Chua, N. Kato, T. Yoshida, H. Taguchi, Hydrolysis of cellulose by cellulases of Sporotrichum cellulophilum in an ultrafilter membrane reactor, *Enzyme Microb. Technol.*, **8**, 691–695, 1986.
17. K. Bélafi-Bakó, A. Koutinas, N. Nemestóthy, L. Gubicza, C. Webb, Continuous enzymatic cellulose hydrolysis in a tubular membrane bioreactor, *Enzyme Microb. Technol.*, **38**, 155–161, 2006.
18. Q. Gan, S. Allen, G. Taylor, Design and operation of an integrated membrane reactor for enzymatic cellulose hydrolysis, *Biochem. Eng. J.*, **12**, 223–229, 2002.

19. P. Andrić, A.S. Meyer, P.A. Jensen, K. Dam-Johansen, Effect and modeling of glucose inhibition and in situ glucose removal during enzymatic hydrolysis of pretreated wheat straw, *Appl. Biochem. Biotechnol.*, **160**, 280–297, 2010.
20. P. Andrić, A.S. Meyer, P.A. Jensen, K. Dam-Johansen, Reactor design for minimizing product inhibition during enzymatic lignocellulose hydrolysis: II. Quantification of inhibition and suitability of membrane reactors, *Biotechnol. Adv.*, **28**, 407–425, 2010.
21. J.J. Stickel, J.S. Knutsen, M.W. Liberatore, W. Luu, D.W. Bousfield, D.J. Klingenberg, C.T. Scott, T.W. Root, M.R. Ehrhardt, T.O. Monz, Rheology measurements of a biomass slurry: an inter-laboratory study, *Rheol. Acta*, **48**, 1005–1015, 2009.
22. C.M. Roche, C.J. Dibble, J.S. Knutsen, J.J. Stickel, M.W. Liberatore, Particle concentration and yield stress of biomass slurries during enzymatic hydrolysis at high-solids loadings, *Biotechnol. Bioeng.*, **104**, 290–300, 2009.
23. H. Jørgensen, J.B. Kristensen, C. Felby, Enzymatic conversion of lignocellulose into fermentable sugars: challenges and opportunities, *Biofuels, Bioprod. Biorefining*, **1**, 119–134, 2007.
24. J.S. Knutsen, R.H. Davis, Combined sedimentation and filtration process for cellulase recovery during hydrolysis of lignocellulosic biomass, *Appl. Biochem. Biotechnol.*, **98–100**, 1161–1172, 2002.
25. W.D. Mores, J.S. Knutsen, R.H. Davis, Cellulase recovery via membrane filtration, *Appl. Biochem. Biotechnol.*, **91–93**, 297–310, 2001.
26. F. Carstensen, A. Apel, M. Wessling, In situ product recovery: Submerged membranes vs. external loop membranes, *J. Memb. Sci.*, **394–395**, 1–36, 2012.
27. M. Nakajima, T. Shoji, H. Nabetani, Protease hydrolysis of water soluble fish proteins using a free enzyme membrane reactor, *Process Biochem.*, **27**, 155–160, 1992.
28. R.A. Korus, A.C. Olson, Use of glucose isomerase in hollow fiber reactors, *J. Food Sci.*, **42**, 258–260, 1977.
29. F. Sannier, J.-M. Piot, P. Dhulster, D. Guillochon, Stability of a mineral membrane ultrafiltration reactor for peptide hydrolysis of hemoglobin, *J. Chem. Technol. Biotechnol.*, **61**, 43–47, 1994.
30. P. Pasta, G. Carrea, E. Monzani, N. Gaggero, S. Colonna, Chloroperoxidase-catalyzed enantioselective oxidation of methyl phenyl sulfide with dihydroxyfumaric acid/oxygen or ascorbic acid/oxygen as oxidants, *Biotechnol. Bioeng.*, **62**, 489–493, 1999.
31. S. Bouhallab, C. Touzé, Continuous hydrolysis of caseinomacropeptide in a membrane reactor: kinetic study and gram-scale production of antithrombotic peptides, *Lait*, **75**, 251–258, 1995.
32. M. Prata-Vidal, S. Bouhallab, G. Henry, P. Aimar, An experimental study of caseinomacropeptide hydrolysis by trypsin in a continuous membrane reactor, *Biochem. Eng. J.*, **8**, 195–202, 2001.
33. V. Cauwenberg, P. Vergossen, A. Stankiewicz, H. Kierkels, Integration of reaction and separation in manufacturing of pharmaceuticals: membrane-mediated production of S-ibuprofen, *Chem. Eng. Sci.*, **54**, 1473–1477, 1999.
34. D. Start, U. von Stockar, In situ product removal (ISPR) in whole cell biotechnology during the last twenty years, *in Process Integration in Biochemical Engineering*, U. von Stockar and L. van der Wielen (Eds), Advances in Biochemical Engineering/Biotechnology 80, Springer, Berlin/Heidelberg, 2003.

35. L. Greiner, D.H. Müller, E.C.D. van den Ban, J. Wöltinger, C. Wandrey, A. Liese, Membrane aerated hydrogenation: enzymatic and chemical homogeneous catalysis, *Adv. Synth. Catal.*, **345**, 679–683, 2003.
36. Y. Isono, M. Nakajima, H. Nabetani, Solvent-free esterification of oleic acid and oleyl alcohol using membrane reactor and lipase-surfactant complex, *J. Ferment. Bioeng.*, **86**, 138–140, 1998.
37. L.-E. Shi, G.-Q. Ying, Z.-X. Tang, J.-S. Chen, W.-Y. Xiong, H. Wang, Continuous enzymatic production of 5′-nucleotides using free nuclease P1 in ultrafiltration membrane reactor, *J. Memb. Sci.*, **345**, 217–222, 2009.
38. L. Wang, T. Khan, K. Mohanty, R. Ghosh, Cascade ultrafiltration bioreactor-separator system for continuous production of $F(ab')_2$ fragment from immunoglobulin G, *J. Memb. Sci.*, **351**, 96–103, 2010.
39. D.L. Grzenia, S.R. Wickramasinghe, D.J. Schell, Fermentation of reactive-membrane-extracted and ammonium-hydroxide-conditioned dilute-acid-pretreated corn stover, *Appl. Biochem. Biotechnol.*, **166**, 470–478, 2012.
40. S.S. Kulkarni, E.W. Funk, N.N. Li, Ultrafiltration, *in Membrane handbook*, W.S.W. Ho, K.K. Sirkar (Eds), van Nostrand Reinhold, New York, 1992.
41. B. Van der Bruggen, J. Geens, Nanofiltration, *in Advanced Membrane Technology and Applications*, N.N. Li, A.G. Fane, W.S.W. Ho, T. Matsuura (Eds), Hoboken, NJ, 2008.
42. Y.H. Weng, H.J. Wei, T.Y. Tsai, T.H. Lin, T.Y. Wei, G.L. Guo, C.P. Huang, Separation of furans and carboxylic acids from sugars in dilute acid rice straw hydrolyzates by nanofiltration, *Bioresour. Technol.*, **101**, 4889–4894, 2010.
43. H.H. Himstedt, K.M. Marshall, S.R. Wickramasinghe, pH-responsive nanofiltration membranes by surface modification, *J. Memb. Sci.*, **366**, 373–381, 2011.
44. H.H. Himstedt, H. Du, K.M. Marshall, S.R. Wickramasinghe, X. Qian, pH responsive nanofiltration membranes for sugar separations, *Ind. Eng. Chem. Res.*, **52**, 9259–9269, 2013.
45. B. Qi, J. Luo, X. Chen, X. Hang, Y. Wan, Separation of furfural from monosaccharides by nanofiltration, *Bioresour. Technol.*, **102**, 7111–7118, 2011.
46. S.K. Maiti, Y. Lukka Thuyavan, S. Singh, H.S. Oberoi, G.P. Agarwal, Modeling of the separation of inhibitory components from pretreated rice straw hydrolysate by nanofiltration membranes, *Bioresour. Technol.*, **114**, 419–427, 2012.
47. Dow Filmtec product information for NF290, December 2012.
48. Alfa Laval product information for RO90, November 2012.
49. Alfa Laval product information for RO98, November 2012.
50. Alfa Laval product information for RO99, November 2012.
51. M. Malmali, J.J. Stickel, S.R. Wickramasinghe, Sugar concentration and detoxification of clarified biomass hydrolysate by nanofiltration, *Sep. Purif. Technol.*, **132** 655–665, 2014.
52. M.L. Cazetta, M.A.P.C. Celligoi, J.B. Buzato, I.S. Scarmino, Fermentation of molasses by *Zymomonas mobilis*: effects of temperature and sugar concentration on ethanol production, *Bioresour. Technol.*, **98**, 2824–2828, 2007.
53. S. Brethauer, C.E. Wyman, Review: Continuous hydrolysis and fermentation for cellulosic ethanol production, *Bioresour. Technol.*, **101**, 4862–4874, 2010.
54. X. Qian, J. Lei, S.R. Wickramasinghe, Novel polymeric solid acid catalysts for cellulose hydrolysis, *RSC Adv.*, **3**, 24280–24287, 2013.

5

Integrated Membrane Processes for the Preparation of Emulsions, Particles and Bubbles

Goran T. Vladisavljević
Chemical Engineering Department, Loughborough University, Leicestershire, United Kingdom

5.1 Introduction

Synthetic membranes are mainly used for separation and chemical or biochemical conversion. In membrane separation processes, the feed stream is divided into two product streams of different compositions: retentate and permeate (Figure 5.1a) as a result of selective permeation through the membrane (Mulder, 1996). A shear rate is maintained at the retentate/membrane interface to limit concentration polarisation and accumulation of the rejected solids on the high-pressure side of the membrane, thereby improving the transmembrane flux. For the last two decades, microfluidic applications of membranes have been gaining in popularity, as a result of rising global interest in microfluidic technologies and significant progress in microfabrication processes. Membrane microfluidic processes can be classified into two groups: (i) formation of dispersed systems (gas–liquid, liquid–liquid, and solid–liquid) (Figure 5.1b) and (ii) treatment of dispersions (homogenisation of dispersed systems, phase separation in dispersed systems, and phase inversion). In a membrane dispersion process (Figure 5.1b), phase I is injected through a microporous membrane into phase II for the purpose of: (i) mixing two miscible solvents, usually followed by nanoprecipitation of solute(s) initially dissolved in phase I or (ii) forming droplets or bubbles of phase I into phase II. Membrane treatment of dispersions (Figure 5.1c) involves injection

Integrated Membrane Systems and Processes, First Edition. Edited by Angelo Basile and Catherine Charcosset.

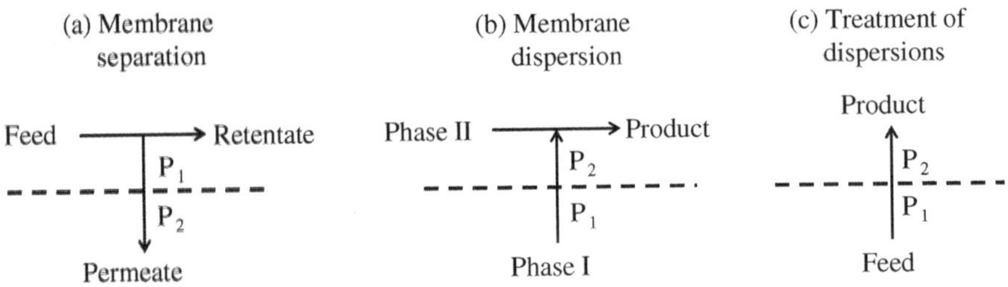

Figure 5.1 Comparison between pressure-driven membrane separation and membrane microfluidic processes, where $P_1 > P_2$. In a membrane separation process (a), feed stream (a dispersion or solution) is split into two product streams of different composition (Mulder, 1996). In a membrane dispersion process (b), two streams (miscible or immiscible) are combined together to form a product dispersion. Membrane treatment of dispersions (c) involves passing the feed dispersion through the membrane, which results in the modification of the particle size distribution in the original dispersion and/or phase inversion.

of the feed dispersion through the membrane, which results in the physicochemical and mechanical interactions between the dispersed phase passing through the pores and the surface of the pore walls leading to the modification of the original particle-size distribution and/or phase separation.

5.1.1 Membrane Dispersion Processes

Membrane dispersion processes include direct membrane emulsification (DME) (Nakashima et al., 2000), membrane-based gas dispersion (formation of micro and nanobubbles (Kukizaki and Goto, 2007; Kukizaki and Goto, 2006), and membrane micromixing (Chen et al., 2004). A shear is applied at the membrane/continuous phase interface to improve mixing efficiency or facilitate the detachment of the dipsersed phase from the membrane surface. In DME, one liquid (a dispersed phase) is injected through a microporous membrane into another immiscible liquid (the continuous phase) (Nakashima et al., 1991) leading to the formation of droplets at the membrane/continuous phase interface (Figure 5.2a). Hydrophobic membranes are needed to produce water-in-oil (W/O) emulsions (Cheng et al., 2008; Jing et al., 2006), and hydrophilic membranes are required to prepare oil-in-water (O/W) emulsions (Figure 5.2a). In membrane microbubbling, a pressurised gas is forced through a hydrophilic membrane into aqueous surfactant solution, leading to the formation of microbubbles (1 μm $< d_{bubble} <$ 1 mm) or nanobubbles (1 nm $< d_{bubble} <$ 1 μm), depending on the pore size of the membrane and operating conditions (Figure 5.2b). Micromixing is interpenetration of miscible solutions at the molecular level and it is a crucial step in any homogeneous reaction (Okhonin et al., 2011). Membrane micromixing is interpenetration of miscible solutions via a microporous membrane (Figure 5.2c). Usually, the solution passing through the membrane is an organic solution containing water-miscible volatile organic solvent and particle-forming material or an aqueous salt solution. Membrane micromixing is often combined with nanoprecipitation to produce inorganic (Chen et al., 2004) or organic (Laouini et al., 2011) nanoparticles.

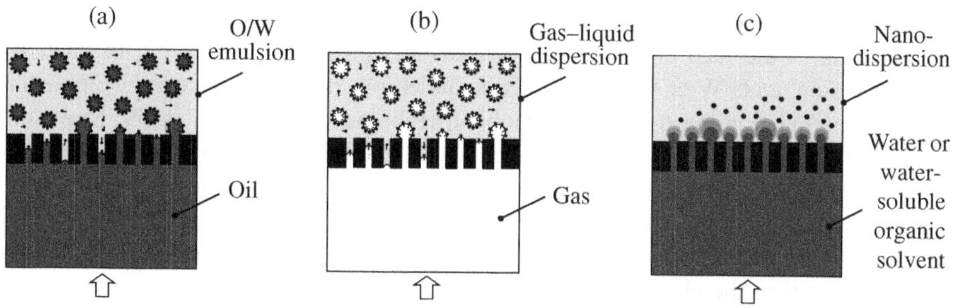

Figure 5.2 *Membrane dispersion processes with hydrophilic membrane. (a) Production of liquid–liquid dispersion (O/W emulsion) by DME (Nakashima et al., 2000). (b) Production of gas–liquid dispersion by microbubbling (Kukizaki and Goto, 2007c) or nanobubbling (Kukizaki and Goto, 2006). (c) Production of solid–liquid (nanoparticle) dispersion by membrane micromixing/nanoprecipitation method (Chen et al., 2004).*

5.1.2 Membrane Treatment of Dispersions

Membrane processes used to modify particle-size distribution of dispersions can be classified into four groups: (a) simple premix membrane emulsification (PME) (Suzuki et al., 1996); (b) PME with phase inversion (Suzuki et al., 1996); (c) membrane demulsification (Kukizaki and Goto, 2008); and (d) homogenisation of suspensions (e.g. coarse vesicle suspension) by extrusion through the membrane (Olson et al., 1979). In PME (Figure 5.3a), a pre-emulsion is forced through a microporous membrane (Suzuki et al., 1996) or a packed bed of uniform particles (van der Zwan et al., 2008; Yasuda et al., 2010; Laouini et al., 2014). As in DME, hydrophobic and hydrophilic membranes are needed to produce W/O and O/W emulsions, respectively. If the transmembrane pressure is lower than the capillary pressure in a pore, the membrane will reject the droplets, while allowing a pure continuous phase liquid to pass through, which will lead to the separation of the feed emulsion into a droplet-free continuous phase and concentrated emulsion (Koltuniewicz et al., 1995).

If the dispersed phase of the feed emulsion wets the membrane wall, the rate of droplet coalescence in the membrane pores becomes faster than the rate of droplet breakup, which

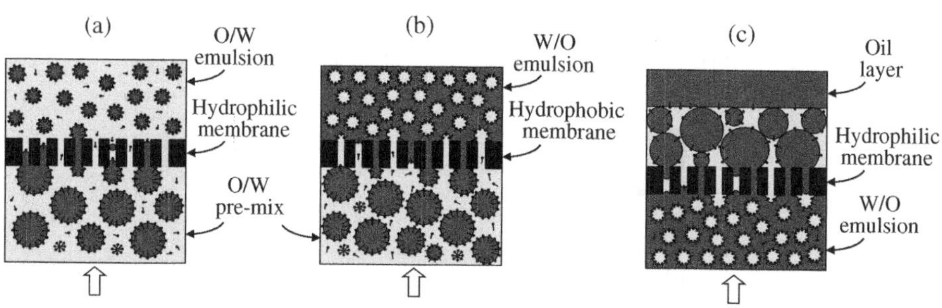

Figure 5.3 *Treatment of emulsions using membranes: (a) Production of O/W emulsion by PME (Suzuki et al., 1996); (b) Production of W/O emulsion by PME with phase inversion (Suzuki et al., 1999); (c) Demulsification of W/O emulsion (Kukizaki and Goto, 2008).*

leads to inversion of phases in the emulsion passing through the membrane (Figure 5.3b) or separation of the feed emulsion into two distinct phases (Figure 5.3c). In PME with phase inversion, an O/W or W/O/W emulsion undergoes inversion into a W/O emulsion as a result of permeation through a hydrophobic membrane (Suzuki et al., 1999; Kawashima et al., 1991). Similarly, a W/O or O/W/O emulsion can be inverted into an O/W emulsion by permeation through a hydrophilic membrane. A successful phase inversion requires feed emulsion to contain a blend of surfactants with a low and high hydrophilic–lipophilic balance (HLB) number (Suzuki et al., 1999) or otherwise, the emulsion breaking is more likely to occur than the phase inversion.

5.1.3 Comparison of Membrane and Microfluidic Drop Generation Processes

Generation of droplets and bubbles in microfluidic devices usually involves injection of one fluid through a single microchannel (MC) into a stream of second immiscible fluid flowing through another perpendicular channel (T-junction) (Thorsen et al., 2001) or break-up of coaxial jets of immiscible fluids flowing through a narrow orifice (flow-focusing device) (Anna et al., 2003). The droplets and bubbles generated in microfluidic devices are highly uniform in size, with a variation of droplet sizes in the dripping regime of 3% or less (the relative span 0.25 or less), and the drop generation frequency can exceed 10,000 Hz (Yobas et al., 2006). However, the volume flow rate of the dispersed phase in microfluidic devices is very low, usually 10^{-2}–10 mL h^{-1}, because there is typically only one injection channel, thus only one droplet generation unit (DGU). In contrast, a membrane has a multitude of pores that are DGUs in membrane emulsification (ME). Membrane pores resemble massively parallel T-junctions but there are some distinct differences in the geometry between the two DGUs, as outlined below.

The main difference between the DME and T-junction is that the D_2/D_1 ratio is greater than 10^2 in DME and less than 10 in T-junction, where D_2 and D_1 is the hydraulic diameter of the continuous phase and dispersed phase channel, respectively (Figure 5.4). Consequently in a T-junction, an emerging droplet occupies almost the entire cross section of the main

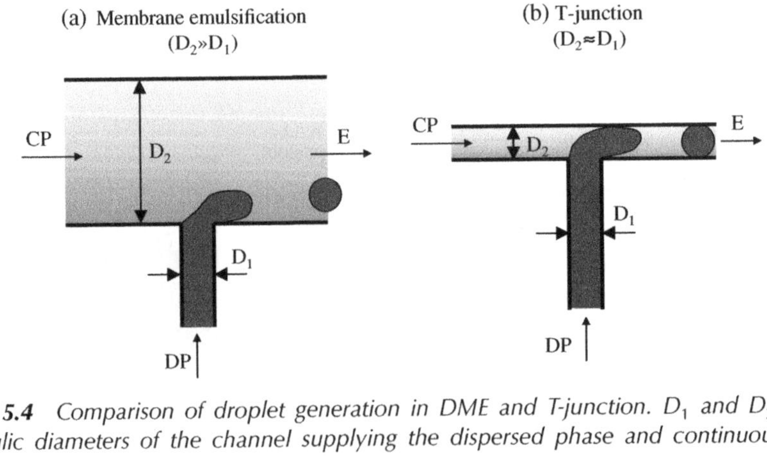

Figure 5.4 Comparison of droplet generation in DME and T-junction. D_1 and D_2 are the hydraulic diameters of the channel supplying the dispersed phase and continuous phase, respectively.

channel at low capillary numbers of the continuous phase ($Ca_c < 0.015$) and droplet breakup is mainly caused by the pressure gradient across the droplet while it is protruding into the main channel, which is known as the plugging–squeezing mechanism (Seemann et al., 2012; Vladisavljević et al., 2012). In DME, the cross-sectional area available for flow of the continuous phase is much greater than the cross-sectional area of the droplet and a pressure gradient across the emerging droplet is negligible, preventing the occurrence of the squeezing regime. In addition, at high flow rates of the continuous phase ($Ca_c > 0.3$), T-junction operates in the jetting regime characterised by the formation of long threads of the dispersed phase prior to breakup. In DME, the continuous phase channel is so large that viscous shear force from the surrounding aqueous phase is not high enough to pull the dispersed phase into a long jet. Instead, at high dispersed phase flow rates DME operates in the continuous outflow regime (Kobayashi et al., 2003) characterised by the formation of big droplets that retain its spherical shape during the entire growth phase without formation of a long thread, as evidenced by real-time microscopic observations of drop generation (Vladisavljević et al., 2007, 2011).

The main advantage of ME devices compared to microfluidic drop generators is in higher droplet productivity. The volume flow rate of dispersed phase in PME is above 10^4 mL h^{-1} for the product emulsion with 10 μm oil droplets and the membrane area below 4 cm^2 (Vladisavljević et al., 2004b). Considering that commercial membrane modules have much larger membrane areas, while the scale-up of microfluidic devices is challenging due to significant pressure drop in microfluidic channels and difficulties of controlling the flow rate of individual fluid streams in long channel networks, it is clear that membrane devices are more suitable for high throughput applications. On the other hand, the variation of droplet sizes in ME is above 10%, which corresponds to the relative span of above 0.4. Another advantage of microfluidic over membrane devices is in their ability to produce droplets with a complex morphology and to manipulate individual droplets with high precision after production. In addition, in ME processes only one type of dispersed phase flows through all the pores and all the droplets are collected in a single bulk channel. Therefore, the droplets cannot be individually manipulated after production and multiple emulsions can only be formed via a two-step emulsification process. In microfluidic drop generators, the continuous phase channel is so narrow that the droplets often form a single layer in the outlet channel with a constant inter-droplet distance and can easily be manipulated (sorted, split, merged with another droplets or continuous liquid streams and loaded with a controlled amount of encapsulated material) as they are moving through the outlet channel. Furthermore, different fluids can be simultaneously supplied through different entry channels, enabling single-step generation of complex droplets, such as core–shell droplets, droplets with constant controlled number of inner droplets, droplets containing inner droplets of two distinct phases and bifacial (Janus) droplets.

5.1.4 Comparison of Membrane and Conventional Homogenisation Processes

Compared to high shear rotor–stator devices, high-pressure valve homogenisers, ultrasonic and static mixers, membrane dispersion devices operate under mild shear stress conditions, allowing constant temperature operation and high yields of inner droplets in multiple emulsion production (Surh et al., 2007; Vladisavljević and Williams, 2008; Dragosavac et al., 2012). Conventional emulsification techniques are not suitable when preparing emulsions with shear sensitive ingredients, because they apply high-energy inputs per unit volume and high shear rates to disrupt droplets (Karbstein and Schubert, 1995). In DME, the shear rate on the membrane surface is in the range of (1–50) × 10^3 s^{-1} but uniform droplets can be produced even in the absence of shear by exploiting the "push-off force" arising from the droplet–droplet interactions before detachment (Kukizaki, 2009; Kukizaki and Goto, 2009; Kosvintsev et al., 2008). A shear rate in rotor–stator devices such as high-shear in-line mixers and colloid mills is (1–2) × 10^5 s^{-1} and can reach 10^7 s^{-1} in a microfluidisers®. In PME, the energy input per unit volume is typically 0.1–5 kJ dm^{-3}, while it is 10–200 kJ dm^{-3} in high-pressure valve homogenisers. In addition, energy input in conventional dispersion devices is not spatially uniform. For example, in rotor–stator devices, shear forces are high in close proximity to a rotor and low in "dead zones", leading to the production of polydispersed emulsions. On the other hand, in the majority of membrane dispersion processes, shear is uniformly distributed over the membrane surface.

Another advantage of ME compared to conventional emulsification devices is that membrane systems allow integration of emulsification step and emulsion post-processing to achieve simultaneous drop generation and chemical/biochemical conversion or physicochemical transformation. The examples include integration of ME with liquid–liquid extraction (Chen et al., 2004c, Xu et al., 2005), biphasic enzymatic transformation (Li and Sakaki, 2008; Mazzei et al., 2010), pervaporation (Chang and Hatton, 2012), and complex coacervation (Piacentini et al., 2013).

5.2 Membranes for Preparation of Emulsions and Particles

Membranes used for preparation of dispersions should have the following properties: (i) uniform pores with a wide range of available mean pore sizes to suit different applications; (ii) low hydrodynamic resistance; (iii) high mechanical strength and thermal and chemical resistance; (iv) high tolerance to organic solvents used in formulations; (v) ease of surface modification of membrane (variation of contact angle, charge, permeability, etc.); (vi) low fabrication costs per unit membrane area. In this chapter, only membranes most commonly used in membrane dispersion processes will be discussed.

5.2.1 SPG Membrane

Shirasu porous glass (SPG) meets the majority of the above-mentioned criteria and it is the most widely used microporous membrane in membrane dispersion processes. Advantages of SPG membrane over microengineered membranes are in higher porosity, more versatile surface chemistry that can be used to modify the pore walls, broader range of pore sizes available, and lower fabrication costs.

5.2.1.1 Fabrication of SPG Membrane

SPG membrane is fabricated from Na_2O–CaO–Al_2O_3–B_2O_3–SiO_2 or Na_2O–CaO–MgO–Al_2O_3–B_2O_3–SiO_2 type mother glass through phase separation by spinodal decomposition (Nakashima and Kuroki, 1981; Nakashima and Shimizu, 1986; Kukizaki and Nakashima, 2004). The mother glass is prepared by mixing and melting raw materials (Shirasu, limestone, and boric acid) at about 1350°C. Typical mixing ratios of raw materials are given in

Table 5.1 *Typical mixing ratios of raw materials in the production of SPG from Na_2O–CaO–Al_2O_3–B_2O_3–SiO_2 mother glass (Nakashima, 2002)*.*

	wt%
Shirasu	51
Limestone	23
Boric acid	22
Soda ash	4

*MgO (≈5 wt%) can also be added.

Table 5.1. Soda ash (Na_2CO_3) and sometimes MgO and ZrO_2 are added to molten glass to adjust the rate and temperature of phase separation and alkaline durability of the glass. Shirasu is a volcanic ash deposit from southern Kyushu, which contains 72–77 wt% SiO_2, 10–15 wt% Al_2O_3, and small amounts of other inorganic oxides (Table 5.2). Molten mother glass is shaped into tubes or flat discs by blowing and then heat treated at 650–750°C for the period ranging from several hours to several tens of hours. The thermal treatment causes a homogeneous glass melt to separate into an acid-insoluble (Al_2O_3–SiO_2 rich) phase and acid-soluble (CaO–B_2O_3 rich) phase (Figure 5.5). The phase-separated glass is then immersed into a hydrochloric acid solution to dissolve CaO–B_2O_3 rich phase, which results in the formation of porous skeleton, whose composition is shown in Table 5.2. The porosity of SPG membrane is determined by the volume fraction of the acid-soluble phase in the phase separated mother glass and ranges between 50% and 60% (Vladisavljević et al., 2005). If the fraction of acid-soluble phase is too low or too high, separation takes place by the nucleation and growth mechanism. The nucleation and growth mechanism occurs in the metastable region of the phase diagram shown in Figure 5.6, between the spinodal and binodal lines, and leads to the formation of discrete spherical particles of one phase embedded in a continuous matrix of the other. This morphology is undesirable in the fabrication of SPG membrane and must be avoided.

Table 5.2 *Composition of primary glass*, SPC* and porous Vycor glass and Pyrex glass (Nakashima et al., 1992; Nakashima, 2002).*

	Primary glass for SPG (wt%)	SPG (wt%)	Vycor® glass (wt%)	Pyrex® glass (wt%)
SiO_2	49	69	94–99.5	81
Al_2O_2	10	13	0–0.5	2
CaO	17	2	–	–
B_2O_3	16	7	0.2–6.0	13
Na_2O	5	5	<0.1	4
K_2O	2	4	–	–
Fe_2O_3	1	0.4	–	–

*Based on the proportions of raw materials given in Table 5.1.

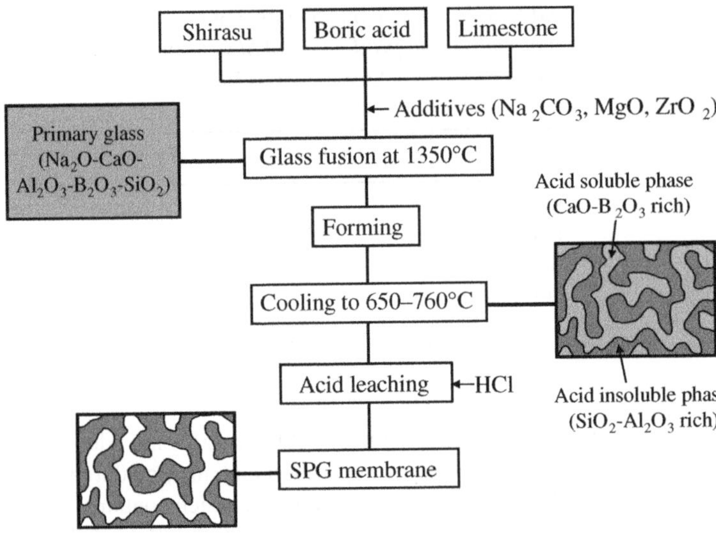

Figure 5.5 *A flow diagram of different steps involved in the fabrication of SPG membrane. Reprinted from reference Vladisavljević et al. (2012) with permission from Springer.*

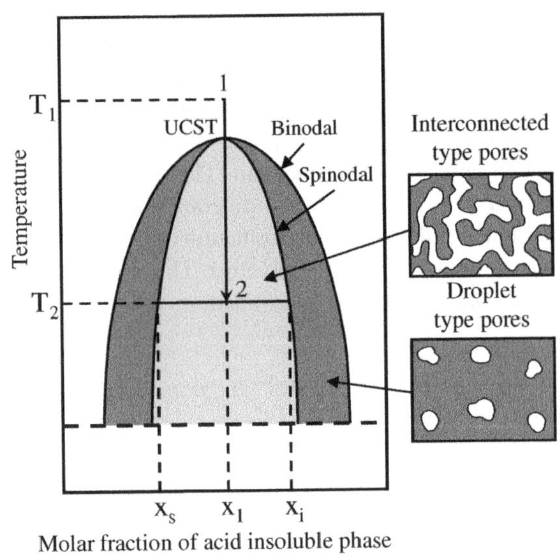

Figure 5.6 *Spinodal decomposition of glass induced by cooling mother glass from an initial temperature T_1 to temperature T_2 lying in the spinodal region (within the spinodal line). To prevent phase separation via nucleation, a transition from the stable to the spinodal region of the phase diagram must proceed quickly or through the upper critical solution temperature (UCST).*

Figure 5.6 depicts spinodal decomposition induced by cooling a homogeneous glass melt from a temperature T_1 (~1350°C) at which all components are miscible in all proportions to a temperature T_2, which lies within the spinodal (unstable) region. A homogeneous glass with composition of x_1 is separated into two immiscible phases with compositions of x_s and x_i. The ratio of acid-soluble to acid-insoluble phase (x_s/x_i) can be found by the lever rule and is equal to $(x_i-x_1)/(x_1-x_s)$. The mean pore diameter d_p of SPG membrane can be controlled by the time, t and temperature, T_2 of the heat treatment process (Kukizaki, 2010):

$$d_p = 4K^{1/2}(V_p/m_m)t^{1/2}\exp[-E_a/(2RT_2)] \quad (5.1)$$

where K is a constant depending on the composition of mother glass, $E_a = 400$–600 kJ mol^{-1} is the activation energy for diffusion during phase separation (Nakashima, 2002; Kukizaki, 2010), $R = 8314$ kJ kmol^{-1} K^{-1} is the universal gas constant, and V_p/m_m is the total pore volume per unit mass of dry membrane. Therefore, the mean pore diameter of SPG membrane is proportional to the square root of the heating time at constant temperature, whereas a logarithm of the mean pore diameter is inversely proportional to $1/T_2$ at constant heating time. It means that fine pores can be obtained by keeping a high temperature during phase separation for a short time.

5.2.1.2 *Properties of SPG Membrane*

SPG membrane is available from SPG Technology Ltd. (Sadowara, Japan) with a mean pore sizes ranging from 0.040 to 40 µm (Table 5.3). The membrane has a uniform internal microstructure, as confirmed by x-ray microtomography (XMT) (Vladisavljević et al., 2007), characterised by interconnected cylindrical pores with a tortuosity factor of $\xi \approx 1.3$. On SEM and XMT images, the pores have a non-cylindrical shape (Figure 5.7), because they extend in all directions and pore junctions can also be seen on the surface. The number of pores per unit cross-sectional area of SPG membrane is given by (Vladisavljević et al., 2005);

$$N/A_m = 0.56/d_p^2 \quad (5.2)$$

Table 5.3 *Properties of commercial isotropic (symmetric) SPG membrane (Vladisavljević et al., 2005; Nakashima, 2002; Kukizaki, 2009b; Nakashima et al., 1992).*

Shape	Tubes or flat discs
Thickness (δ_m)	0.4–1 mm
Compressive strength	200–280 MPa
Pore diameter (d_p)	0.04–40 µm
Porosity (ε)	50–60%
True density	2000–2500 kg m^{-3}
Zeta potential at pH = 3–10 and C_{NaCl} = 1–100 mol m^{-3}	−15 to −45 mV
Pore tortuosity (ξ)	1.25–1.4
Number of pores per unit cross-sectional area (N/A_m)	10^9–10^{14} m^{-2}
Specific pore volume (V_p/m_m)	0.5–0.6 dm^3 kg^{-1}
Hydraulic resistance ($R_{m,sym}$)	10^8–10^{12} m^{-1}

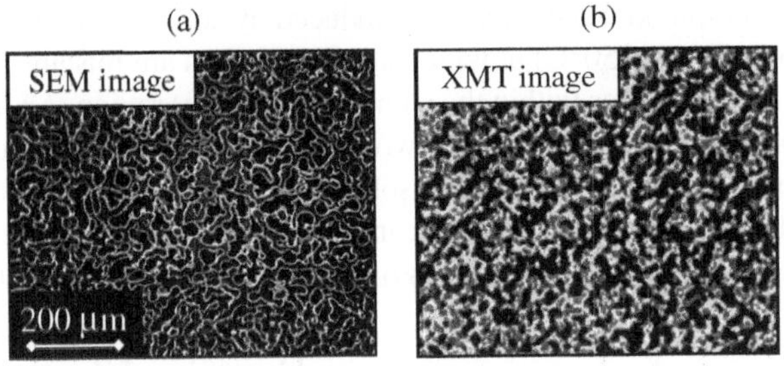

Figure 5.7 (a) Scanning electron micrograph of the surface of SPG membrane polished with diamond paste and used for visualisation of ME by metalographic microscope; (b) XMT image of SPG membrane. Reprinted from Vladisavljević et al. (2007), with permission from Elsevier.

where N/A_m and d_p are in m^{-2} and m, respectively. The hydraulic resistance of isotropic SPG membrane is given by (Vladisavljević et al., 2005):

$$R_{m,sym} = 32\xi^2\delta_m/\left(d_p^2\varepsilon\right) \tag{5.3}$$

where δ_m is the membrane thickness and ε is the membrane porosity. The hydraulic resistance of isotropic SPG membrane is relatively high (Table 5.3), due to its substantial thickness of 400–1000 μm, but can be reduced if the membrane is fabricated with anisotropic structure (Kukizaki and Goto, 2007b). Assuming that the pore tortuosity and porosity, ξ and ε, are independent on the pore size. The hydraulic resistance of anisotropic SPG membrane is given by (Kukizaki and Goto, 2007b):

$$R_{m,asym} = 32\xi^2\delta_{skin}/\left(d_{p,skin}^2\varepsilon\right) + 32\xi^2\delta_{sup}/\left(d_{p,sup}^2\varepsilon\right) \tag{5.4}$$

where δ_{skin} and δ_{sup} are the thicknesses of the skin and support layer, respectively, and $d_{p,skin}$ are $d_{p,sup}$ their mean pore diameters. According to Kukizaki and Goto (2007b), the thickness of the skin layer is 6% of the overall membrane thickness and the ratio of the pore diameters in the skin and support layer is around 7, so it can be written:

$$R_{m,asym} = 32(\xi^2/\varepsilon)\left(\delta_{skin}/d_{p,skin}^2 + \delta_{sup}/d_{p,sup}^2\right) = R_{m,sym}(0.06/1 + 0.94/7^2) \approx 0.08R_{m,sym}$$

Therefore, the hydraulic resistance of asymmetric SPG membrane is just 8% of that for symmetric membrane of the same total thickness.

A comparison of chemical composition of SPG and common porous glasses is shown in Table 5.2. SPG is more stable in water and alkaline solutions than Porous Vycor® glass, because it contains less SiO_2 and more Al_2O_3. However, the durability of both membranes at high pH is poor, due to the attack of hydroxide ions on siloxane (Si–O–Si) bonds:

$$\equiv \text{Si–O–Si} \equiv + \text{OH}^- \rightarrow \equiv \text{Si–O}^- + \equiv \text{Si–OH}$$

Alkaline durability of SPG can be improved by incorporating about 3 mol% ZrO_2 into the glass skeleton, which results in to the formation of stable Zr–O–Si bonds in the silicate network (Kukizaki, 2010). The compressive strength of SPG of 200–280 MPa is much

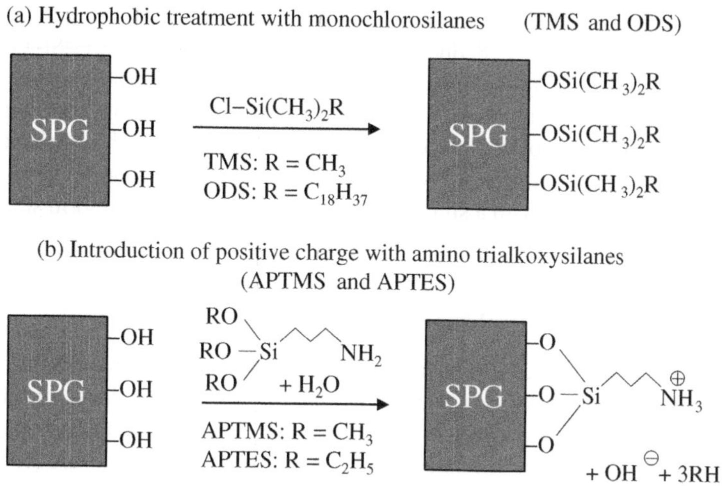

Figure 5.8 *Chemical modification of SPG surface by treatment with organosilane compounds: (a) Hydrophobic treatment with monochlorosilanes (Kai et al., 2006); (b) Introduction of amino groups by amino trialkoxysilanes to render the surface positively charged.*

higher than that of porous alumina or zirconia of the same porosity (Nakashima et al., 1992), because SPG is made up of a continuous glass skeleton with very few defects, while porous alumina or zirconia is composed of skeletal grains joined together discontinuously via grain boundaries.

5.2.1.3 Surface Modification of SPG Membrane

The surface of SPG membrane can be hydrophobised by chemical reaction with organosilane compounds such as chlorosilanes (Kukizaki and Wada, 2008) or physical coating with silicone resin (Vladisavljević et al., 2005). Monochlorosilanes such as trimethylchlorosilane (TMS) and octadecyldimethylchlorosilane (ODS) are suitable agents for hydrophobisation because they contain only one chlorine atom, which means that no polymerisation between silane molecules can occur while they react with a silanol group on the pore surface (Figure 5.8a) (Kai et al., 2006). The longer the carbon chain length in the organosilane compound, more hydrophobic the membrane surface becomes (Kukizaki and Wada, 2008). The membrane hydrophobicity can be enhanced by depositing silica nanoparticles onto the surface of SPG membrane prior to treatment with TMS (Meng et al., 2013). The surface of SPG membrane can also be made with thermoresponsive hydrophilic–hydrophobic properties by depositing silica nanoparticles containing poly(N-isopropylacrylamide) (PNIPAM) brushes grafted on their surface (Meng et al., 2010). The porosity and hydraulic resistance of SPG membrane can be modified over a wide range by synthesising dextran macromolecules within the pores via *in-situ* enzymatic reaction between dextransucrase and sucrose (Kawakita et al., 2009; Seto et al., 2011). A reversible change in the hydraulic resistance of dextran-loaded SPG membrane is a consequence of reversible extension and shrinkage of solvent-responsive dextran chains inside the pores initiated by washing the membrane with water or organic solvent, respectively.

The surface of untreated SPG surface has a negative zeta potential between −15 and −45 mV within a pH range of 2–8, due to dissociation of silanol groups (≡Si–OH D ≡SiO$^-$ + H$^+$) (Kukizaki, 2009). A positive charge on the membrane surface can be induced by treating the membrane with amino trialkoxysilanes, such as (3-aminopropyl)-trimethoxysilane (APTMS) and (3-aminopropyl)-triethoxysilane (APTES) (Figure 5.8b). Amino trialkoxysilanes undergo hydrolysis in aqueous solution resulting in the formation of silanol groups that can be then condensed with a silanol group on the SPG surface to form stable siloxane bond (Si–O–Si).

5.2.2 Microengineered Membranes

Microengineered membranes are used in ME to achieve high transmembrane fluxes at low transmembrane pressures (Wagdare et al., 2010). However, because of a low surface area, the dispersed phase flow rate through a microengineered membrane is often lower than that in SPG membrane modules. Microengineered membranes are microfiltration membranes with a controlled pore geometry and spatial arrangement manufactured by semiconductor fabrication methods (Brans et al., 2006). Typical microsieves used in ME are nickel microengineered membranes manufactured using UV-LIGA process (Nazir et al., 2011; Schadler and Windhab, 2006; Egidi et al., 2008), silicon nitride Aquamarijn^{TM} microsieves fabricated by reactive ion etching (RIE) (van Rijn et al., 1997), stainless steel membranes fabricated by pulsed laser drilling (Dowding et al., 2001; Vladisavljević and Williams, 2006; Geerken et al., 2008) or end-milling (Kobayashi et al., 2008), and single crystal silicon MC arrays fabricated by deep reactive ion etching (DRIE) (Kobayashi et al., 2003). Microengineered membranes are ultra-thin foils with rectilinear pores and very low internal pore area and porosity, so they are less prone to fouling by emulsion ingredients than highly tortuous and porous SPG membranes. It is especially important in PME, where a whole emulsion, rather than a pure dispersed phase liquid, is injected through the membrane.

LIGA is a German acronym for *LIthographie, Galvanoformung, Abformung*, which means lithography, electroplating, and moulding. The fabrication of nickel microengineered membranes by the UV-LIGA process consists of three main steps: UV lithography (UVL), nickel electroplating and membrane release (Figure 5.9a). UVL starts with spin coating positive photoresist (typically SU-8) on an electrically conductive substrate. The photoresist is exposed to UV light through a mask that determines the geometry of the pores. The photoresist is then developed to remove the exposed parts of the photoresist, rinsed in deionised water and dried. The membrane is then electroformed in a nickel electroplating bath by depositing a nickel film in the voids left by the removed photoresist. The nickel foil will be perforated at the regions on the substrate masked by the photoresist. Starting from a particular size of the photoresist areas on the substrate, the final perforation diameter can be adjusted as small as desired by overgrowth of nickel over the masked areas in the electroplating operation. The membrane is released from the substrate using an ultrasound bath. Pure nickel membrane is hydrophobic, but can be made hydrophilic by oxygen plasma treatment, coating with diamond like carbon, and other hydrophilic agents. Nickel microsieves can be produced with a variety of different pore morphologies, including slotted pores (Holdich et al., 2006) and cylindrical micronozzles protruding above the membrane surface (Geerken, 2006). The pore size can be reduced by plasma enhanced chemical vapour deposition (CVD) (Schadler and Windhab, 2006).

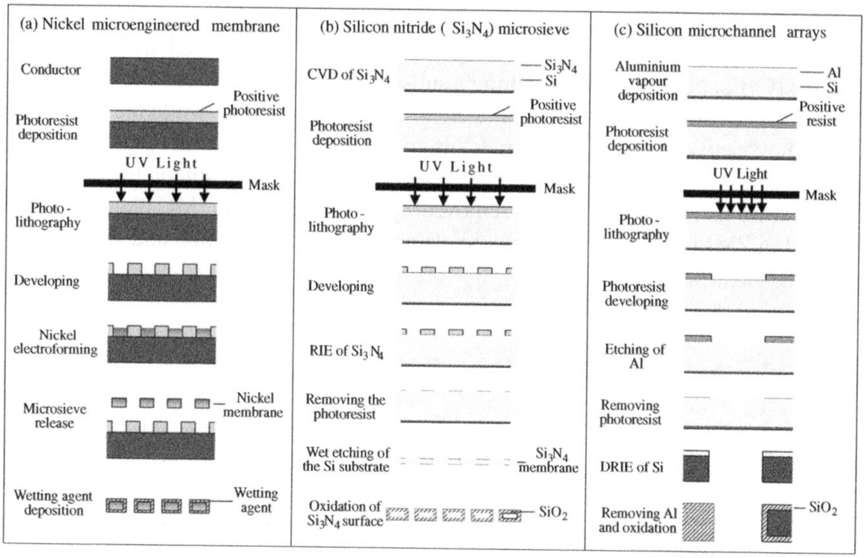

Figure 5.9 *Fabrication of nickel, silicon nitride, and single crystal silicon microengineered membranes used in integrated membrane dispersion processes. Source: Reprinted from Vladisavljević et al. (2012) with permission from Springer Science and Business Media.*

The fabrication process used for Aquamarijn® microsieves is described in Figure 5.9b. It consists of four main steps: CVD of silicon nitride film on a single crystal silicon substrate, photolithography, RIE of silicon nitride film, and anisotropic wet etching of the substrate. First, a layer of silicon nitride is deposited on the surface of the substrate by CVD, which involves a reaction between dichlorosilane (SiH_2Cl_2) and ammonia at elevated temperature and low pressure. A positive photoresist is then applied to the surface of the silicon nitride film and exposed to UV light through the mask. After exposure, the photoresist is developed in a dilute NaOH solution to remove the exposed photoresist and form the photoresist mask having openings. The silicon nitride exposed in these openings is etched by RIE in CHF_3/O_2 plasma to form a pattern of pores corresponding in locations to the openings in the photoresist mask. After dry etching of silicon nitride, the silicon underneath the membrane is etched by anisotropic wet etching using a KOH solution. Finally, the microsieve is treated with air plasma to grow hydrophilic silicon dioxide layer on the membrane surface. The thickness of Aquamarijn® microsieves is only 0.1–5 μm because the RIE process is not suitable for larger membrane thickness. Because of the low membrane thickness, it is more difficult to obtain uniform droplets, because the capillary pressure is very low and very small fluctuations in pressures may cause large differences in transmembrane flux. Single crystal silicon membranes with a thickness of about 100 μm produced by DRIE are very suitable for a robust monodisperse droplet generation.

The fabrication process for single crystal silicon membranes consists of the four main steps shown in Figure 5.9c: (i) CVD of aluminium film on a single crystal silicon substrate, (ii) photolithography, (iii) aluminium film etching, and (iv) DRIE of silicon substrate. The DRIE process patented by Robert Bosch GmbH involves three steps: (a) etching a shallow trench into silicon substrate using sulphur hexafluoride (SF_6) plasma; (b) passivating that newly formed cavity with fluorocarbon polymer ($-(CF_2)_x-$), created with octafluorocyclobutane (C_4F_8) plasma, and (c) etching a subsequent and deeper trench with SF_6 plasma. Passivation with polymer prevents lateral etching of the sidewalls, while the hole becomes deeper. The reactive species (neutral radicals and ions such as F^* and SF_n^+), formed by the collision of SF_6 molecules with high-energy electrons, react with silicon forming SiF_4 gas (e.g. $Si + 4F^* \rightarrow SiF_4$), that can be removed by a vacuum pump. MC plates can be also fabricated in PMMA by x-ray lithography using synchotron radiation and subsequent wet etching (Kobayashi et al., 2008b).

5.3 Production of Emulsions Using SPG Membrane

SPG membrane was widely used for production of macroemulsions by DME (Vladisavljević et al., 2004; Vladisavljević and Schubert, 2002) and PME (Vladisavljević et al., 2004b, 2006a, 2006b). The advantages of PME over DME are in smaller droplet sizes (Figure 5.10) and higher transmembrane fluxes that can be achieved for any given pore size. On the other hand, in PME a more severe membrane fouling and broader particle-size distribution can be expected, compared to DME.

Various membrane modules have been used in DME: (i) cross-flow module with tubular SPG membrane with a length of the membrane tube up to 500 mm; (ii) short SPG membrane tube with a length of 7–15 mm immersed in a stirred continuous phase (internal or external pressure microkit), and (iii) rotating SPG membrane tube immersed in an otherwise stagnant continuous phase. In a batch cross-flow system, a continuous phase liquid circulates from a storage tank through the bore of the membrane tube, and back to the storage tank (Figure 5.11). A dispersed phase-forming liquid stored in a pressure vessel is fed to the outside of the membrane tube and penetrates through the membrane under the pressure difference which is 1.1–5 times higher than the capillary pressure (Vladisavljević and Schubert, 2003a). The apparatus is operated until a desired dispersed phase concentration is achieved in the product emulsion. The recirculation of emulsion at high flow rate can

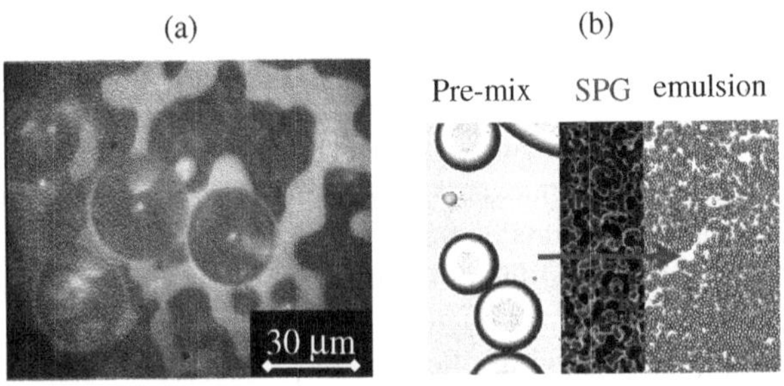

Figure 5.10 (a) Micrograph of droplets formed on the surface of SPG membrane in DME (Vladisavljević et al., 2007); (b) Micrographs of droplets before PME and after passing five times through 8-µm SPG membrane Reprented from (Vladisavljević et al. 2007; 2013), with permission from Elsevier.

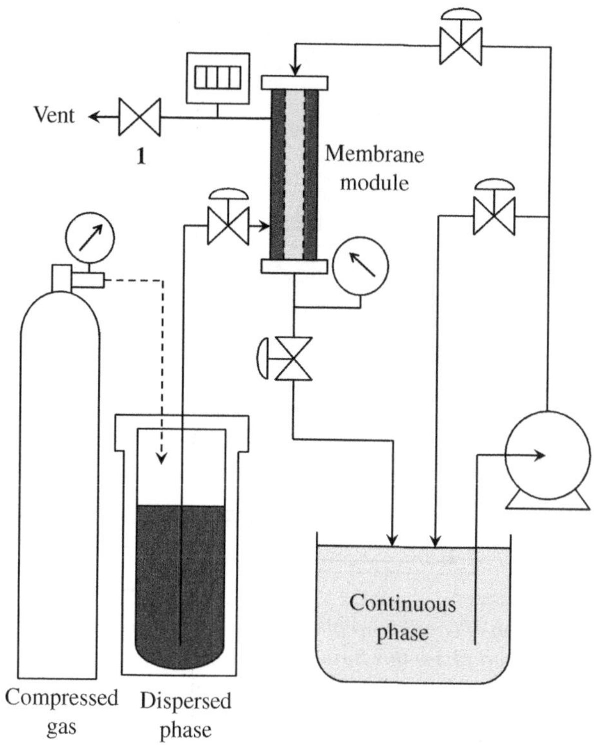

Figure 5.11 Apparatus for cross-flow DME using tubular SPG membrane. During start-up, valve 1 must be open to remove any trapped air from the module (Nakashima et al., 1994).

lead to uncontrolled droplet breakup. Two methods have been used to decrease a flow rate of the continuous phase in the recirculation loop, while keeping the same shear stress at the membrane surface: insertion of static turbulence promoters in the lumen of the membrane tube (Koris et al., 2011) and introduction of back-and-forward pulsations in the continuous phase flow (Piacentini et al., 2013b). The particle-size distribution of the product emulsion with a dispersed phase content of 30% (v/v) was compared by operating the process for 8.5 hours using conventional cross-flow and pulsed cross-flow. When the emulsion was produced by the pulsed cross-flow, the mean particle size and size distribution were constant during the whole time while emulsion became polydispersed after 3 hours of conventional cross-flow operation due to the recirculation effect.

Transmembrane flux in cross-flow DME should be kept below a certain critical value (1–30 L m^{-2} h^{-1}) to obtain uniform droplets with a relative span factor of 0.25–0.45. Therefore, the flow rate ratio (Q_d/Q_c) is very low and the product stream must be recirculated to achieve a reasonably high dispersed phase concentration. Transmembrane flux can be increased by two orders of magnitude by introducing the continuous phase into SPG tube radially, as shown in Figure 5.12. A tangential introduction of the continuous phase generates spiral streamlines in the axial direction (swirl flow) that exert a strong centrifugal

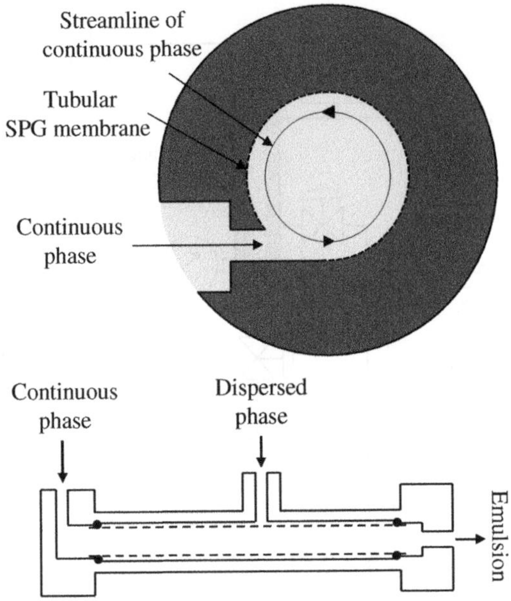

Figure 5.12 Introduction of continuous phase in cross-flow module from the tangential direction to improve a dispersed phase flux through the membrane (Shimoda et al., 2011).

force onto the inner surface of the membrane helping to sweep away droplets from the membrane surface (Shimoda et al., 2011). At the swirl-flow velocity of 0.85–5.4 m s^{-1} and the transmembrane flux of 0.3–3 m^3 m^{-2} h^{-1}, a relative span factor of 0.45–0.64 was achieved with an oil phase/water phase volume ratio in single-pass operation of up to 0.4 (Shimoda et al., 2011).

Cross-flow systems are easy to scale up and offer a constant shear stress along the membrane surface. However, at least several hundred millilitres of the continuous phase must be used in the system to provide recirculation. SPG test kit shown in Figure 5.13a. requires less than 50 mL of the continuous phase and can be operated with a very low hold-up volume of both phases, which is useful for expensive samples, such as medical preparations (Higashi and Setoguchi, 2000). In this kit, the continuous phase is kept under agitation by a magnetic stir bar or impeller, while the dispersed phase is injected through the membrane tube from outside to inside. The membrane tube also serves as a draft tube, which results in more effective circulation of the continuous phase than in the internal pressure SPG kit.

In addition to DME with static SPG membrane, where shear stress is controlled by providing fluid flow over the membrane surface, dynamic SPG membrane has been developed, where shear is controlled by rotating the membrane within a static continuous phase (Pawlik and Norton, 2012, 2013). Rotating SPG membrane can be operated batchwise or continuously. In a continuous flow operation, surface shear is decoupled from the cross-flow velocity, which means that sufficient shear on the membrane surface can be achieved no matter how small the flow rate of the continuous phase is. Therefore, emulsions with a

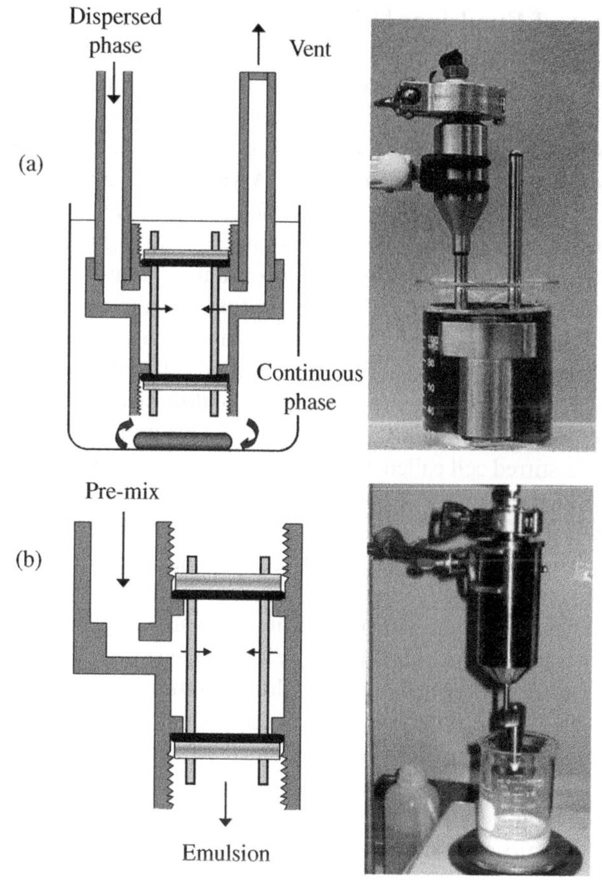

Figure 5.13 *External pressure type microkits available by SPG Technology Co., Ltd (Sadowara, Japan) for (a) DME and (b) PME. The kits are supplied with SPG membrane tube with an effective length of 10–15 mm.*

high dispersed phase concentration can be produced without emulsion recirculation, which can help to prevent damage to shear sensitive material and secondary breakup of the drops formed by the membrane.

SPG membrane rig for PME is shown in Figure 5.13b. A pressurised pre-mix from a pressure vessel is passing through the membrane tube from outside to inside under the driving force of pressure difference ranging from several bars (for a 10-μm membrane) to more than 10 bar (for 1-μm membrane) and up to 50 bar for the membrane with sub-micron pore sizes. The product emulsion is collected inside the membrane tube and is discharged by gravity from the bottom of the tube. In addition, for reducing the droplet size and improving droplet size uniformity, product emulsion can be passed repeatedly through the membrane (Vladisavljević et al., 2004b, 2006a, 2006b). Membrane homogenisation using repeated cycles was first developed by Olson et al. (1979) and used for homogenisation of lipid vesicles using track-etch polycarbonate filters.

5.4 Production of Emulsions Using Microengineered Membranes

Microengineered membranes have been widely used in production of macroemulsions. A shear on the surface of microengineered membrane has been produced using cross-flow (Abrahamse et al., 2002), paddle stirrer (Kosvintsev et al., 2005), rotating (spinning) membrane (Vladisavljević and Williams, 2006; Aryanti et al., 2006), vibrating membrane (Holdich et al., 2010), and pulsed cross-flow (Holdich et al., 2013). Kosvintsev et al. (2005) developed a constant-shear membrane dispersion cell with a geometry similar to 'plate and cone' rheometer by replacing normally impervious plate underneath the cone with a nickel microengineered membrane. The dispersed phase from underneath the membrane was injected into the continuous phase under constant shear provided by a rotating cone. However, it has been shown that a simple paddle-stirred cell with non-uniform shear profile on the membrane surface provides the same degree of drop size uniformity as the modified Weissenberg rheometer with a constant shear-stress operation (Kosvintsev et al., 2005). This simple paddle-stirred cell called dispersion cell (Figure 5.14a) is increasingly used for batch production of emulsions (the batch size ~100 mL) (Dragosavac et al., 2008, 2012; Egidi et al., 2008).

In order to achieve a continuous production, the cross-flow velocity must be very low and hence, shear stress on the membrane surface must be controlled by alternative methods. The shear was decoupled from the cross-flow velocity by rotating or vibrating microengineered membrane within a static continuous phase (Vladisavljević and Williams, 2006; Holdich et al., 2010; Zeng et al., 2013) or introducing flow pulsations in the continuous phase flow (Holdich et al., 2013). The rotating stainless steel membrane with laser drilled pores is shown in Figure 5.14b. The membrane is mounted on an overhead stirrer to control the rotation speed. The dispersed phase is introduced inside the membrane tube and passes through the pores by a hydrostatic head difference. The droplets are detached from the membrane surface due to combined action of centrifugal force and viscous shear force at low angular velocities, but Taylor vortices play a significant role at high angular velocities. The oscillating ME system developed by Micropore Technologies Ltd. is shown in Figure 5.14c. The membrane is a hollow cylinder with conical stainless steel bottom or in the form of a hollow ring. The oscillation signal generated by an audio generator is amplified by a power amplifier and used to drive the electromechanical oscillator on which the inlet manifold and membrane tube are mounted. The droplet size was found to depend on the peak shear stress on the membrane surface but was independent on the individual values of the frequency or amplitude (Holdich et al., 2010). The same conclusion was confirmed by Zeng et al. (2013) using oscillatory woven metal micro-screen. Pulsed back-and-forward cross-flow is another method of obtaining emulsions with a high dispersed to continuous phase ratio continuously, without recirculation of the product stream. The potential benefits of pulsed flow over membrane vibration are: (i) liquid pulsations require less energy than membrane vibrations because liquids have a lower inertia than solids; (ii) the consumption of energy to maintain pulsed flow is independent on the membrane size, whereas in a vibrating ME system the energy consumption is proportional to the membrane surface area, and (iii) pulsed flow in the continuous phase can be extended to a baffled reactor, connected in series to the membrane module, to achieve simultaneous drop generation and reaction in the produced emulsion (Piacentini et al., 2013).

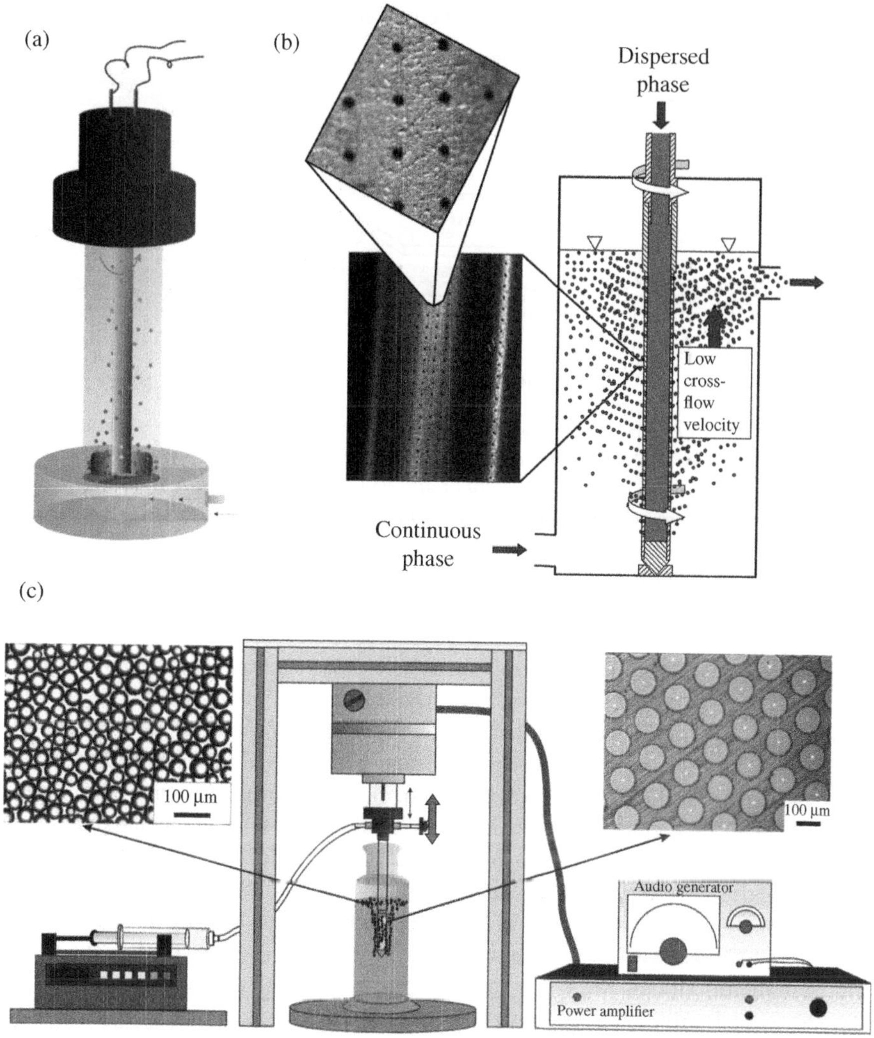

Figure 5.14 (a) A batch stirred cell for membrane emulsification commercialised by Micropore Technologies (Dragosavac et al., 2008). (b) A continuous flow membrane emulsification system consisting of rotating stainless steel membrane with 100 μm laser drilled pores (Vladisavljević and Williams, 2006). (c) vibrating membrane emulsification system consisting of a nickel membrane with a 10 μm pore size and 180 μm pore spacing oscillating with a frequency of 10–90 Hz and an amplitude of 1–5 mm (Holdich et al., 2010).

Drop generation in straight-through MC arrays developed in Nakajima Lab is shown in Figure 5.15. These membranes are used in cross-flow modules containing a transparent cover glass and the height of the cross-flow channel between the membrane surface and the cover glass is less than 1 mm, so that the drop generation process can be visually controlled using a microscope video system. Symmetric membranes shown in Figure 5.15a. have a

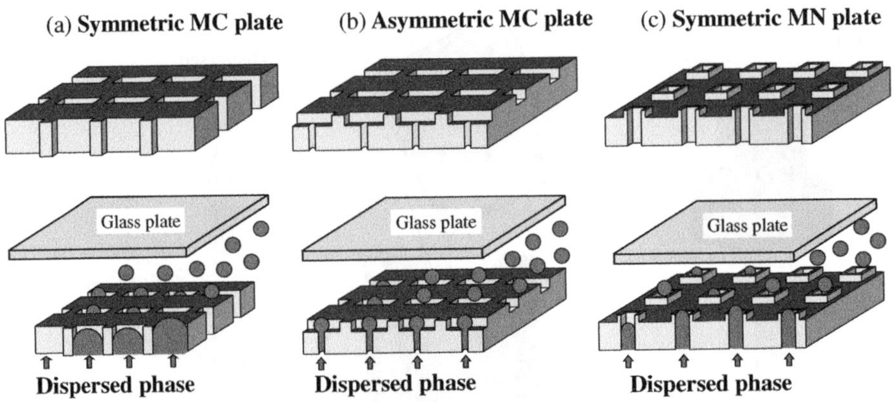

Figure 5.15 *Single crystal silicon membrane systems with different pore geometries developed in Nakajima Lab at National Food Research Institute, Japan. (a) Symmetric membrane with slotted pores on both sides (Kobayashi et al., 2002). (b) Asymmetric membrane with circular pores on upstream side and slotted pores on downstream side (Kobayashi et al., 2005). (c) Symmetric membrane with circular micronozzles extended above the surface (Sugiura et al., 2005).* Source: *Reprinted from Vladisavljević et al. (2012) with permission from Springer Science and Business Media.*

constant geometry of the channels over the whole cross section. The membrane size can range from 15 × 15 to 40 × 40 mm and the total number of MCs can exceed 200,000 (Kobayashi et al., 2005b). Rectangular MCs with an aspect ratio of at least three are more suitable for monodispersed drop generation than circular MCs (Kobayashi et al., 2004, 2009a; van Dijke et al., 2010). Asymmetric membrane shown in Figure 5.15b. contains circular channels on the upstream (bottom) side of the membrane and rectangular channels on the downstream (top) side (Kobayashi et al., 2005; Vladisavljević et al., 2008, 2011). This asymmetric geometry is particularly useful for generation of uniform droplets when the droplet phase viscosity is 1 mPa s or less, for example, when the dispersed phase is water or a volatile hydrocarbon (Kobayashi et al., 2005, 2009b). The size of droplets produced using asymmetric silicon MCs can range from just several microns (Kobayashi et al., 2008c) to several millimetres (Kobayashi et al., 2008a). Single crystal silicon membranes with circular micronozzle array shown in Figure 5.15c. consist of circular micronozzles having outlets extending above the plate surface. They can be used to increase the velocity of the continuous phase at the channel outlet, which could be useful if the viscosity of the dispersed phase is high (Sugiura et al., 2005).

5.5 Factors Affecting Droplet Size in DME

The size distribution of droplets produced in DME depends on many factors, that can be divided into membrane parameters (surface wettability and charge, porosity, pore size, pore morphology and spatial arrangement), formulation parameters (viscosity of the dispersed and continuous phase, type and concentration of surfactants and additives) and process parameters (shear stress generated on the membrane surface and transmembrane flux) (Joscelyne and Trägårdh, 2000).

5.5.1 Effect of Transmembrane Pressure and Flux

The minimum transmembrane pressure for driving the dispersed phase through the pores is known as the capillary pressure, P_{cap}, and is given by the Young–Laplace equation:

$$P_{cap} = \frac{4\gamma_{wo}\cos\theta}{d_p} \tag{5.5}$$

where γ_{wo} is the equilibrium interfacial tension between the dispersed and continuous phase, θ is the contact angle, that is, the angle formed by the dispersed phase at the three-phase boundary where the continuous phase, dispersed phase and membrane intersect (Figure 5.16). A hydrophilic membrane ($\theta < 90°$) is used in the production of O/W emulsion, and thus $P_{cap} > 0$ and $P_o > P_w$. A hydrophobic membrane ($\theta > 90°$) is used in the production of W/O emulsion, and thus $P_{cap} < 0$ and $P_o < P_w$, that is, the water phase pressure should be higher than the oil phase pressure by P_{cap} to drive the water phase through the membrane.

Two mechanisms of drop formation in DME are: (a) shear-controlled detachment as a result of shear stress on the membrane surface and (b) spontaneous detachment driven by interfacial tension (Sugiura et al., 2002). Shear-controlled detachment occurs in membranes with circular rectilinear pores (Kosvintsev et al., 2005), while spontaneous droplet detachment exists in membranes with tortuous and non-circular pores, such as SPG membrane and single crystal silicon membranes shown in Figure 5.15. In shear-controlled droplet formation process, bigger droplets are formed at higher transmembrane flux (line 1 in Figure 5.17). It was usually explained by the assumption that the detachment of a droplet

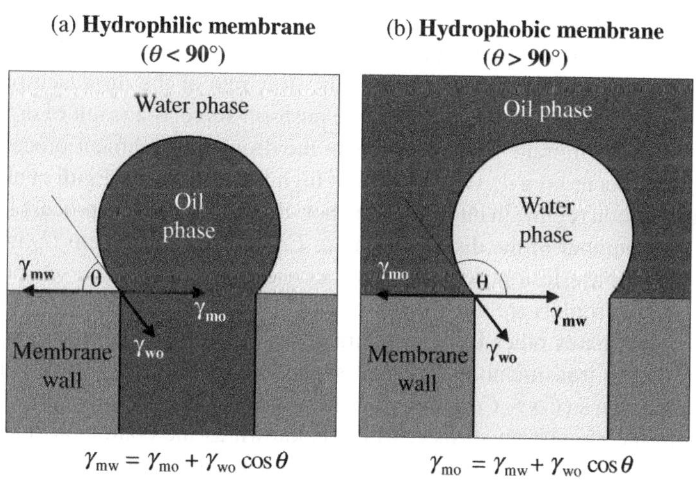

Figure 5.16 *Typical contact angles through the water phase and phase pressures encountered in membrane emulsification: (a) Production of O/W emulsion ($\theta < 90°$, $P_o > P_w$); (b) Production of W/O emulsion ($\theta > 90°$, $P_o < P_w$). The contact angle θ is the angle measured through the water phase, where a liquid/liquid interface meets a membrane surface. γ_{mw}, interfacial tension between the membrane and water phase, γ_{mo}, interfacial tension between the membrane and oil phase, γ_{wo}, interfacial tension between the water and oil phase.*

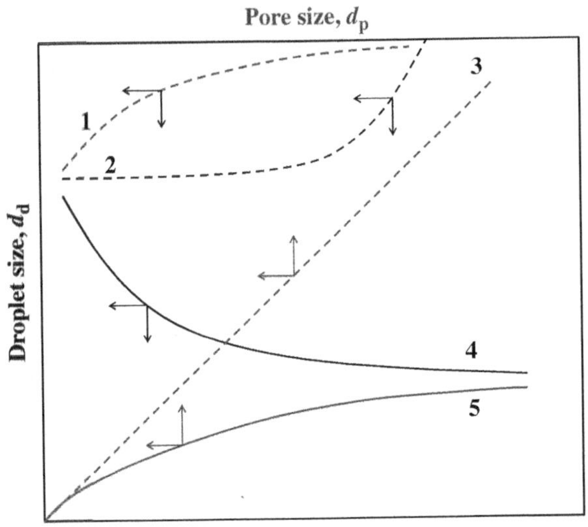

Figure 5.17 Mean droplet size, d_d in DME (dashed lines) and PME (solid lines) as a function of mean pore size, d_p and transmembrane flux, J. (1) d_d versus J in shear-controlled DME. (2) d_d versus J in interfacial-tension-driven DME. (3) d_d versus d_p in interfacial-tension-driven DME in dripping regime. (4) d_d versus J in PME. (5) d_d versus d_p in PME.

from the membrane surface is not instantaneous but requires a finite time, the necking time, during which an additional amount of dispersed phase will flow into the droplet (van der Graaf et al., 2006). An inflow of the dispersed phase into a forming droplet during the detachment phase increases with increasing the flow rate of the dispersed phase causing the droplet size to increase. At high fluxes, the push-off force as a result of droplet–droplet interaction on the membrane surface assists in the droplet detachment process, causing a plateau region to occur on a d_d versus J plot at high fluxes (line 1) (Egidi et al., 2008).

Droplet generation regime in interfacial-tension-driven detachment process is determined by the capillary number of the dispersed phase: $Ca = U_d\eta_d/\gamma_{wo}$, where U_d is the velocity of the dispersed phase in a pore and μ_d is the viscosity of the dispersed phase. At low fluxes, $Ca < Ca_{cr}$, droplets are formed in the dripping regime. In this regime, the interfacial tension force dominates other force (Sugiura et al., 2002) and the droplet size is almost independent on the transmembrane flux or shear stress (line 2 in Figure 5.17). At high transmembrane fluxes ($Ca > Ca_{cr}$), droplets grow to large sizes ($d_d > 10d_p$) before being detached from the membrane surface, which is known as the continuous outflow regime (Kobayashi et al., 2003). In this regime, the viscous force dominates the interfacial tension force and the droplet size sharply increases with increasing the transmembrane flux. The transmembrane flux at which the transition from dripping to continuous outflow regime occurs is independent on the pore size (Kobayashi et al., 2011) and increases with decreasing the viscosity of the dispersed phase (Vladisavljević et al., 2011). DME must be conducted in the dripping regime, since the emulsions produced in the continuous outflow regime are polydisperse, due to the random nature of the droplet formation process. The transition from dripping to continuous outflow does not occur simultaneously for all the pores, leading to bimodal particle size distribution of the droplets as a result of the large differences in the droplet size in the dripping and continuous outflow regime.

5.5.2 Influence of Pore (Channel) Size and Shear Stress on the Membrane Surface

For interfacial-tension-driven droplet formation, a linear correlation between the mean droplet size and the mean pore size of SPG membrane exists in dripping regime: $d_d = K'd_p$ (line 3 in Figure 5.17), where $K' = 2.8$–3.5 for SPG membrane (Kukizaki and Goto, 2009, 2007c; Nakashima et al., 1991; Vladisavljević et al., 2006). For MC arrays shown in Figure 5.15: $d_d = K'D_1$, where D_1 is the hydraulic diameter of the dispersed phase channels and $K' \approx 3$. A slope of d_d versus d_p line slightly increases with decreasing the shear stress on the membrane surface, but even without any shear, $K' = 3.3$ for 1% Tween-80 stabilised O/W emulsions produced using SPG membrane (Kukizaki and Goto, 2009).

In shear-controlled droplet formation process, the mean droplet size is determined by a balance between the shear force exerted on the liquid–liquid interface by the continuous phase, F_d and the capillary force, F_{ca} (Kosvintsev et al., 2005):

$$F_{ca} = \pi d_p \gamma \tag{5.6}$$

$$F_d = 9\pi\tau_w d_d \sqrt{(d_d/2)^2 - r_p^2} \tag{5.7}$$

where r_p is the pore radius and τ_w is the shear stress on the membrane surface. At $J \to 0$, the effect of transmembrane flux can be neglected and the droplet breakup occurs when $F_{ca} = F_d$:

$$d_d = \frac{\sqrt{18\tau_w^2 r_p^2 + 2\sqrt{81\tau_w^4 r_p^4 + 4r_p^2\tau_w^2\gamma^2}}}{3\tau_w} \tag{5.8}$$

The mean drop diameter decreases with increasing shear stress on the membrane surface until it reaches $2r_p$ at very high shear stresses. The shear stress in different DME systems is given in Table 5.4 and the variations of the shear stress with time or position along the membrane surface are shown in Figure 5.18 (Laouini et al., 2013). In a pulsed cross-flow system, flow oscillations are superimposed on a low cross-flow velocity, but the contribution of the cross-flow to the shear stress can be neglected (Piacentini et al., 2013).

5.5.3 Influence of Surfactant

The role of surfactant in ME is to rapidly adsorb to the newly formed oil–water interface to reduce the interfacial tension and facilitate droplet detachment, thus stabilising the formed droplets against coalescence. The effect of kinetics of adsorption of surfactant at oil–aqueous interface during DME on the droplet size has been investigated by Schröder et al. (1998), van der Graaf et al. (2004), and Rayner et al. (2005). As a rule, the faster the surfactant molecules adsorb to the newly formed interface, the smaller the droplet size becomes. Surfactant molecules must not adsorb to the membrane surface, since otherwise the dispersed phase will spread over the membrane surface instead of being

Table 5.4 *Shear stress on the membrane surface in DME for different membrane shapes and shear generation techniques.*

Shear generation technique	Membrane shape	Shear stress on membrane surface
Cross-flow	Cylindrical	$\tau_w = \dfrac{8\eta_c U_c}{d_{mi}}$
Cross-flow	Flat	$\tau_w = \dfrac{3Q_c\eta_c}{2h^2W}$
Rotating	Cylindrical	$\tau_w = \dfrac{2\eta_c r_{mo}^2\omega}{r_b^2 - r_{mo}^2}$
Paddle stirrer	Flat	$\tau_{w,max} = \dfrac{0.825\eta_c\omega r_{trans}}{\sqrt{\eta_c/(\rho_c\omega)}}$ *
Oscillating	Any	$\tau_{w,max} = (2\pi A f)^{3/2}(\eta_c\rho_c)^{1/2}$

$^{*}r_{trans} = 1.23\frac{D}{2}\left(0.57 + 0.35\frac{D}{D_{cell}}\right)\left(\frac{b}{T}\right)^{0.036} n_b^{0.116}\mathrm{Re}/(1000 + 1.43\mathrm{Re})$

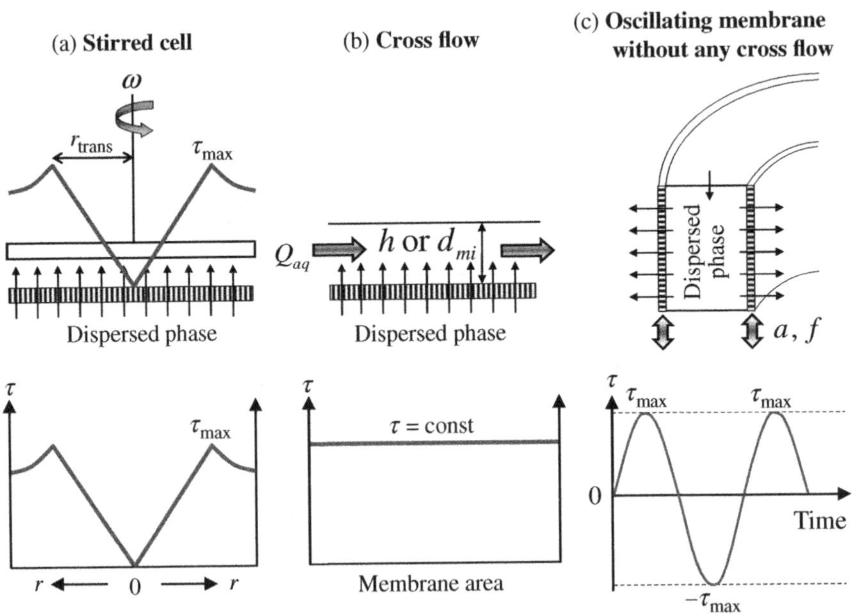

Figure 5.18 *Spatial or temporal variations of the shear stress at the membrane surface in different DME systems: (a) stirred cell; (b) cross flow; (c) oscillating membrane without any cross flow. The variation of the shear stress with time for pulsed flow of the continuous phase is exactly the same as in oscillation membrane system shown in (c).*

confined within the pores, which leads to uncontrolled droplet generation. This means that the functional groups of surfactant molecules must not carry a positive charge to avoid electrostatic deposition onto the negatively charged surface of SPG membrane (Nakashima et al., 1993). The use of cationic surfactants, for example, alkyltrimethylammonium salts such as cetyltrimethylammonium bromide (CTAB) leads to polydispersed O/W emulsions with $d_d/d_p > 20$ (Nakashima et al., 1993). The use of zwitterionic surfactants must also be avoided, even when they carry a net negative charge. It was found that lecithin at pH 3 fouls SPG membrane due to electrostatic interactions between positively charged groups ($-N(CH_3)_3^+$ and $-NH_3^+$) on phospholipid molecules and negatively charged silanol groups on SPG surface, although at pH 3 the net charge on lecithin molecules is negative (Surh et al., 2008). To produce cationic droplets using SPG membrane, the membrane must be pre-treated with amino trialkoxysilanes to induce a positive charge on the membrane surface (Figure 5.8b) or the charge of anionic droplets must be altered after ME by surfactant displacement (Vladisavljević and McClements, 2010).

5.6 Factors Affecting Droplet Size in PME

The mean droplet size in PME depends on several parameters such as the mean pore size of SPG membrane, transmembrane pressure, number of passes through the membrane, viscosity of the continuous and dispersed phase and interfacial tension (Nazir et al., 2010). The mean droplet size is a non-linear function of the mean pore size (line 5 in Figure 5.17):

$$d_d = K(d_p)^n \tag{5.9}$$

where $n < 1$. The d_d/d_p ratio decreases with increasing the mean pore size and ranges from 1 to 1.5 for SPG membrane with $d_p = 5$–20 μm at the shear stress on the pore wall of 200 Pa (Vladisavljević et al., 2006a). The critical pressure in PME is given by (Park et al., 2001):

$$P_{cap} = \frac{\gamma[2 + 2a^6/\sqrt{2a^6 - 1} \times \arccos(1/a^3) - 4a^2]}{a + \sqrt{a^2 - 1}} \tag{5.10}$$

where $a = d_{pm}/d_p$ and d_{pm} is the mean droplet size in pre-mix. If $a \gg 1$, the capillary pressure is given by Eq. (5). In PME, the optimum transmembrane pressure is 10–50 times larger than P_{cap} (Vladisavljević et al., 2004b). The mean droplet size decreases with increasing the mean shear stress on the pore walls, given by:

$$\tau_{w,p} = 8\eta_e J\xi/(\varepsilon d_p) \tag{5.11}$$

where η_e is the viscosity of emulsion inside the pores. Therefore, in PME smaller droplets can be produced at higher transmembrane pressures, as shown by line 4 in Figure 5.17. In repeated PME, the droplet size is reduced by increasing the number of passes through the membrane (Vladisavljević et al., 2004b; Laouini et al., 2014).

5.7 Integration of ME with Solid/Semi-Solid Particle Fabrication

In the early 1990s, ME was used for the preparation of O/W or W/O emulsions with a narrow particle size distribution and controlled mean particle size (Nakashima et al., 1991). Since that time, the application of ME have been extended to the production of multiple emulsions, such as solid-in-oil-in-water (S/O/W) (Kukizaki, 2009c), oil-in-water-in-oil (O/W/O) (Wei et al., 2013; Cho et al., 2005) and water-in-oil-in-water (W/O/W) (Surh et al., 2007), nano- and micro-emulsions (Koga et al., 2010; Oh et al., 2011; Laouini et al., 2012; Choi et al., 2012; Pradhan et al., 2013; Oh et al., 2013), emulsions with droplets laminated with multilayered biopolymer films (Vladisavljević and McClements, 2010; Gudipati et al., 2010; Nazir et al., 2012), microbubbles (Kukizaki and Goto, 2007), nanobubbles (Kukizaki and Goto, 2006), micro- and nanoparticles (Vladisavljević and Williams, 2005, 2010), and vesicles (liposomes and niosomes) (Hwang et al., 2011; Pham et al., 2012).

Examples of particles fabricated by DME or PME using SPG membrane are given in Table 5.5. Emulsion droplets were transformed into solid particles by implementing different chemical reactions or physicochemical processes within the droplets, such as crosslinking of hydrogel forming polymers (Wei et al., 2013), polymerisation of monomer mixtures (Omi et al., 2005), solidification from a melt (Kukizaki and Goto, 2007c), polymer precipitation induced by solvent evaporation or extraction (Liu et al., 2005), redox reaction (Kakazu et al., 2010), complex coacervation (Kage et al., 1997), and thermal coagulation (El-Mahdy et al., 1998).

5.7.1 Integration of ME and Crosslinking of Gel-forming Polymers

Hydrogels are three-dimensional networks composed of crosslinked hydrophilic polymers. Hydrogels are insoluble in water but can absorb and hold large amount of water due to their hydrophilic nature (Hoare and Kohane, 2008). A three-dimensional network can be formed by chemical gelation (free-radical polymerisation or polycondensation) or physical gelation, which can involve heating (heat-setting gels), cooling (cold-setting gels) or electrostatic interaction (ionotropic gelation). Ionotropic gelation can be used if the gel-forming polymer contains charged functional groups, such as amino groups of chitosan or carboxylic group of alginate, that can interact with oppositely charged divalent or polyvalent crosslinkers. The main strategies used for ME/ionotropic gelation are internal gelation, external gelation and droplet coalescence (Figure 5.20).

5.7.1.1 Internal Gelation

In the internal gelation method (Figure 5.20a), the dispersed phase contains a gel-forming polymer and a crosslinking agent in a non-dissociated (inactive) form (e.g. Ca^{2+} in the form of $CaCO_3$ particles). The beads are formed by adding a species (e.g. H^+) that diffuses into the droplets and triggers the release of the crosslinking agent in its active form. In the case of $CaCO_3/H^+$/alginate system, the released Ca^{2+} ions bind to the L-guluronate residues of alginate, crosslinking the polymer and causing the droplets to gel. The H^+ ions can be added by dropping glacial acetic acid into the resultant W/O emulsion under stirring, as shown in Table 5.5A. The size of $CaCO_3$ particles should be at least one order of magnitude smaller than the pore size to prevent pore blockage. When the dispersed phase contained $CaCO_3$

Table 5.5 Microparticles fabricated by integration of DME and PME using SPG membrane and various secondary reactions/processes.

Product	Example	Secondary reaction/process after DME or PME	References
Ceramic particles	Silica nano- or microparticles	Polymerisation of silicic acids by interfacial or internal reaction	Kandori et al. (1992)
Liquid crystal particles	Thermochromic liquid crystal particles	Melt crystallisation in O/W emulsion	Segura et al. (2013)
Carbon particles	Carbon cryogel	sol-gel polycondensation followed by freeze-drying and carbonisation	Yamamoto et al. (2010)
Metal particles	Solder metal microparticles	Solidification of liquid metal in M/W or M/O emulsion	Torigoe et al. (2011)
	Silver nanoparticles	Reduction of silver ions in W/O microemulsions	Kakazu et al. (2010)
Solid lipid particles	W/S microcarrier	Melt crystallisation in W/O/W emulsion	Kukizaki and Goto (2007c)
	S/S microcarrier	Melt crystallisation in S/O/W emulsion	Kukizaki (2009c)
	Coherent particles	Melt crystallisation in O/W emulsion	D'oria et al. (2009); Li et al. (2011)
Gel micro- and nanoparticles	Ca-alginate	Crosslinking of sodium alginate with Ca^{2+} in W/O emulsion	Liu et al. (2003); You et al. (2001); Akamatsu et al. (2011)
	Chitosan	Crosslinking of chitosan with GA in W/O emulsion	Wang et al. (2005); Wei et al. (2010); Yue et al. (2011); Akamatsu et al. (2012)
		Crosslinking of chitosan with GA in O/W/O emulsion	Wei et al. (2013)
	HTCC/GP	Thermal gelation in W/O emulsion	Wu et al. (2008)
	Alginate/chitosan	Coalescence of Na-alginate droplets with Ca^{2+} droplets and particle coating with chitosan	Zhang et al. (2011)
	Agarose	Helix-coil transition induced by cooling	Zhou et al. (2007; 2008; 2009)

(continued)

Table 5.5 *(Continued)*

Product	Example	Secondary reaction/process after DME or PME	References
Protein microspheres	Albumin	Heat or chemical denaturation of albumin in W/O emulsion	El-Mahdy et al. (1998); Muramatsu and Kondo (1995); Muramatsu and Nakauchi (1998)
Composite organic–inorganic particles	Polymer particles with embedded TiO_2/Fe_3O_4 nanoparticles or quantum dots	Solvent evaporation from oil phase in S/O/W emulsion	Supsakulchai et al. (2002; 2002b); Omi et al. (2001); Wang et al. (2013); Yang et al. (2010); Zhou et al. (2012)
	Polymeric particles coated with silica nanoparticles	Solvent evaporation followed by electrostatic layer-by-layer deposition	Ito et al. (2010)
Coherent polymeric micro-or nanospheres	PSt, P(St-*co*-DVB), P(St-*co*-MMA), PUU-VP, etc.	One-stage suspension polymerisation in O/W emulsion	Yuyama et al. (2000); Omi et al. (1994); Nuisin et al. (2000); Ma et al. (2003);
	PSt-PAAm composite	One-stage suspension polymerisation in W/O/W emulsion	Ma et al. (2004)
	P(AAm-*co*-AA) and PNaAMPS hydrogel	One-stage suspension polymerisation in W/O emulsion	Nagashima et al. (1998); Hu et al. (2011)
	PMMA microspheres and large P(St-*co*-DVB) spheres	Two-stage suspension polymerisation in O/W emulsion	Omi et al. (1995, 1997)
	PUU, PSt-PMMA,	Solvent evaporation from oil phase droplets in O/W emulsion	Yuyama et al. (2000b); Ma et al. (1999a, b, c)
Synthetic biodegradable polymer particles	Coherent PLA and PLGA spheres	Solvent evaporation from oil phase droplets in O/W emulsion	Ito et al. (2011); Yue et al. (2012); Kanakubo et al. (2010)
	PLA or PLGA capsules for HAIs, DFB loaded PLA capsules	Solvent evaporation from oil phase in W/O/W emulsion	Liu et al. (2005a, b); Doan et al. (2011); Hou et al. (2009)
	mPEG-PLA capsules for HAIs	Solvent extraction from oil phase in W/O/W emulsion	Wei et al. (2008, 2011)

Core/shell and hollow particles	P(St-*co*-DMAEMA), P(St-*co*-DVB), PDVB	One-stage suspension polymerisation and internal phase separation in O/W emulsion	Ma et al. (2001, 2002, 2003b); Lee et al. (2010); Hao et al. (2009)
	Polymer-supported palladium catalyst	One-stage suspension polymerisation, internal phase separation and ligand exchange	Liu et al. (2010a, 2010b)
	P(St-*co*-DVB-*co*-MAA)	Two-stage suspension polymerisation and internal phase separation in O/W emulsion	Wang et al. (2012)
	ENB-P(M-*co*-U-*co*-F) core–shell capsules	*In-situ* polymerisation	Liu et al. (2011)
	Chitosan	Crosslinking of chitosan shell onto alginate particles and core dissolution	Akamatsu et al. (2010)
	Molecularly imprinted P(MMA-*co*-EDMA) particles	Molecular imprinting using CAP as a template molecule	Kou et al. (2012)
	PGPR-PE2CA core–shell particles	Interfacial polymerisation followed by solvent evaporation	Lee et al. (2009)
	Hollow porous silica nanocapsules loaded with Fe_3O_4 nanoparticles	One-stage suspension polymerisation, followed by sol-gel process and calcination	Kong et al. (2010, 2012, 2013)
Thermo-responsive capsules	Porous PA shells with P(NIPAM) gates in the pores	Interfacial polymerisation	Chu et al. (2002; 2003)
	P(NIPAM-*co*-AA) capsules	Suspension polymerisation in W/O emulsion	Si et al. (2011); Wang et al. (2013)
Janus particles	PS/PPC	Solvent pervaporation and internal phase separation	Chang and Hatton (2012)
	PMMA/P(S-BIEM)-*g*-PDMAEMA or PS/P(MMA-CMS)-*b*-PDMAEMA	Solvent evaporation, followed by internal phase separation and atom transfer radical polymerisation	Tanaka et al. (2010); Ahmad (2008)
	PS/PMMA	Solvent evaporation followed by internal phase separation	Yamashita et al. (2012)
Complex coacervate microcapsules	Gelatin/acacia microcapsules	Complex coacervation in O/W emulsion	Kage et al. (1997)
Non-spherical particles	Hemispherical polymer particles	Cleavage of Janus particles	Yamashita et al. (2012)
3D colloidal assemblies	Clusters containing silica-encapsulated magnetite nanoparticles	Solvent pervaporation and coating of clusters with silica	Chang and Hatton (2012)

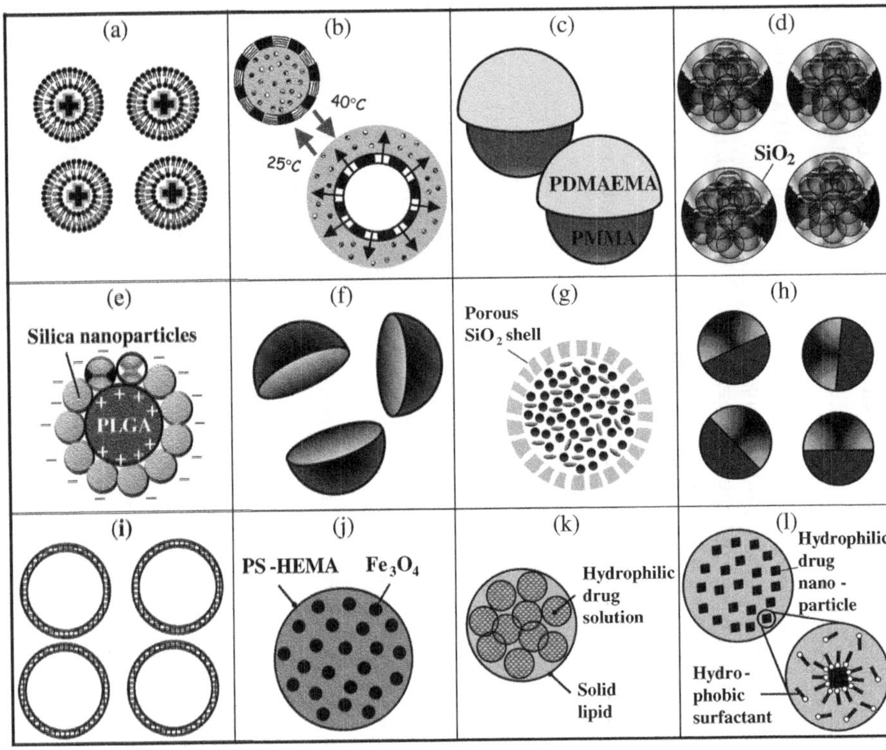

Figure 5.19 *Microparticles fabricated by integration of SPG membrane emulsification and post-processing. (a) DOX-loaded liposomes prepared by the film-hydration method combined with repeated PME and remote loading of DOX (Hwang et al., 2011). (b) Porous thermoresponsive capsules with PMIPAM gates prepared by DME, interfacial polymerisation and plasma-graft pore-filling polymerisation (Chu et al., 2002). (c) Mushroom-like Janus particles prepared by DME, internal phase separation and ATRP (Tanaka et al., 2010). (d) Silica-encapsulated magnetite nanoparticle clusters prepared by DME, solvent pervaporation and sol-gel coating (Chang and Hatton, 2012). (e) PLGA particles coated with silica nanoparticles, prepared by DME and layer-by-layer electrostatic deposition of PAH and silica nanoparticles onto PLGA particles (Ito et al., 2010). (f) hemispherical particles produced by cleavage of Janus particles fabricated by PME (Yamashita et al., 2012). (g) Porous silica shells loaded with magnetic nanoparticles and anticancer drug prepared by DME, polymerisation of styrene droplets, silica sol-gel coating of PS particles, removing PS core by thermal treatment and drug loading (Kong et al., 2010). (h) Janus PMMA/PS particles produced by DME and evaporation of toluene from homogeneous PMMA/PS/toluene droplets (Yamashita et al., 2012). (i) Chitosan shells prepared by coating chitosan onto alginate particles produced by DME, followed by crosslinking the shell and dissolution of the alginate core (Akamatsu et al., 2010a). (j) Magnetic polymer microspheres prepared by PME, chemical coprecipitation of Fe3O4 within the inner water phase of W/O/W emulsion and solvent evaporation (Yang et al., 2010). (k) W/O solid lipid carrier prepared by temperature-controlled DME and cooling of W/O/W emulsion (Kukizaki and Goto, 2007c). (l) S/O solid lipid carrier containing surfactant-coated hydrophilic drug nanoparticles embedded in solid lipid matrix, prepared by temperature-controlled PME and cooling of S/O/W emulsion (Kukizaki, 2009).*

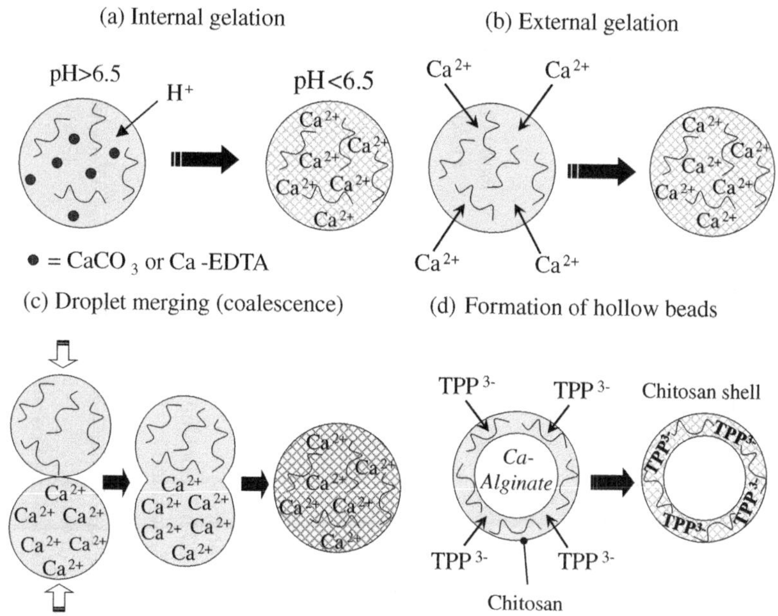

Figure 5.20 *Strategies used to form spherical or hollow gel microbeads by integration of membrane emulsification and ionotropic gelation.*

particles with a mean size of 2.3 μm, blockage of microengineered membrane occurred when 10 μm pore size was used, but no blockage was observed when using a 20 μm pore size membrane (Hanga and Holdich, 2014). For SPG membrane, blockage is even more likely to occur due to tortuous and interconnected pore structure. Another disadvantage of internal gelation is that it can lead to non-uniform distribution of Ca^{2+} in the dispersed phase and may leave grains of non-dissolved crosslinking agent in the beads (Ogończyk et al., 2011).

Alginate gelation with Ca^{2+} is a reversible process and the beads can be solubilised in an aqueous solution containing monovalent ions due to exchange of Ca^{2+} ions with non-crosslinkable monovalent ions. Irreversible gelation can be achieved using alginate with phenol moieties (Alg-Ph), which can be crosslinked via oxidative C–C and C–O coupling of phenol moieties in the side alginate chains with hydrogen peroxide and peroxidase (Sakai et al., 2007). In this process, the dispersed phase is an aqueous Alg-Ph solution containing horseradish peroxidase and the continuous phase is H_2O_2 dissolved in an oil phase. H_2O_2 can penetrate into alginate droplets causing enzymatic crosslinking of Alg-Ph.

5.7.1.2 External Gelation

In the external gelation method (Figure 5.20b), the aqueous phase droplets contain only a polymer and the crosslinking agent is added once the W/O emulsion is formed. Typical crosslinking agents used in this process are Ca^{2+} for alginate beads and glutaraldehyde (GA) and tripolyphosphate (TPP) for chitosan beads (Table 5.5B). In the case of calcium

Table 5.5 *Gel microbeads fabricated by integration of ME and a physical gelation process.*

A. Internal gelation

Process	Product	Formulation	References
DME, cross-flow nickel membrane, $d_p = 2.9–5.2\ \mu m$	Calcium alginate beads ($d_{part} = 44–146\ \mu m$)	DP: 0.5–2% (w/v) sodium alginate and 25 mM $CaCO_3$ in 0.9% (w/v) NaCl in water CP: 0.5–2% (v/v) Span 80 in paraffin oil XL: Glacial acetic acid	Liu et al. (2003)
DME, stirred cell nickel microengineered membrane, $d_p = 5–20\ \mu m$	P(NIPAM)-coated alginate beads ($d_{part} = 55–650\ \mu m$)	DP: 1.5 wt% sodium alginate + 0.5–5 wt% $CaCO_3$ in water CP: 1% wt Span 80 in Miglyol 840 XL: Glacial acetic acid	Hanga and Holdich (2014)

B. External gelation

Process	Product	Formulation	References
DME, internal pressure SPG microkit $d_p = 2.9\ \mu m$	Calcium alginate beads ($d_{part} = 4\ \mu m$) loaded with lidocaine HCl, 4-acetamidophenol or sodium salicylate	DP: 2 wt% sodium alginate + drug (drug:alginate = 0.1–1:1) in water CP: 6 vol% Span 80 in isooctane XL: Aqueous $CaCl_2$ solution	You et al. (2001)
DME using 0.5–19.6 μm SPG membrane, external pressure microkit	Chitosan beads ($d_{part} = 0.4–14\ \mu m$)	DP: 1–2 wt% chitosan + 1 wt% acetic acid + 0.9 wt% NaCl in water CP: 4 wt% PO-500 in liquid paraffin/petroleum ether (7:5, v/v) XL: Toluene saturated by GA	Wang et al. (2005)

DME using 1.1–11.9 μm SPG membrane, internal pressure microkit	Chitosan beads (d_{part} = 0.75–9.5 μm)	DP: 1 wt% chitosan + 2 wt% acetic acid in water CP: 1 wt% TGCR in kerosene XL: 25 wt% GA in water	Akamatsu et al. (2010b)
Repeated PME using 1.4–9.0 μm SPG membrane	Chitosan beads (d_{part} = 0.3–1.85 μm)	DP: 0.3–1 wt% chitosan + 1 wt% acetic acid in water CP: 2–8 wt% PO-500 in liquid paraffin/petroleum ether (1:2, v/v) XL: Toluene saturated by GA	Lv et al. (2009); Ma et al. (2010)

C. Cooling of emulsion to room temperature

Process	Product	Formulation	References
DME using 4.7–19.6 μm SPG membrane, external pressure microkit	Agarose beads (d_{part} = 15–60 μm)	DP: 4 wt% agarose + 0.9 wt% NaCl in water CP: 1–6 wt% PO-500 in liquid paraffin/petroleum ether (7/5, v/v) Emulsification temperature: 55–70°C	Zhou et al. (2007)
Repeated PME using 10.2 μm SPG external pressure microkit	Agarose beads (d_{part} = 5–6 μm)	DP: 2–14 wt% agarose + 0.9 wt% NaCl in water CP: 4 wt% PO-500 in liquid paraffin/petroleum ether (7/5, v/v) Emulsification temperature: 60°C	Zhou et al. (2008)
Repeated PME, external pressure SPG microkit, d_p = 5.7–19 μm or PE membrane, d_p = 11.8–25.6 μm	Agarose beads (d_{part} = 3–9 μm)	DP: 10 wt% agarose + 0.9 wt% NaCl in water CP: 4 wt% PO-500 in liquid paraffin/petroleum ether (7/5, v/v) Emulsification temperature: 60°C	Zhou et al. (2009)

(continued)

Table 5.6 *(Continued)*

Process	Product	Formulation	References
D. Mixing two emulsions and droplet merging			
DME using 1.1–11.9 μm SPG membrane, internal pressure microkit	Hybrid chitosan/alginate beads (d_{part} = 1–4.4 μm)	DP1: 1 wt% chitosan + 1 wt% $CaCl_2$ + 2 wt% acetic acid in water DP2: 2 wt% sodium alginate in water CP: 1 wt% TGCR in kerosene	Akamatsu et al. (2010b)
DME using 1.1–5.4 μm SPG membrane, internal pressure microkit	Chitosan beads (d_{part} = 0.84–1.5 μm)	DP1: 1 wt% chitosan + 2 wt% acetic acid in water DP2: 10 wt% sodium hydroxide in water CP: 1 wt% TGCR in kerosene	Akamatsu et al. (2010b); Park et al. (2004)
DME using 12 μm SPG membrane, internal pressure microkit	Hollow chitosan beads (d_{part} = 2.1–4.4 μm) with very thin shell	DP1: alginate beads coated with chitosan gel in water DP2: 3 wt% TPP in water CP: 3 wt% Span 85 in hexane	Akamatsu et al. (2010)
E. Thermal gelation			
DME using 2.8–15.4 μm SPG membrane, external pressure microkit	quaternised chitosan/ glycerophosphate (HTCC/GP) beads (d_{part} = 7–53 μm)	DP: 3.5 wt% HTCC + 8wt% GP in 0.1 M lactic acid solution CP: 1–4 wt% PO-500 in liquid paraffin/petroleum ether (7/5, v/v)	Wu et al. (2008)
PME using 8 μm SPG membrane	W/O/W emulsions with gelled inetrnal water droplets (d_{part} = 10.5 μm after five passes)	DP: W/O emulsion containing 15 wt% WPI in 5 mM phosphate buffer as the dispersed and 8% PGPR in corn oil as the oil phase CP: 0.5 wt% Tween 20 in 5 mM phosphate buffer	Surh et al. (2007)

DP, dispersed phase, CP, continuous phase, XL, crosslinking agent, d_p, pore diameter, d_{part}, particle diameter.

alginate and chitosan/TPP beads, crosslinking involves electrostatic interactions, but in the case of GA-crosslinked chitosan, the gelling occurs due to covalent bonding between amino groups of chitosan and aldehyde groups of GA and formation of a Schiff-base type bond. In order to preserve the original droplet size distribution, toluene saturated with GA was used for crosslinking reaction, rather than aqueous GA solution (Wang et al., 2005). The oil phase used in the external gelation process was usually a mixture of oil-soluble surfactant and low-viscosity organic solvent, such as isooctane, kerosene or 7:5 (v/v) paraffin/petroleum ether mixture (Table 5.5B).

5.7.1.3 Gel Formation by Droplet Merging (Coalescence)

In this method (Figure 5.20c), the gelation is achieved by mixing two W/O emulsions: emulsion 1, containing droplets of polymer solution and emulsion 2, containing droplets of the crosslinking agent. The crosslinking mechanism involves merging the two droplets (1 and 2) and rapid mixing of their contents resulting in a crosslinking reaction within the merged droplet. However, a stochastic nature of the coalescence process inevitably leads to the formation of polydisperse beads even if the original droplets are monodisperse (Sugiura et al., 2005). This strategy was used to create chitosan beads as a result of merging acidic chitosan drops with sodium hydroxide drops. The gelling was a result of precipitation of chitosan in a basic environment formed within the merged droplets (Table 5.5D). Hybrid chitosan/alginate beads were formed as a result of merging chitosan and alginate drops and subsequent electrostatic interaction between polycation (chitosan) and polyanion (alginate). Hollow chitosan beads were formed using the following procedure: (i) alginate beads were obtained by mixing W/O emulsions containing sodium alginate (emulsion 1) and $CaCl_2$ (emulsion 2) (ii) alginate beads were then coated with chitosan by mixing the resultant alginate beads with a W/O emulsion containing chitosan in the dispersed phase; (iii) the chitosan layer was crosslinked by mixing the chitosan-coated alginate beads with a W/O emulsion containing TPP in the dispersed phase. Polyphosphate has twofold action: it is a crosslinker for chitosan and Ca^{2+} catcher for calcium alginate, which results in the dissolution of alginate core and formation of hollow chitosan particles.

5.7.1.4 Gel Formation by Cooling or Heating

Polymers like gelatine and agarose can form hydrogels upon cooling from an aqueous solution. The gel formation is due to coil-to-helix transition: coil $\rightarrow$ helix $\rightarrow$ gel (Viebke et al., 1994). The first step in this process is a transition from a random coil to helix confirmation of macromolecules and the second step is association and branching of the helices into a 3D network. This method was used to produce agarose beads by DME or PME at an elevated temperature (55–70 °C) followed by cooling of the resultant W/O emulsion containing agarose. In PME, the ratio of the mean diameter of agarose beads to the mean pore size of SPG membrane was 0.49 μm in the investigated range of pore sizes (Zhou et al., 2009). In DME, the mean size of agarose beads was three times the mean pore size (Zhou et al., 2007) and no difference in size was found between the droplets formed at elevated temperature and the produced beads. The droplets were much more uniform when low-viscosity mixture of liquid paraffin and petroleum ether oil (7:5 v/v) was used as the continuous phase compared to high-viscosity oils such as pure paraffin and vegetable oil (Zhou et al., 2007).

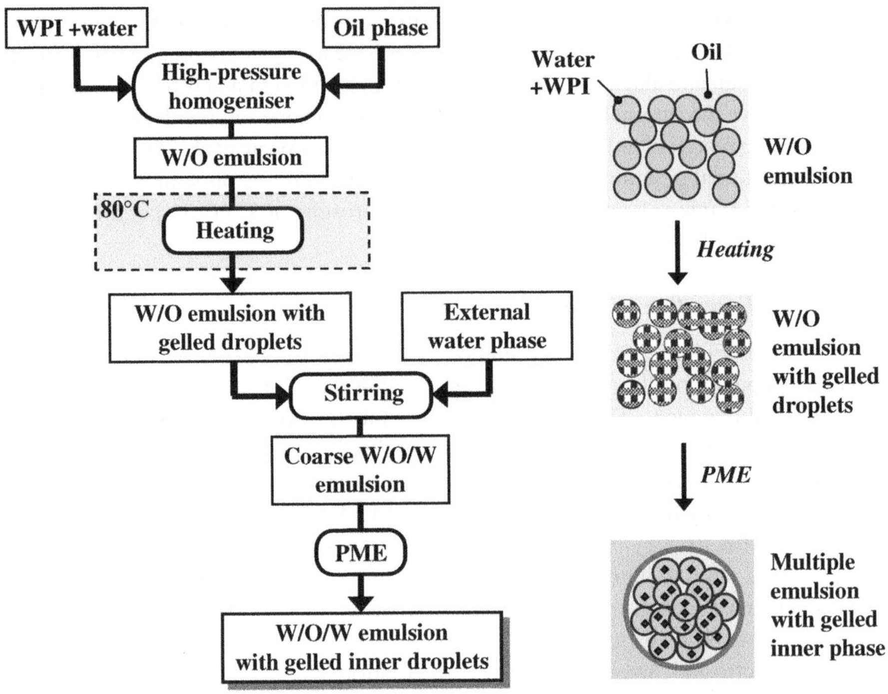

Figure 5.21 Production of W/O/W emulsions containing biopolymer-gelled water droplets by integration of PME and thermal gelation within inner water phase (Surh et al., 2007).

Hydrogel beads were also formed by thermal gelation within the droplets. Chitosan beads were obtained by reverse thermal gelation at a temperature close to 37°C of the droplets containing quaternised chitosan and glycerol-2-phosphate (GP) (Table 5.5E). The gelation occurred due to hydrophobic interactions between chitosan chains that are prevented at low temperatures due to protective hydration of the chains in the presence of GP (Wu et al., 2008). W/O emulsions containing gelled water droplets were formed by irreversible thermal gelation of whey protein isolate within the aqueous phase droplets at 80°C (Surh et al., 2007). The thermal gelation of whey protein isolate is a two-step process involving unfolding protein molecules and covalent crosslinking of unfolded protein molecules, most probably due to the formation of new disulphide (S–S) bonds. These W/O emulsion were used to form a W/O/W emulsion with improved stability and controlled release properties (Figure 5.21).

5.7.2 Integration of ME and Melt Solidification

ME/melt solidification involves ME above the melting point T_m of the dispersed phase followed by cooling the emulsion below the melting point. This approach was used for fabrication of solid lipid particles for oral drug delivery (Kukizaki, 2009c), low-melting-point metal particles for soldering microcomponents in microelectronics (Torigoe et al., 2011) and thermochromic liquid crystal particles for digital particle tracking thermography/

velocimetry (Segura et al., 2013). Solid lipid microparticles are particles made from lipids that are solid at room temperature. The commonly used solid lipids in ME formulations were mono-, di-, and triesters of glycerol or polyethylene glycol (PEG), as well as saturated fatty acids, such as palmitic acid, stearic acid and behenic acid (Table 5.7). The melting point of these lipids is between 44°C and 65°C and ME must be carried out at the temperature which is at least 5°C higher than the melting point of the lipid to prevent membrane fouling.

Three different melt dispersion methods were used for the preparation of solid lipid particles depending on the type of the template emulsion: O/W, W/O/W or S/O/W (Table 5.7). The O/W melt dispersion method was used for production of homogeneous (coherent) solid lipid particles for encapsulation of lipophilic active ingredients (LAIs), such as vitamin E (Laouini et al., 2014). LAI is first dissolved in the hot melted lipid and the solution is then injected through the membrane into a hot aqueous surfactant solution. The produced O/W emulsion is allowed to cool either at room temperature or in an ice bath. The W/O/W melt dispersion method was used for production of solid lipid particles containing inner water droplets for encapsulation of hydrophilic active ingredients (HAIs), such as vitamin B12 (Figure 5.19k). Following oral administration, HAI is gradually released in the small intestine due to degradation of the solid lipid matrix by lipase. Kukizaki and Goto (2007c) developed a two-stage DME process to prepare first a W/O and then a W/O/W emulsion using SPG membranes with the pore sizes of 0.3 and ≥1 μm, respectively. The droplet to pore size ratio was 3.3–3.4 in both cases. Finally, the W/O/W emulsion was cooled down to crystallise the lipid phase and convert the emulsion into a W/S/W dispersion (Figure 5.22). W/S particles contain a considerable amount of water (up to 40 wt%), because water droplets within the carrier remain intact after crystallisation of the oil phase. To improve a microbiological stability of the solid lipid particles, the inner water was removed by vacuum evaporation at 60°C to form nanoparticles of HAI suspended in the oil phase (Figure 5.23). A layer of lipophilic surfactant (PGPR) adsorbed on the droplet/particle surface (Figure 5.19 l) inhibited droplet coalescence and particle aggregation during dehydration of the water droplets, so the final mean particle size in the S/O dispersion was around 130 nm. This S/O dispersion was then dispersed in the aqueous surfactant solution by PME to form a hot S/O/W emulsion (Figure 5.23). The ratio of mean particle diameter to the mean pore diameter was 1–1.5 (Kukizaki, 2009). Finally, the S/O/W emulsion was cooled to obtain an S/S/W dispersion containing solid lipid particles with embedded nanoparticles of HAI.

5.7.3 Integration of ME and Polymerisation

Non-biodegradable polymeric particles were produced by combining ME and suspension polymerisation within the emulsified droplets (Omi et al., 1994). Suspension polymerisation was performed in O/W (Ma et al., 2003), W/O (Hu et al., 2011) or W/O/W emulsion (Ma et al., 2004) and can be combined with droplet swelling technique (which is known as the two-stage suspension polymerisation) to produce microspheres from hydrophilic monomers (Omi et al., 1997). Hollow polymeric particles were produced by combining SPG ME with interfacial polymerisation (Chu et al., 2003), internal phase separation (Liu et al., 2010), molecular imprinting (Kou et al., 2012), and coating polymer particles by a sol-gel process (Kong et al., 2013) or interfacial crosslinking (Akamatsu et al., 2010a) followed by core disintegration by calcination or chemical dissolution. A detailed discussion of integration

Table 5.7 *Microparticles fabricated by integration of ME integrated with solidification of melted droplets.*

Process	Product	Formulation	References
DME, cross-flow SPG, $d_p = 0.2–1$ μm	Gelucire 44/14 ($T_m = 44$°C, $d = 50–130$ nm) or Compritol 888 ($T_m = 69–74$°C, $d_{part} =$ 560–760 nm) particles	DP: Gelucire 44/14 melt at 65°C or Compritol 888 melt at 80°C CP: 0.125 wt% Tween 20 or 1.26 wt% Pluronic F68 in water Method: O/W melt dispersion	D'oria et al. (2009)
PME, external pressure SPG microkit, $d_p = 5.4–14.8$ μm	Trimyristin particles ($T_m =$ 56–57°C, $d_{part} = 55–650$ μm) with embedded solid nanoparticles	DP: S/O dispersion at 60°C (S: vitamin B_{12} nanoparticles (d = 132 nm), O: 5 wt% PGPR in trimyristin melt) CP: 1% wt Tween 40 in water Method: S/O/W melt dispersion	Kukizaki (2009c)
DME using 0.3–9.9 μm SPG membrane	Tripalmitin particles ($T_m = 65$°C, $d_{part} = 3.1–32.8$ μm) with embedded water droplets	DP: W/O emulsion at 70°C (W: 1 wt% vitamin B_{12} + 0.22 % (w/v) NaCl in water, O: 10 wt% TGCR in tripalmitin melt) CP: 1 wt% DGCS + 0.22 % (w/v) NaCl in water Method: W/O/W melt dispersion	Kukizaki and Goto (2007c)
PME, packed bed of glass beads, $d_{bead} = 30–90$ μm	Glyceryl palmitostearate ($T_m =$ 56°C, $d_{part} = 1.5–2.1$ μm) particles	DP: 7 wt% vitamin E in glyceryl palmitostearate melt at 65°C CP: 2 wt% Tween 80 in water	Laouini et al. (2014)
DME, external pressure SPG microkit, $d_p = 5.5$ μm	Termochromic liquid crystal particles ($T_m = 45–47$°C, $d_{part} = 16$ μm)	DP: Termochromic liquid crystal melt at 55°C CP: 5% poloxamer 188 in water	Segura et al. (2013)
PME, external pressure SPG microkit, $d_p = 5.5–20.2$ μm	Solder particles containing 4.7% Bi, 22.6% Pb, 8.3% Sn, 5.3% Cd, and 19.1% In ($T_m =$ 46.8°C, $d_{part} = 4.8–11.5$ μm)	DP: Alloy melt at 55°C CP: 5 wt% TGCR in toluene Method: M/O melt dispersion	Torigoe et al. (2011)
PME, external pressure SPG microkit, $d_p = 20.2$ μm	Solder particles containing 63% Sn and 37% Pb ($T_m =$ 183°C, $d_{part} = 9.3$ μm)	DP: 63Sn/Pb melt at 200°C CP: 5 wt% TGCR in lubricating oil Method: M/O melt dispersion	Torigoe et al. (2011)

DP, dispersed phase; CP, continuous phase; d_p, pore diameter; d_{part}, particle diameter.

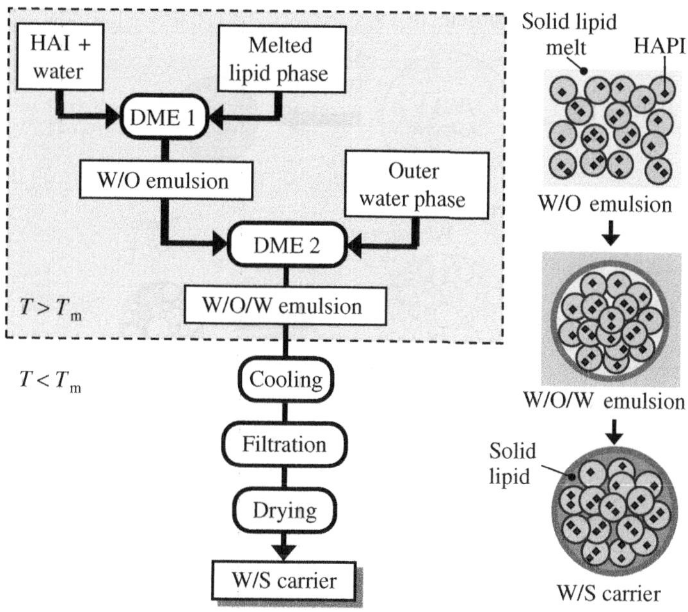

Figure 5.22 *Production of water-in-solid lipid (W/S) carrier by integration of two-stage DME at $T > T_m$ and crystallisation of oil phase in the obtained W/O/W emulsion at $T < T_m$.*

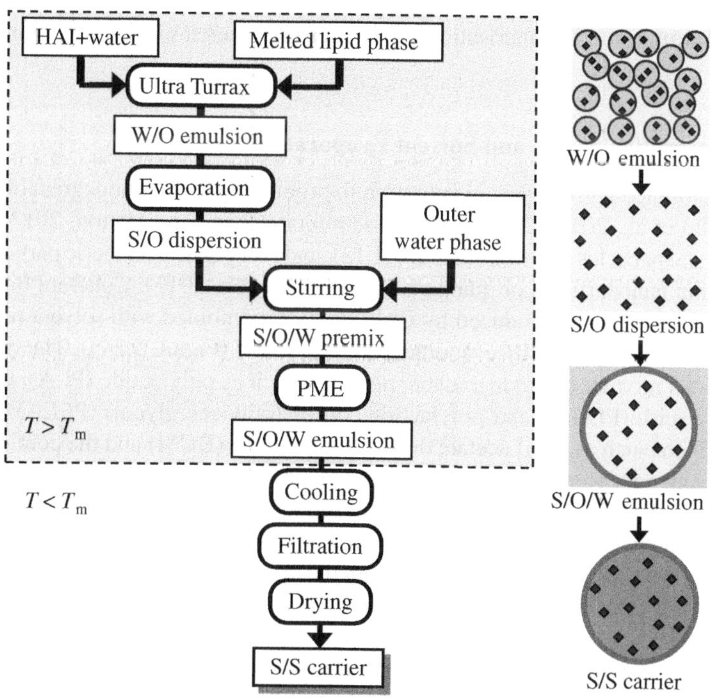

Figure 5.23 *Production of solid-in-solid lipid (S/S) carrier by PME at $T > T_m$ and crystallisation of oil phase in the obtained S/O/W emulsion at $T < T_m$.*

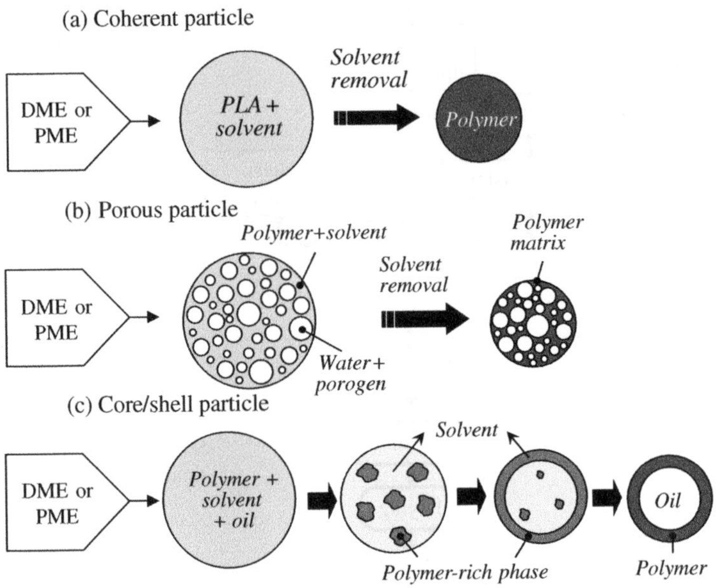

Figure 5.24 Production of particles with different morphologies from biodegradable synthetic polymers by combination of membrane emulsification and solvent evaporation.

of ME and suspension polymerisation is provided elsewhere (Vladisavljević and Williams, 2005).

5.7.4 Integration of ME and Solvent Evaporation/Extraction

ME was combined with solvent evaporation to produce synthetic biodegradable polymeric particles (Ito et al., 2011), 3D colloidal assemblies (Chang and Hatton, 2012), Janus and hemispherical particles (Yamashita et al., 2012), and composite polymeric particles containing inorganic nanoparticles or quantum dots (Wang et al., 2013). Coherent biodegradable polymeric particles were produced by DME or PME combined with solvent removal from the prepared O/W droplets by evaporation or extraction (Figure 5.24a). The oil phase is a homogeneous mixture of biodegradable polymer, such as polylactide (PLA), poly(lactide-co-glycolic acid) (PLGA), and polylactide-co-poly(ethylene glycol) (PELA) in a volatile organic solvent such as ethyl acetate or dichloromethane (DCM) and the continuous phase is typically an aqueous solution of polyvinyl alcohol (Table 5.8A). A hydrophobic active ingredient can be added into the oil phase to produce a controlled delivery system for the hydrophobic active. The final particle size can be predicted from the mass balance equation by assuming a complete solvent removal from the droplet phase, because the particle porosity is negligible after solvent evaporation. In the case of PME through SPG membrane, the mean particle size can be nearly 10 times smaller than the mean droplet size (Table 5.8A).

Porous matrix-type polymeric particles were obtained by combining ME and solvent evaporation from a W/O/W emulsion (Figure 5.24b). Porous particles can find applications as a controlled drug delivery system for HAIs and ultrasound contrast agents (UCAs).

Table 5.8 *Microparticles fabricated by integration of ME and solvent evaporation.*

Process	Product	Formulation	References
A. Coherent particles			
Repeated ($n = 3$) PME, external pressure SPG microkit, $d_p = 5.9–19.9$ μm	RIF-loaded PLGA particles ($d_{part} = 0.64–5.51$ μm)	DP: 3–30 wt% PLGA + 1–3 wt% RIF in EA CP: 3% (w/v) PVA + 8.5 vol% EA, pH = 4	Doan et al. (2011)
B. Porous particles			
Repeated ($n = 3$) PME, external pressure SPG microkit, $d_p = 5.9–19.9$ μm	PLGA particles loaded with RIF and RIF-HPβCD complex ($d_{part} = 2.08–8.43$ μm)	W_1: 18 mg mL^{-1} RIF + 0.066 M HPβCD in 50 mM borate buffer, pH 9 O: 3–30 wt% PLGA + 2 wt% RIF in EA W_2: 3% (w/v) PVA + 8.5 vol% EA, pH = 4	Doan et al. (2011)
DME, cross-flow SPG membrane, $d_p = 1.1$ μm	PLA-based UCA particles ($d_{part} = 1.99–3.58$ μm)	W_1: 2.5 wt% of ammonium bicarbonate in water O: 2.5 % (w/v) PLA + 1 % (w/v) Span 80 in DCM W_2: 1% PVA (w/v) + 0.5% (w/v) SDS in water	Hou et al. (2009)
C. Core/shell particles			
Repeated ($n = 1–15$) PME, glass fiber syringe membrane, $d_p = 1$ μm	Hollow PLLA particles ($d_{part} = 0.35–5$ μm)	DP: 0.6 wt% PLLA + 9.1 wt% DD in DCM CP: 0.3% PVA + (30 wt% MeOH or 30 wt% EtOH or 25 wt% PrOH) in water	Sawalha et al. (2008)

d_p, pore diameter; d_{part}, particle diameter.

UCAs are gas-filled hollow microspheres with a diameter of less than 7 μm containing a biodegradable lipid, protein or synthetic polymer shell or matrix. They can be injected into the blood flow to increase the backscattered signal from blood when insonified by ultrasound waves in contrast-enhanced power Doppler sonography (Cosgrove, 2006). Targeted UCAs are those containing adhesion ligands incorporated into their external surface, which allows them to bind to a specific cell type (Klibanov, 2007). Targeted UCAs can be retained on the endothelium at the site of pathology and used for molecular imaging or ultrasound-triggered targeted drug release (Böhmer et al., 2009).

Core/shell UCA particles were produced by combining PME and internal phase separation induced by solvent evaporation (Sawalha et al., 2008). As shown in Figure 5.24c, the process starts with the removal of volatile solvent from the drops containing a mixture of shell-forming polymer, such as PLA, a volatile water-immiscible solvent such as DCM and relatively non-volatile water-immiscible non-solvent ('oil', e.g. a C6–C20 hydrocarbon). As the solvent evaporates, the drop gradually shrinks and the polymer precipitates in the non-solvent forming droplets of polymer-enriched phase dispersed in the original drop. As the polymer is more hydrophilic than the non-solvent, polymer-enriched droplets tend to migrate to the interface with the aqueous phase and engulf the original drop. Further solvent removal leads to the creation of coherent polymer shell around the oil. UCAs can be produced through a serious of subsequent downstream operations including particle coating with a hydrophilic layer, transferring the particles in a vial, freeze-drying to remove the oil from the interior of the particles and particle loading with a suitable gas. It was found that the presence of alcohols (25–30 wt%) in the continuous phase during DCM evaporation increased the solvent removal rate from the droplets, as DCM is miscible with pure alcohols (Sawalha et al., 2008). As a result, the polymer solidifies more quickly and smaller and more uniform PLA particles were obtained by PME. However, at higher alcohol concentrations in the continuous phase (>50% for methanol, >45% for ethanol, or >35% for 2-propanol), the dispersed phase would become completely soluble in the alcohol–water mixture, which would result in nanoprecipitation of synthetic polymer during pre-mix formation (Sawalha et al., 2008). This phenomenon occurs for all alcohols, but at different concentrations, and leads to the formation of polymeric nanoparticles as explained in Section 10.

5.8 Integration of Membrane Permeation and Gas Dispersion

Microbubbles or nanobubbles can be produced by injecting gas phase through a hydrophilic SPG membrane into an aqueous surfactant solution (direct injection method) or by loading porous particles fabricated by SPG ME with a suitable gas (Hou et al., 2009). Monodispersed microbubbles with a relative span factor of about 0.5 were generated when the contact angle at membrane/water/air interface was in the range of $0° < \theta < 45°$ and the bubble-to-pore size ratio was 7.9 (Kukizaki and Wada, 2008). Nanobubbles are bubbles with a diameter between 1 nm and 1 μm. They have a long-term stability in the liquid phase as a result of the fact that the buoyancy force is smaller than any current in the liquid and the contact line pinning as a nanobubble gets smaller (Zimmerman et al., 2011). Nanobubbles with a mean diameter of 360–720 nm and relative span factor of 0.45–0.48 were produced by injecting air through SPG membranes with a mean pore diameter of 43–85 nm into 0.05–0.5 wt% sodium dodecyl sulphate (SDS) solution (Kukizaki and Goto, 2006). The mean size of nanobubbles was 8.6 times larger than the mean pore size and unaffected by the flow velocity of air in the pores within a range of 0.5–3.7 $m \cdot s^{-1}$ (Kukizaki and Goto, 2006). Microbubbles generated by SPG membranes can find applications in the production of aerated food products (Zúñiga and Aguilera, 2008), UCAs for ultrasonic examinations (Hou et al., 2009) and aerobic wastewater treatment (Liu et al., 2012, 2013), which can be combined with UV irradiation (Tasaki et al., 2009) or activated sludge process (Liu et al., 2012b).

5.9 Integration of Membrane Micromixing and Nanoprecipitation

Membrane micromixing can be combined with nanoprecipitation to achieve better control over the size of the nanoparticles. Precipitation of inorganic compounds involves membrane dispersion of aqueous solution of water-soluble solute **A** into an aqueous solution of water-soluble solute **B**. Nanoparticles are formed as a result of chemical reaction between the two solutes:

$$\mathbf{A} + \mathbf{B} \rightarrow \mathbf{C} + \mathbf{D} \tag{5.12}$$

where one of the products (**C** or **D**) is sparingly soluble in water and precipitates from the resultant mixture in the form of nanoparticles (Table 5.9). Precipitation of organic compounds requires dispersion of water-miscible organic solvent containing particle-forming organic compounds into an aqueous phase (anti-solvent). In this case, precipitation occurs as a result of lower solubility of the particle-forming solutes in the aqueous phase and self-assembly of aggregated polymer chains. In pharmaceutical nanodispersions, particle-forming organic compounds are active principle ingredient (API) and excipient(s). Typical excipients used in membrane micromixing/nanoprecipitation are phospholipids and synthetic biodegradable polymers or copolymers and typical nanostructures that have been fabricated are liposomes, niosomes, micelles and polymeric nanoparticles (Table 5.10). The size of nanoparticles was found to increase with increasing the membrane pore size, but many other parameters play an important role such as pore spacing, aqueous-to-organic volume phase ratio, transmembrane flux, shear stress on the membrane surface and chemical composition of the two phases (Laouini et al., 2013).

Table 5.9 *Formation of inorganic nanoparticles by membrane micromixing/nanoprecipitation method.*

	Salt A	Salt B	Membrane and pore size	References
$BaSO_4$ nanoparticles (d = 20–200 nm)	0.1–0.3M $BaCl_2$	0.1M $NaSO_4$	5 μm stainless steel, 0.2–0.9 μm Ni	Chen et al. (2004)
Anatase-TiO_2 nanoparticles (d = 9–20 nm)	0.03–0.15M $Ti(SO_4)_2$	0.1–0.3M NH_4HCO_3	0.2 μm Ni	Chen et al. (2004b)
ZnO nanoparticles (d = 9.4–14 nm)	0.2–1.2M $ZnSO_4$	2.25M NH_4HCO_3	5 μm stainless steel	Wang et al. (2010)

Table 5.10 *Formation of organic nanoparticles by membrane micromixing/nanoprecipitation method.*

Product	Excipients	Solvent and API	Membrane and pore size	References
BDP-loaded liposomes (d = 60–200 nm)	20–60 mg mL^{-1} Lipoïd® E80 + 4–12 mg mL^{-1} Chl	EtOH + 0.4 mg mL^{-1} BDP	0.4–10.2 μm SPG	Jaafar-Maalej et al. (2011)
SPL-loaded liposomes (d = 110–190 nm)	20–80 mg mL^{-1} Lipoïd® E80 + 4–16 mg mL^{-1} Chl	EtOH + 3 mg mL^{-1} SPL	40 nm PP hollow fibre	Laouini et al. (2011)
Vitamin E-loaded liposomes (d = 59–308 nm)	20 or 50 mg mL^{-1} Lipoïd® E80 or POPC + 5–12.5 mg mL^{-1} Chl, SA or CB	EtOH + 0–5 mg mL^{-1} vitamin E	5–40 μm nickel microengineered membrane	Laouini et al. (2013b)
Vitamin E-loaded micelles (d = 82–165 nm)	1 mg mL^{-1} PEG-PCL	THF or acetone + 0–2.5 mg mL^{-1} vitamin E	5–40 μm nickel microengineered membrane	Laouini et al. (2013c)
Vitamin E-loaded PCL nanoparticles (d = 250–350 nm)	5 mg mL^{-1} PCL	Acetone + 4 mg mL^{-1} vitamin E	0.2–10.2 μm SPG	Khayata et al. (2012)
caffeine and SPL-loaded niosomes (d = 111–115 nm)	105 mM Tween 60 + 105 mM Chl + 23.3 mM DCP	EtOH + 10 mg mL^{-1} caffeine or 3 mg mL^{-1} SPL	0.9 μm SPG	Pham et al. (2012)

5.10 Conclusions

Microporous membranes are increasingly being used in microfluidic applications aiming at generating uniform micro- and nanodroplets, -bubbles, and -particles. They have also been used for modification of emulsions (phase inversion, phase separation and homogenisation) and micromixing. Membranes can overcome low-throughput limitations of conventional microfluidic junctions and flow-focusing devices by providing numerous drop generation units – pores that serve as massively parallel T-junctions. However, there are some distinct differences in geometry and operational features between the membrane pores and T-junctions; the most notable difference is that the size of the outlet channel in a T-junction is of the same order of magnitude as the size of the inlet channel causing the droplet to squeeze in the channel at low flow rates of the continuous phase or form a jet at high velocities of the continuous phase. In DME, the mean droplet diameter is roughly three to five times larger than the mean pore size, while in PME the ratio of the mean droplet size to the mean pore size is lower than 1.5 and can be below unity. To form microparticles, DME or PME can be combined with a variety of physiochemical or chemical processes, that can be applied individually or in combination, such as polymerisation, ionic crosslinking, solvent evaporation, layer-by-layer electrostatic deposition, internal phase separation, coagulation, calcination, carbonisation, sol-gel processing and crystallisation. Organic and inorganic nanoparticles were formed by membrane micromixing/nanoprecipitation and spontaneous emulsification processes.

List of Acronyms

Chemicals and Materials: AA, acrylic acid; APTES, (3-aminopropyl)-triethoxysilane; APTMS, (3-aminopropyl)-trimethoxysilane; API, active principle ingredient; CB, cocoa butter; DMAEMA, dimethylaminoethyl methacrylate; CAP, chloramphenicol; Chl, cholesterol; CMS, chloromethylstyrene; BDP, beclomethasone dipropionate; DCM, dichloromethane; DCP, dicetyl phosphate; DD, dodecane; DFB, decafluorobutane; DGCS, decaglycerin condensed stearic acid ester; DOX, Doxorubicin, DVB, divinylbenzene; EA, ethyl acetate; EDMA, ethylene dimethacrylate; ENB 5-ethylidene-2-norbornene; EtOH, ethanol; HAI, hydrophilic active ingredient; HPβCD, (2-hydroxypropyl-β-cyclodextrin; HTCC, *N*-[(2-hydroxy-3-trimethylammonium) propyl] chitosan chloride; GP, α-β-glycerophosphate; Lipoïd® E80, egg yolk lecithin manufactured by Lipoïd GmbH; MAA, methacrylic acid; MeOH, methanol; LAI, lipophilic active ingredients; MMA, methyl methacrylate; mPEG, poly(monomethoxy polyethylene glycol); NIPAM, N-isopropylacrylamide; PAAm, polyacrylamide; PAH, poly(allylamine hydrochloride); ODS, octadecyldimethylchlorosilane; PCL, polycaprolactone; PE2CA,

	poly(ethyl 2-cyanoacrylate); PGPR, polyglycerol polyricinoleate; PLA, polylactic acid or polylactide; PLGA, poly(lactic-*co*-glycolic acid); PLLA, poly(L-lactic acid); P(M-*co*-U-*co*-F), PMMA, poly(methyl methacrylate); Poly(melamine-*co*-urea-*co*-formaldehyde); PNaAMPS, poly(sodium 2-(acrylamido)-2-methylpropanesulfonate); PO-500; POPC, 1-palmitoyl-2-oleoyl-sn-glycero-3-phosphocholine; hexaglycerin penta ester; PP, polypropylene; PPC, poly(propylene carbonate); PrOH, 2-propanol; PS, polystyrene; P(S-BIEM), poly(styrene-2-(2-bromoisobutyryloxy)ethyl methacrylate; PUU, polyurethaneurea; PVA, polyvinylalcohol; SA, stearic acid; Span 80, sorbitan monooleate; Span 85, Sorbitane trioleate; RIF, rifampicin; SPG, Shirasu porous glass; SDS, sodium dodecyl sulfate; SPL, spironolactone; St, styrene; SU-8, negative photoresist from from Shell Chemical; TGCR, tetraglycerol condensed ricinoleate; TMS, trimethylchlorosilane; TPP, tripolyphosphate; Tween 20, polyoxyethylene (20) sorbitan monolaurate; Tween 60, polyoxyethylene (20) sorbitan monostearate; Tween 80, Polyoxyethylene (20) sorbitan monooleate; UCA, ultrasound contrast agent; VP, vinyl polymer; WPI, whey protein isolate.
Emulsions/dispersions	M/O, metal-in-oil; M/W, O/W, oil-in-water, metal-in-water; O/W/O, oil-in-water-in-oil; S/O, solid-in-oil; S/O/W, solid-in-oil-in-water; S/S, solid-in-solid; W/O, water-in-oil, W/O/W, water-in-oil-in-water.
Other	ATRP, atom transfer radical polymerisation; CVD, chemical vapour deposition; DGU, droplet generation unit; DME, direct membrane emulsification; DRIE, deep reactive ion etching; LIGA, Ger. LIthographie, Galvanoformung, Abformung (lithography, electroplating and moulding), ME, membrane emulsification; PME, premix membrane emulsification; RIE, reactive ion etching; UCST, upper critical solution temperature; UVL, ultraviolet lithography; XMT, X-ray microtomography.

Symbols

A_{m}	Cross-sectional area of membrane
A	Amplitude of membrane oscillation
A	Reactant in reactive nanoprecipitation, see Equation (5.12)
a	Ratio of droplet size in pre-mix and pore size
b	Height of stirrer blade
B	Reactant in reactive nanoprecipitation, see Equation (5.12)
C	Product in reactive nanoprecipitation, see Equation (5.12)
D	Product in reactive nanoprecipitation, see Equation (5.12)
Ca	Capillary number
D	Stirrer diameter

D_{cell}	Internal diameter of stirred cell
D_1	Hydraulic diameter of dispersed phase channel
D_2	Hydraulic diameter of continuous phase channel
d_d	Droplet diameter
d_{pm}	Droplet diameter in pre-mix
d_{mi}	Inner diameter of membrane tube
d_p	Pore diameter
d_{part}	Particle diameter
E_a	Activation energy for diffusion during phase separation
F_{ca}	Capillary force
F_d	Drag force
f	Frequency of membrane oscillation
h	Height of cross-flow channel
J	Transmembrane flux
K	Constant in Equation (5.1)
K'	Proportionality constant between pore size and droplet size
K''	Proportionality constant in Equation (5.9)
m_m	Total mass of dry membrane
N	Total number of pores in membrane
n_b	Number of stirrer blades
Q	Volume flow rate
P	Pressure
P_{cap}	Capillary pressure
R	Universal gas constant (8314 kJ kmol^{-1} K^{-1})
Re	Rotating Reynolds number of continuous phase
R_m	Hydraulic resistance of membrane
r_b	Inner radius of a cylinder in which the membrane is rotating
r_{mo}	Outer radius of membrane tube
r_p	Pore radius
r_{trans}	Radial distance from the axis of rotation at which the shear is maximal
T	Temperature
T_m	Melting point
T_1	Initial temperature of mother glass
T_2	Temperature of phase separation of glass
t	Time
V_p	Total volume of all pores in membrane
W	Width of cross-flow channel
γ	Interfacial tension between water and oil phase
δ	Membrane thickness
ε	Membrane porosity
η	Viscosity
ρ_c	Density of continuous phase
ξ	Pore tortuosity
θ	Contact angle
τ_w	Shear stress on surface
ω	Angular velocity of membrane or stirrer

Subscripts

c Continuous phase
d Dispersed phase
e Emulsion
o Oil phase
w Water phase
skin Skin layer
sup Support layer
sym Symmetric
p Pore

References

Abrahamse, A.J., van Lierop, R., van der Sman, R.G.M., van der Padt, A. and Boom, R.M. 2002 'Analysis of droplet formation and interactions during cross-flow membrane emulsification', *J. Membr. Sci.*, **204**: 125–137.

Ahmad, H., Saito, N., Kagawa, Y. and Okubo, M. 2008 'Preparation of micrometer-sized, monodisperse "Janus" composite polymer particles having temperature-sensitive polymer brushes at half of the surface by seeded atom transfer radical polymerization', *Langmuir*, **24**: 688–691.

Akamatsu, K., Chen, W., Suzuki, Y., Ito, T., Nakao, A., Sugawara, T., Kikuchi, R. and Nakao, S. 2010a 'Preparation of monodisperse chitosan microcapsules with hollow structures using the SPG membrane emulsification technique', *Langmuir*, **26**: 14854–14860.

Akamatsu, K., Kaneko, D., Sugawara, T., Kikuchi, R. and Nakao, S. 2010b 'Three preparation methods for monodispersed chitosan microspheres using the Shirasu porous glass membrane emulsification technique and mechanisms of microsphere formation', *Ind. Eng. Chem. Res.*, **49**: 3236–3241.

Akamatsu, K., Maruyama, K., Chen, W., Nakao, A. and Nakao, S. 2011 'Drastic difference in porous structure of calcium alginate microspheres prepared with fresh or hydrolyzed sodium alginate', *J. Colloid Interface Sci.*, **371**: 46–51.

Akamatsu, K., Ikeuchi, Y., Nakao, A. and Nakao, S. 2012 'Size-controlled and monodisperse enzyme-encapsulated chitosan microspheres developed by the SPG membrane emulsification technique', *J. Colloid Interface Sci.*, **363**: 707–710.

Anna, S.L., Bontoux, N. and Stone, H.A. 2003 'Formation of dispersions using "flow focusing" in microchannels', *Appl. Phys. Lett.*, **82**: 364–366.

Aryanti, N., Williams, R.A., Hou, R. and Vladisavljević, G.T. 2006 'Performance of rotating membrane emulsification for O/W production', *Desalination*, **200**: 572–574.

Böhmer, M.R., Klibanov, A.L., Tiemann, K., Hall, C.S., Gruella, H. and Steinbach, O.C. 2009 'Ultrasound triggered image-guided drug delivery', *Eur. J. Radiol.*, **70**: 242–253.

Brans, G., Kromkamp, J., Pek, N., Gielen, J., Heck, J., van Rijn, C.J.M., van der Sman, R.G.M, Schroën, C.G.P.H. and Boom, R.M. 2006 'Evaluation of microsieve membrane design', *J. Membr. Sci.*, **278**: 344–348.

Chang, E.P. and Hatton, T.A. 2012 'Membrane emulsification and solvent pervaporation processes for the continuous synthesis of functional magnetic and Janus nanobeads', *Langmuir*, **28**: 9748–9758.

Chen, G.G., Luo, G.S., Xu, J.H. and Wang, J.D. 2004a 'Membrane dispersion precipitation method to prepare nanopartials', *Powder Technol.*, **139**: 180–185.

Chen, G., Luo, G., Yang, X., Sun, Y. and Wang, J. 2004b 'Anatase-TiO_2 nano-particle preparation with a micro-mixing technique and its photocatalytic performance', *Mater. Sci. Eng. A*, **380**: 320–325.

Chen, G.G., Luo, G.S., Sun, Y., Xu, J.H. and Wang, J.D. 2004c 'A ceramic microfiltration tube membrane dispersion extractor', *AIChE J.*, **50**: 382–387.

Cheng, C.J., Chu, L.Y., Xie, R. and Wang, X.W. 2008 'Preparation of highly monodisperse W/O emulsions with hydrophobically modified SPG membranes', *Chem. Eng. Technol.*, **31**: 377–383.

Cho, Y.H., Lee, J.J., Park, I.B., Huh, C.S., Baek, Y.J. and Park, J. 2005 'Protective effect of microencapsulation consisting of multiple emulsification and heat gelation processes on immunoglobulin in yolk', *J. Food Sci.*, **70**: E148–E151.

Choi, Y.K., Poudel, B.K., Marasini, N., Yang, K.Y., Kim, J.W., Kim, J.O., Choi, H.G. and Yong, C.S. 2012 'Enhanced solubility and oral bioavailability of itraconazole by combining membrane emulsification and spray drying technique', *Int. J. Pharm.*, **434**: 264–271.

Chu, L.Y., Park, S.H., Yamaguchi, T. and Nakao S. 2002 'Preparation of micron-sized monodispersed thermoresponsive core-shell microcapsules', *Langmuir*, **18**: 1856–1864.

Chu, L.Y., Rui, X., Zhu, J.H., Chen, W.M., Yamaguchi, T. and Nakao, S. 2003 'Study of SPG membrane emulsification processes for the preparation of monodisperse core-shell microcapsules', *J. Colloid Interface Sci.*, **265**: 187–196.

Cosgrove, D. 2006 'Ultrasound contrast agents: An overview', *Eur. J. Radiol.*, **60**: 324–330.

van Dijke, K, Kobayashi, I., Schroën, K., Uemura, K., Nakajima, M. and Boom, R. 2010 'Effect of viscosities of dispersed and continuous phases in microchannel oil-in-water emulsification', *Microfluid. Nanofluid.*, **9**: 77–85.

Doan, T.V.P., Couet, W. and Olivier, J.C. 2011 'Formulation and in vitro characterization of inhalable rifampicin-loaded PLGA microspheres for sustained lung delivery', *Int. J. Pharm.*, **414**: 112–117.

Dowding, P.J., Goodwin, J.W. and Vincent, B. 2001 'Production of porous suspension polymer beads with a narrow size distribution using a cross-flow membrane and a continuous tubular reactor', *Colloid. Surf. A*, **180**: 310–309.

D'oria, C., Charcosset, C., Barresi, A.A. and Fessi, H. 2009 'Preparation of solid lipid particles by membrane emulsification—Influence of process parameters', *Colloids Surf., A*, **338**: 114–118.

Dragosavac, M.M., Sovilj, M.N., Kosvintsev, S.R., Holdich, R.G. and Vladisavljević, G.T. 2008 'Controlled production of oil-in-water emulsions containing unrefined pumpkin seed oil using stirred cell membrane emulsification', *J. Membr. Sci.*, **322**: 178–188.

Dragosavac, M.M., Holdich, R.G., Vladisavljević, G.T. and Sovilj, M.N. 2012 'Stirred cell membrane emulsification for multiple emulsions containing unrefined pumpkin seed oil with uniform droplet size', *J. Membr. Sci.*, **392–393**: 122–129.

Egidi, E., Gasparini, G., Holdich, R.G., Vladisavljević, G.T. and Kosvintsev, S.R. 2008 'Membrane emulsification using membranes of regular pore spacing: Droplet size and uniformity in the presence of surface shear', *J. Membr. Sci.*, **323**: 414–420.

El-Mahdy, M., Ibrahim, E.S., Safwat, S., el-Sayed, A., Ohshima, H., Makino, K., Muramatsu, N. and Kondo, T. 1998 'Effects of preparation conditions on the monodispersity of albumin microspheres', *J. Microencapsulation*, **15**: 661–673.

van der Graaf, S., Schroën, C.G.P.H., Van der Sman, R.G.M. and Boom, R.M. 2004 'Influence of dynamic interfacial tension on droplet formation during membrane emulsification', *J. Colloid Interface Sci.*, **277**: 456–463.

van der Graaf, S., Nisisako, T., Schroën, C.G.P.H., van der Sman, R.G.M. and Boom, R.M. 2006 'Lattice Boltzmann simulations of droplet formation in a T-shaped microchannel', *Langmuir*, **22**: 4144–4152.

Geerken, M.J. 2006 'Emulsification with micro-engineered devices', PhD Thesis, University of Twente.

Geerken, M.J., Groenendijk, M.N.W., Lammertink, R.G.H. and Wessling, M. 2008 'Microfabricated metal nozzle plates used for water-in-oil and oil-in-water emulsification', *J. Membr. Sci.*, **310**: 374–383.

Gudipati, V., Sandra, S., McClements, D.J. and Decker, E.A. 2010 'Oxidative stability and in vitro digestibility of fish oil-in-water emulsions containing multilayered membranes', *J. Agric. Food Chem.*, **58**: 8093–8099.

Hanga, M.P. and Holdich, R.G. 2014 'Membrane emulsification for the production of uniform poly-N-isopropylacrylamide-coated alginate particles using internal gelation', *Ind. Eng. Chem. Res.*, **2**: 1664–1673.

Hao, D.X., Gong, F.L., Hu, G.H., Lei, J.D., Ma, G.H. and Su, Z.G. 2009 'The relationship between heterogeneous structures and phase separation in synthesis of uniform PolyDVB microspheres', *Polymer*, **50**: 3188–3195.

Higashi, S. and Setoguchi, T. 2000 'Hepatic arterial injection chemotherapy for hepatocellular carcinoma with epirubicin aqueous solution as numerous vesicles in iodinated poppy-seed oil microdroplets: clinical application of water-in-oil-in-water emulsion prepared using a membrane emulsification technique', *Adv. Drug Deliv. Rev.*, **45**: 57–64.

Hoare, T.R. and Kohane, D.S. 2008 'Hydrogels in drug delivery: Progress and challenges', *Polymer*, **49**: 1993–2007.

Holdich, R., Kosvintsev, S., Cumming, I. and Zhdanov, S. 2006 'Pore design and engineering for filters and membranes', *Phil. Trans. R. Soc. A*, **364**: 161–174.

Holdich, R.G., Dragosavac, M.M., Vladisavljević, G.T. and Kosvintsev, S.R. 2010 'Membrane emulsification with oscillating and stationary membranes', *Ind. Eng. Chem. Res.*, **49**: 3810–3817.

Holdich, R.G., Dragosavac, M., Vladisavljević, G.T. and Piacentini, E. 2013 'Continuous membrane emulsification with pulsed oscillatory) flow', *Ind. Eng. Chem. Res.*, **52**: 507–515.

Hou, Z., Lin, C. and Zhang, Q. 2009 'Preparation and characterization of PLA ultrasound contrast agents by combining an ultrasound method and a Shirasu porous glass (SPG) membrane emulsification technique', *3rd International Conference on Bioinformatics and Biomedical Engineering*, 2009. ICBBE.

Hu, J., Hiwatashi, K., Kurokawa, T., Liang, S.M., Wu, Z.L. and Gong, J.P. 2011 'Microgel-reinforced hydrogel films with high mechanical strength and their visible mesoscale fracture structure', *Macromolecules*, **44**: 7775–7781.

Hwang, T., Park, T.J., Koh, W.G., Cheong, I.W., Choi, S.W. and Kim, J.H. 2011 'Fabrication of nano-scale liposomes containing doxorubicin using Shirasu porous glass membrane', *Colloids Surf., A*, **392**: 250–255.

Ito, F., Uchida, Y. and Murakami, Y. 2010 'Facile technique for preparing organic–inorganic composite particles: Monodisperse poly(lactide-co-glycolide) (PLGA) particles having silica nanoparticles on the surface', *Colloids Surf., A*, **361**: 109–117.

Ito, F., Kanakubo, Y. and Murakami, Y. 2011 'Rapid preparation of monodisperse biodegradable polymer nanospheres using a membrane emulsification technique under low gas pressure', *J. Polym. Res.*, **18**: 2077–2085.

Jaafar-Maalej, C., Charcosset, C. and Fessi, H. 2011 'A new method for liposome preparation using a membrane contactor', *J. Liposome Res.*, **21**: 213–220.

Jing, C., Chin, C.Y. and Xie, R. 2006 'Preparation of highly monodisperse W/O emulsions with hydrophobically modified SPG membranes', *J. Colloid Interface Sci.*, **300**: 375–382.

Joscelyne, S.M. and Trägårdh, G. 2000 'Membrane emulsification – a literature review', *J. Membr. Sci.*, **169**: 107–117.

Kage, H., Kawahara, H., Ogura, H. and Matsuno, Y. 1997 'Microencapsulation of mono-dispersed droplets by complex coacervation method and membrane thickness of generated capsules', *Kagaku Kogaku Ronbunshu*, **23**: 652–658.

Kai, T., Suma, Y., Ono, S., Yamaguchi and Nakao, S. 2006 'Effect of the pore surface modification of an inorganic substrate on the plasma-grafting behavior of pore-filling-type organic/inorganic composite membranes', *J. Polym. Sci., Part A-1: Polym. Chem.*, **44**: 846–856.

Kakazu, E., Murakami, T., Akamatsu, K., Sugawara, T., Kikuchi, R. and Nakao, S. 2010 'Preparation of silver nanoparticles using the SPG membrane emulsification technique', *J. Membr. Sci.*, **354**: 1–5.

Kanakubo, Y., Ito, F. and Murakami, Y. 2010 'Novel one-pot facile technique for preparing nanoparticles modified with hydrophilic polymers on the surface via block polymer-assisted emulsification/evaporation process', *Colloids Surf., B*, **78**: 85–91.

Kandori, K., Kishi, K. and Ishikawa T. 1992 'Preparation of uniform silica hydrogel particles by SPG filter emulsification method', *Colloids Surf.*, **62**: 259–262.

Karbstein, H. and Schubert H. 1995 'Developments in the continuous mechanical production of oil-water macro-emulsions', *Chem. Eng. Process.*, **34**: 205–211.

Kawakita, H., Hamamoto, K., Seto, H., Ohto, K., Harada, H. and Inoue, K. 2009 'Porosity estimation of a membrane filled with dextran produced by immobilized dextransucrase', *AIChE J.*, **55**: 275–278.

Kawashima, Y., Hino, T., Takeuchi, H., Niwa, T. and Horibe, K. 1991 'Shear-induced phase inversion and size control of water/oil/water emulsion droplets with porous membrane', *J. Colloid Interface Sci.*, **145**: 512–523.

Khayata, N., Abdelwahed, W., Chehna, M.F. Charcosset, C. and Fessi, H. 2012 'Stability study and lyophilization of vitamin E-loaded nanocapsules prepared by membrane contactor' *Int. J. Pharm.*, **439**: 254–259.

Klibanov, A.L. 2007 'Ultrasound molecular imaging with targeted microbubble contrast agents', *J. Nucl. Cardiol.*, **14**: 876–884.

Kobayashi, I., Nakajima, M., Chun, K., Kikuchi, Y. and Fujita, H. 2002 'Silicon array of elongated through-holes for monodisperse emulsion droplets', *AIChE J.*, **48**: 1639–1644.

Kobayashi, I., Nakajima, M. and Mukataka, S. 2003 'Preparation characteristics of oil-in-water emulsions using differently charged surfactants in straight-through microchannel emulsification', *Colloid. Surf. A*, **229**: 33–41.

Kobayashi, I., Mukataka, S. and Nakajima, M. 2004 'Effect of slot aspect ratio on droplet formation from silicon straight-through microchannels', *J. Colloid Interface Sci.*, **279**: 277–280.

Kobayashi, I., Mukataka, S. and Nakajima, M. 2005a 'Novel asymmetric through-hole array microfabricated on a silicon plate for formulating monodisperse emulsions', *Langmuir*, **21**: 7629–7632.

Kobayashi, I., Mukataka, S. and Nakajima, M. 2005b 'Production of monodisperse oil-in-water emulsions using a large silicon straight-through microchannel plate', *Ind. Eng. Chem. Res.*, **44**: 5852–5856.

Kobayashi, I., Wada, Y., Uemura, K. and Nakajima, M. 2008a 'Generation of uniform drops via through-hole arrays micromachined in stainless-steel plates', *Microfluid. Nanofluid.*, **5**: 677–687.

Kobayashi, I., Hirose, S., Katoh, T., Zhang, Y., Uemura, K. and Nakajima, M. 2008b 'High-aspect-ratio through-hole array microfabricated in a PMMA plate for monodisperse emulsion production', *Microsyst. Technol.*, **14**: 1349–1357.

Kobayashi, I., Takano, T., Maeda, R., Wada, Y., Uemura, K. and Nakajima, M. 2008c 'Straight-through microchannel devices for generating monodisperse emulsion droplets several microns in size', *Microfluid. Nanofluid.*, **4**: 167–177.

Kobayashi, Y. Murayama, T. Kuroiwa, K. Uemura, M. and Nakajima, M. 2009a 'Production of monodisperse water-in-oil emulsions consisting of highly uniform droplets using asymmetric straight-through microchannel arrays', *Microfluid. Nanofluid.*, **7**: 107–119.

Kobayashi, I., Murayama, Y., Kuroiwa, T., Uemura, K. and Nakajima, M. 2009b 'Production of monodisperse water-in-oil emulsions consisting of highly uniform droplets using asymmetric straight-through microchannel arrays', *Microfluid. Nanofluid.*, **7**: 107–119.

Koga, K., Takarada, N. and Takada, K. 2010 'Nano-sized water-in-oil-in-water emulsion enhances intestinal absorption of calcein, a high solubility and low permeability compound' *Eur. J. Pharm. Biopharm.*, **74**: 223–232.

Kobayashi, I., Vladisavljević, G.T., Uemura, K. and Nakajima, M 2011 'CFD analysis of microchannel emulsification: Droplet generation process and size effect of asymmetric straight flow-through microchannels', *Chem. Eng. Sci.*, **66**: 5556–5565.

Koltuniewicz, A.B., Field, R.W. and Arnot, T.C. 1995 'Cross-flow and dead-end microfiltration of oily-water emulsion. Part I: Experimental study and analysis of flux decline', *J. Membr. Sci.*, **102**: 193–207.

Kong, S.D., Zhang, W., Lee, J.H., Brammer, K., Lal, R., Karin, M. and Jin, S. 2010 'Magnetically vectored nanocapsules for tumor penetration and remotely switchable on-demand drug release', *Nano Lett.*, **10**: 5088–5092.

Kong, S.D., Zhang, W., Lee, J.H., Choi, C., Khamwannah, J., Karin, M. and Jin, S. 2012 'Externally triggered on-demand drug release and deep tumor penetration', *J. Vac. Sci. Technol., B: Microelectron. Nanometer Struct.*, **30**: 02C102–02C102-7.

Kong, S.D., Choi, C., Khamwannah, J. and Jin, S. 2013 'Magnetically vectored delivery of cancer drug using remotely on–off switchable nanocapsules', *IEEE Trans. Magn.*, **49**: 349–352.

Koris, A., Piacentini, E., Vatai, G., Bekassy-Molnar, E., Drioli, E. and Giorno, L. 2011 'Investigation on the effects of a mechanical shear-stress modification method during cross-flow membrane emulsification', *J. Membr. Sci.*, **371**: 28–36.

Kosvintsev, S.R., Gasparini, G., Holdich, R.G., Cumming, I.W. and Stillwell, M.T. 2005 'Liquid-liquid membrane dispersion in a stirred cell with and without controlled shear', *Ind. Eng. Chem. Res.*, **44**: 9323–9330.

Kosvintsev, S.R., Gasparini, G. and Holdich, R.G. 2008 'Membrane emulsification: droplet size and uniformity in the absence of surface shear', *J. Membr. Sci.*, **313**: 182–189.

Kou, X., Li, Q., Lei, J., Geng, L., Deng, H., Zhang, G., Ma, G., Su, Z. and Jiang, Q. 2012 'Preparation of molecularly imprinted nanospheres by premix membrane emulsification technique', *J. Membr. Sci.*, 417–418: 87–95.

Kukizaki, M. and Nakashima, T. 2004 'Acid leaching process in the preparation of porous glass membranes from phase-separated glass in the Na_2O–CaO–MgO–Al_2O_3–B_2O_3–SiO_2 system', *Membrane*, **29**: 301–308.

Kukizaki, M. and Goto, M. 2006 'Size control of nanobubbles generated from Shirasu-porous-glass (SPG) membranes', *J. Membr. Sci.*, **281**: 386–396.

Kukizaki, M. and Goto, M. 2007a 'Spontaneous formation behavior of uniform-sized microbubbles from Shirasu porous glass (SPG) membranes in the absence of water-phase flow', *Colloids Surf., A*, **296**: 174–181.

Kukizaki, M. and Goto, M. 2007b 'Preparation and characterization of a new asymmetric type of Shirasu porous glass (SPG) membrane used for membrane emulsification', *J. Membr. Sci.*, **299**: 190–199.

Kukizaki, M. and Goto, M. 2007c 'Preparation and evaluation of uniformly sized solid lipid microcapsules using membrane emulsification', *Colloids Surf., A*, **293**: 87–94.

Kukizaki, M. and Goto, M. 2008 'Demulsification of water-in-oil emulsions by permeation through Shirasu-porous-glass (SPG) membranes', *J. Membr. Sci.*, **322**: 196–203.

Kukizaki, M. and Wada, T. 2008 'Effect of the membrane wettability on the size and size distribution of microbubbles formed from Shirasu-porous-glass (SPG) membranes', *Colloids Surf., A*, **317**: 146–154.

Kukizaki, M. and Goto, M. 2009 'A comparative study of SPG membrane emulsification in the presence and absence of continuous-phase flow', *J. Chem Eng. Jpn*, **42**: 520–530.

Kukizaki, M. 2009a 'Shirasu porous glass (SPG) membrane emulsification in the absence of shear flow at the membrane surface: Influence of surfactant type and concentration, viscosities of dispersed and continuous phases, and transmembrane pressure', *J. Membr. Sci.*, **327**: 234–243.

Kukizaki, M. 2009b 'Relation between salt rejection and electrokinetic properties on Shirasu porous glass (SPG) membranes with nano-order uniform pores', *Sep. Sci. Technol.*, **69**: 87–96.

Kukizaki, M. 2009c 'Preparation of solid lipid microcapsules via solid-in-oil-in-water dispersions by premix membrane emulsification', *Chem. Eng. J.*, **151**: 387–396.

Kukizaki, M. 2010 'Large-scale production of alkali-resistant Shirasu porous glass (SPG) membranes: Influence of ZrO_2 addition on crystallization and phase separation in

$Na_2O–CaO–Al_2O_3–B_2O_3–SiO_2$ glasses; and alkali durability and pore morphology of the membranes', *J. Membr. Sci.*, **360**: 426–435.

Laouini, A., Jaafar-Maalej, C., Sfar, S., Charcosset, C. and Fessi, H. 2011 'Liposome preparation using a hollow fiber membrane contactor – application to spironolactone encapsulation', *Int. J. Pharm.*, **415**: 53–61.

Laouini, A., Charcosset, C., Georgiadou, S., Fessi, H., Holdich, R.G. and Vladisavljević, G.T. 2013a 'Preparation of liposomes: A novel application of microengineered membranes – From laboratory scale to large scale', *Colloids Surf., B*, **112**: 272–278.

Laouini, A., Charcosset, C., Georgiadou, S., Fessi, H., Holdich, R.G. and Vladisavljević, G.T. 2013b 'Preparation of liposomes: a novel application of microengineered membranes – investigation of the process parameters and application to the encapsulation of vitamin E', *RSC Advances*, **3**: 4985–4994.

Laouini, A., Koutroumanis, K.P., Charcosset, C., Georgiadou, S., Fessi, H., Holdich, R.G. and Vladisavljević, G.T. 2013c 'pH-Sensitive micelles for targeted drug delivery prepared using a novel membrane contactor method', *ACS Appl. Mater. Interfaces*, **5**: 8939–8947.

Lee, S.H., Baek, H.H., Kim, J.H. and Choi, S.W. 2009 'Core-shell poly(d,l-lactide-co-glycolide)/poly(ethyl 2-cyanoacrylate) microparticles with doxorubicin to reduce initial burst release', *Macromol. Res.*, **17**: 1010–1014.

Lee, J., Hwang, D.R. and Shim, S.E. 2010 'Controlling morphology of polymer microspheres by Shirasu porous glass (SPG) membrane emulsification and subsequent polymerization: from solid to hollow', *Macromol. Res.*, **18**: 1142–1147.

Li, N. and Sakaki, K. 2008 'Performance of an emulsion enzyme membrane reactor combined with premix membrane emulsification for lipase-catalyzed resolution of enantiomers', *J. Membr. Sci.*, **314**: 183–192.

Li, Y., Fessi, H. and Charcosset, C. 2011 'Preparation of indomethacin-loaded lipid particles by membrane emulsification', *Adv. Sci. Lett.*, **4**: 591–595.

Liu, X.D., Bao, D.C., Xue, W.M., Xiong, Y., Yu, W.T., Yu, X.J., Ma, X.J. and Yuan, Q. 2003 'Preparation of uniform calcium alginate gel beads by membrane emulsification coupled with internal gelation', *J. Appl. Polym. Sci.*, **87**: 848–852.

Liu, R., Ma, G.H., Meng, F.T. and Su, Z.G. 2005a 'Preparation of uniform-sized PLA microcapsules by combining Shirasu Porous Glass membrane emulsification technique and multiple emulsion-solvent evaporation method', *J. Controlled Release*, **103**: 31–43.

Liu, R., Ma, G.H., Wan, Y.H. and Su, Z.G. 2005b 'Influence of process parameters on the size distribution of PLA microcapsules prepared by combining membrane emulsification technique and double emulsion-solvent evaporation method', *Colloids Surf., B*, **45**: 144–153.

Liu, Y., Feng, X.J., Bao, D.C., Li, K.X. and Bao, M. 2010a 'Preparation of microcapsule-supported palladium catalyst using SPG (Shirasu Porous Glass) emulsification technique', *Chin. Chem. Lett.*, **21**: 979–982.

Liu, Y., Feng, X.J., Bao, D.C., Li, K. and Bao, M. 2010b 'A new method for the preparation of microcapsule-supported palladium catalyst for Suzuki coupling reaction', *J. Mol. Catal. A: Chem.*, **323**: 16–22.

Liu, X., Lee, J.K. and Kessler, M.R. 2011 'Microencapsulation of self-healing agents with melamine-urea-formaldehyde by the Shirasu porous glass (SPG) emulsification technique', *Macromol. Res.*, **19**: 1056–1061.

Liu, C., Tanaka, H., Zhang, L., Zhang, J., Huang, X., Ma, J. and Matsuzawa, Y. 2012a 'Fouling and structural changes of Shirasu porous glass (SPG) membrane used in aerobic wastewater treatment process for microbubble aeration', *J. Membr. Sci.*, **421–422**: 225–231.

Liu, C., Tanaka, H., Ma, J., Zhang, L., Zhang, J., Huang, X. and Matsuzawa, Y. 2012b 'Effect of microbubble and its generation process on mixed liquor properties of activated sludge using Shirasu porous glass (SPG) membrane system', *Water Res.*, **46**: 6051–6058.

Liu, C., Tanaka, H., Zhang, J., Zhang, L., Yang, J., Huang, X. and Kubota, N. 2013 'Successful application of Shirasu porous glass (SPG) membrane system for microbubble aeration in a biofilm reactor treating synthetic wastewater', *Sep. Purif. Technol.*, **103**: 53–59.

Laermer F and Urban A 2011 'Through-silicon vias Using Bosch DRIE process technology' in Burghartz, J.N. (ed.), *Ultra-thin Chip Technology and Applications*, Springer: New York, pp. 81–92.

Laouini, A., Fessi, H. and Charcosset, C. 2012 'Membrane emulsification: A promising alternative for vitamin E encapsulation within nano-emulsion', *J. Membr. Sci.*, **423–424**: 85–96.

Laouini, A., Charcosset, C., Fessi, H., Holdich, R.G. and Vladisavljević, G.T. 2013 'Preparation of liposomes: a novel application of microengineered membranes: From laboratory scale to large scale', *Colloids Surf. B*, **112**: 272–278.

Laouini, A., Charcosset, C., Fessi, H. and Schroen, K. 2014 'Use of dynamic membranes for the preparation of vitamin E-loaded lipid particles: An alternative to prevent fouling observed in classical cross-flow emulsification', *Chem. Eng. J.*, **236**: 498–505.

Lv, P.P., Wei, W., Gong, F.L., Zhang, Y.L., Zhao, H.Y., Lei, J.D., Wang, L.Y. and Ma, G.H. 2009 'Preparation of uniformly sized chitosan nanospheres by a premix membrane emulsification technique'. *Ind. Eng. Chem. Res.*, **48**: 8819–8828.

Ma, G.H., Nagai, M. and Omi, S. 1999a 'Study on preparation and morphology of uniform artificial polystyrene–poly(methyl methacrylate) composite microspheres by employing the SPG (Shirasu Porous Glass) membrane emulsification technique', *J. Colloid Interface Sci.*, **214**: 264–282.

Ma, G.H., Nagai, M. and Omi, S. 1999b 'Effect of lauryl alcohol on morphology of uniform polystyrene-poly(methyl methacrylate) composite microspheres prepared by Porous glass membrane emulsification technique', *J. Colloid Interface Sci.*, **219**: 110–128.

Ma, G.H., Nagai, M. and Omi, S. 1999c 'Preparation of uniform poly(lactide) microspheres by employing the shirasu porous glass (SPG) emulsification technique', *Colloids Surf. A*, **153**: 383–394.

Ma, G.H., Nagai, M. and Omi, S. 2001 'Study on preparation of monodispersed poly(styrene-co-N-dimethylaminoethyl methacrylate) composite microspheres by SPG (shirasu porous glass) emulsification technique', *J. Appl. Polym. Sci.*, **79**: 2408–2424.

Ma, G.H., Omi, S., Dimonie, V.L., Sudol, E.D. and El-Aasser, M.S. 2002 'Study of the preparation and mechanism of formation of hollow monodisperse polystyrene microspheres by SPG (Shirasu Porous Glass) emulsification technique', *J. Appl. Polym. Sci.*, **85**: 1530–1543.

Ma, G.H., An, C.J., Yuyama, H., Su, Z.G. and Omi, S. 2003a 'Synthesis and characterization of polyurethaneurea-vinyl polymer (PUU-VP) uniform hybrid microspheres by SPG emulsification technique and subsequent suspension polymerization', *J. Appl. Polym. Sci.*, **89**: 163–178.

Ma, G.H., Chen, A.Y., Su, Z.G. and Omi, S. 2003b 'Preparation of uniform hollow polystyrene particles with large voids by a glass-membrane emulsification technique and a subsequent suspension polymerization', *J. Appl. Polym. Sci.*, **87**: 244–251.

Ma, G.H., Sone, H. and Omi, S. 2004 'Preparation of uniform-sized polystyrene-polyacrylamide composite microspheres from a W/O/W emulsion by membrane emulsification technique and subsequent suspension polymerization', *Macromolecules*, **37**: 2954–2964.

Ma, G.H., Yang, J., Lv, P.P., Wang, L.Y., Wei, W., Tian, R., Wu, J. and Su, Z.G. 2010 'Preparation of uniform microspheres and microcapsules by modified emulsification process' *Macromol. Symp.*, **288**: 41–48.

Mazzei, R., Drioli, E. and Giorno, L. 2010 'Biocatalytic membrane reactor and membrane emulsification concepts combined in a single unit to assist production and separation of water unstable reaction products', *J. Membr. Sci.*, **352**: 166–172.

Meng, T., Xie, R., Chen, Y.C., Cheng, C.J., Li, P.F., Ju, X.J. and Chu, L.Y. 2010 'A thermo-responsive affinity membrane with nano-structured pores and grafted poly(N-isopropylacrylamide) surface layer for hydrophobic adsorption', *J. Membr. Sci.*, **349**: 258–267.

Meng, T., Xie, R., Ju, X.J., Cheng, C.J., Wang, S., Li, P.F., Liang, B. and Chun, L.Y. 2013 'Nano-structure construction of porous membranes by depositing nanoparticles for enhanced surface wettability', *J. Membr. Sci.*, **427**: 63–72.

Mulder, M. 1996 *Basic Principles of Membrane Technology*, Dordrecht: Kluwer Academic Publishers.

Muramatsu, N. and Kondo, T. 1995 'An approach to prepare microparticles of uniform size', *J. Microencapsulation*, **2**: 129–136.

Muramatsu, N. and Nakauchi, K. 1998 'A novel method to prepare monodisperse microparticles', *J. Microencapsulation*, **15**: 715–723.

Nagashima, S., Ando, S., Tsukamoto, T., Ohshima, H. and Makino, K. 1998 'Preparation of monodisperse poly(acrylamide-co-acrylic acid) hydrogel microspheres by a membrane emulsification technique and their size-dependent surface properties', *Colloids Surf., B*, **11**: 47–56.

Nakashima, T. and Kuroki, Y. 1981 'Effect of composition and heat treatment on the phase separation of NaO–B_2O_3–SiO_2–Al_2O_3–CaO glass prepared from volcanic ashes', *Nippon Kagaku Kaishi*, **8**: 1231–1238.

Nakashima T. and Shimizu M. 1986 'Porous glass from calcium alumino boro-silicate glass', *Ceramics Japan*, **21**: 408–412.

Nakashima T., Shimizu M. and Kukizaki, M. 1991 'Membrane emulsification by microporous glass', *Key Eng. Mater.*, **61–62**: 513–516.

Nakashima, T., Shimizu, M. and Kukizaki M. 1992 'Mechanical strength and thermal resistance of porous glass', *J. Ceram. Soc. Jpn. Int. Ed.*, **100**: 1389–1393.

Nakashima, T., Shimizu, M. and Kukizaki M. 1993 'Effect of surfactant on production of monodispersed O/W emulsion in membrane emulsification', *Kag. Kog. Ronbunshu*, **19**: 991–997.

Nakashima, T., Shimizu, M. and Kukizaki M. 1994 'Monodisperse single and double emulsions and method of producing same', *US Patent* **5**,326,484.

Nakashima, T., Shimizu, M. and Kukizaki, M. 2000 'Particle control of emulsion by membrane emulsification and its applications', *Adv. Drug Deliv. Rev.*, **45**: 47–56.

Nakashima, T. 2002 'Porous glass material and its recent applications', paper presented at 38th International SPG Forum on Membrane and Particle Science and Technology in Food and Medical Care in Sadowara, Japan.

Nazir, A., Schroën, K. and Boom, R. 2010 'Premix emulsification: A review', *J. Membr. Sci.*, **362**: 1–11.

Nazir, A., Schroën, K. and Boom, R. 2011 'High-throughput premix membrane emulsification using nickel sieves having straight-through pores', *J. Membr. Sci.*, **383**: 116–123.

Nazir, H., Wang, L., Lian, G., Zhu, S., Zhang, Y., Liu, Y. and Ma, G. 2012 'Multilayered silicone oil droplets of narrow size distribution: Preparation and improved deposition on hair', *Colloids Surf., B*, **100**: 42–49.

Nuisin, R., Ma, G.H., Omi, S. and Kiatkamjornwong, S.J. 2000 'Dependence of morphological changes of polymer particles on hydrophobic/hydrophilic additives', *J. Appl. Polym. Sci.*, **77**: 1013–1028.

Ogończyk, D., Siek, M. and Garstecki, P. 2011 'Microfluidic formulation of pectin microbeads for encapsulation and controlled release of nanoparticles', *Biomicrofluidics*, **5**: 013405.

Oh, D.H., Balakrishnan, P., Oh, Y.K., Kim, D.D., Yong, C.S. and Choi, H.G. 2011 'Effect of process parameters on nanoemulsion droplet size and distribution in SPG membrane emulsification', *Int. J. Pharm.*, **404**: 191–197.

Oh, D.H., Din, F.U., Kim, D.W., Kim, J.O., Yong, C.S. and Choi, H.G. 2013 'Flurbiprofen-loaded nanoparticles prepared with polyvinylpyrrolidone using Shirasu porous glass membranes and a spray-drying technique: nano-sized formation and improved bioavailability', *J. Microencapsulation*, **30**: 674–680. DOI: 10.3109/02652048.2013.774447.

Okhonin V., Petrov, A.P., Krylova, S.M. and Krylov, S.N. 2011 'Quantitative characterization of micromixing based on uniformity and overlap', *Angew. Chem. Int. Ed.*, **50**: 11999–12002.

Olson, F., Hunt, C.A. and Szoka, F.C. 1979 'Preparation of liposomes of defined size distribution by extrusion through polycarbonate membranes', *Biochim. Biophys. Acta*, **557**: 9–23.

Omi, S., Katami, K., Yamamoto, A. and Iso, M. 1994 'Synthesis of polymeric microspheres employing SPG emulsification technique', *J. Appl. Polym. Sci.*, **51**: 1–11.

Omi, S., Katami, K., Taguchi, T., Kaneko, K. and Iso, M. 1995 'Synthesis of uniform PMMA microspheres employing modified SPG (Shirasu Porous Glass) emulsification technique', *J. Appl. Polym. Sci.*, **57**: 1013–1024.

Omi, S., Taguchi, T, Nagai, M. and Ma, G.H. 1997 'Synthesis of 100 μm uniform porous spheres by SPG emulsification with subsequent swelling of the droplets', *J. Appl. Polym. Sci.*, **63**: 931–942.

Omi, S., Kanetaka, A., Shimamori, Y., Supsakulchai, A., Nagai, M. and Ma, G.H. 2001 'Magnetite (Fe_3O_4) microcapsules prepared using a glass membrane and solvent removal', *J. Microencapsulation*, **19**: 749–765.

Pham, T.T., Jaafar-Maalej, C., Charcosset, C. and Fessi, H. 2012 'Liposome and niosome preparation using a membrane contactor for scale-up', *Colloids Surf., B*, **94**: 15–21.

Park, S.H., Yamaguchi, T. and Nakao, S. 2001 'Cut-off of dilute O/W emulsions through a microfiltration membrane', *J. Membr. Sci.*, **190**: 167–178.

Pawlik, A.K. and Norton, I.T. 2012 'Encapsulation stability of duplex emulsions prepared with SPG cross-flow membrane, SPG rotating membrane and rotor-stator techniques—A comparison', *J. Membr. Sci.*, **415–416**: 459–468.

Pawlik, A.K. and Norton, I.T. 2013 'SPG rotating membrane technique for production of food grade emulsions', *J. Food Eng.*, **114**: 530–537.

Piacentini, E., Giorno, L., Dragosavac, M.M., Vladisavljević, G.T. and Holdich, R.G. 2013a 'Microencapsulation of oil droplets using cold water fish gelatine/gum arabic complex coacervation by membrane emulsification', *Food Res. Int.*, **53**: 362–372.

Piacentini, E., Drioli, E. and Giorno, L. 2013b 'Pulsed back-and-forward cross-flow batch membrane emulsification with high productivity to obtain highly uniform and concentrate emulsions', *J. Membr. Sci.*, **453**: 119–125.

Pradhan, R., Lee, D.W., Choi, H.G., Yong, C.S. and Kim, J.O. 2013 'Fabrication of a uniformly sized fenofibrate microemulsion by membrane emulsification', *J. Microencapsulation*, **30**: 42–48.

Rayner, M., Trägårdh, G. and Trägårdh, C. 2005 'The impact of mass transfer and interfacial expansion rate on droplet size in membrane emulsification processes', *Colloids Surf., A*, **266**: 1–17.

van Rijn, C.J.M. and Elwenspoek, M.C. 1995 'Micro filtration membrane sieve with silicon micro machining for industrial and biomedical applications', Proceedings of the Micro Electro Mechanical Systems (MEMS), Amsterdam, The Netherlands, January 29–February 2 (1995), p. 83.

van Rijn, C., van der Wekken, M., Nijdam, W. and Elwenspoek, M. 1997 'Deflection and maximum load of microfiltration membrane sieves made with silicon micromachining', *J. Microelectromech. Syst.*, **6**: 48–54.

Sakai, S., Hashimoto, I., Ogushi, Y. and Kawakami, K. 2007 'Peroxidase-catalyzed cell encapsulation in subsieve-size capsules of alginate with phenol moieties in water-immiscible fluid dissolving H_2O_2', *Bomacromolecules*, **8**: 2622–2626.

Sawalha, H., Fan, Y., Schroën, K. and Boom, R. 2008 'Preparation of hollow polylactide microcapsules through premix membrane emulsification—Effects of nonsolvent properties', *J. Membr. Sci.*, **325**: 665–671.

Schadler, V. and Windhab, E.J. 2006 'Continuous membrane emulsification by using a membrane system with controlled pore distance', *Desalination*, **189**: 130–135.

Schröder, V., Behrend, O. and Schubert H. 1998 'Effect of dynamic interfacial tension on the emulsification process using microporous, ceramic membrane', *J. Colloid Interface Sci.*, **202**: 334–340.

Seemann, R., Brinkmann, M., Pfohl, T. and Herminghaus, S. 2012 'Droplet based microfluidics', *Rep. Prog. Phys.*, **75**: 016601 (41pp).

Seto, H., Ohto, K. and Kawakita, H. 2011 'Reversible extension and shrinkage of solvent-responsive dextran chains produced by enzymatic reaction', *J. Membr. Sci.*, **370**: 76–81.

Segura, R., Cierpka, C., Rossi, M., Joseph, S., Bunjes, H. and Kähler, C. 2013 'Non-encapsulated thermo-liquid crystals for digital particle tracking thermography/velocimetry in microfluidics', *Microfluid. Nanofluid.*, **14**:445–456.

Shimoda, M., Miyamae, H., Nishiyama, K., Yuasa, T., Noma, S. and Igura, N. 2011 'Swirl-flow membrane emulsification for high throughput of dispersed phase flux through Shirasu porous glass (SPG) membrane', *J. Chem. Eng. Jpn*, **44**: 1–6.

Si, T., Wang, Y., Wei, W., Lv, P., Ma, G. and Su, Z. 2011 'Effect of acrylic acid weight percentage on the pore size in poly(N-Isopropyl acrylamide-co-acrylic acid) microspheres', *React. Funct. Polym.*, **71**: 728–735.

Sugiura, S., Nakajima, M., Kumazawa, N., Iwamoto, S. and Seki, M. 2002 'Characterization of spontaneous transformation-based droplet formation during microchannel emulsification', *J. Phys. Chem. B*, **106**: 9405–9409.

Sugiura, S., Oda, T., Izumida, Y., Aoyagi, Y., Satake, M., Ochiali, A., Ohkohchi, N. and Nakajima, M. 2005 'Size control of calcium alginate beads containing living cells using micro-nozzle array', *Biomaterials*, **26**: 3327–3331.

Supsakulchai, A., Ma, G.H., Nagai, M. and Omi, S. 2002a 'Microencapsulation of fine titanium dioxide powders from (S/O)/W emulsion with subsequent solvent evaporation', *ACS Symp. Ser.*, **801**: 260–275.

Supsakulchai, A., Ma, G.H., Nagai, M. and Omi, S. 2002b 'Uniform titanium dioxide (TiO_2) microcapsules prepared by glass membrane emulsification with subsequent solvent evaporation', *J. Microencapsulation*, **19**: 425–449.

Surh, J., Vladisavljević, G.T., Mun, S. and McClements, D.J. 2007 'Preparation and characterization of water/oil and water/oil/water emulsions containing biopolymer-gelled water droplets', *J. Agric. Food Chem.*, **55**: 175–184.

Surh, J., Jeong, Y.G. and Vladisavljević, G.T. 2008 'On the preparation of lecithin-stabilized oil-in-water emulsions by multi-stage premix membrane emulsification', *J. Food Eng.*, **89**: 164–170.

Suzuki, K., Shuto, I. and Hagura, Y. 1996 'Characteristics of the membrane emulsification method combined with preliminary emulsification for preparing corn oil-in-water emulsions', *Food Sci. Technol. Int. Tokyo*, **2**: 43–47.

Suzuki, K., Fujiki, I. and Hagura, Y. 1999 'Preparation of high concentration of O/W and W/O emulsions by the membrane phase inversion emulsification using PTFE membranes', *Food Sci. Technol. Int. Tokyo*, **5**: 234–238.

Tanaka, T., Okayama, M., Kitayama, Y., Kagawa, Y. and Okubo, M. 2010 'Preparation of "mushroom-like" janus particles by site-selective surface-initiated atom transfer radical polymerization in aqueous dispersed systems, *Langmuir*, **26**: 7843–7847.

Tasaki, T., Wada, T., Fujimoto, K., Kai, S. and Ohe, K., Oshima, T., Baba, Y. and Kukizaki, M. 2009, 'Degradation of methyl orange using short-wavelength UV irradiation with oxygen microbubbles', *J. Hazard. Mater.*, **162**: 1103–1110.

Thorsen, T., Roberts, R.W., Arnold, F.H. and Quake, S.R. 2001 'Dynamic pattern formation in a vesicle-generating microfluidic device', *Phys. Rev. Lett.*, **86**: 4163–4166.

Torigoe, K., Shimizu, M., Yamamoto, K., Mizozoe, M., Takahashi, H., Suzuki, T. and Murase, M. 2011 'Method and apparatus for manufacturing low melting point metal fine particles', US patent 7,976,608, 12 July 2011.

Viebke, V., Piculell, L. and Nilsson, S. 1994 'On the mechanism of gelation of helix-forming biopolymers', *Biomacromolecules*, **27**: 4160–4166.

Vladisavljević, G.T. and Schubert, H. 2002 'Preparation and analysis of oil-in-water emulsions with a narrow droplet size distribution using Shirasu-porous-glass (SPG) membranes', *Desalination*, **144**: 167–172.

Vladisavljević, G.T. and Schubert, H. 2003a 'Preparation of emulsions with a narrow particle size distribution using microporous α-alumina membranes', *J. Dispersion Sci. Technol.*, **24**: 811–819.

Vladisavljević, G.T. and Schubert, H. 2003b 'Influence of process parameters on droplet size distribution in SPG membrane emulsification and stability of prepared emulsion droplets', *J. Membr. Sci.*, **225**: 15–23.

Vladisavljević, G.T., Lambrich, U., Nakajima M. and Schubert H. 2004 'Production of O/W emulsions using SPG membranes, ceramic α-Al_2O_3 membranes, microfluidizer and a microchannel plate: a comparative study', *Colloids Surf., A*, **232**: 199–207.

Vladisavljević, G.T., Shimizu, M. and Nakashima, T. 2004b 'Preparation of monodisperse multiple emulsions at high production rates by multi-stage premix membrane emulsification', *J. Membr. Sci.*, **244**: 97–106.

Vladisavljević, G.T., Shimizu, M. and Nakashima, T. 2005 'Permeability of hydrophilic and hydrophobic Shirasu-porous-glass (SPG) membranes to pure liquids and its microstructure', *J. Membr. Sci.*, **250**: 69–77.

Vladisavljević, G.T. and Williams, R.A. 2005 'Recent developments in manufacturing emulsions and particulate products using membranes', *Adv. Colloid Interface Sci.*, **113**: 1–20.

Vladisavljević, G.T. and Williams, R.A. 2006 'Manufacture of large uniform droplets using rotating membrane emulsification', *J. Colloid Interface Sci.*, **299**: 396–402.

Vladisavljević, G.T., Shimizu, M. and Nakashima, T. 2006a 'Production of multiple emulsions for drug delivery systems by repeated SPG membrane homogenization: Influence of mean pore size, interfacial tension and continuous phase viscosity', *J. Membr. Sci.*, **284**: 373–383.

Vladisavljević, G.T., Surh, J. and McClements, D.J. 2006b 'Effect of emulsifier type on droplet disruption in repeated Shirasu porous glass membrane homogenization', *Langmuir*, **22**: 4526–4533.

Vladisavljević, G.T., Kobayashi, I., Nakajima, M., Williams, R.A., Shimizu, M. and Nakashima, T. 2007 'Shirasu Porous Glass membrane: Characterisation of microstructure by high resolution x-ray microtomography and visualisation of droplet formation in real time', *J. Membr. Sci.*, **302**: 243–253.

Vladisavljević, G.T. and Williams, R.A. 2008 'Recent developments in manufacturing particulate products from double-emulsion templates using membrane and microfluidic devices', in Aserin, A. (ed.), *Multiple Emulsions: Technology and Applications*, Hoboken: John Wiley & Sons, Inc.

Vladisavljević, G.T., Kobayashi, I. and Nakajima, M. 2008 'Generation of highly uniform droplets using asymmetric microchannels fabricated on a single crystal silicon plate: Effect of emulsifier and oil types', *Powder Technol.* **183**: 37–45.

Vladisavljević, G.T. and Williams, R.A. 2010 'Recent developments in manufacturing nanoparticles from emulsion droplets', in Starov, V. (ed.), *Nanoscience: Colloidal and Interfacial Aspects*, Boca Raton, FL: CRC Press, pp. 437–491.

Vladisavljević, G.T. and McClements, D.J. 2010 'Modification of interfacial characteristics of monodisperse droplets produced using membrane emulsification by surfactant displacement and/or polyelectrolyte electrostatic deposition', *Colloids Surf., A*, **364**: 123–131.

Vladisavljević, G.T., Kobayashi, I. and Nakajima, M. 2011 'Effect of dispersed phase viscosity on maximum droplet generation frequency in microchannel emulsification using asymmetric straight-through channels', *Microfluid. Nanofluid.*, **10**: 1199–1209.

Vladisavljević, G.T., Kobayashi, I. and Nakajima, M. 2012 'Production of uniform droplets using membrane, microchannel and microfluidic emulsification devices', *Microfluid. Nanofluid.*, **13**: 151–178.

Wagdare, N.A., Marcelis, A.T.M., Ho, O.B., Boom, R.M. and van Rijn, C.J.M. 2010 'High throughput vegetable oil-in-water emulsification with a high porosity micro-engineered membrane', *J. Membr. Sci.*, **347**: 1–7.

Wang, L.Y., Ma, G.H. and Su, Z.G. 2005 'Preparation of uniform sized chitosan microspheres by membrane emulsification technique and application as a carrier of protein drug', *J. Controlled Release*, **106**: 62–75.

Wang, Y., Zhang, C., Bi, S. and Luo, G. 2010 'Preparation of ZnO nanoparticles using the direct precipitation method in a membrane dispersion micro-structured reactor', *Powder Technol.*, **202**: 130–136.

Wang, G., Dou, H. and Sun, K. 2012 'Facile synthesis of hollow polymeric microparticles possessing various morphologies via seeded polymerization', *Colloid Polym. Sci.*, **290**: 1867–1877.

Wang, G., Leng, Y., Dou, H., Wang, L., Li, W., Wang, X., Sun, K., Shen, L., Yuan, X., Li, J., Sun, K., Han, J., Xiao, H. and Li, Y. 2013 'Highly efficient preparation of multiscaled quantum dot barcodes for multiplexed hepatitis B detection', *ASC Nano*, **7**: 471–481.

Wang, Y., Qin, J., Yi, W., Li, C. and Ma, G. 2013 'Preparation strategies of thermo-sensitive P(NIPAM-co-AA) microspheres with narrow size distribution', *Powder Technol.*, **236**: 107–113.

Wei, Q., Wei, W., Tian, R., Wang, L.Y., Su, Z.G. and Ma, G.H. 2008 'Preparation of uniform-sized PELA microspheres with high encapsulation efficiency of antigen by premix membrane emulsification', *J. Colloid Interface Sci.*, **323**: 267–273.

Wei, W., Ma, G.H., Wang, L.Y., Wu, J. and Su, Z.G. 2010 'Hollow quaternized chitosan microspheres increase the therapeutic effect of orally administered insulin', *Acta Biomater.*, **6**: 205–209.

Wei, Y., Wang, Y., Wang, L., Hao, D. and Ma, G.H. 2011 'Fabrication strategy for amphiphilic microcapsules with narrow size distribution by premix membrane emulsification', *Colloids Surf., B*, **87**: 399–408.

Wei W., Lv, P.P., Chen, X.M., Yue, Z.G., Fu, Q., Liu, S.Y., Yue, H. and Ma, G.H. 2013 Codelivery of mTERT siRNA and paclitaxel by chitosan-based nanoparticles promoted synergistic tumor suppression', *Biomaterials*, **34**: 3912–3923.

Wu, J., Wei, W., Wang, L.Y., Su, Z.G. and Ma, G.H. 2008 'Preparation of uniform-sized pH-sensitive quaternized chitosan microsphere by combining membrane emulsification technique and thermal-gelation method', *Colloids Surf., B*, **63**: 164–175.

Xu, J.H., Luo, G.S., Chen, G.G. and Tan, B. 2005 'Mass transfer performance and two-phase flow characteristic in membrane dispersion mini-extractor', *J. Membr. Sci.*, **249**: 75–81.

Yamamoto, T., Ohmori, T. and Kim, Y.H. 2010 'Synthesis of monodisperse carbon cryogel microspheres using membrane emulsification of a phenol–formaldehyde solution', *Carbon*, **48**: 912–928.

Yamashita, N., Konishi, N., Tanaka, T. and Okubo, M. 2012 'Preparation of hemispherical polymer particles by cleavage of a Janus poly(methyl methacrylate)/polystyrene composite particle, *Langmuir*, **28**: 12886–12892.

Yang, J., Hao, D.X., Bi, C.X., Su, Z.G., Wang, L.Y. and Ma, G.H. 2010 'Rapid synthesis of uniform magnetic microspheres by combing premix membrane emulsification and in situ formation techniques', *Ind. Eng. Chem. Res.*, **49**: 6047–6053.

Yasuda, M., Goda, T., Ogino, H., Glomm, W.R. and Takayanagi, H. 2010 'Preparation of uniform monomer droplets using packed column and continuous polymerization in tube reactor', *J. Colloid Interface Sci.*, **349**: 392–410.

Yobas, L., Martens, S., Ong, W.L. and Ranganathan, N. 2006 'High-performance flow-focusing geometry for spontaneous generation of monodispersed droplets', *Lab Chip*, **6**: 1073–1079.

You, J.O., Park, S.B., Park, H.Y., Haam, S., Chung, C.H. and Kim, W.S. 2001 'Preparation of regular sized Ca-alginate microspheres using membrane emulsification method', *J. Microencapsulation*, **18**: 521–532.

Yue, Z.G., Wei, W., Lv, P.P., Yue, H., Wang, L.Y., Su, Z.G. and Ma, G.H. 2011 'Surface charge affects cellular uptake and intracellular trafficking of chitosan-based nanoparticles', *Biomacromolecules*, **12**: 2440–2446.

Yue, H., Wei, W., Fan, B., Yue, Z., Wang, L., Ma, G. and Su, Z. 2012 'The orchestration of cellular and humoral responses is facilitated by divergent intracellular antigen trafficking in nanoparticle-based therapeutic vaccine, *Pharmacol. Res.*, **65**: 189–197.

Yuyama, H., Hashimoto, T., Ma, G.H., Nagai, M. and Omi, S. 2000a 'Mechanism of suspension polymerization of uniform monomer droplets prepared by glass membrane (Shirasu Porous Glass) emulsification technique', *J. Appl. Polym. Sci.*, **78**: 1025–1043.

Yuyama, H., Yamamoto, K., Shirafuji, K., Nagai, M. and Ma, G.H. 2000b 'Preparation of polyurethaneurea (PUU) uniform spheres by SPG membrane emulsification technique', *J. Appl. Polym. Sci.*, **77**: 2237–2245.

Zeng, W., Gomaa, H.G., Liu, J. and Zhu, J. 2013 'Intensification of production of O/W emulsions using oscillatory woven metal micro-screen (WMMS)', *Chem. Eng. Process.*, **73**: 111–118.

Zhang, Y., Wei, W., Lv, P., Wang, L. and Ma, G. 2011 'Preparation and evaluation of alginate–chitosan microspheres for oral delivery of insulin', *Eur. J. Pharm. Biopharm.*, **77**: 11–19.

Zhou, Q.Z., Wang, L.Y., Ma, G.H. and Su, Z.G. 2007 'Preparation of uniform-sized agarose beads by microporous membrane emulsification technique', *J. Colloid Interface Sci.*, **311**: 118–127.

Zhou, Q.Z., Wang, L.Y., Ma, G.H. and Su, Z.G. 2008 'Multi-stage premix membrane emulsification for preparation of agarose microbeads with uniform size', *J. Membr. Sci.*, **322**: 98–104.

Zhou, Q.Z., Ma, G.H. and Su, Z.G. 2009 'Effect of membrane parameters on the size and uniformity in preparing agarose beads by premix membrane emulsification', *J. Membr. Sci.*, **326**: 694–700.

Zhou, Q., Zhang, M.C., Shuang, C.D., Li, Z.Q. and Li, A.M. 2012 'Preparation of a novel magnetic powder resin for the rapid removal of tetracycline in the aquatic environment', *Chin. Chem. Lett.*, **23**: 745–748.

Zimmerman, W.B., Tesař, V. and Bandulasena, H.C.H. 2011 'Towards energy efficient nanobubble generation with fluidic oscillation', *Curr. Opin. Colloid Interface Sci.*, **16**: 350–356.

Zúñiga, R.N. and Aguilera, J.M. 2008 'Aerated food gels: fabrication and potential applications', *Trends Food Sci. Technol.*, **19**: 176–187.

van der Zwan, E.A., Schroën, C.G.P.H. and Boom, R.M. 2008 'Premix membrane emulsification by using a packed layer of glass beads', *AIChE J.*, **54**: 2190–2197.

6

Nanofiltration in Integrated Membrane Processes

Bart Van der Bruggen
Department of Chemical Engineering, ProcESS – Process Engineering for Sustainable Systems, KU Leuven, Belgium
Faculty of Engineering and the Built Environment, Tshwane University of Technology, Private Bag X680, Pretoria 0001, South Africa

6.1 Introduction

Nanofiltration has become a reference separation process for drinking water production, for process water treatment, and even for wastewater recycling applications [1]. Even though not all questions are answered, research on nanofiltration is today more critical than ever before, because the process is now to be considered mature. Therefore, unexpected outcomes need explanations, so that the performance of a nanofiltration unit can be fully understood and predicted. This allows to apply nanofiltration as a reliable part of a treatment chain, and to integrate it with other (separation) processes. In 2009, Van der Bruggen et al. outlined six challenges for nanofiltration [2]: (1) avoiding membrane fouling, and possibilities to remediate, (2) improving the separation between solutes that can be achieved, (3) further treatment of concentrates, (4) chemical resistance and limited lifetime of membranes, (5) insufficient rejection of pollutants in water treatment, and (6) the need for modelling and simulation tools. These are today as relevant as they were in 2009. It is instructive to observe how the various challenges were (are) handled in research. For membrane fouling, the classical approach is directed to membrane synthesis, by modifying the surface structure of existing membranes [3], by using additives such as hydrophilic nanoparticles [4–6], through layer-by-layer assembly [7,8] or by surface coating [9]. These

Integrated Membrane Systems and Processes, First Edition. Edited by Angelo Basile and Catherine Charcosset.

efforts, even though interesting, have not resulted in a dramatic improvement of the membrane performance with respect to fouling reduction. It is obvious that each of these novel membranes have some benefits; unfortunately, it seems not easy to scale up or commercialize such membranes, for various reasons (of which the membrane performance is probably not the most prominent one). Researchers tirelessly keep trying to improve the membrane performance on laboratory scale in the quest for the ideal non-fouling membrane. However, it may be more worthwhile to consider engineering aspects of membrane separations, which leads directly to the development of integrated processes as discussed in this chapter. Such approach may intrinsically solve the fouling problem at least partly. It is evident that this requires a case-by-case approach, since each application is different with respect to membrane fouling. Therefore, identification of foulants [10, 11] and cleaning procedures [12] is of vital importance. Repetitive membrane fouling and chemical cleaning is typically required to maintain a high level of water production, but may also influence the membrane performance negatively. Fouling influences the rejection by pore blocking, modification of the membrane surface charge, and cake-enhanced concentration polarization; however, not all foulants were removed from the membrane surface by a caustic cleaning procedure. The membrane surface hydrophobicity and zeta potential were observed to change, and a temporary enlargement of the membrane pores due to the caustic cleaning was found to have an impact on the rejection, mainly for charged compounds. These findings again underline the need for integrated solutions making use of nanofiltration membranes combined with complementary processes able to reduce fouling.

A further issue is the need for fractionation. Challenges related to fractionation may not be realistically solved by a single nanofiltration membrane. A classical benefit of a nanofiltration membrane is the differential rejection of monovalent and multivalent salts, and the gradual rejection of organic compounds related to the molar mass or molar size [13]. This may be of interest, for example, lowering the rejection in general when the purity requirements do not need reverse osmosis membranes. However, fractionation is still a further challenge that cannot be adequately solved by a single nanofiltration membrane because the rejection as a function of molar size is far from a step function, and the differences between charged solutes are insufficient for a one-step application. This prompts for engineered solutions, which again can be found in integration of membrane processes, using the approach of membrane cascades. Keurentjes (1992) suggested that when separation of solvents by distillation is difficult (referring to the formation of an azeotropic mixture), membrane processes can be an alternative; a cascade of membranes has to be applied when it is difficult or impossible to achieve a complete separation in a single-membrane separation process [14]. Gunderson et al. (2007) further elaborated this concept, extending the idea of using diafiltration for the fractionation of solute pairs [15]. The insufficient rejection of trace compounds in drinking water production may also find a solution in a similar way; this comes down to a fractionation problem in which the minerals should remain in the water apart from some undesired ionic species such as nitrate, boron, fluoride, and arsenic, to name but a few, and a wide range of organic micropollutants ranging from pesticides over chemicals to pharmaceuticals and personal care products [16].

Several other challenges may be resolved as well by integration, hybridization or cascading of processes. Concentrate treatment is an obvious aspect of any membrane treatment process and should be further considered. The classical approach is to consider concentrates as waste fractions, and to develop methods for treatment of liquid waste such as incineration, direct or indirect discharge in groundwater or in surface water (for aqueous concentrates), and discharge on a landfill [17]. Alternatively, one may consider reuse of concentrates, or further removal of contaminants by other separation processes. More advanced solutions would gradually shift to the integration of nanofiltration with other membrane or non-membrane processes.

Process integration is an opportunity for membrane processes to gain recognition in chemical engineering not only as miraculous black box solutions, but as well-controlled, engineered separation processes that can be operated in connection with reactors or with other separation units. This requires more extended tools for modelling and simulation. Such tools are generally not well integrated with simulation packages, mostly because of the versatility of membrane properties, which requires specific data for any given membrane, or any given application. Standardization of membranes and performance parameters would help in solving this problem.

Membranes, and more specifically nanofiltration membranes, have been considered in various ways as parts of an integrated process. This ranges from their use as pre- or post-treatment processes, which would have a loose connection with other processes in a treatment train, to a fully integrated unit in which all flows are intrinsically connected and in which each sub-process should be operated in such a way that not only the process itself, but also the overall flowsheet is optimized. This results in a non-uniform terminology and some ambiguity with respect to what should be understood as process integration. In the strictest sense, process integration would refer only to those processes in which there is an intrinsic link of the performance of both connected devices. This, however, is not a rigid definition since the influence of one process to another can be a minimal influence through for example, the variation of the composition of the feed flow to the second process, or changing flow dynamics. Such processes could be considered as integrated, even though they are rather loosely connected without a recirculation stream. If a recirculation is applied, the integration is evident; such processes are denoted as hybrids. Unfortunately, the word 'cascade' is used in the literature for any combination of processes, whereas in the strict chemical engineering sense it refers to a specific hybridization of a process with itself, yielding a superior performance due to the repeated self-integration. The classical example is a distillation column [18]. However, membranes (and nanofiltration membranes in particular) can also be used in a similar approach, which yields a membrane cascade, which is a hybridization of a nanofiltration unit with itself. Many references to membrane cascades in the literature would in fact consider a sequence of processes, which may be denoted (with some flexibility) as an integration of processes. It is strongly recommended to use the correct terminology in this context, because an inappropriate use of the word 'cascade' would automatically lead to a devaluation of the terminology related to process integration and, consequently, confusion in the interpretation of process descriptions.

A further extension of the concept of process integration is the consideration of membrane reactors using nanofiltration membranes. This would have particular benefits for the performance of the reactor. A typical example that is often applied is a membrane bioreactor, which can be used for wastewater treatment not only with microfiltration or ultrafiltration membranes but also with nanofiltration membranes in an aerobic or an anaerobic configuration [19,20], but also in many bioproductions of chemical specialities, where the product inhibits the bioreaction [21]. Integration can be done in an external module for membrane extraction or by an extraction membrane module immersed directly in the bioreactor.

In this chapter, an overview is given of the potential of integrated separation processes based on, or making using of, nanofiltration. A gradual approach is used in which the integration aspect is more emphasized step by step. First, the need for pretreatment processes prior to nanofiltration is considered. Second, nanofiltration is considered as the pretreatment unit, serving a further process in the treatment train. This is further extended to the use of nanofiltration in processes in series, followed by the integration of processes. The distinction between these two is a very thin line, balancing on the specification of the degree of interaction between two processes; although the two concepts may also be indicated on a greyscale with only small nuances of difference. Hybrid processes are described according to the strict definition of a hybrid, requiring (a) only two, and no more, outgoing flows, and (b) back coupling of at least one flow to the first process in the hybrid configuration. The last part of the chapter is reserved for membrane cascades as a particular type of hybrid processes, with specific features that allow to strengthen the separation capability of a single nanofiltration unit.

6.2 Pretreatment for Nanofiltration

It is generally assumed that prior to nanofiltration, pretreatment is necessary to prevent excessive fouling of nanofiltration membranes. Various methods are used to achieve this, many of which are classical treatment processes. Manekar et al. [22] give an overview of pretreatment processes to be applied for the specific case of textile effluents, which can be considered a worst case scenario since such heavily loaded feed stream would require the most extensive pretreatment. The authors include chemical precipitation, electro-coagulation, chlorination, ozonation, catalytic oxidation, bio-oxidation (activated sludge process or a sequential batch reactor), pressure sand filtration, and activated carbon adsorption. This is schematically shown in Figure 6.1. Based on a multicriteria decision-making tool considering capital cost, operation and maintenance cost, footprint, performance, and sludge generation, the authors concluded that an activated sludge process preceded and followed by chemical precipitation (Module IV in Figure 6.1) is to be preferred. However, rather than considering only the outcome (which may be even slightly subjective), it is more instructive to evaluate the type and range of candidate processes, and the criteria used in decision making. Processes like electro-coagulation, catalytic oxidation, and ozonation were denoted as advanced pretreatment methods, with high capital cost and operational cost, and a moderate-to-low performance. The dominance of cost issues shifts the outcome to conventional, chemicals-based processes with little potential for integration in an overall membrane approach. Such pretreatment would often be operated as feed conditioning, linked with, but not integrated with a subsequent nanofiltration unit.

A higher degree of integration may be obtained by including microfiltration membranes in the pretreatment. Tahri et al. [23] studied dye baths with a combination of microfiltration and nanofiltration, and concluded that the use of microfiltration as a pretreatment prior to nanofiltration improves the treatment effectiveness by increasing the operating time and the permeate flux. Van der Bruggen et al. [24] studied the effect of a microfiltration pretreatment for synthetic solutions of fluorescent monodisperse latex particles of 0.1 and 1 μm diameter and an organic foulant (benzyl alcohol) in water, and concluded that the beneficial effect depends on the size and hydrophobicity of foulant particles, and the roughness of the nanofiltration membrane.

Sari and Chellam [25] extended this approach by studying four different advanced pretreatment processes to reduce fouling of a commercially available thin-film

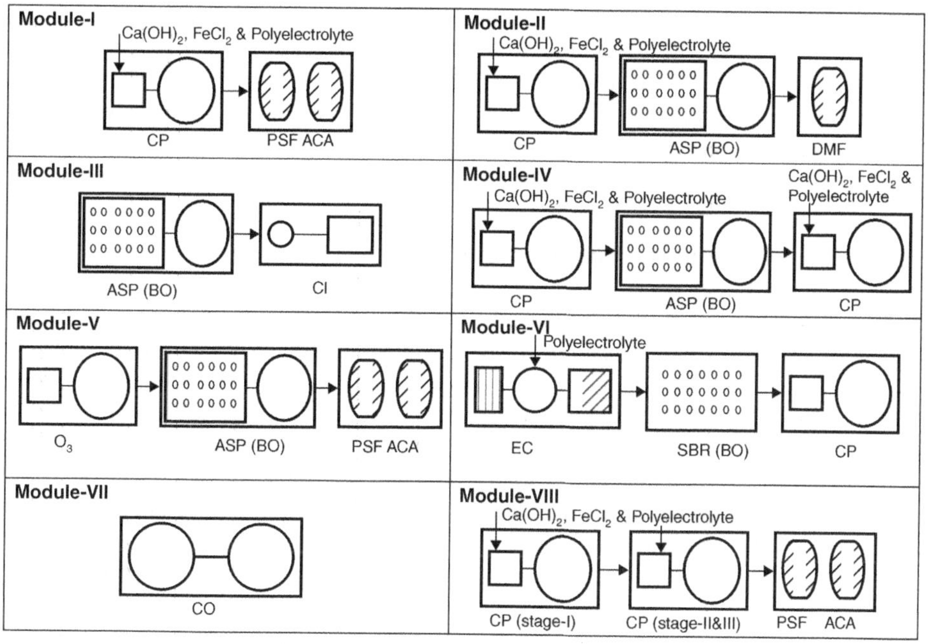

Figure 6.1 *Possible pretreatment methods prior to membrane filtration for textile effluents. CP, chemical precipitation; BO, bi-oxidation; ASP, activated sludge process; EC, electro-coagulation; SBR, sequential batch reactor; CO, catalytic oxidation; O_3, ozonation; Cl, chlorination; DFM, dual media filter; PSF, pressure sand filter; ACC, activated carbon adsorption.* Source: *Reprinted from Reference 22 with permission from Elsevier.*

composite nanofiltration membrane during surface water treatment, that is, microfiltration (alone, chemical (alum) coagulation–microfiltration, electrocoagulation–microfiltration, and electroflotation–microfiltration). It was found that the latter process yielded the best performance of the nanofiltration membrane, which suggests that a combination of processes is necessary. Stoller et al. [26] relate the performance of the pretreatment method to the concept of the threshold flux and suggest that the performance of a nanofiltration unit should be understood in a close relation with preceding processes.

A further intensification of the pretreatment method could be in the use of a membrane bioreactor combined with nanofiltration [27], which may be related to the effect of effluent organic matter on membrane fouling. The beneficial effect of such operation remains unclear, because the bioreactor would produce large amounts of polysaccharide-like substances with small molecular weight causing fouling of nanofiltration membranes. Furthermore, the hydrophilic fraction was found to foul the membrane, even though it is generally assumed that the hydrophobic fraction is the main cause of fouling. Biodegradation would shift the composition of foulants to the hydrophilic fraction, so that the eventual outcome when using a bioreactor is rather unclear.

The ultimate pretreatment method prior to nanofiltration would be ultrafiltration. This would offer the maximal protection of the nanofiltration membrane against fouling and yield the best performance. Ultrafiltration allows an efficient removal of algae, bacteria, parasites, and viruses for clarification and disinfection purposes, suspended particles, colloids, and turbidity. This reduces all types of fouling in the nanofiltration unit, including biofouling, to a minimum. However, the use of ultrafiltration is seldom practiced as a mere pretreatment to nanofiltration because of the additional process cost. In some applications, the integration of both processes is of interest, but in such cases the function of the ultrafiltration unit surpasses that of a pretreatment technology (discussed in Sections 6.4. and 6.5).

6.3 Nanofiltration as a Pretreatment Method

In the same philosophy of pretreatment processes, nanofiltration can be used prior to enhanced separation by, for example, reverse osmosis. This is of interest for desalination, because the nanofiltration unit is more economical in terms of energy consumption and removes the scalants that cause problems in the reverse osmosis unit (calcium, magnesium, sulphates) with some selectivity. This is applied in one of the commercial seawater reverse osmosis plants in Ummlujj, Saudi Arabia, which is in operation since September 2000 [28]. The nanofiltration unit was operated at 65% recovery at pH 6 at a feed pressure below 25 bar, which led to increase in production of the desalination plant by 42%. A scheme for this plant is shown in Figure 6.2. The concept was developed at the turn of the century,

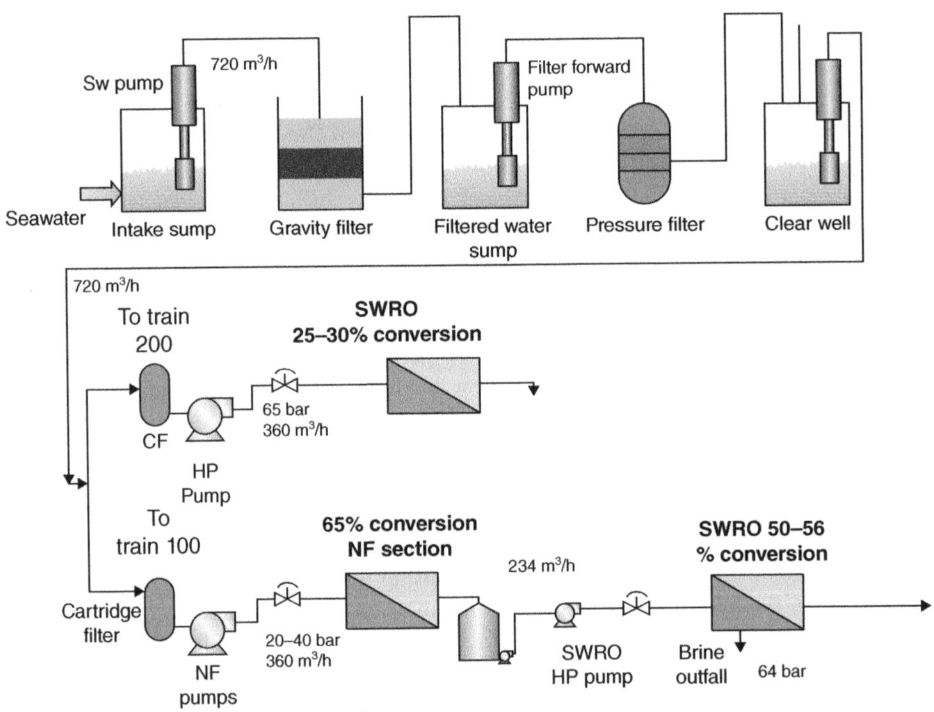

Figure 6.2 *Nanofiltration followed by reverse osmosis for desalination, operated since 2000 in Ummlujj, Saudi Arabia. NF, nanofiltration; SWRO, seawater reverse osmosis; SW, seawater.* Source: *Reprinted from Reference 28 with permission from Elsevier.*

aiming at reverse osmosis as well as multi-stage flash (MSF) as desalination technology. The use of nanofiltration as a pretreatment to thermal desalination may allow to increase the top brine temperature (the maximum temperature in the first stage of the process, limiting the productivity to up to 160°C [29]. A techno-economical analysis concluded that the integration of nanofiltration in an existing desalination plant using MSF as technology enables to increase the top brine temperature to 130°C, so that the production can be increased by 19%. However, the cost analysis showed that the capital cost is 5.4% higher than for a conventional MSF system (operated at 110°C) due to the additional capital cost of the nanofiltration system [30]. When reverse osmosis is used as the desalination technology, the rejection of scale forming ions is focused on sulphate and divalent cations. Overall encouraging results were found for a range of nanofiltration membranes for synthetic and real seawater, obtaining a sulphate rejection above 90% for most membranes, yielding an effective scaling prevention [31].

In addition to the salt removal potential of nanofiltration membranes, other aspects such as fouling by nanoparticles are currently of interest. Removal of colloidal nanoparticles may help in a successful operation of reverse osmosis membranes [32], which can be achieved with nanofiltration.

It should be noted that in contrast to a pretreatment by microfiltration or ultrafiltration, a nanofiltration unit has a direct impact on the reverse osmosis unit or the thermal desalination unit, since it increases the recovery and may also reduce the energy consumption of the overall process.

In a different context, nanofiltration was studied as a means of seawater softening for offshore oilfield water and polymer flooding [33]. Furthermore, Esmi et al. [34] investigated the use of nanofiltration as a pretreatment step in a zero rejection process for wastewater treatment, in which the nanofiltration mainly served for retaining metal cations. Benito-Alcazar [35] considered different pretreatments applied to a petrochemical secondary effluent to produce a suitable feed water quality for reverse osmosis treatment, among which nanofiltration; other options included granulated activated carbon filtration, ultrafiltration, and granulated activated carbon filtration coupled with nanofiltration. They concluded that although the permeate obtained with ultrafiltration was suitable for reverse osmosis, nanofiltration, and granulated activated carbon filtration coupled with nanofiltration provided a better effluent quality for reverse osmosis than the other pretreatments considered in the study. Purkait et al. [36] focused on a similar approach for treatment of leather plant effluent. They observed that a combined operation of nanofiltration, using a membrane with a cut-off of 400 Da, followed by a reverse osmosis unit, has a better performance than a single operation of nanofiltration with a 200-Da cut-off membrane; the BOD and COD values of the treated effluent were well within the permissible limits.

This suggests a new trend following the use of combined nanofiltration and reverse osmosis in desalination, which aims at nearly zero discharge of wastewater and an almost complete recovery of water. Until today, this has not been extensively studied, because of the enhanced cost of such two-step system, which cannot (yet) be justified for wastewater treatment, in contrast to seawater desalination. This, however, is entirely related to the value of the end product and not to any technical aspect of the application. Since this is due to change, one can also expect more applications of such two-step processes. Nevertheless, it may then be necessary to consider this in a broader context comprising not only the by-now classical challenge of water reuse, but also recycling of valuable compounds such as nutrients. This requires an enhanced fractionation of the feed, and would typically lead to a more complex treatment train by using processes in series or integrated/hybrid systems. These will be discussed in the following sections.

6.4 Processes in Series

In many cases, the use of nanofiltration preceded or followed by another process is not sufficient, especially for difficult separations or when multiple objectives have to be met. There is confusion on the correct terminology to be used, which may make it difficult to clearly understand systems comprising more than one process. Many applications would fit in this section, as they are operated as processes in series, but can be found described in the literature as integrated systems, dual processes, hybrid processes or even membrane cascades, all of which is incorrect terminology. An example of a processes-in-series approach is the removal of oil and grease from bilge wastewaters accumulating in the lower recesses of ships [37]. The purpose was not only in purification, but also in minimizing concentrate production (since the concentrate has to be stored on board this is a very important issue). The approach was denoted as a membrane cascade, but as will be further explained, it was in reality a series of processes of which each had its own performance. Tubular ceramic microfiltration and ultrafiltration membranes were used in the first stage, where the concentration of particulate matter is high, and cross-flow cells representing spiral wound modules in the second stage. In this second stage, flat sheet ultrafiltration (1–100 kDa cut-off) and nanofiltration membranes were used. This system was found to produce water containing below detectable levels of hexane extractable material, although the authors observed that nanofiltration was less beneficial in this case because the (unnecessary) ion retention and the smaller pores resulted in lower fluxes, whereas tight ultrafiltration membranes would yield a permeate of sufficient quality as well.

A similar approach was used for treatment of produced water with high silica content [38]. Nanofiltration membranes were used to remove divalent cations from produced water, and the resulting permeate was fed to reverse osmosis membranes after increasing the pH to 10.0 to enhance silica solubility. To reduce the fouling potential of the nanofiltration and reverse osmosis membranes, dissolved air flotation, ceramic ultrafiltration, MYCELX cartridges, and organoclay filters were tested as pretreatment alternatives. This is shown in Figure 6.3.

It was found that pretreatment processes were effective for turbidity and oil and grease removal but were not for retaining organic matter, primarily protein-like and polysaccharide-like material, which eventually fouled the first pass nanofiltration membranes, whereas the reverse osmosis membranes were predominantly scaled by silica. This limited the overall feed water recovery of the processes in series to a maximum of 70%. Thus, the application of a two-pass configuration met the discharge limits for most of the contaminants in produced water, but a more stringent pretreatment process for selective removal of organics and silica is essential to operate the membrane systems at higher recoveries (above 85% is suggested). Such high recoveries were also an objective for using nanofiltration as part of a treatment train in surface treatment. The overall system comprised of riverbank filtration, perforated plate aeration, dual media filtration, aeration, rapid sand filtration, and nanofiltration, in which the nanofiltration may be followed by or replaced by granular activated carbon

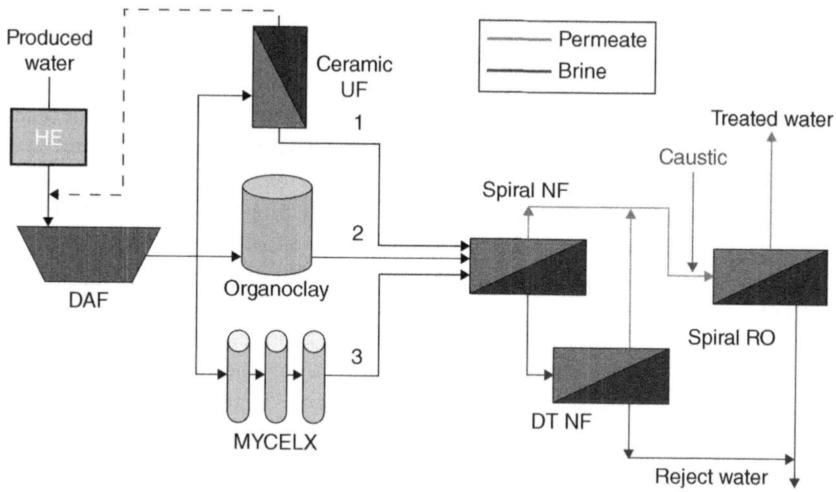

Figure 6.3 Processes in series used for treatment of produced water with high silica content. DAF, dissolved air flotation; UF, ultrafiltration; DT, disc tube; NF, nanofiltration. Source: Reprinted from Reference 38 with permission from Taylor & Francis.

filtration. This is schematically shown in Figure 6.4, representing the serial character of the overall system.

Interestingly, this system was denoted as an 'integrated multi-objective membrane system'; even though the integrative function may be questioned, the multi-objective aspect is highly important in the context of processes in series. The study aimed at the removal of colour, hardness, and a number of synthetic organic chemicals originating from the river. Furthermore, the water recovery was maximized and fouling was minimized by using a combination of treatment processes. Such serial process has some level of integration due to the fact that each process is important in maintaining the overall performance of the plant.

Processes in series can also be found in other applications. Turek [40] further elaborated the combination of nanofiltration with MSF and/or reverse osmosis for desalination mentioned before, and extended this approach by using the highly concentrated brine as a

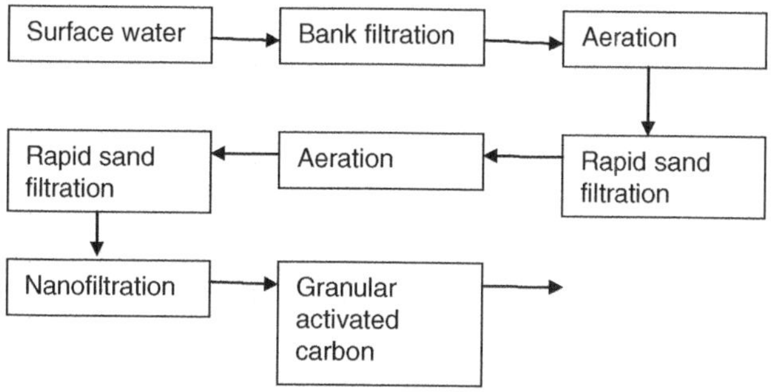

Figure 6.4 Sequence of treatment processes proposed for surface water treatment [39].

possible by-product. The outcome is a dual-purpose desalination–salt production system. Two arrangements were considered: ultrafiltration–nanofiltration–MSF–crystallization and ultrafiltration–nanofiltration–reverse osmosis–MSF–crystallization. This resulted in a calculated water cost that is competitive compared to potable water produced in thermal or RO seawater plants. One decade later one may expect that benefits have gone up and costs have gone down. It is of interest to reconsider the suggestions made by Turek using some adjustments, such as the use of a membrane crystallizer for the proposed production of salts [41].

Processes in series based on nanofiltration are also of interest for bioconversions. Qi et al. [42] studied a sequence of ultrafiltration and nanofiltration for recycling cellulase and concentrating glucose from enzymatic hydrolyzate of steam exploded wheat straw. Ultrafiltration allowed the recovery of 73.9% of the cellulase protein present in the hydrolyzate suspension, while allowing free transmission of glucose. The permeate was then fed to a nanofiltration unit, so that the glucose concentration in the ultrafiltered hydrolyzate increased from 30.2 to 110.2 g/L. Recycling cellulase by ultrafiltration reduced the cost of hydrolysis of the lignocellulosic feedstock, while concentrating glucose by nanofiltration improved the fermentation efficiency of lignocellulosic hydrolyzate and lowered the separation and purification cost. Thus, the use of processes in series has a clear advantage in such applications in which multiple objectives should be met, in which each individual process has a well-defined and independent function. With the current trend of aiming at enhanced fractionation for optimized production or recycling, this approach will be applied more and more in the near future.

6.5 Integrated Processes

A further integration of processes used in combination with nanofiltration can be obtained when functionalities of the individual processes are shared. This yields an enhanced interaction between the processes, because the influence of the performance of one process would directly influence the performance of other processes and therefore, the general performance of the overall system. This would go beyond the approach of processes in series, where the influence of a given process on a subsequent process is limited to either the separation of a specific compound or fraction, or to condition the flow so that further treatment is secured (this is typically related to a reduction of the fouling potential). Such integrated approach requires that the individual processes have at least partly the same functionality. A recycle stream, however, is not necessary.

An example is the combination of nanofiltration and solar photo-Fenton treatment of water containing micropollutants [43]. Five different pharmaceuticals (carbamazepine, flumequine, ibuprofen, ofloxacin, and sulfamethoxazole) dissolved in natural water were preconcentrated by nanofiltration and treated by solar photo-Fenton, considering the removal of the pharmaceuticals and the hydrogen peroxide consumption as the parameters to evaluate and optimize the process. An influence of the nanofiltration unit on the hydrogen peroxide consumption and on the photo-Fenton treatment time was observed, which led the authors to the suggestion of using combined nanofiltration and photo-Fenton treatment for wastewater containing extremely low concentrations of micropollutants. Berberidou et al. [44] combined photocatalytic oxidation (heterogeneous photocatalytic oxidation, and photo-Fenton oxidation, in the presence of UV-A or visible light) and membrane filtration using nanofiltration or reverse osmosis for decolouration, reduction of the concentration of organic matter and of salinity for a synthetic dyestuff effluent containing the dye Reactive Black 5, with the objective of producing water that can be reused for dyeing purposes. The integration of either photocatalytic methods with a reverse osmosis unit, leads to complete decolourization, as well as to a more than 95% reduction of the initial organic content and salinity.

The same idea can be applied to drinking water production [45]. The integration of nanofiltration with direct and indirect UV photolysis for drinking water treatment was studied in order to guarantee effective removal of different hormones with endocrine disruption capabilities (mestranol, octylphenol, nonylphenol, progesterone, estrone, estriol, 17 alpha-ethynylestradiol, and beta-estradiol were studied) from real surface water matrices. The integrated process comprising nanofiltration and direct or indirect photolysis was shown to be remarkably efficient to remove all the target hormones from water. Other oxidation processes may have the same effect; ozone was used as the oxidant in an integrated nanofiltration oxidation process for the degradation (and removal) of natural organic matter [46]. Dudziak [47] followed a similar approach with an integration of ozonation and nanofiltration for mycoestrogen removal; the efficiency was higher than that observed for single ozonation while the effectiveness of nanofiltration in the integrated system and as a single process was similar.

The idea of process integration can be further elaborated for wastewater applications by considering enhanced fractionation of the various constituents. This was proposed by Van der Bruggen et al. [47] for textile finishing wastewater. The original scheme is shown in Figure 6.5.

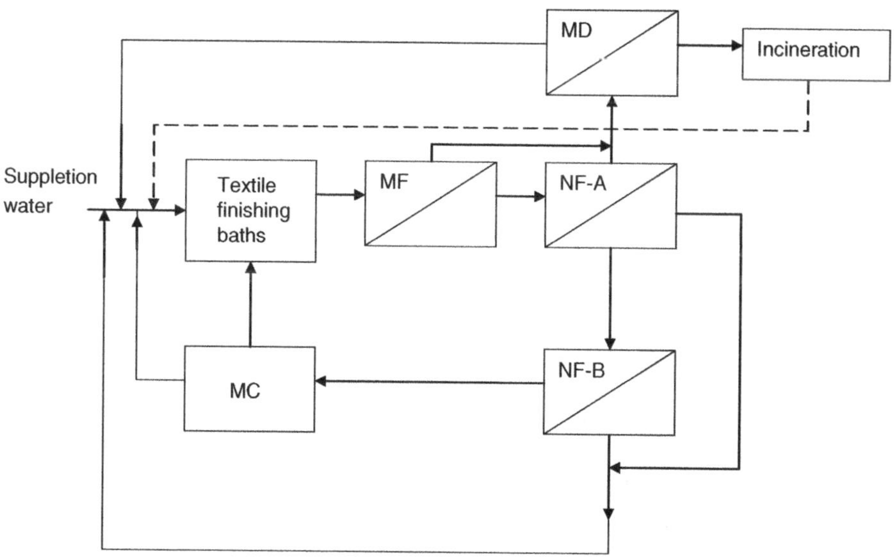

Figure 6.5 *Integrated membrane processes based on nanofiltration for wastewater fractionation in the textile finishing industry. Full line, water stream; dashed line, energy stream. MD, membrane distillation; NF, nanofiltration; MC, membrane crystallizer.* Source: *Reprinted from Reference 48 with permission from Elsevier.*

The objective in this approach is a full fractionation, partially based on a sequence of separations including a pretreatment by microfiltration and a further treatment by membrane distillation. Such approach would fit well in the concept of processes in series. However, the dual passage over nanofiltration membranes and therefore also the remaining membrane crystallization represent an integration of processes in which all processes have an intrinsically related performance in terms of separation. The nanofiltration units here are different from one another, the first one aiming at rejection of organic solutes and the second one aiming at salt rejection. However, it is obvious that they cannot be operated as independent processes; the overall performance depends on each of the individual sub-processes. More recently [49], it was noted that the concept of process integration is well known, but apart from some exceptions, not yet applied in wastewater treatment, although (or perhaps because) it is new, unusual and challenging. Vergili et al. [50] have taken up the challenge in an attempt to achieve a zero liquid discharge in the treatment of textile dye bath wastewater by using integrated membrane processes. The integrated process proposed by these authors includes various combinations of ultrafiltration, loose nanofiltration, tight nanofiltration, and reverse osmosis and was considered both technically feasible and economically viable. A remaining problem indicated by the authors is the incineration cost associated with the concentrate, which limits the return on investment.

Van der Bruggen [49] emphasizes the need for new separation or transformation processes to achieve such ambitions. Such processes should offer a separation more selective than can be achieved by conventional processes, should not generate waste by-products and should allow full integration of all process streams including recycle streams. This leads to the development of three crucial processes for integrated physico-chemical wastewater treatment, which are nanofiltration, membrane distillation, and membrane crystallization [49]. Tun and Groth [51] integrated a nanofiltration unit with a membrane crystallizer with the aim of resource recovery, and compared this configuration with membrane distillation and crystallizer for the concentration of a reverse osmosis brine, which was a secondary reject from an industrial wastewater treatment plant. The intention of these systems was to achieve zero liquid discharge. An overall water recovery of 80% in the first case, and 95% in the second case, was achieved. Several more examples integrating membrane crystallization with nanofiltration can be found. Curcio et al. [52] integrated nanofiltration with membrane crystallization for the removal of sodium sulphate from aqueous wastes originating from the production of base raw materials (Ni-H) for special rechargeable batteries. Nanofiltration was used in this case in view of the high rejection of sodium sulphate (>99%). Such combination seems a step forward in a sustainable use of nanofiltration including the consideration of the concentrate that is generated. This is confirmed by Drioli et al. [53], who studied the use of membrane contactors (membrane distillation/crystallization), operated in conjunction with nanofiltration or reverse osmosis. This increases the water recovery to 92.8% without a significant increase of the costs if waste thermal energy is available. In addition, the sale of salt crystals grown in highly concentrated brines might reduce the overall desalination cost. Interestingly, the integrated system appeared to score also well in an energy and exergy analysis.

Applications of integrated membrane processes are not limited to wastewater treatment. For clarified thin sugar juice pre-heating and concentration, Gul and Harasek [54] combined nanofiltration with a multiple-effect evaporator. Both processes have essentially the same function, but play a complementary role. In that sense, the process is more than a serial

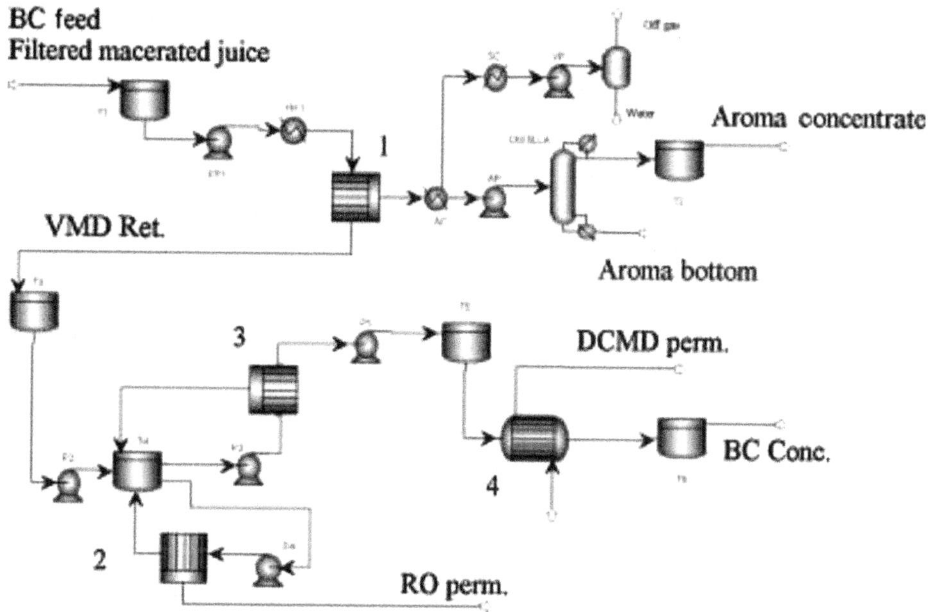

Figure 6.6 Integrated membrane process using nanofiltration for aroma recovery and juice concentration. 1, aroma recovery by vacuum membrane distillation followed by distillation; 2, preconcentration by reverse osmosis; 3, preconcentration with nanofiltration and 4, final concentration by direct contact membrane distillation. Source: Reprinted from Reference 55 with permission from Elsevier.

application of nanofiltration, although it is not a hybrid process either (see Section 6.6). In the study, the thin sugar juice was concentrated from 15 to 50 wt% with nanofiltration or reverse osmosis, and from 50 to 70 wt% with the multiple-effect evaporator. This reduces the energy consumption needed for concentration of thin juice significantly, and reduces the heat transfer area of heat exchangers for preheating of the thin juice prior to evaporation by 70%. Also in the food industry, Sotoft et al. [55] developed an integrated process for blackcurrant juice concentrate production, comprising aroma recovery with vacuum membrane distillation and water removal by reverse osmosis, nanofiltration and direct contact membrane distillation to replace traditional multiple step evaporators and aroma recovery (see Figure 6.6). An economical evaluation for a full-scale plant yielded a saving in operational costs by 43% compared to the traditional approach. More processes are needed, which may increase the investment cost, but for membrane systems the capital cost tends to be lower for less intensified traditional processes.

6.6 Hybrid Processes

Many processes are denoted as hybrid processes in the literature, but not all of them are hybrid processes in reality. A hybrid (separation) process is an integrated process configuration in which subprocesses are connected in two directions, that is, there should be a recycle stream to the first separation unit. The overall system would also have only two outgoing flows. Classical combinations of processes usually do not fit in this concept, but the current trend is towards more integration, stepping off the approach of a separation as a unit operation but rather considering it as a tool in an integral approach. Nanofiltration membranes are recent additions to the 'hybrid toolbox'. Hybrid configurations would even go beyond separation as such: the combination of reaction and separation has been proven very beneficial in this context. The development of nanofiltration in organic solvents has boosted applications for this, since most reactions would need non-aqueous solvents. Schmidt et al. [56] demonstrate this for the ligand-assisted hydroformylation of cyclooctene to cyclooctanecarbaldehyde. The technical performance depends on the design and performance of the nanofiltration membrane, and this can be easily adjusted by, for example, using an appropriate surface area for a given flux of the membrane. Difficulties would rather be related to membrane stability, lifetime and declining performance, all of which seem to have been resolved with the new generation of solvent-resistant nanofiltration membranes [57]. This can be further extended towards reactive and membrane-assisted distillation [58]. Such processes are believed to be important technologies for retrofitting existing processes and for incorporation into future processes for efficient and flexible (bio)chemical production, and organic solvent nanofiltration is one of the promising processes to achieve this, along with vapour permeation and pervaporation. The main attraction for nanofiltration would then be its role as a highly efficient separator of homogenous catalysts. This was applied by Rabiller-Baudry et al. [59] in olefin metathesis for a model ring closing metathesis reaction involving a commercially available pre-catalyst, Grubbs–Hoveyda II, and a commercial polyimide membrane (Starmem 122) in cross-flow filtration and in semi-continuous or continuous mode. A suggestion from this study is to look for a compromise between the high retention of the pre-catalyst and a high transmission of the product in order to limit the following diafiltration steps, which were both time and solvent consuming. A similar study was made by Fahrenwaldt et al. [60] for the enantioselective Henry reaction of ethyl pyruvate and nitromethane catalyzed by benzoyl cupreine. Peeva et al. [61] describe a one-pot, long-term continuous Heck coupling reaction performed in dimethylformamide (DMF) at 80°C and organic base concentrations >0.9 mol/L, where the Pd catalyst is retained by a polymeric membrane. This is shown in Figure 6.7. They achieved a stable performance for

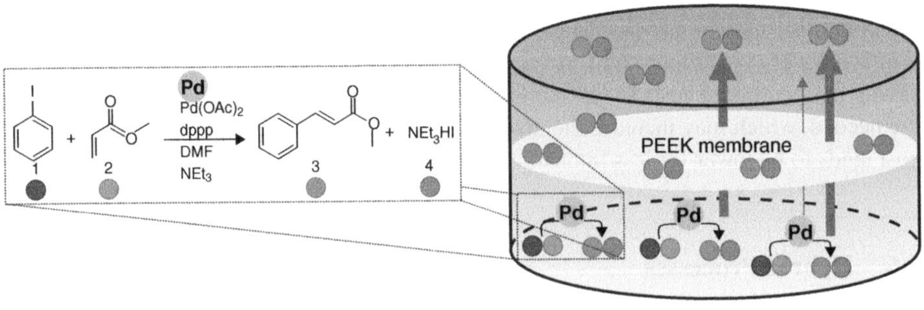

Figure 6.7 *Hybrid reaction and separation using organic solvent nanofiltration.* Source: *Reprinted with permission from reference 61 Copyright 2013 American Chemical Society.*

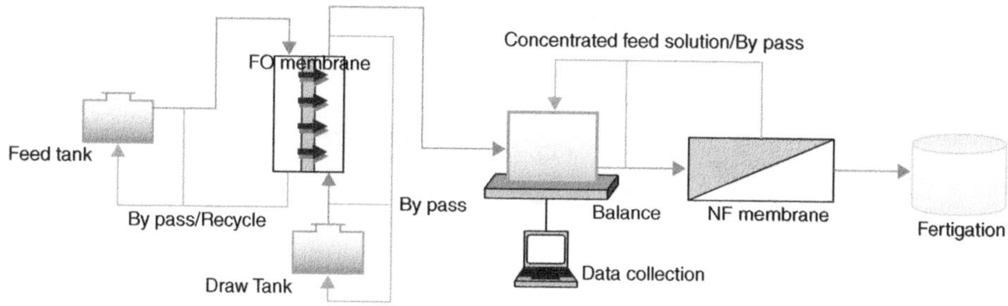

Figure 6.8 *Integration of forward osmosis and nanofiltration for fertigation.* Source: *Reprinted from Reference 66 with permission from Taylor & Francis.*

more than 1000 hours at conversion rates higher than 85%, and with an average product contamination in the continuous process integrated with membrane separation that was 20 times lower than in a simple batch process with the same catalyst loading.

Many other applications can be found in the production of fine chemicals and in bioprocessing. Nimmig and Kaspereit [62] consider a process combining single-column chromatography, racemization and solvent removal by membrane filtration to produce single enantiomers at high yields. Other examples are given by Ormerod et al. [63] and Kajetanowicz et al. [64], among others.

Another area in which hybrid systems related to nanofiltration appear is in applications of forward osmosis. The design of this process requires intrinsically a hybrid approach; nanofiltration is the logical candidate because it may allow to be operated as a method for the recycling of the draw solution, or otherwise be involved in preconditioning or postconditioning. Many different designs can be found in the literature which may or may not make sense. Two areas seem of specific interest in the hybridization and application of forward osmosis: fertilizer-drawn forward osmosis, and hybrid wastewater treatment and desalination. The former application solves the intrinsic problem of recycling the draw solution since it is a useful product. Nanofiltration serves then to reduce the nutrient concentrations in the final product water and thereby allow direct use of the product water for fertigation without further dilution [65]. This was also the approach followed by Kim et al. [66], shown in Figure 6.8. Even though this is not a hybridization of forward osmosis and nanofiltration according to the definition above, it could be operated as such, and seems a promising approach for fertigation.

The alternative approach is to integrate seawater desalination and wastewater reclamation [67]. This makes sense since it reduces both energy requirements and environmental impacts by osmotic dilution of the seawater and concentrated brine with an impaired low salinity stream, such as treated wastewater effluent. This can also be applied with a different objective, shown in Figure 6.9 [68]. Here, the draw solution is not a brine but seawater; the diluted seawater is then subjected to desalination, in which a lower pressure can be used (and membranes with lower rejections, such as low-pressure reverse osmosis membranes or even nanofiltration membranes).

Other attempts to integrate forward osmosis and nanofiltration are focused on the regeneration of the draw solution, in various ways. For example, Ling and Chung [69] used draw solutions of surface-dissociated nanoparticles with regeneration through an

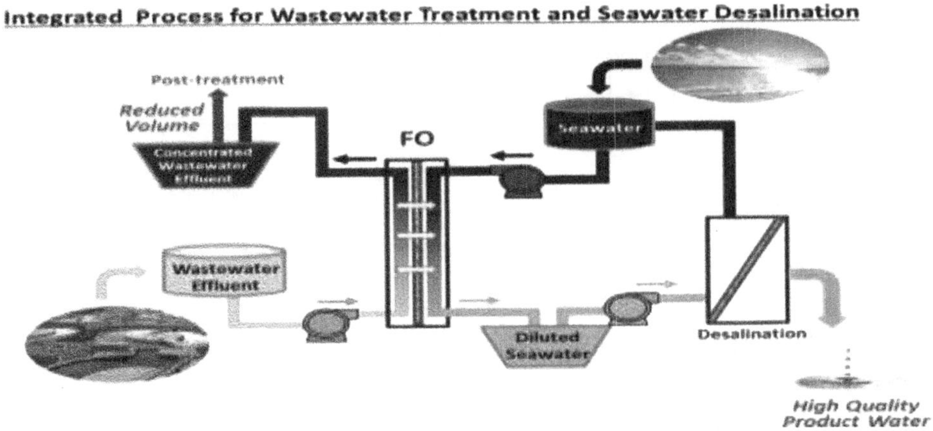

Figure 6.9 *Integration of forward osmosis and low-pressure reverse osmosis for a hybrid system of desalination and wastewater treatment.* Source: *Reprinted from Reference 68 with permission from Elsevier.*

integrated electric field–nanofiltration system for water reclamation, while Ge et al. [70] used polyelectrolytes as draw solutes, which can be separated relatively easily by nanofiltration. Using a divalent draw solute (e.g. Na_2SO_4 or $MgSO_4$) may offer a solution as well, given the high rejection that can be achieved with nanofiltration membranes for such salts [71].

A final example of a hybrid based on nanofiltration and forward osmosis is given by Cho et al. [72]. They use nanofiltration and high-flux reverse osmosis membranes to selectively separate aqueous organic acids from pretreated fermentation feed solutions; other organics and salts are rejected by the membranes by varying pH conditions in the feed. A low-energy-consuming forward osmosis process was applied for dewatering of the aqueous organic acid solutions to concentrate organic acid. The concentrated organic acid was successfully obtained by using conventional desalination and/or commercial forward osmosis membranes.

6.7 Nanofiltration Cascades

Nanofiltration membranes were developed in the 1980s for aqueous applications with the objective of lowering the rejection of (monovalent) salts, thus aiming at a separation of monovalent ions from multivalent ions and organic solutes. Nanofiltration has been applied on industrial scale since the mid-1990s. However, a (complete) separation between inorganic and organic solutes is not possible, and separation of organic solutes from one another is incomplete. Therefore, nanofiltration is only useful when the final concentrations are not critical, on condition that they are below a given standard. A quantitative fractionation is beyond the possibilities of nanofiltration as a stand-alone process. Integration and hybridization, as explained above, may help to overcome this limitation. However, solutions may lie within the process itself. The idea of a membrane cascade is to hybridize a nanofiltration

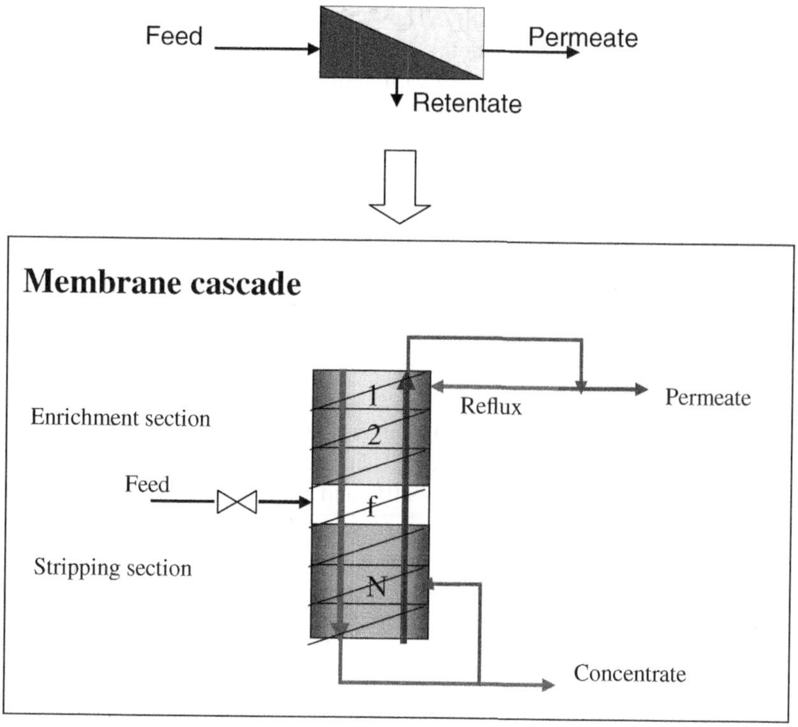

Figure 6.10 Principle of a membrane cascade in analogy with a distillation column.

unit with itself, which yields a new separation tool that allows for a new generation of separations not yet possible with the available membrane technologies. Membrane cascades are derived in analogy to separations based on vapour–liquid equilibria. Using a single step, in a flash drum, the separation that can be attained is limited by the thermodynamic equilibrium. This limitation disappears when different steps are combined and integrated by connecting all streams to the adjacent steps. By doing this, a new theoretical frame is obtained that allows separating any two compounds with a difference in vapour pressure. The technological translation of this principle is a distillation column, the most important separation tool in the classical process technology. In analogy with separations based on vapour–liquid equilibria, membrane separations (i.e. nanofiltration) can be combined and integrated. This is schematically shown in Figure 6.10.

Membrane cascades for liquid separations are still recent, although they have been used much longer in gas separation. Aqueous applications of nanofiltration cascades were considered first by Caus et al. [73], who assessed the potential of integrated countercurrent nanofiltration cascades for advanced separation of individual organic components in aqueous solution by means of single-stage filtration experiments using xylose and maltose and cascade simulations. When applied to drinking water production, this allows to remove micropollutants (pesticides) from water completely while removing only a small fraction of salts [74]. Vanneste et al. [75] studied three different sugar separations: raffinose–sucrose, fructose–glucose and xylose–glucose using a nanofiltration membrane cascade and simulated moving bed chromatography. They concluded that membrane cascades seem most promising for large-scale continuous processes for producing pure but depending on the selectivity not ultrapure products. This limitation is an economic one, because a more challenging separation would require more stages (and be therefore more expensive) but technically still viable. Patil et al. [76] came to a similar conclusion: three- and five-stage ideal cascades show a significant improvement in separation with diafiltration compared to single-stage systems.

Nevertheless, it is attractive to use the principle of a cascade in applications with high added value, because the flexibility of using a larger number of stages is there higher. Abejon et al. [77] used cascades for ultrapurification of high-purity electronic grade hydrogen peroxide for industrial production, using high pressure membranes. More applications can be found for cascades making using of organic solvent nanofiltration. Kim et al. [78] developed a simplified-control cascade process to minimize the product loss and to maximize the purity, considering fractionation of an organic solution of polyethylene glycol 400 and polyethylene glycol 2000 in acetonitrile. Using this process, it was possible to increase the final yield of PEG-2000 from 59% to 94%, and the solvent-consumption/productivity ratio was also reduced. In further work, a two-stage cascade configuration coupled to an adsorptive solvent recovery unit was validated on purification of active pharmaceutical ingredients from genotoxic impurities using organic solvent nanofiltration [79]. The process yield was increased from 58% to 95% while maintaining less than 5 ppm of genotoxic impurities in the final solution.

Vanneste et al. [80] illustrate how powerful this method is, by studying the separation of an intermediate I (MW 221 g/mol) from an impurity, ethylene bromide (MW 188 g/mol). The potential of nanofiltration cascades is compared to diafiltration by Siew et al. [81]. Solute fractionation using a cascade where the more permeable solute is stripped can reduce solvent usage significantly and increase separation yield with minimal stages added. Such cascade is more productive and effective in solute fractionation than constant volume diafiltration, where only the recovery and not the purity is optimized, while both parameters can be optimized simultaneously in a nanofiltration membrane cascade.

Practical demonstrations of the applicability of nanofiltration cascades remain scarce to date, despite the vast potential. Nevertheless, the viability has been proven, for example, in the study of Siew et al. [82], who used a three-stage cascade for the concentration of a dilute API product solution and concurrent solvent recovery downstream of a chromatographic process. They report that the cascade was able to achieve an effective rejection of 80% compared to a single pass rejection of 55%, and that control of the cascade was simple and its operation was stable. Thus, it can be expected that this is an area of significant growth and a high economical value.

6.8 Conclusions

Nanofiltration membranes can be used in different ways in integrated processes. The level of integration, however, may differ significantly, and the terminology is not always correctly used in the literature. This chapter has made a systematic classification according to the method of integration. The lowest level of integration is achieved by considering pretreatment processes to nanofiltration, which serves for protection of the nanofiltration unit and for maintaining an optimal performance. The use of nanofiltration as a pretreatment for further purification represents a higher level of integration since the interaction is typically more intensive (with seawater desalination as the reference example). This approach can be further extended in processes in series in which every part of the overall system has a different function, which may however go beyond the role of a pretreatment. Integrated processes, in which separation objectives can be shared among different processes in the system, are often developed for multi-objective applications.

Hybrid processes are sometimes mentioned when a simple sequence is meant. Hybrid processes should be more than an integration of various processes; they should be operated in a synergetic way so that there is an intrinsic link in two directions between the different processes. This has a large potential particularly in organic solvent nanofiltration and for aqueous applications, in the hybridization with forward osmosis.

Lastly, a hybridization of nanofiltration with itself leads to a new era of separations, which opens vast possibilities in applications with high added value, such as in the production of fine chemicals. The technical viability of nanofiltration cascades has been demonstrated for aqueous as well as for non-aqueous applications.

It is safe to say that integration, hybridization and cascading are among the most prominent future trends for nanofiltration. Engineering of membrane applications has been quite overlooked until recently, and scientific research and applications seem to catch up with this gap step by step. This is a crucial evolution, which will bring membrane science and technology to a higher level of development.

List of Abbreviations

DMF	Dimethylformamide
MSF	Multi-stage flash
MW	Molecular weight
RO	Reverse osmosis

References

1. A.I. Schäfer, A.G. Fane, T.D. Waite (Eds.), *Nanofiltration: Principles and Applications*. Elsevier, Oxford, 2005.
2. Van der Bruggen, B.; Mänttäri, M.; Nyström, M. Drawbacks of applying nanofiltration and how to avoid them: A review. *Separ. Purif. Technol.* 2008, **63**, 251–263
3. Abu Seman, N.M.; Hilal, N.; Khayet, M. UV-photografting modification of NF membrane surface for NOM fouling reduction. *Desal. Water Treat.* 2013, **51** (25–27), 4855–4861.
4. Balta, S.; Sotto, A.; Kim, J.; Luis, P.; Benea, L.; Van der Bruggen, B. A new outlook on membrane enhancement with nanoparticles: The alternative of ZnO. *J. Membr. Sci.* 2012, **389**, 155–161.
5. Sotto, A.; Boromand, A.; Balta, S.; Kim, J.; Van der Bruggen, B. Doping of polyethersulfone nanofiltration membranes: Antifouling effect observed at ultralow concentrations of TiO_2 nanoparticles. *J. Mater. Chem.* 2011, **21** (28), 10311–10320.

6. Ahmad, A.L.; Abdulkarim, A.A.; Ooi, B.S.; Ismail, S. Recent development in additives modifications of polyethersulfone membrane for flux enhancement. *Chem. Eng. J.* 2013, **223**, 246–267.
7. Ishigami, T.; Amano, K.; Fujii, A.; Ohmukai, Y.; Kamio, E.; Maruyama, T.; Matsuyama, H. Fouling reduction of reverse osmosis membrane by surface modification via layer-by-layer assembly. *Sep. Purif. Technol.* 2012, **99**, 1–7.
8. Kim, J.H.; Sotto, A.; Chang, J.; Nam, D.; Boromand, A.; Van der Bruggen, B. Embedding and surface coating by layer-by-layer deposition of TiO_2 nanoparticles on nanoporous polymeric films. *Micropor. Mesopor. Mater.* 2013, **173**, 121–128.
9. Madaeni, S.S.; Falsafi, M.; Ghaemi, N. A novel method for preparation of low-fouling membranes: Surface coating by extracted wax from leafy cabbage. *Desalination* 2011, **283**, 148–155.
10. Alzahrani, S.; Mohammad, A.W.; Hilal, N.; Abdullah, P.; Jaafar, O. Identification of foulants, fouling mechanisms and cleaning efficiency for NF and RO treatment of produced water. *Sep. Purif. Technol.* 2013, **118**, 324–341.
11. Saravia, F.; A.; Richard, A.Z.; Frimmel, F.H. Identification of nanofiltration fouling layer constituents. *Desal. Water Treat.* 2013, **51** (37–39), 6921–6928.
12. Simon, A.; Price, W.E.; Nghiem, L.D. Changes in surface properties and separation efficiency of a nanofiltration membrane after repeated fouling and chemical cleaning cycles. *Sep. Purif. Technol.* 2013, **113**, 42–50.
13. Van der Bruggen, B.; Schaep, J.; Wilms, D.; Vandecasteele, C. Influence of molecular size, polarity and charge on the retention of organic molecules by nanofiltration. *J. Membr. Sci.* 1999, **156**, 29–41.
14. Keurentjes, J.T.F.; Linders, L.J.M.; Beverloo, W.A.; Vantriet, K. Membrane cascades for the separation of binary mixtures. *Chem. Eng. Sci.* 1992, **47** (7), 1561–1568.
15. Gunderson, S.S.; Brower, W.S.; O'Dell, J.L.; Lightfoot, E.N. Design of membrane cascades. *Sep. Sci. Technol.* 2007, **42** (10), 2121–2142.
16. Van der Bruggen, B.; Verliefde, A.; Braeken, L.; Cornelissen, E.R.; Moons, K.; Verberk, J.C.J.Q.; Van Dijk, J.C.; Amy, G. Assessment of a semi-quantitative method for estimation of the rejection of organic compounds in aqueous solution in nanofiltration. *J. Chem. Technol. Biotechnol.* 2006, **81** (7), 1166–1176
17. Van der Bruggen, B.; Lejon, L.; Vandecasteele, C. Reuse, treatment and discharge of the concentrate of pressure driven membrane processes. *Environ. Sci. Technol.* 2003, **37** (17), 3733–3738.
18. Seader, J.D.; Henley, E.J.; Roper, D.K. *Separation Process Principles*. John Wiley & Sons, 2011.
19. Pal, P.; Kumar, R. Treatment of Coke wastewater: A critical review for developing sustainable management strategies. *Sep. Purif. Rev.* 2014, **43** (2), 89–123.
20. Ozgun, H.; Dereli, R.K.; Ersahin, M.E.; Kinaci, C.; Spanjers, H.; Van Lier, J. A review of anaerobic membrane bioreactors for municipal wastewater treatment: Integration options, limitations and expectations. *Sep. Purif. Technol.* 2013, **118**, 89–104.
21. Mihal, M.; Kristofikova, L.; Markos, J. Production of 2-phenylethanol in hybrid system using airlift reactor and immersed hollow fiber membrane module. *Chem. Eng. Proc.* 2013, **72**, 144–152.
22. Manekar, P.; Nandy, T.; Sargaonkar, A.; Rathi, B.; Karthik, M. FAHP ranking and selection of pretreatment module for membrane separation processes in textile cluster. *Biores. Technol.* 2011, **102** (2), 558–566.

23. Tahri, N.; Masmoudi, G.; Ellouze, E.; Jrad, A.; Drogui, P.; Ben Amar, R. Coupling microfiltration and nanofiltration processes for the treatment at source of dyeing-containing effluent. *J. Clean. Prod.* 2012, **33**, 226–235.
24. Van der Bruggen, B.; Kim, J.H.; DiGiano, F.A.; Geens, J.; Vandecasteele, C. Influence of MF pretreatment on NF performance for aqueous solutions containing particles and an organic foulant. *Separ. Purif. Technol.* 2004, **36** (3), 203–213.
25. Sari, M.A.; Chellam, S. Surface water nanofiltration incorporating (electro) coagulation-microfiltration pretreatment: Fouling control and membrane characterization. *J. Membr. Sci.* 2013, **437**, 249–256.
26. Stoller, M.; Bravi, M.; Chianese, A. Threshold flux measurements of a nanofiltration membrane module by critical flux data conversion. *Desalination* 2013, **315**, 142–148.
27. Chon, K.; Cho, J.; Shon, H.K. Fouling characteristics of a membrane bioreactor and nanofiltration hybrid system for municipal wastewater reclamation. *Biores. Technol.* 2013, **130**, 239–247.
28. Al-Hajouri, A.A.; Al-Amoudi, A.S.; Farooque, A.M. Long term experience in the operation of nanofiltration pretreatment unit for seawater desalination at SWCC SWRO plant. *Desal. Water Treat.* 2013, **51** (7–9), 1861–1873.
29. Al-Sofi, M.A.K. Seawater desalination – SWCC experience and vision. *Desalination* 2001, **135** (1–3), 121–139.
30. Mabrouk, A.N.A.; Fath, H.E.S. Techno-economic analysis of hybrid high performance MSF desalination plant with NF membrane. *Desal. Water Treat.* 2013, **51** (4–6), 844–856.
31. Llenas, L.; Ribera, G.; Martinez-Llado, X.; Rovira, M.; de Pablo, J. Selection of nanofiltration membranes as pretreatment for scaling prevention in SWRO using real seawater. *Desal. Water Treat.* 2013, **51** (4–6), 930–935.
32. Park, J.Y.; Lim, S.; Park, K. A new approach for determination of fouling potential by colloidal nanoparticles during reverse osmosis (RO) membrane filtration of seawater. *J. Nanopart. Res.* 2013, **15** (4), UNSP 1548.
33. Su, B.; Dou, M.W.; Gao, X.L.; Shang, Y.W.; Gao, C.J. Study on seawater nanofiltration softening technology for offshore oilfield water and polymer flooding. *Desalination* 2012, **297**, 30–37.
34. Esmi, C.F.; Schrive, L.; Barre, Y.; Palmeri, J.; Deratani, A. Using nanofiltration in a "zero rejection" process: the removal of Ni_2^+ and Co_2^+ from salty wastewater. *Desal. Water Treat.* 2013, **51** (1–3), 476–484.
35. Benito-Alcazar, C.; Vincent-Vela, M.C.; Gozalvez-Zafrilla, J. M.; Lora-Garcia, J. Study of different pretreatments for reverse osmosis reclamation of a petrochemical secondary effluent. *J. Haz. Mater.* 2010, **178** (1–3), 883–889.
36. Purkait, M.K.; Kumar, V.D.; Maity, D. Treatment of leather plant effluent using NF followed by RO and permeate flux prediction using artificial neural network. *Chem. Eng. J.* 2009, **151** (1–3), 275–285.
37. Peng, H.; Tremblay, A.Y. The selective removal of oil from wastewaters while minimizing concentrate production using a membrane cascade. *Desalination* 2008, **229** (1–3), 318–330.
38. Subramani, A.; Schlicher, R.; Long, J.; Yu, J.; Lehman, S.; Jacangelo, J.G. Recovery optimization of membrane processes for treatment of produced water with high silica content. *Desal. Water Treat.* 2011, **36** (1–3), 297–309.

39. Van Paassen, J.A.M.; Kruithof, J.C.; Bakker, S.M.; Kegel, F.S. Recovery optimization of membrane processes for treatment of produced water with high silica content. *Desalination* 1998, **118** (1–3), 239–248.
40. Turek, M. Seawater desalination and salt production in a hybrid membrane-thermal process. *Desalination* 2003, **153** (1–3), 173–177.
41. Luis, P.; Van Aubel, D.; Van der Bruggen, B. Technical viability and exergy analysis of membrane crystallization: closing the loop of CO_2 sequestration. *Int. J. Greenh. Gas. Con.* 2013, **12**, 450–459.
42. Qi, B.K.; Luo, J.Q.; Chen, G.Q.; Chen, X.R.; Wan, Y.H. Application of ultrafiltration and nanofiltration for recycling cellulase and concentrating glucose from enzymatic hydrolyzate of steam exploded wheat straw. *Biores. Technol.* 2012, **104**, 466–472.
43. Miralles-Cuevas, S.; Arques, A.; Maldonado, M.I.; Sanchez-Perez, J.A.; Malato Rodriguez, S. Combined nanofiltration and photo-fenton treatment of water containing micropollutants. *Chem. Eng. J.* 2013, **224**, 89–95.
44. Berberidou, C.; Avlonitis, S.; Poulios, I. Dyestuff effluent treatment by integrated sequential photocatalytic oxidation and membrane filtration. *Desalination* 2009, **249** (3), 1099–1106.
45. Pereira, V.J.; Galinha, J.; Crespo, M.T.B.; Matos, C.T.; Crespo, J.G. Integration of nanofiltration, UV photolysis, and advanced oxidation processes for the removal of hormones from surface water sources. *Sep. Purif. Technol.* 2012, **95**, 89–96.
46. Van Geluwe, S.; Degrève, J.; Vinckier, C.; Braeken, L.; Creemers, C.; Van der Bruggen, B. Kinetic study and scale-up of the oxidation of nanofiltration retentates by O_3. *Ind. Eng. Chem. Res.* 2012, **51** (20), 7056–7066.
47. Dudziak, M. Combined ozonation – Nanofiltration for mycoestrogens removal from water. *Arch. Environ. Prot.* 2012, **38** (2), 103–109.
48. Van der Bruggen, B.; Curcio, E.; Drioli, E. Process intensification in the textile industry: the role of membrane technology. *J. Environ. Manage.* 2004, **73** (3), 267–274.
49. Van der Bruggen, B. Integrated membrane separations for recycling of valuable wastewater streams: nanofiltration, membrane distillation and membrane crystallizers revisited. *Ind. Eng. Chem. Res.* 2013, **52** (31), 10335–10341.
50. Vergili, I.; Kaya, Y.; Sen, U.; Gonder, Z.B.; Aydiner, C. Techno-economic analysis of textile dye bath wastewater treatment by integrated membrane processes under the zero liquid discharge approach. *Resour. Conservat. Recycl.* 2012, **58**, 25.
51. Tun, C.M.; Groth, A.M. Sustainable integrated membrane contactor process for water reclamation, sodium sulfate salt and energy recovery from industrial effluent. *Desalination* 2011, **283**, 187–192.
52. Curcio, E.; Ji, X.; Quazi, A.M.; Barghi, S.; Di Profio, G.; Fontananova, E.; Macleod, T.; Drioli, E. Hybrid nanofiltration-membrane crystallization system for the treatment of sulfate wastes. *J. Membr. Sci.* 2010, **360** (1–2), 493–498.
53. Drioli, E.; Curcio, E.; Di Profio, G.; Macedonio, F.; Criscuoli, A. Integrating membrane contactors technology and pressure-driven membrane operations for seawater desalination – Energy, exergy and costs analysis. *Chem. Eng. Res. Des.* 2006, **84** (A3), 209–220.
54. Gul, S.; Harasek, M. Energy saving in sugar manufacturing through the integration of environmental friendly new membrane processes for thin juice pre-concentration. *Appl. Therm. Ener.* 2012, **43**, 128–133.

55. Sotoft, L.F.; Christensen, K.V.; Andresen, R.; Norddahl, B. Full scale plant with membrane based concentration of blackcurrant juice on the basis of laboratory and pilot scale tests. *Chem. Eng. Proc.* 2012, **54**, 12–21.
56. Schmidt, P.; Becker, M.; Priske, M.; Hamers, B.; Kreis, P.; Gorak, A. Process design of integrated reaction and membrane separation by organic solvent nanofiltration using evolutionary algorithms. *Comput. Aided Chem. Eng.* 2012, **30**, 727–731.
57. Buekenhoudt, A.; Beckers, H.; Ormerod, D.; Bulut, M.; Vandezande, P.; Vleeschouwers, R. Solvent based membrane nanofiltration for process intensification. *Chem. Ing. Tech.* 2013, **85** (8), 1243–1247.
58. Lutze, P.; Gorak, A. Reactive and membrane-assisted distillation: Recent developments and perspective. *Chem. Eng. Res. Des.* 2013, **91** (10), 1978–1997.
59. Rabiller-Baudry, M.; Nasser, G.; Renouard, T.; Delaunay, D.; Camus, M. Comparison of two nanofiltration membrane reactors for a model reaction of olefin metathesis achieved in toluene. *Sep. Purif. Technol.* 2013, **116**, 46–60.
60. Fahrenwaldt, T.; Grosseheilmann, J.; Erben, F.; Kragl, U. Organic solvent nanofiltration as a tool for separation of quinine-based organocatalysts. *Org. Proc. Res. Dev.* 2013, **17** (9), 1131–1136.
61. Peeva, L.; Arbour, J.; Livingston, A. On the potential of organic solvent nanofiltration in continuous heck coupling reactions. *Org. Process Res. Dev.*, 2013, **17** (7), 967–975.
62. Nimmig, S.; Kaspereit, M. Continuous production of single enantiomers at high yields by coupling single column chromatography, racemization, and nanofiltration. *Chem. Eng. Proc.* 2013, **67**, 89–98.
63. Ormerod, D.; Bongers, B.; Porto-Carrero, W.; Giegas, S.; Vijt, G.; Lefevre, N.; Lauwers, D.; Brusten, W.; Buekenhoudt, A. Separation of metathesis catalysts and reaction products in flow reactors using organic solvent nanofiltration. *RSC Adv.* 2013, **44** (3), 21501–21510.
64. Kajetanowicz, A.; Czaban, J.; Krishnan, G.R.; Malinska, M.; Wozniak, K.; Siddique, H.; Peeva, L.G.; Livingston, A.G.; Grela, K. Batchwise and continuous nanofiltration of POSS-tagged Grubbs-Hoveyda-type olefin metathesis catalysts. *Chem. Sus. Chem.* 2013, **6** (1), 182–192.
65. Phuntsho, S.; Hong, S.; Elimelech, M.; Shon, H.K. Forward osmosis desalination of brackish groundwater: Meeting water quality requirements for fertigation by integrating nanofiltration. *J. Membr. Sci.* 2013, **436**, 1–15.
66. Kim, J.E.; Phuntsho, S.; Shon, H.K. Pilot-scale nanofiltration system as post-treatment for fertilizer-drawn forward osmosis desalination for direct fertigation. *Desal. Water Treat.* 2013, **51** (31–33), 6265–6273.
67. Boo, C.; Elimelech, M.; Hong, S. Fouling control in a forward osmosis process integrating seawater desalination and wastewater reclamation. *J. Membr. Sci.* 2013, **444**, 148–156.
68. Linares, R.V.; Li, Z.Y.; Abu-Ghdaib, M.; Wei, C.H.; Amy, G.; Vrouwenvelder, J.S. Water harvesting from municipal wastewater via osmotic gradient: An evaluation of process performance. *J. Membr. Sci.* 2013, 447, 50–56.
69. Ling, M.M.; Chung, T.S. Surface-dissociated nanoparticle draw solutions in forward osmosis and the regeneration in an integrated electric field and nanofiltration system. *Ind. Eng. Chem. Res.* 2012, **51** (47), 15463–15471.

70. Ge, Q.C.; Su, J.C.; Amy, G.L.; Chung, T.S. Exploration of polyelectrolytes as draw solutes in forward osmosis processes. *Water Res.* 2012, **46** (4), 1318–1326.
71. Zhao, S.F.; Zou, L.; Mulcahy, D. Brackish water desalination by a hybrid forward osmosis-nanofiltration system using divalent draw solute. *Desal. Water Treat.* 2012, **284**, 175–181.
72. Cho, Y.H.; Lee, H.D.; Park, H.B. Integrated membrane processes for separation and purification of organic add from a biomass fermentation process. *Ind. Eng. Chem. Res.* 2012, **51** (30), 10207–10219.
73. Caus, A.; Braeken, L.; Boussu, K.; Van der Bruggen, B. The use of integrated counter-current nanofiltration cascades for advanced separations. *J. Chem. Technol. Biotechnol.* 2009, **84** (3), 391–398.
74. Caus, A.; Vanderhaegen, S.; Braeken, L.; Van der Bruggen, B. Integrated nanofiltration cascades with low salt rejection for complete removal of pesticides in drinking water production. *Desalination* 2009, **241** (1–3), 111–117.
75. Vanneste, J.; De Ron, S.; Vandecruys, S.; Soare, S.A.; Darvishmanesh, S.; Van der Bruggen, B. Techno-economic evaluation of membrane cascades relative to simulated moving bed chromatography for the purification of mono- and oligosaccharides. *Separ. Purif. Technol.* 2011, **80** (3), 600–609.
76. Patil, N.V.; Janssen, A.E.M.; Boom, R.M. The potential impact of membrane cascading on downstream processing of oligosaccharides. *Chem. Eng. Sci.* 2014, **106**, 86–98.
77. Abejon, R.; Garea, A.; Irabien, A. Integrated countercurrent reverse osmosis cascades for hydrogen peroxide ultrapurification. *Comp. Chem. Eng.* 2012, **41**, 67–76.
78. Kim, J.F.; da Silva, A.M.F.; Valtcheva, I.B.; Livingston, A.G. When the membrane is not enough: A simplified membrane cascade using organic solvent nanofiltration (OSN). *Sep. Purif. Technol.* 2013, **116**, 277–286.
79. Kim, J.F.; Szekely, G.; Valtcheva, I.B.; Livingston, A.G. Increasing the sustainability of membrane processes through cascade approach and solvent recovery-pharmaceutical purification case study. *Green Chem.* 2014, **16** (1), 133–145.
80. Vanneste, J.; Ormerod, D.; Theys, G.; Van Gool, D.; Van Camp, B.; Darvishmanesh, S.; Van derBruggen, B. Towards high resolution membrane-based pharmaceutical separations. *J. Chem. Technol. Biotechnol.* 2013, **88** (1), 98–108.
81. Siew, W.E.; Livingston, A.G.; Ates, C.; Merschaert, A. Continuous solute fractionation with membrane cascades – A high productivity alternative to diafiltration. *Sep. Purif. Technol.* 2013, **102**, 1–14.
82. Siew, W.E.; Livingston, A.G.; Ates, C.; Merschaert, A. Molecular separation with an organic solvent nanofiltration cascade – Augmenting membrane selectivity with process engineering. *Chem. Eng. Sci.* 2013, **90**, 299–310.

7

Seawater, Brackish Waters, and Natural Waters Treatment with Hybrid Membrane Processes

Maxime Pontié[1] *and Catherine Charcosset*[2]
[1]*L'UNAM, Laboratory GEPEA UMR CNRS 6144, Nantes University, Nantes, France*
[2]*Laboratoire d'Automatique et de Génie des Procédés, CNRS, Université Lyon 1, Villeurbanne Cedex, France*

7.1 Introduction

Water is necessary to all human activities on earth. Resources decrease in some places of the globe and the significant increase in human population intensifies the need for water and the availability issue. Desalination is an increasingly important solution to the rising water scarcity afflicting many of the world's regions. In addition, in several places around the world, natural waters are reported to contain several contaminants with severe potential health problems, such as organic matter, arsenic, fluorides, and iron. Therefore, ground waters are increasingly treated to obtain drinking water.

Hybrid or integrated membrane processes may be classified into three categories: (1) the removal mechanism such as adsorption or reaction is included in the membrane itself. It may be the case of membrane reactors, ion-exchange membranes, adsorptive membranes, and others; (2) other hybrid or integrated membrane processes combine several membrane separation steps, one step being dependent of the former one, in a multistage configuration. This may be the case of pretreatment using ultrafiltration (UF) or microfiltration (MF) before reverse osmosis (RO); (3) finally, the hybrid or integrated membrane processes may associate a different separation technique to membrane filtration, for example, activated

Integrated Membrane Systems and Processes, First Edition. Edited by Angelo Basile and Catherine Charcosset.

carbon adsorption combined to UF and MF. In this case, the membrane retains the carbon particles in a continuous process for water treatment.

In seawater, brackish water desalination, natural water treatment, and several examples of hybrid membrane processes have been reported and some of them are applied at an industrial scale. In desalination, the multistage membrane process is mainly used to remove fouling species by UF, MF, or nanofiltration (NF) before RO. For natural water treatment, several hybrid membrane processes have been proposed, which usually combine UF or MF to another separation technique like activated carbon adsorption, advanced oxidation, or precipitation. Multistage NF has also been applied to groundwater treatment.

7.2 Desalination Market

7.2.1 Growth of Desalination Capacity Worldwide

Desalination technologies and their applications have grown substantially over the last 50 years [1]. Today, some countries depend on desalination technologies for the purpose of meeting their freshwater requirements. In particular, water desalination is a vital and dependable freshwater resource in the Middle East, in countries such as Saudi Arabia, United Arab Emirates, and Kuwait. Overall, it was estimated in 2007 that over 75 million people worldwide obtain freshwater by desalinating seawater or brackish water. The IDA Desalting Inventory 2007 Report shows that at the end of 2006, installed and contracted brackish and seawater desalination plants worldwide totalized 12,791 desalination plants. The world desalination plant capacity was 43 million m^3 per day and the annual average growth rate for the last 10 years was 12%. Today, desalination plants operate in approximately 125 countries, with seawater desalination plants contributing 60% of the total worldwide desalination capacity and 22% with brackish water. More than 17,000 units are installed in the world for a production of 100 million m^3 per day (see Figure 7.1).

A projection from the data in Figure 7.1 predicts a production of 124 million m^3 per day in 2016. Some arid regions depend heavily on desalination for their water supply. The Middle East countries, mainly the Gulf Cooperation Council states, are the biggest users of desalination technology with 50% of the world's capacity, followed by America, then Europe, as shown in Figure 7.2.

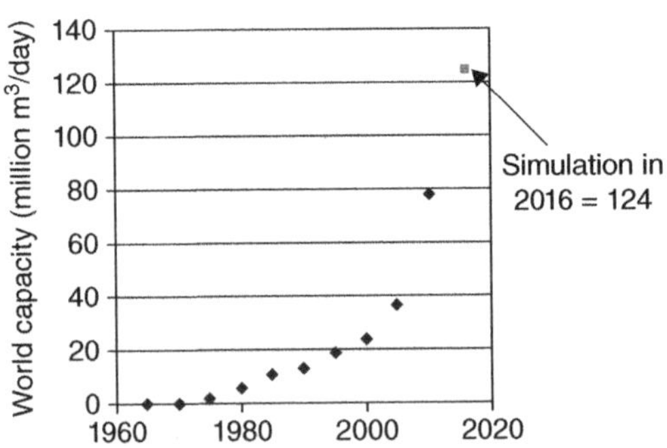

Figure 7.1 Desalination world capacity per year installed in the world [2].

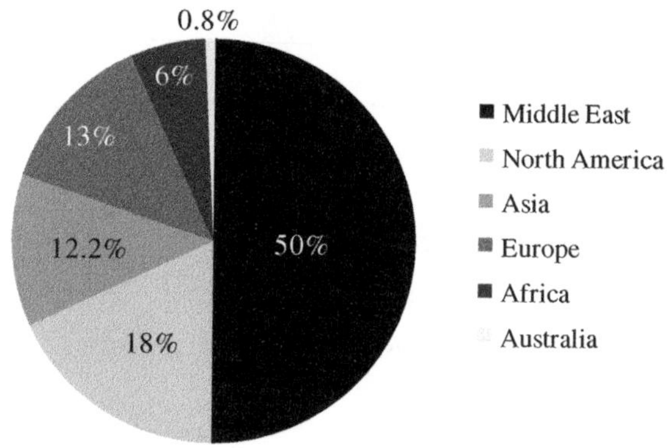

Figure 7.2 *Chart showing fraction of the worldwide capacity of desalination plants by region [2].*

7.2.2 Desalination Technologies

Many different desalination technologies exist to separate dissolved salts from water. The desalinated water is recovered for consumption, whereas the salts are concentrated in a stream of water called the brine reject, disposed either to the sea or to a saline aquifer, or in evaporation ponds. A typical flow diagram of the process with inputs and outflows is shown in Figure 7.3.

The water desalination processes require significant quantities of energy to achieve salt separation and get freshwater. The amount and type of the energy required differs according to the technique used.

The choice of the technology used for water desalination depends on a number of site-specific factors, including source water quality, the intended use of the water produced, plant size, capital costs, energy costs, and the potential for energy reuse. The commercially tested desalination processes are given in Figure 7.4.

Water desalination can be accomplished by different processes that can be classified into two categories: thermal and membrane processes [3, 4]. The thermal processes can be subdivided into the following processes: (i) multistage flash evaporation, (ii) multiple effect distillation, and (iii) vapor compression. The principle of thermal water desalination is the elimination of saline content by evaporating water in distillation systems. The membrane

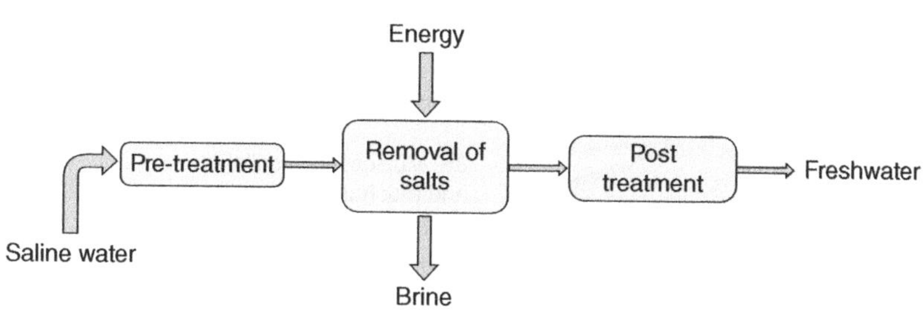

Figure 7.3 *Typical desalination process.*

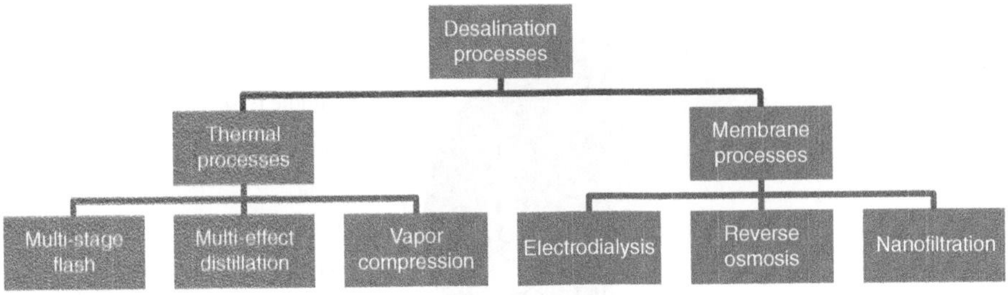

Figure 7.4 *Industrial desalination processes.*

processes are subdivided into: (i) RO, (ii) electrodialysis (ED), and (iii) NF. RO is the principally used technique for salted water desalination (see Figure 7.5) [2, 5].

7.3 Seawater and Brackish Waters Composition

7.3.1 Seawater Composition

Table 7.1 reports the mineral composition of Pacific Ocean sea water. Other seawater compositions are reported elsewhere [6, 7].

Two ranges of total dissolved solid (TDS) are usually considered: lower than 15,000 ppm (15,000 mg/L) and higher than 25,000 ppm (25,000 mg/L). Water under lower salinity is denoted brackish water as others are assimilated to seawater. According to the World Health Organization (WHO) [9], TDS should be less than 1000 mg/L in drinking water based on taste considerations. By comparison, seawater has an average TDS of about 35,000 mg/L.

7.3.2 Brackish Water versus Seawater

The desalination industry makes a distinction between seawater and brackish water. A classification of source waters according to the quantity of dissolved solids is given in Table 7.2.

Seawater typically has a salt concentration in the order of 35,000 mg/L. More than 70 elements are dissolved in seawater, but only two elements (chloride and sodium) make up greater than 85% by weight of all the TDS. Seawater is a saline solution of nearly constant

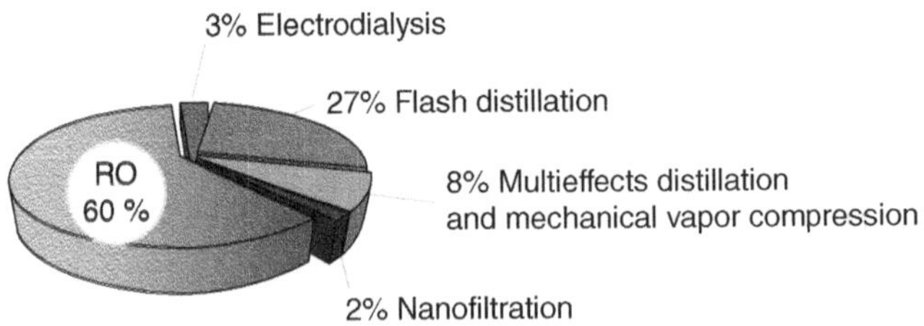

Figure 7.5 *Percentage of membrane operations (RO, NF, ED) versus thermal water desalination processes installed units in the world [2].*

Table 7.1 Pacific ocean water quality [8].

Constituent	Value
Calcium	405
Magnesium	1.334
Sodium	11.041
Potassium	383
Ammonia (as N)	0.05
Barium	<0.025
Strontium	7.7
Bicarbonate (as $CaCo_3$)	115
Sulfate	2.537
Chloride	19.080
Bromide	64
Boron	3.8
Nitrate (as N)	<25
Fluoride	0.9
Silica	<10
TDSs	34.490
pH	8.1
TOC	1.2
Temperature (°F)	60–75

All values in mg/L, except pH and temperature.

composition. Brackish water contains less TDS than seawater but more than freshwater. The TDS concentrations in brackish water can range between 1000 and 15,000 mg/L; most brackish water environments are dynamic and TDS levels in these environments fluctuate spatially and temporally. The salinity of brackish surface water near the coast can vary depending on the tide, the amount of freshwater entering the system as rain or river flows, and the rate of evaporation. Brackish water also occurs in coastal aquifers. Some deep groundwater aquifers contain brackish water that occurs under natural conditions. In coastal aquifers, excessive groundwater withdrawals may cause the seawater to move into freshwater aquifers (a phenomenon known as saltwater intrusion) and create brackish water in the aquifer.

Table 7.2 Classification of source water according to quantity of dissolved solids.

Water source	Total dissolved solids (mg/L)
Potable water	<1000
Mildly brackish waters	1000–5000
Moderately brackish waters	5000–15,000
Heavily brackish waters	15000–35,000
Average sea water	35,000

Note: Some seas and evaporative lakes can show wide variability in TDS; for example, the Arabian Gulf has an average TDS of 48 g/L, and the Mediterranean Sea 38 g/L.

7.3.3 Product Water Specification

An important consideration for any desalination system is the quality required of the product water. Some systems provide very pure water for particular industrial processes, but the majority is designed to provide drinking water. It is clearly important to establish an appropriate upper limit for the concentration of the water delivered to the consumer. In the desalination literature, it is often stated that the WHO recommends a limit of 500 mg/L TDS, but the exact origin of this figure is unclear. There is no evidence of a 500 mg/L TDS limit within the WHO guidelines. Regarding TDS, the WHO guidelines states that no health-based guideline value is proposed. However, the presence of high levels of TDS in drinking water may be objectionable to the consumer. Water with extremely low concentrations of TDS may also be unacceptable because of its flat, insipid taste [9].

Moreover, the first edition of the WHO guidelines in 1984 did suggest a limit of 1000 mg/L for TDS, based on the taste. This limit was removed in the second edition. There is no mention of any limit at 500 mg/L. The WHO guidelines indicated that the palatability of water with a TDS level of less than 600 mg/L is generally considered to be good; drinking water becomes significantly and increasingly unpalatable at TDS levels greater than about 1000 mg/L [9].

In summary, the acceptable concentration of salt in water is primarily a matter of taste, not health. The acceptable taste will depend greatly on who the consumers are, what they are accustomed to, and what their alternatives are. A product concentration of 500 mg/L remains a useful target, but in areas with no alternative supply, 1000 mg/L may well be perfectly acceptable, especially if this can be produced in greater quantity or with improved reliability.

7.4 Desalination with Integrated Membrane Processes

The most reported hybrid membrane systems for desalination consist of seawater pretreatment before RO, using UF, MF, or NF. These techniques have been tested from the beginning of the 1990s and are now well implemented at several desalination plants. These processes are presented in the following sections, with special attention to a desalination plant built in Saudi Arabia with NF as a pretreatment to RO.

Other hybrid membrane processes have been reported in the field of desalination, which will not be detailed in this chapter. For example, various hybrid membrane processes have been proposed for boron removal from seawater, such as ion-exchange resins associated to UF/MF [10]. In addition, submerged hybrid systems coupled with physico-chemical processes (coagulation and adsorption) have been recently used as a pretreatment to seawater RO [11]. Small molecular species that are not rejected by the MF alone can be absorbed by the powder activated carbon (PAC) and coagulated by the $FeCl_3$ in the membrane hybrid systems. Another hybrid membrane system reported in desalination application is membrane distillation to concentrate RO brines and increase the global recovery of the process [12, 13].

7.4.1 MF/UF–RO

UF or MF seawater pretreatment prior to RO has been reported from the beginning of the 1990s [14]. With decreasing membrane costs and increasing requirements on water quality, these techniques have become suitable alternatives to conventional chemical pretreatments [3]. Both UF and MF membranes are able to remove suspended particles, colloidal materials, microalgae, bacteria, viruses and pathogens, microorganisms from raw water, reducing RO membranes fouling. In addition, UF or MF pretreatment usually requires less chemical addition than conventional pretreatments and significantly less space. They are less sensitive to fluctuations of feedwater quality and supply the RO stage with superior water quality for long-term operation. Membrane pretreatment therefore might be able to increase RO membrane life and ensure stable operation even under adverse conditions and could thus lead to overall cost reductions.

At laboratory scale, UF and MF of natural seawater from San Diego Bay were compared in dead-end filtration before RO experiments [15]. The seawater was first filtered through a 1 µm prefilter, then using 0.1 µm MF, 20 kDa, or 100 kDa UF. MF pretreatment was found more effective in reducing RO fouling than conventional filtration. UF pretreatment with the 100 kDa membrane did not decrease fouling compared to the MF membrane. The 20 kDa UF membrane was the most effective in reducing fouling but operated at higher pressures for the same flux as the MF and 100 kDa UF membranes.

In several desalination plants, UF/MF pretreatment has been reported to be very effective. At Jeddah Port on the Red Sea (Saudi Arabia), Pearce et al. [16] operated UF pretreatment to RO desalination for a 6-month period as an alternative to its conventional pretreatment facility, which could not meet targeted feedwater quality during periods of algal bloom and storms. An average filtrate silt density index (SDI) of 2.2 was obtained, approximately two units better than the existing conventional pretreatment. In a desalination plant in the Yellow Sea in China, Xu et al. [17] utilized UF hollow fiber membranes as pretreatment prior to RO. During the experimental periods, the UF–RO system was run stably and successfully, without any chemicals required for disinfection, flocculation, enhanced chemical backwash, and cleaning. The efficacy of MF pretreatment was also confirmed in a pilot study conducted from October 2007 to November 2010 to manage salt levels in Central Arizona Project water [18]. Pretreatments consisting of MF and slow sand filtration were compared based on turbidity removal, SDI, and volume treated between cleaning events. Although the economic costs of MF and slow sand filtration pretreatments were similar; MF was preferred based on the quality of treated water and stability of downstream RO operation.

Algal blooms are reported as a major cause of RO membrane fouling. Since 10 years, seawater quality has been decreasing due to anthropic activities and the consequence is a dramatic increase in the algal blooms frequency all over the world. An example of a toxic microalgae encountered on the Atlantic coast, *Alexandrium minutum*, is shown in Figure 7.6 [19].

The main effect of bloom development is the increase of the dissolved organic carbon (DOC) from 16% to 50% [20]. Specific pretreatments are needed and membrane processes are today very well identified, notably UF and/or MF, as reported recently by Castaing et al. [19, 21]. In these studies, submerged MF/UF membranes were investigated for the removal of various dinoflagellate suspensions (*Heterocapsa triquetra*, *A. minutum*, and *Prorocentrum lima*). Three membranes were compared (10 kDa, 300 kDa, and 0.2 µm) by filtration of 30,000 cells/mL of *H. triquetra* suspension (equivalent to a major natural phytoplanktonic bloom). The influence of microalgal species and its concentrations (1000 and 30,000 cells/mL) on filtration yield and membrane fouling mechanisms were evaluated. The dissolved organic substances and particulate size distribution were shown to be important factors not only affecting internal but also external fouling. These results were

Figure 7.6 Example of a toxic microalgae encountered on the Atlantic coast, *Alexandrium minutum*. Source: Reprinted from Reference 18 with permission from Elsevier.

mainly dedicated for the pretreatment of seawater for little units in the shellfish industry; however, it can be foreseen that these results can be applied to MF and/or UF pretreatment prior to RO desalination.

7.4.2 NF versus RO

The history of membrane demineralization using RO dates from the 1960s [8]. However, the high pressures traditionally used in RO resulted in considerable energy consumption and reduced RO development. By the second half of the 1980s, low-pressure RO membranes, called NF membranes, with lower energy consumption, lower rejections of dissolved components, but with higher water permeability, were dedicated to partial demineralization of brackish waters [22].

NF offers several advantages in comparison to RO such as low operation pressure, high flux, high retention of multivalent anion salts and organics compounds with molecular weight situated between 100 and 1000 Da, relatively low investment and low operation/maintenance costs. Because of these advantages, the applications of NF worldwide have increased. Today 15% of the brackish waters treatment in the world is realized using NF membranes. While NF is a relatively new membrane process, it is already widely used

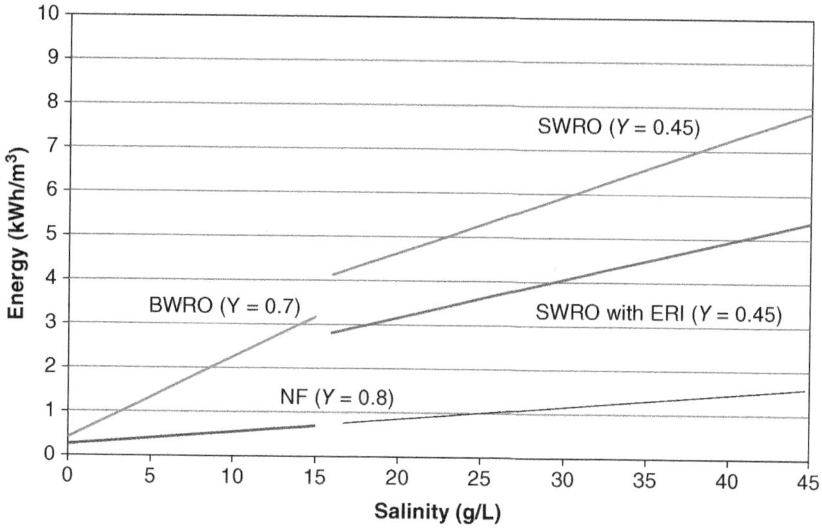

Figure 7.7 *Energy consumption in RO/NF versus water salinity [23]. Y, flow yield ratio defined as the ratio between permeate flow versus feed flow; BW, brackish water; SW, seawater; ERI, pressure exchanger from Energy Recovery Inc.*

for water treatment in different parts of Europe, Israel, and the United States. Striving toward improved quality, efficiency, and applicability, research is continuing in an attempt to understand and model the varying parameters involved in NF. A technique that is often used for the evaluation of membranes is the flux and rejection behavior of uncharged and charged solutes. However, many membranes have to be screened before finding a suitable one.

The actual membrane desalination market is dominated by RO (51%) and only 4% by NF (see Figure 7.5). Energy consumption of water desalination by both processes can be expressed as a function of salinity (see Figure 7.7).

Considering the theoretical osmotic pressure, brackish water requires less pressure to be desalinated than seawater. So NF or brackish water RO (BWRO) is required for brackish waters desalination as seawater RO (SWRO) is required for seawater desalination.

The life cycle electricity consumption is the total electricity consumed during the treatment plant construction and operation phases. Electricity consumption at the potable water plant is calculated using site measurements and the previously described equipment models [23]. Electricity consumption for off-site chemicals production is calculated using ecoinvent datasets. Electricity consumption for plant construction, also calculated using ecoinvent datasets, is found to be negligible in comparison with the electricity consumption for plant operation, as reported elsewhere [23]. The life cycle electricity consumption of the RO plant is about 4.6 kWh/m^3 of potable water, as reported on Figure 7.8. Electricity consumption for the RO process represents most of the life cycle electricity consumption with 84% of total electricity consumption.

The comparison between RO and NF plants, with similar intake and potable water distribution conditions, estimates that the NF plant needs 1.3 kWh/m^3 of potable water.

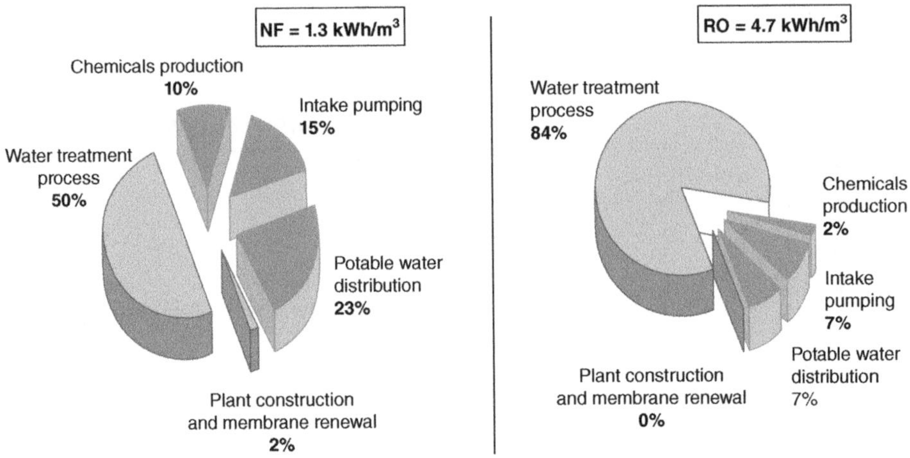

Figure 7.8 *Life cycle electricity consumption of NF and RO plants.*

Electricity for the water treatment process represents 50% of the life cycle electricity consumption, whereas, drinkable water distribution and chemicals production represent both one-third of life cycle electricity consumption. Concerning the electricity cost of membrane renewal, a very low cost is observed, identical between RO and NF with 0.02 kWh/m^3 of seawater combining RO and NF.

7.4.3 NF–RO

The use of NF as pretreatment before RO is also a possible alternative as water hardness can thus be strongly reduced. Indeed, NF membrane is usually negatively charged and preferentially rejects divalent ions which are prone to scaling [24, 25]. In addition, NF can retain 10–50% monovalent species depending on their molecular weight, hydration, and electrostatic interactions with the membrane. As a consequence, the osmotic pressure of RO feed and retentate streams is decreased, thus allowing the system to operate at high water recovery factors. In addition to the ability of NF membranes to reduce scale-forming hardness ions, NF can also reduce TDS and remove microorganisms, turbidity, and organic matter [26]. The first industrial plant pretreating seawater by NF in replacement of the first stage RO was built in Umm Lujj (Saudi Arabia) at the beginning of the 2000s by the Saline Water Conversion Corporation [27]. The design of the plant is schematized in Figure 7.9.

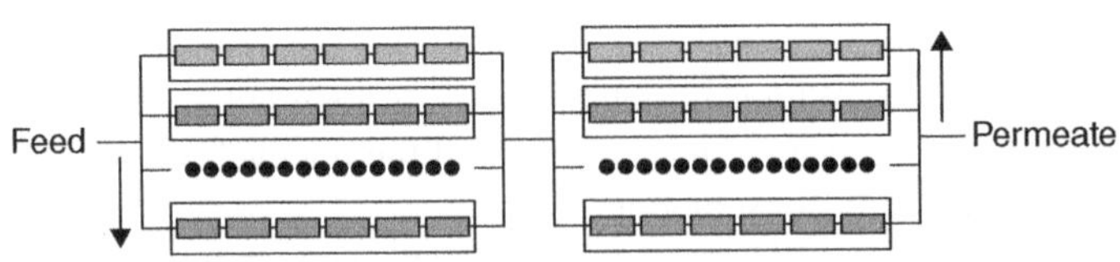

Figure 7.9 *Design of Umm Lujj NF–RO plant, with stages one and two.*

Table 7.3 *Flow yield ratio, energy gains, and TDS retention for the two different stages configurations based on Umm Lujj design.*

Membrane	Gain of flow yield (%)	Gain of energy (%)	Rejection of TDS (%)
NF90–SW30	16	24	99.6
NF200–SW30	30	29	99.1
NF-OI Umm Lujj	38	39	99.2

Based on data operated at the Umm Lujj plant and using the commercial software ROSA® (Dow Chemical Company), different configurations of NF associated to RO were compared in terms of gain of energy, water quality, and salt rejection (Table 7.3) [28]. The following different NF–RO configurations were considered: NF200-SW30 and NF90-SW30 in reference to the SW30-SW30 were classically used.

The NF200-SW30 gave the best energy gain with 29% and the permeate obtained presented a TDS of 324 ppm. Furthermore, this configuration showed a gain of 16% for the recovery. But as observed for Umm Lujj, the configuration NF-OI gave a gain in recovery of 8. In the case of NF200 simulation, the convective flow through the NF membrane was not taken into account by the software ROSA®, then a higher recovery was expected, but a lower quality of TDS due to convective mass transfer, as illustrated in Figure 7.10.

The gain in recovery was correlated with a loss in water quality. The opposite result with the following configurations NF90-SW30 and XLE-SW30 (Dow Filmtec) are observed where a gain in water quality is associated to a decrease in water productivity. XLE is a RO membrane dedicated to lower pressure conditions; it has a higher hydraulic permeability than the usual SW30 membranes. So, the best balance must be optimized for each situation.

However, simulations using the ROSA™ software are limited, especially for NF, where the active layer presents some microporosities, leading to convective mass transfer which is not taken into account. As a consequence, simulations must be validated by experiments done on real seawaters, at sufficient scale and time.

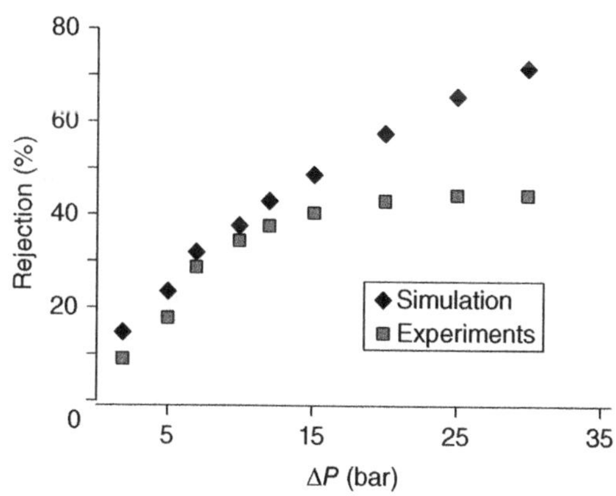

Figure 7.10 *Rejection = f(ΔP) for a first stage NF200 membrane. Comparison between simulation and experimentation with NaCl 0.6 M (35 g/L).*

Table 7.4 *Measurements of SDI3, TOC, and turbidity on seawater for NF200 and NF90 membranes with real seawater from the Atlantic coast (Le Croisic, France).*

	Seawater prefiltered on 1 μm	Permeate NF200	Permeate NF90
SDI_3 (%/min) ± 2	21	4	2.5
TOC (ppm) ± 0.04	0.81	<0.10	0.08
Turbidity (NTU) ± 0.1	2.2	<0.3	0.2

The following parameters have been determined:

- SDI for 3 minutes
- TOC (total organic carbon)
- Turbidity

for the comparison of NF90 and NF200 membranes. The seawater was prefiltered on 1 μm membrane nominal pore size, before NF. The data are presented in Table 7.4.

SDI_{15} was not measurable on natural seawater as reported by Leparc et al. [7]. The SDI_3 was therefore determined, as usually preconized [7]. Initially, a value of 19 ± 2%/min was obtained for untreated seawater, value in the range of other natural seawater in the world with SDI_3 reported between 5 and 35%/min. Furthermore, prefiltration to 1 μm had no influence on the fouling properties of seawater, SDI stayed constant, but the turbidity in the same time decreased from 9.7 to 2.2. Maybe particulates larger than membrane pore size (0.45 μm) are retained but without influencing SDI values. NF200 and NF90 filtration of seawater permits to decrease the SDI_3 to a value of 4 and 2.5, respectively.

Furthermore, due to the very low TOC in both permeates (<0.1 mg/L) the biofouling development on the RO membrane of the second pass should be very limited. But, as a consequence, the durability of NF membranes will depend on the pretreatments developed before this operation.

An intensification of MF/UF seawater pretreatment can be predicted in the future, associated to NF as the first stage desalination before RO, as illustrated in Figure 7.11.

Figure 7.11 resumed the design of future seawater desalination plants positioning the integration of membranes MF or UF and NF as seawater pretreatments operations before RO.

7.5 Natural Water Treatment Using Hybrid Membrane Processes

Natural water sources may contain several contaminants: synthetic organic chemicals such as pesticides, herbicides, industrial solvents, and chemicals; inorganic pollutants such as arsenic, nitrate, and toxic metals; natural organic matter (NOM) and microorganisms such as protozoa, bacteria, and viruses. Over the last decades, various water treatment technologies including adsorption, coagulation, flotation, ozonation, ion-exchange, and pressure-driven membrane processes such as MF, UF, NF, and RO have been thoroughly investigated for natural water treatment. The tightening of water quality regulations and the increased attention given to trace contaminants in drinking water has been favoring the emergence of alternative treatment technologies in order to improve water treatment processes.

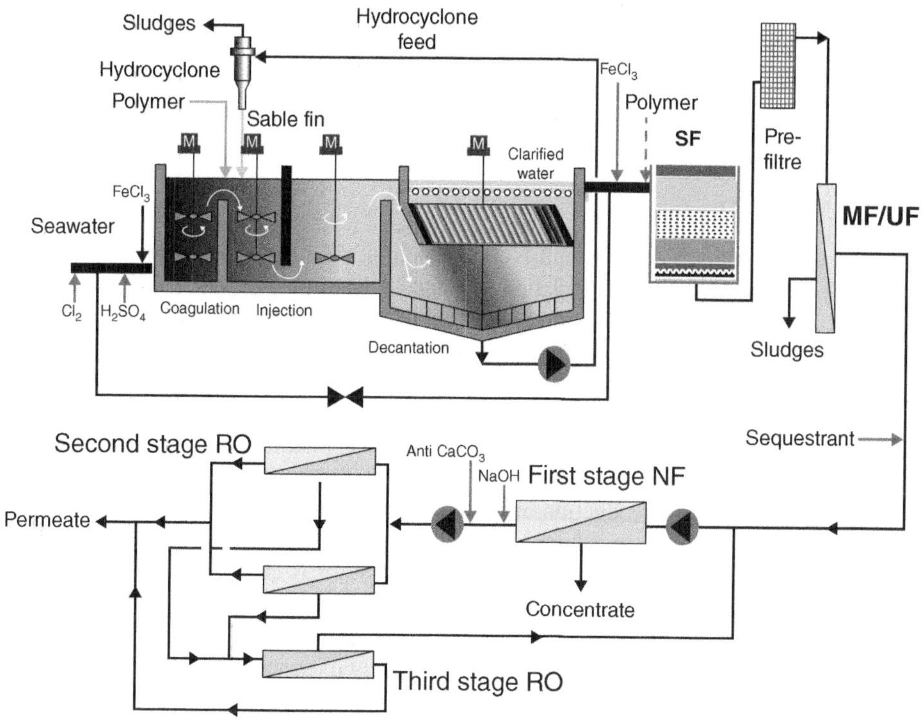

Figure 7.11 Future design of seawater plants integrating membranes.

Recently, hybrid membrane processes for drinking water treatment have been extensively investigated. In these processes, pretreatment is usually added before membrane separation, usually UF or MF. The general idea is to increase the size of the species to be removed so they can be eliminated easily by UF and MF, instead of NF and RO. Several advantages are thus expected: less energy consumption and reduced cost for implementation than NF and RO, improvement of efficiency removal and eventually decreasing membrane fouling compared to UF and MF alone, improvement of efficiency, and easiest implementation at large scale compared to the pretreatment technique alone.

Various hybrid membrane systems have been developed combining pretreatment before membrane processes. These pretreatment techniques include precipitation/coagulation/flocculation, adsorption on activated carbon or other adsorptive media, chemical oxidation or advanced oxidization technique such as ultraviolet (UV) and ozone. Very few applications are related to multistage membrane separation. To our knowledge, the main application concerns multistage NF for natural water treatment. This technique has been applied to demineralization/defluoridation and used in the first world unit built recently in Thiadiaye (Senegal) [29].

Other hybrid membrane techniques for drinking water treatment include, for example, the hybrid membrane bioreactor technology [30], the hybrid ion-exchange/pressure-driven membrane processes [31], and chelating polymer-assisted membrane separation [32], which will not be detailed in this chapter. Example of hybrid membrane processes are given below for removal of organic matter, arsenic, and other species such as iron and fluorides in drinking water treatment. Other trace contaminants such as manganese, nitrate, pesticides, herbicides, and pharmaceuticals will not be considered in this chapter although their presence in natural water is increasingly reported as well as their removal using hybrid membrane technologies.

7.5.1 Natural Organic Matter

Natural waters contain colloids and NOM, which is mainly composed of polydispersed molecules, derived from the degradation of plants and microorganisms. NOM is a cause of color in natural water sources, and is not considered to be harmful to humans. However, NOM is associated with the formation of disinfection by-products (DBPs) during chlorination in drinking water treatment. Due to harmful health effects of DBPs, such as increased cancer risk, a significant amount of NOM should be removed prior to chlorination [33]. For example, Norwegian drinking waters are mainly supplied from surface water sources, generally from lakes with high NOM concentrations. The removal of NOM is therefore a major requirement in the production of potable water where concentrations typically in the 30–80 mg/L Pt true color range have to be reduced to less than 10 mg/L Pt [34].

The conventional treatment process for NOM removal is coagulation/flocculation followed by either traditional rapid sand filtration or direct filtration, where flocculated water is flowed directly onto the filter, forming a porous, low-density cake on the surface that is easily removed by scouring or backwashing [31]. Given more stringent water quality regulations, pressure-driven membrane processes are increasingly used; however, membrane fouling is a major limitation. Previous studies regarding the membrane treatment of natural water sources such as lake and river waters show that NOM is one of the major foulants. To improve the membrane performance as well as the water quality produced by MF or UF, various hybrid membrane systems have been developed, such as powdered activated carbon/UF (PAC/UF), iron oxide particles (IOPs), adsorption/UF, coagulation/UF or MF, iron oxide adsorption/UF, photocatalysis/MF, ozonation/UF, and electrocoagulation (EC)/MF. A major fraction of NOM present in surface or ground waters is composed of humic substances (HA). HA was found to have a high fouling potential and thus is used as a model compound of NOM in many studies.

7.5.1.1 Powdered Activated Carbon/UF

Hybrid membrane processes using activated carbon treatment for drinking water have been reviewed recently [35]. The different configurations were classified into three categories:

- Activated carbon treatment prior to UF or MF (pretreatment configuration, see Figure 7.12)
- An integrated activated carbon treatment and UF/MF (integrated treatment configuration, see Figure 7.13)
- Activated carbon treatment after UF/MF (post-treatment configuration, see Figure 7.14).

The carbon contactor can either be an adsorptive granular activated carbon (GAC) filter or a biological GAC, a fluidized GAC suspension or a PAC contactor, where PAC is maintained in suspension. The pretreatment configuration is the most reported at the industrial scale. The UF/MF system can be pressurized or immersed and operated in both cases either under

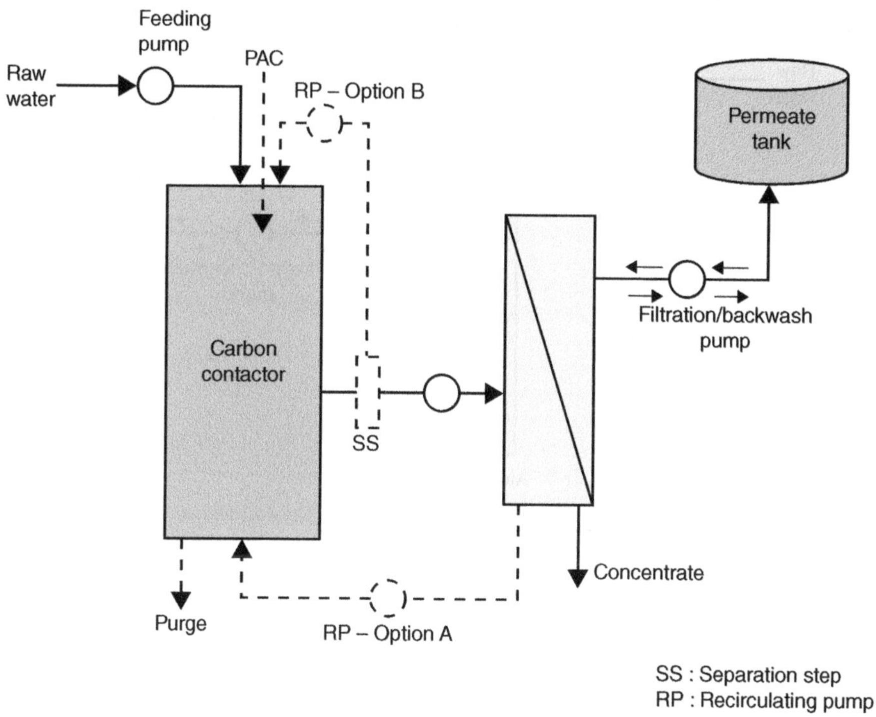

Figure 7.12 *Schematic representation of the hybrid membrane process with activated carbon pretreatment. When PAC is used, the concentrate can be either recirculated in the carbon contactor (option A) or separated using a process before being recirculated (option B).* Source: *Reprinted from Reference 35 with permission from Elsevier.*

an outside-in or an inside-out mode. Pressurized systems can be operated with or without crossflow.

The configuration consisting of a PAC contactor, UF, and a recirculation loop was used at an industrial scale in France under the commercial name CRISTAL® process. Two plants were reported in France, at Vigneux (28,000 m^3/d), L'Apié (55,000 m^3/d) and in Angers (France) in 2004 (110,000 m^3/d) [36].

In the removal of organic substances, PAC/UF systems were found to be effective for both low and high molecular weight compounds such as phenols and low molecular HA fractions. For example, Tomaszewska and Mozia [37] reported that approximately 90% of HAs and 100% of phenol were removed at initial concentration of 10 and 1 mg/L, respectively, at a PAC dosage of 100 mg/L. The main advantage of PAC/UF system is to remove organic compounds, which cannot be removed by UF alone. However, the addition of PAC into feedwater containing HAs may cause a decrease in permeate flux depending on parameters such as membrane materials and natural water composition. The high drop in the permeate flux measured during UF of HA solutions through cellulose acetate (CA) and polysulfone (PSF) membranes was explained by the membrane surface chemistry [38]. Negatively charged functional groups present on the membrane surface, such as carboxyl groups (CA membrane) and sulfone groups (PSF membrane), interact with HA resulting in

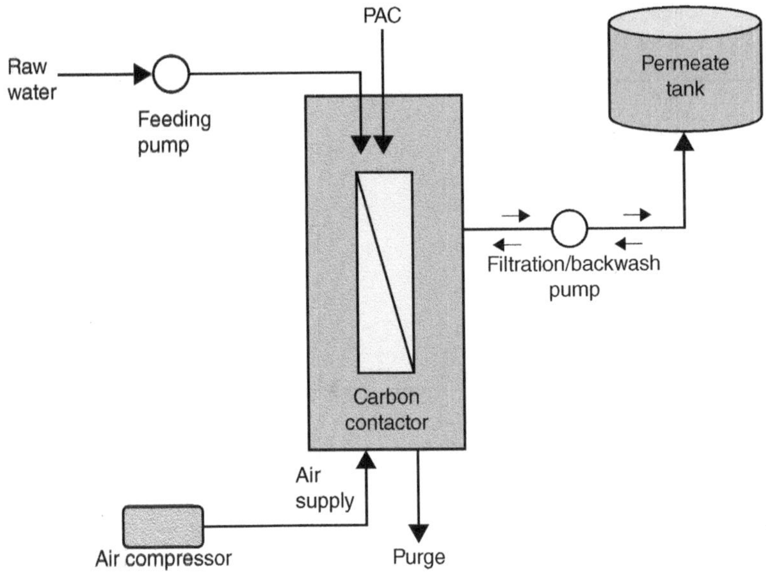

Figure 7.13 *Schematic representation of the hybrid membrane process with integrated activated carbon treatment.* Source: *Reprinted from Reference 35 with permission from Elsevier.*

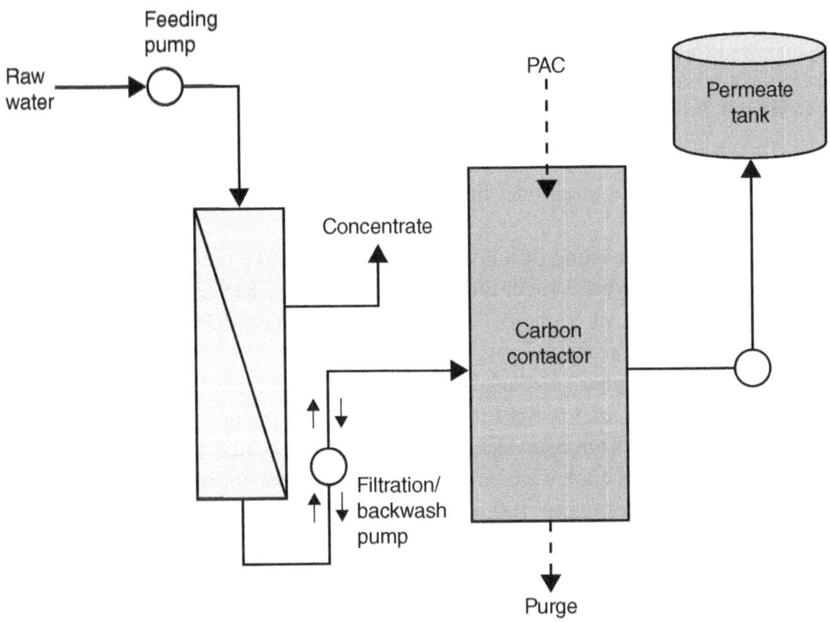

Figure 7.14 *Schematic representation of the hybrid membrane process with activated carbon post-treatment.* Source: *Reprinted from Reference 35 with permission from Elsevier.*

coating of the membrane surface. Such a phenomenon represents a drawback for the use of PAC/UF systems. Also, the disposal of saturated PAC particles containing toxic chemicals or nonbiodegradable compounds is relatively costly.

7.5.1.2 *Iron Oxide Particles Adsorption/UF*

IOPs are an alternative adsorbent to PAC, which could be less expensive and easily regenerated. IOPs added to UF improved both NOM removal and membrane flux, which could be attributed to the adsorption of NOM by IOPs and the subsequent protection of the membrane [39]. NOM-loaded IOPs were easily reversed by backwashing, and were regenerated in alkaline pH conditions. Whether IOPs were suspended in aqueous solutions or attached to the membrane surface, they had no negative effect on the membrane flux, unless a substantially significant amount of IOPs was added. Similar results were reported by Lee et al. [40] for the treatment of water from the Nakdong River by a hybrid process combining IOPs adsorption/UF. IOPs addition to UF improved both NOM removal and membrane flux even at high IOP concentration of up to 500 mg/L. This was explained by the adsorptive removal of NOM by IOPs and subsequent fouling reduction, sufficiently compensating a possible adverse effect on flux that may arise from the formation of a thicker IOP cake layer at the membrane surface. The better control of fouling by IOP than PAC was confirmed by Kang and Choo [41]. In this study, two adsorbents, IOPs adsorbent ferrihydrite (FH) and PAC were compared in terms of the degree of fouling control under the condition where the same amount of DOC was removed. The main reason why FH was better than PAC with respect to fouling control was suggested to be the selective adsorption of large NOM molecules and the formation of more porous cakes with FH particles.

7.5.1.3 *Coagulation/UF or MF*

Several research groups have investigated coagulation pretreatment before UF or MF as an alternative to reduce and control membrane fouling by NOM. Polymeric membranes have been mostly used, although inorganic membranes have also been reported. For example, Leiknes et al. [34] investigated the feasibility and potential of using inorganic metal MF membranes in a submerged membrane configuration with coagulation pretreatment. Coagulation pretreatment with polyaluminum chloride (PAX-16, Kemira Chemicals AS, Norway) of raw water with a color of 50 mg/L Pt revealed that a specific aluminum dosage of 5 mg/L aluminum removed >95% of true color, ~87% of UV-absorbing compounds, and 65–75% of DOC). A high permeate flux was achieved for all operating modes investigated.

Hydrolyzed metal salts are often used in the coagulation pretreatment. Iron-based coagulants can be preferred to conventional aluminum-based coagulants which are reported to have a possible link with Alzheimer's disease. Polymer flocculants, which are commonly used in the conventional treatment process to enhance floc sedimentation, have been also adopted in the coagulation/MF process by many treatment plants. However, Wang et al. [42] underlined that the application of polymers in MF pretreatment needs to be carefully evaluated. In a very complete study, these authors investigated the effects of three types of polymer flocculants on MF permeate water quality and membrane fouling using a hollow fiber membrane under two different operation modes, coagulation/flocculation–sedimentation MF (CFSM) and coagulation/flocculation–MF (CFM). The authors recommended that the use of anionic and neutral polymer flocculants at concentrations higher than 0.5 mg/L should be avoided considering the severe membrane fouling and the small improvement in water quality. Cationic polymer flocculants may be used at low dosage in the CFSM mode to enhance NOM removal as well as to reduce membrane fouling.

7.5.1.4 Photocatalysis/MF

Photocatalytic degradation in combination with membranes has been reported in many studies as an alternative for the removal of NOM or toxic organics in water [33]. In these photocatalytic membrane reactors, the membranes are used to confine the photocatalysts inside the reactor. TiO_2 is one of the most commonly used photocatalyst, due to its high photocatalytic activity and environmentally benign properties. Once illuminated, TiO_2 particles simultaneously adsorb NOM and efficient photons, creating a redox environment able to oxidize toxic chemicals and NOM into harmless compounds or mineralize into carbon dioxide and water. For example, Le-Clech et al. [43] used an experimental setup composed of a light-protected slurry reactor, a cross-flow filtration cell, and a UV lamp. The hybrid membrane/photocatalysis process was used to treat waters containing small concentration of NOM. The optimization of the process indicated that operating conditions should be slightly acidic (pH around 4–5) and contain a small concentration of TiO_2 (0.1 g/L). These operating values depended on the feedwater composition and would require adjustment when variation in the water composition occurs. In other studies, TiO_2 particles were coated at the membrane surface. For example, Song et al. [44] prepared modified polyvinylidene fluoride (PVDF) membranes by adding different amounts of polyethylene glycol (PEG) and TiO_2 particles. NOM removal and flux decline were measured in both dead-end and cross-flow filtration experiments. Results indicated that appropriate TiO_2-doped modification on PVDF membrane could be an effective way for better removal of NOM and reduction of membrane fouling at the same time. During the UF on TiO_2-doped PVDF membrane, equilibrium could be reached between the foulants deposited on the membrane surface and photocatalytic degradation of the foulants during cross-flow UF coupled with photocatalysis, showing self-cleaning ability.

7.5.1.5 Ozonation/UF

In the past few years, several research groups have coupled ceramic membranes with catalytic ozonation processes to reduce membrane fouling during NOM removal for drinking water treatment. Catalysts, such as metal oxides, can degrade ozone to generate ·OH radicals, which are very strong oxidants that can degrade many contaminants. This reduction is primarily due to the changes of NOM characteristics resulting from the preferential oxidation of electron-rich moieties containing carbon–carbon double bonds and aromatic compounds, which are largely responsible for membrane fouling. For example, Park et al. [45] realized NOM removal using a hybrid process that combined ozonation with a ceramic UF membrane and with a ceramic reactive membrane (RM) incorporating iron oxide nanoparticles (IONs). Their results indicated that the RM-ozonation process enhanced NOM removal and reduced membrane fouling by generating hydroxyl radicals from the catalytic ozonation.

7.5.1.6 Electrocoagulation/MF

Unlike chemical flocculation in which the coagulants are added to the water as salts, in the EC process the coagulants (iron or aluminum) are added to the solution by dissolving the anode in an electrochemical cell. The ions lead to aggregation of particles, which are then removed by sedimentation or filtration. EC has several advantages compared to chemical coagulation (CC), including easy operation, lower quantities of produced sludge, avoidance of chemical usage, and the fact that no anions such as chloride or sulfate need to be added to the solution. Several research groups have observed increasing NOM removal efficiency using EC as a pretreatment to MF. For example, Ben-Sasson et al. [46] studied the performances of a hybrid process that combined EC and dead-end MF. Both iron- and aluminum-based EC pretreatment reduced NOM fouling and improved NOM removal efficiency. At optimal conditions, a 36-fold shorter filtration time and a 20% increase in NOM removal efficiencies were obtained with the EC–MF hybrid process as compared to 100 kDa UF alone. The authors concluded that the EC–MF hybrid process could be a suitable method for NOM removal, due to its ability to obtain high fluxes, low fouling, and high colloidal/NOM removal. The disadvantages may be related to the operation and cost of the additional treatment by EC.

7.5.2 Arsenic

Arsenic is a toxic compound that is present in drinking water supplies worldwide due to industrial pollution sources or natural geochemical activity. Prolonged consumption of arsenic-contaminated water is the cause of various types of human diseases, including respiratory diseases, gastrointestinal, liver and cardiovascular problems, and cancer. High arsenic concentrations are found worldwide, for example, in Bangladesh and West Bengal, India, where many groundwater sources contain arsenic above the maximum contaminant level (MCL) of these countries (50 ppb). The MCL recommended by the WHO, and regulated by the United States and European Union is 10 ppb [47].

Arsenic exists in groundwater in two predominant species: trivalent arsenite As(III) (H_3AsO_3) and pentavalent arsenate As(V) (H_3AsO_4) [48]. Both As(III) and As(V) are found in groundwaters regardless of the environmental conditions. A variety of technologies have been used for arsenic removal, including oxidation, coagulation–precipitation, adsorption, ion-exchange, and membrane separation [49]. Coagulation and adsorption processes provide good arsenic removal efficiency, although their major drawbacks are the requirement of multiple chemical processes, regeneration of medium, and treatment of wet solid sludge. Ion exchange can also be used for arsenic removal, although it has limitations due to its operational instability, resin regeneration, and the required treatment of used resin. Membrane processes such as UF, NF, RO have been used due to their simple process, reduced maintenance and operational requirements, and the development of high flux membranes. RO ensures the highest removal efficiency for As(V) from groundwater along with other contaminants. However, it also involves the highest cost due to RO membranes as well as pumping at high pressure. Integrated membrane systems and hybrid technologies coupled with membrane filtration have recently been proposed to reduce the concentration of arsenic in natural waters for drinking water treatment.

Under normal pH conditions, As(III) behaves as a neutral molecule which can be separated by steric effects, whereas As(V) being negatively charged can be separated by both the Donnan and steric mechanisms. Therefore, for waters containing As(III), pre-oxidation prior to treatment is often required for efficient removal of arsenic. As(III) oxidation techniques include chemical oxidation ($KMnO_4$, H_2O_2, OCl^-, and $S_2O_2{}^{-8}$), Fenton

[Fe(II)/H_2O_2], and ozone (O_3). As(III) oxidation has also been realized by UV, UV/catalyst as well as coupling of ultrasound (US) to UV. The produced sludge usually consists of small particles (0.1–1 μm) [50]. In order to remove these particles, a coagulation–flocculation step is necessary to generate sufficiently large particles that can be removed in a settling basin. However, careful control in the dosing of chemicals and long settling times make the technique difficult. MF is an alternative to coagulation–flocculation settling, where particles can be removed by using a low-pressure pump without chemicals and need for large space.

7.5.2.1 *Coagulation/Flocculation*

The addition of a coagulant followed by the formation of a floc can be used to remove arsenic from groundwater. Positively charged cationic coagulants reduce the negative charge of the colloids and as a result, larger particles are formed due to aggregation of particles. Coagulation is usually performed using ferric chloride and ferric sulfate. Wickramasinghe et al. [51] used coagulation followed by tangential flow MF. The technique was successfully applied to the treatment of groundwater from a city in southern Colorado in the United States and from Sonargaon in Bangladesh. However, the actual efficiency of arsenic removal was pointed out to be highly dependent on the quality of the groundwater. Since a lower pH leads to increased arsenate removal, pH adjustment should be necessary in order to reduce the ferric ion dose. EC is also a possible technique for removal of arsenic from drinking water. Mólgora et al. [52] compared the hybrid process EC/MF to CC (using $FeCl_3$) coupled with MF (CC/MF) for removal of arsenic (C_0 = 100 μg/L) from drinking water. The best arsenic removal (97%) was obtained using the CC–MF process at operating conditions at a pH of 7.0 and a concentration of 4.0 mg/L of Fe^{3+}. The performance of the EC/MF decayed more quickly than that of CC–MF. It was suggested that during EC, the resulting effluent was not enriched with anions and salts content did not increase compared to CC where the effluent was enriched with chloride ions.

7.5.2.2 *Pre-oxidation*

Chemical pre-oxidation with $KMnO_4$ for conversion of As(III) to As(V) can be used prior to membrane filtration. For example, arsenic was separated from contaminated groundwater by using composite polyamide NF membranes in a flat sheet cross-flow membrane module [53]. With pre-oxidation of As(III), arsenic removal increased from 50–63% to 97–100%. The same authors addressed the issue of disposal of arsenic rejects for safe disposal [47]. A membrane hybrid treatment system was developed for continuous removal of arsenic from contaminated groundwater with simultaneous stabilization of arsenic rejects (Figure 7.15). Both As(III) and As(V) were removed by cross-flow NF following a chemical pre-oxidation step for conversion of As(III) to As(V). Continuous removal of more than 98% arsenic was obtained from groundwater that contained around 190 mg/L of total suspended solid, 205 mg/L of TDS, 0.18 mg/L of arsenic, and 4.8 mg/L of iron at pH 7.2. Stabilization of arsenic rejects was obtained in a solid matrix of iron and calcium.

As(III) removal has also been investigated in a hybrid membrane process using ozonation pre-oxidation followed by UF [49]. To make use of the catalytic effect of iron, a reactive ceramic UF membrane incorporating IONs was prepared. As(III) removal with pre-ozonation followed by UF (92%) was significantly increased compared with that of UF alone (63%). The presence of NOM led to a higher As(III) rejection, whereas the

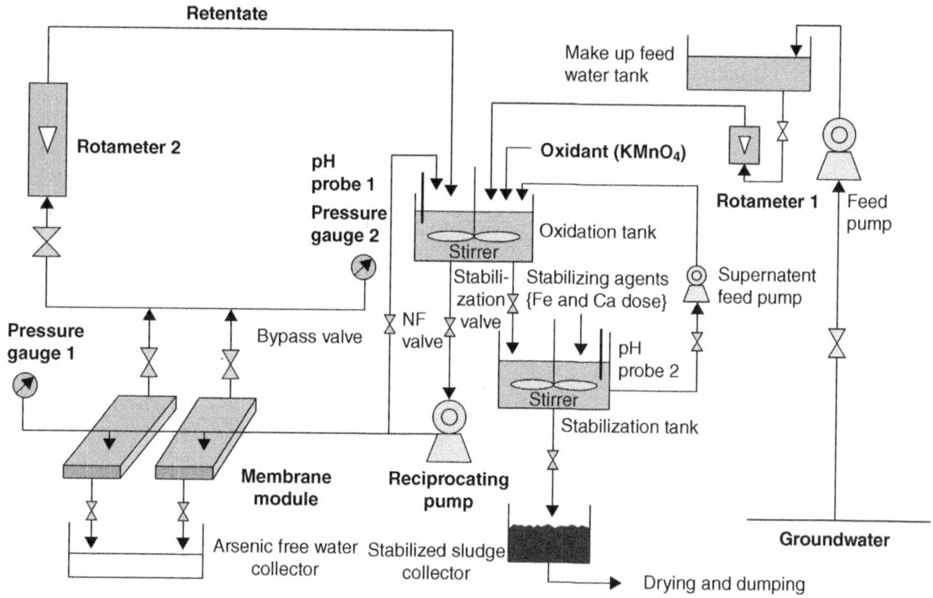

Figure 7.15 *Schematic diagram of the membrane-integrated hybrid process for removal of arsenic from water and its stabilization in solid matrix.* Source: *Reprinted from Reference 48 with permission from Elsevier.*

presence of coexisting ions in the feedwater significantly reduced the As(III) rejection. Finally, a comparison of As(III) rejection in synthetic water and groundwater confirmed the importance of real conditions in the hybrid reactive UF process with continuous ozonation.

7.5.2.3 *Coupling of US to UV/MF*

Arsenic contamination in groundwater is often accompanied by Fe(II) contamination because of release from iron and arsenic-bearing minerals like arsenopyrite (FeAsS). Therefore, together with As, ferrous Fe(II) can be released and should be treated appropriately. Arsenic removal using a pilot system of US and UV followed by MF was proposed by Cui et al. [50]. In water, the oxidation rate of As(III) can be increased by UV irradiation in the presence of oxygen. Compared to the single processes of US or UV, the combined US/UV system was more effective for As(III) oxidation. A high rate constant of As(III) removal was obtained as ferrous Fe(II) ions existed. Like the pseudo-Fenton reaction, Fe(II) species can participate in the production of additional ·OH by reacting with H_2O_2 produced by US, before being oxidized to Fe(III). MF was then used to remove particles produced in the pre-oxidation step.

7.5.2.4 *Anion Exchange Membrane/Coagulation*

Limitations of CC may be related to the direct addition of the coagulant (iron or aluminum) to the water, thus leading to residual levels. A hybrid process consisting of arsenate transport through an anion exchange membrane (Donnan dialysis) followed by coagulation was

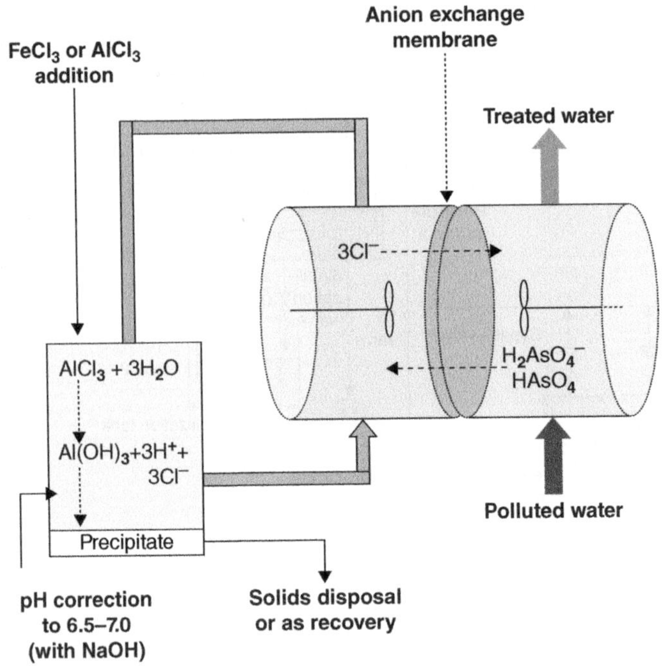

Figure 7.16 *Schematic of arsenate removal through the ion-exchange membrane process.* Source: *Reprinted from Reference 47 with permission from Elsevier.*

proposed recently for arsenic removal from drinking water without secondary contamination by coagulants and pH controlling reagents [47]. In this configuration, the drinking water stream is isolated by an anion exchange membrane barrier, coupled with chemical precipitation to treat the associated brine solution (Figure 7.16). The effects of iron versus aluminum coagulants on the process showed that aluminum was more effective than iron at preventing membrane fouling, likely due to the contrasting charges of the precipitates, which are repelled from the positively charged membrane in the case of aluminum, unlike iron. At optimal conditions, it was found that arsenate was removed below the MCL (6.6 ± 2.8 ppb). The authors concluded that the high quality drinking water obtained and its ease of operation make the process attractive for treating natural water containing arsenic.

7.5.3 Other Species

7.5.3.1 Iron

According to the European standards, the concentration of Fe(II) in drinking water should not exceed 0.3 mg/L. High iron concentrations in potable water produce unpleasant taste and color, and induce the growth of ferrobacteria. Iron usually exists in two oxidation states, reduced soluble divalent ferrous Fe(II) and oxidized trivalent ferric Fe(III). The most commonly used methods for iron removal from water are oxidation–precipitation–filtration, ion-exchange, lime softening, and membrane processes [54].

The oxidation–precipitation of soluble Fe(II) into insoluble iron hydroxides, especially ferric hydroxide $Fe(OH)_3$ is also a standard method for iron removal [55]. Oxidation can be carried out using chemical (hypochlorite, permanganate) or dissolved gas (oxygen, chlorine, ozone). Aeration is usually used for oxidizing ferrous iron in waters containing high iron concentration (>5 mg/L), so that disadvantages of chemicals can be avoided. Stronger oxidants such as ozonation or chlorination are required in case complexed iron forms difficult to remove. Iron hydroxide particles are then removed either by filtration on sand when Fe(II) concentration is lower than 7 mg/L or by decantation at higher concentration.

Compared to NOM and arsenic removal, very few membrane hybrid processes have been proposed for iron removal from drinking water. This could be explained by the relative inconvenience of iron, which is nontoxic for humans. The first technique proposed combines oxidation and MF for iron and manganese removal [56]. Artificial and natural groundwaters containing iron and manganese were first treated using a two-step oxidation procedure. The first step was the oxidation of ferrous iron by air bubbling at pH adjusted to 8 in order to fasten the oxidation of iron. The second step was the oxidation of manganous ion with $KMnO_4$. Oxide particles, the size of which ranging from 1.5 to 50 μm, were then eliminated from water using 0.2 μm cross-flow MF. Both artificial and natural groundwaters exhibited very similar behavior. Iron and manganese were not detected in the permeate indicating that the particles were very efficiently filtered even at relatively high concentrations of iron and manganese (10 and 2 mg/L, respectively). Moreover, the flux was strongly dependent on transmembrane pressure. A significant decrease in flux was observed at higher operating pressures, whereas a constant flux was obtained at lower pressures, that is, below 10 kPa. Under low pressures, MF flux was comparable to slow sand filtration rate at a similar pressure difference but with a much thinner filtering medium.

In a similar process, iron was removed from water by air oxidation followed by MF [57]. Water containing ferrous ion was oxidized by air at a constant flow rate and pH. The suspension was then subjected to cross-flow MF in a recycle configuration. Iron removal efficiency was shown to depend mainly on oxidation process conditions, such as pH. Oxidation at pH 8 followed by MF produced water containing iron at concentrations below 0.3 mg/L, while at pH 7.5 a slightly higher iron concentration was found. Tufekei and Demir [58] performed an integrated air oxidation process by recycling the treated water in the aeration tank. The increase of Fe^{2+} removal in this process was explained by an increase in ferric concentration in the aeration tank as a result of recycling. The higher the Fe(III) concentration in the aeration tank, the better was the Fe(II) removal efficiency up to a Fe(III) concentration of 300 mg/L. This result was explained by the catalytic effect of ferric ions.

7.5.3.2 Fluoride

High fluoride concentrations are found in natural water in several places around the world, like North and East Africa, India, and China [59]. Small amount of fluoride is useful for the development of teeth and bones. Fluorosis, caused by high fluoride intake predominantly through drinking water containing F^- concentrations >2 mg/L, is a chronic disease manifested by mottling of teeth (dental fluorosis) as for >4 mg/L, changes in bone structure, ossification of tendons and ligaments, and neurological damage are reported in severe cases.

The concentration of fluoride ions in drinking water recommended by the WHO varies in a concentration range 0.8–1.5 mg/L at temperatures from 12°C to 25°C.

Water defluoridation techniques include ion exchange, membrane processes, EC, coagulation–precipitation, and adsorption [60]. Among these techniques, ion exchange and membrane process can reduce the fluoride concentrations to suitable values, but they are expensive and require frequent regeneration of resins or membrane cleaning procedures. EC is highly effective, but it consumes energy and soluble anode leads to a secondary pollution by increasing dissolved metals, such as aluminum. The coagulation–precipitation method, also known as the Nalgonda technique in India, is very effective for defluoridation; lime and aluminum being the most commonly used chemicals. Drawbacks associated to this method include the formation of toxic soluble aluminum complexes, an increase of pH and TDSs, so the process needs a supplementary reaction for eliminating the excess of chemicals. Adsorption is also an attractive method for fluoride removal in terms of cost, simplicity of design, and operation.

The first unit dedicated to water defluoridation by multistage NF was built in Thiadiaye, Senegal in 2010. Results obtained in this field as well as the multistage unit are discussed below. Some hybrid membrane processes have also been proposed for fluoride removal including EC/MF and precipitation/MF.

7.5.3.3 Multistage Design

Because of its high and specific membrane selectivity, NF appears to be the best membrane process to remove fluoride from brackish underground waters, as reported for the first time in 1996, at bench-scale laboratory experiments [61]. This was confirmed later by a number of studies [62,63]. The main reason of this success is the high hydration of F^- that facilitates its exclusion by organic NF membranes. A comparison of the performances of RO and NF membranes, respectively, BW30 and NF90 from Dow Filmtech (United States), was carried out at semi-industrial scale, using various initial fluoride contents in Moroccan brackish water spiked with 1, 5, 10, and 15 mg/L NaF concentrations [64]. No influence of the initial fluoride concentration in feedwater on fluoride rejection was observed for the BW30 membrane. The fluoride concentration in the permeate was very low (around 0.1 mg/L) and the beneficial effect of fluoride for drinking water was not attained due to the fact that fluorides were highly rejected by the BW30 membrane. For the NF90 the fluoride concentration in the permeate increased with increasing fluoride concentration in the feedwater, but fluorides were reduced to a satisfactory value for all concentrations (below 1.5 mg/L).

Based on this work, the first unit dedicated to water defluoridation by NF was built in Thiadiaye, Senegal in 2010 [29]. The description of this plant is reported on Figure 7.17. The skid mounted system is a two-stages design plant typical for brackish water treatment, with a water tank of 1 m^3 and a membrane area of 1338 m^2 ($4 \times 6 + 2 \times 6 = 36$ spiral wound modules of the NF90-400 membrane) with a permeate flow production of 35 m^3/h. Cartridge filters (10 μm) are located upstream of the NF unit in order to limit particulates fouling. The interest of the two-stages design is to attain a flow yield ratio of 85%, limiting the volume of the brine solution.

Analysis of feed and nanofiltered waters from Thiadiaye (September 2009) are summarized in Table 7.5. The F^- concentration in the feedwater was 4.7 mg/L. After NF90

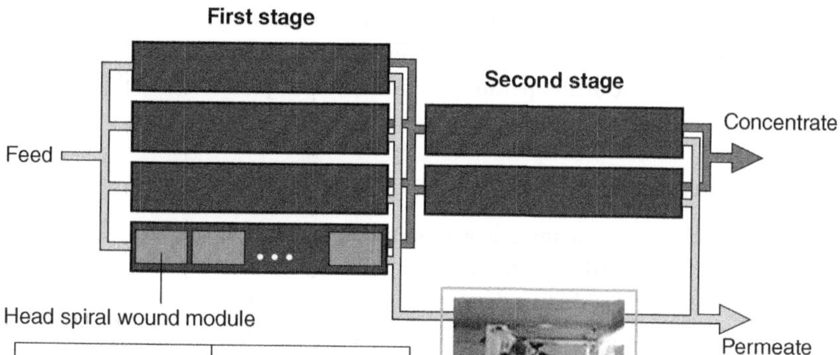

Feed flow (m^3/h)	50
Flow yield	70 %
Membrane area (m^2)	1338
Stages	2
Membrane type (spiral wound)	NF90–400

Figure 7.17 Photography of Thiadiaye's skid (Senegal) and operating conditions. Source: *Personal photography, M. Pontié (2010).*

membrane treatment the value obtained for F^- was drastically decreased to a value of 0.6 mgw/L. Nanofiltered water was also remineralized by percolation across a soil of Neutralite™ and then chlorinated before distribution to inhabitant. The membrane permeability was initially in September 2009 at 3.5 L/h.m^2.bar at 20°C.

7.5.3.4 Electrocoagulation/MF

EC is a possible technique for water defluoridation. During EC, dissolution of electrodes in the form of metallic hydroxides creates flocs which increase the water alkalinity and turbidity. Therefore, the treated water cannot be used for drinking purpose and further treatment is necessary. Therefore, EC has been associated along techniques such as MF. Ghosh et al. [59] reported the use of an EC bath for the treatment of water containing fluoride (10 mg/L). The particles generated during the treatment were in the range of 10 μm. After the completion of EC, an MF treatment was necessary to retain the drinking water quality

Table 7.5 Analysis for feed and nanofiltered waters (Thiadiaye, Senegal, September 2009).

Water parameters	Thiadiaye feedwater	NF water
Fluoride (ppm)	4.7	0.6
Sodium (ppm)	235	25
Chloride (ppm)	280	28
TH (ppm)	11	<1

in terms of pH, conductivity, and TDS. The pH, conductivity, and TDS decreased and were equal to 7.8, 0.2 mS/cm, and 580 mg/L, respectively. However, the high permeate flux decline observed with time at all operating transmembrane pressures, could be an important drawback of the process.

7.5.3.5 *Precipitation/MF*

A hybrid membrane process using adsorption/precipitation followed by MF was proposed by Ben Nasr et al. [65]. Adsorption/precipitation of fluoride was obtained using calcite particles; acetic acid was added to the suspension to increase precipitation and therefore fluoride removal. The suspensions obtained after decantation were then filtered using MF. The method was found suitable for water defluoridation, especially for high fluoride concentration (50 mg/L), the fluoride removal efficiency being very high (>90%). At low fluoride concentration (5 mg/L), the increase in fluoride removal in the presence of acid was mainly attributed to the increase in the available area for adsorption on the particles. At 50 mg/L, supersaturation was high enough for fluorite precipitation to occur. The pH of the treated groundwater remained in an acceptable range. Moreover, acidification of water which could be the main drawback of the method, is already used in water treatment to prevent scaling and reduce the crystallization of calcium carbonate.

7.6 Conclusion

The necessity of water reclamation is growing driven by a stress on water supply and public desire for a significant improvement in water treatment. RO is widely used in seawater, brackish waters, and natural water treatments. MF/UF more dedicated to particles rejection, are well used today as water pretreatment before RO and in hybrid processes with activated carbon or advanced oxidation. NF can be used for partial and/or selective demineralization of water.

There is no doubt that in the near future, membrane processes will play a central role in propagating the success of water desalination. Pretreatment using UF, MF, or NF have been increasingly studied and implemented at industrial scale.

In a fundamental aspect, the following areas can be of particular interest:

- Effects of parameters such as temperature and pH of the solution on NF removal efficiency.
- The removal processes using NF membranes are complex and poorly understood, especially for seawater.
- A full-scale membrane filtration process involves many operating variables including (but not limited to) cross-flow rate, transmembrane pressure, recovery, system arrangement, cleaning frequency, and module design. The pilot study must examine all the variables that might have certain effects on the overall removal efficiency. Therefore, it is essential that the translation of the results obtained from a pilot scale to a real life situation should take various operating variables into careful consideration.
- Fouling is an inevitable phenomenon in most (if not all) of the membrane filtration process. It is known that it strongly influences not only the production of drinking water but also the removal efficiency of the membranes. Numerous dedicated investigations

have been devoted to study the fouling effects on performances of RO membranes. Investigation elucidating the fouling tendency of both membranes (UF, MF, or NF for pretreatment) and RO, will create interesting results and help to better manage membrane processes. It is recommended, therefore, that long-term membrane testing under variable flow conditions should be conducted to establish membranes performances over time for a better scale-up.

For ground water treatment, hybrid membrane processes are increasingly reported in laboratory scale studies. However, to our knowledge, few industrial plants based on these technologies have been developed. To reach an industrial scale, the following points will have to be addressed:

- The hybrid membrane processes are often evaluated at the laboratory scale. The effect of several parameters on removal efficiency is investigated as well as the fouling behavior of the membrane. Before being implemented at large scale, several aspects have to be considered such as the development of a continuous process, with optimization of water flow rate and amount of activated particles, the regeneration of activated particles and membranes, the cost for the implementation and running of the plants compared to other techniques.
- Fouling is very low in a number of studies where the amount of water treated is only few liters. Fouling/biofouling have to be addressed for very large amount of water that will be operated in industrial plants. In addition, fouling and removal efficiency have to be considered for real water conditions such as the presence in the ground water of NOM and microorganisms, which could influence considerably the process performance.

List of Acronyms

BWRO	Brackish water reverse osmosis
CA	Cellulose acetate
CC	Chemical coagulation
CFM	Coagulation/flocculation–MF
CFSM	Coagulation/flocculation–sedimentation
DBPs	Disinfection by-products
DOC	Dissolved organic carbon
EC	Electrocoagulation
FH	Ferrihydrite
GAC	Granular activated carbon
HA	Humic acid
IONs	Iron oxide nanoparticles
IOPs	Iron oxide particles
MCL	Maximum contaminant level
MF	Microfiltration
NF	Nanofiltration
NOM	Natural organic matter
PAC	Powdered activated carbon
PEG	Polyethylene glycol

PSF	Polysulfone
PVDF	Polyvinylidene fluoride
RM	Reactive membrane
RO	Reverse osmosis
SDI	Silt density index
SWRO	Seawater RO
TDS	Total dissolved solid
TOC	Total organic carbon
UF	Ultrafiltration
US	Ultrasound
UV	Ultraviolet
WHO	World Health Organisation

References

1. Balaban, M., Desalination for the Environment: Clean Water and Energy, Congress 11–15 May 2014, Cyprus, Books of Abstract, 2014, European Desalination Society (http:// www.desline.com).
2. Pankratz, T., *IDA Desalination Yearbook 2011–2012*, Media Analytics, Oxford, 2011.
3. Fritzmann, C., J. Löwenberg, T. Wintgens, T. Melin, State-of-the-art of reverse osmosis desalination, *Desalination,* **216,** 2007, 1–76.
4. Charcosset, C., A review on membrane processes and renewable energies for desalination, *Desalination*, **245,** 2009, 214–231.
5. Maurel, A., *Dessalement de l'eau de mer et des eaux saumâtres*, 2nd edition, TEC & DOC, Lavoisier, 2006, 284 p.
6. Gaid, K., Y. Treal, Le dessalement des eaux par osmose inverse: l'expérience de Véolia Water, *Desalination*, **203**, 2007, 1–14.
7. Leparc, J., S. Rapenne, C. Courties, P. Lebaron, J.-P. Croué, V. Jacquemet, G. Turner, Water quality and performance evaluation at seawater reverse osmosis plants through the use of advanced analytical tools, *Desalination,* **203,** 2007, 243–255.
8. Lauer, W.C. (editor) *Desalination of Seawater and Brackish Water*, AWW Trend Series, 1st edition, 2006, 568 p., ISBN 1-58321-328-7.
9. World Health Organisation (WHO), Guidelines for drinking-water quality, 4th edition, 2011 (http://www.who.int/en/).
10. Hilal, N., G.J. Kim, C. Somerfield, Boron removal from saline water: A comprehensive review, *Desalination,* **273,** 2011, 23–35.
11. Jeong, S., S.J. Kim, C.M. Kim, S. Vigneswaran, T.V. Nguyen, H.-K. Shon, J. Kandasamy, I.S. Kim, A detailed organic matter characterization of pretreated seawater using low pressure microfiltration hybrid systems, *Journal of Membrane Science,* **428,** 2013, 290–300.
12. Drioli, E., F. Laganà, A. Criscuoli, G. Barbieri, Integrated membrane operations in desalination processes. *Desalination,* **122,** 1999, 141–145.
13. Mericq, J.-P., S. Laborie, C. Cabassud, Vacuum membrane distillation of seawater reverse osmosis brines. *Water Research,* **44,** 2010, 5260–5273.

14. Okabe, T., M. Hirai, K. Ohta, T. Ishii, T. Goto, Preliminary experiment on seawater pretreatment with UF membranes for RO desalination. *Desalination,* **93,** 1993, 235–244.
15. Kumar, M., S. Adham, W. Pearce, Investigation of seawater reverse osmosis fouling and its relationship to pre-treatment type. *Environmental Science and Technology,* **40,** 2006, 2037–2044.
16. Pearce, G.K., S. Talo, K. Chida, A. Basha, A. Gulamhusein, Pre-treatment options for large scale SWRO plants: Case study of UF trials at Kindasa, Saudi Arabia, and conventional pre-treatment in Spain. *Desalination,* **167,** 2004, 175–189.
17. Xu, J., G. Ruan, X. Chu, Y. Yao, B. Su, C. Gao, A pilot study of UF pretreatment without any chemicals for SWRO desalination in China, *Desalination,* **207,** 2007, 216–226.
18. Corral, A.F., U. Yenal, R. Strickle, D. Yan, E. Holler, C. Hill, W.P. Ela, R.G. Arnold, Comparison of slow sand filtration and microfiltration as pretreatments for inland desalination via reverse osmosis. *Desalination,* **334,** 2014, 1–9.
19. Castaing, J.-B., A. Massé, M. Pontié, V. Séchet, J. Haure, P. Jaouen, Investigating submerged ultrafiltration (UF) and microfiltration (MF) membranes for seawater pre-treatment dedicated to total removal of undesirable micro-algae. *Desalination,* **253,** 2010, 71–77.
20. Meon, B., D. Kirchman, Dynamics and molecular composition of dissolved organic material during experimental phytoplankton blooms. *Marine Chemistry,* **75,** 2001, 185–199.
21. Castaing, J.-B., A. Massé, M. Pontié, V. Séchet, J. Haure, P. Jaouen, Immersed hollow fibres microfiltration (MF) for removing undesirable micro-algae and protecting semi-closed aquaculture basins. *Desalination,* **276,** 2011, 386–396.
22. Conlon, W., Pilot field test data for prototype ultra low pressure reverse osmosis elements. *Desalination,* **56,** 1985, 203–226.
23. Vince, F., E. Aoustin, P. Bréant, F. Maréchal, LCA tool for the environmental evaluation of potable water production. *Desalination,* **220**, 2008, 37–56.
24. Drioli, E., E. Curcio, G. Di Profio, F. Macedonio, A. Criscuoli, Integrating membrane contactors technology and pressure-driven membrane operations for seawater desalination, energy, exergy and costs analysis. *Chemical Engineering Research and Design,* 84, 2006, 209–220.
25. Song, Y., X. Gao, C. Gao, Evaluation of scaling potential in a pilot-scale NF–SWRO integrated seawater desalination system. *Journal of Membrane Science,* **443,** 2013, 201–209.
26. Al-Amoudi, A., A.M. Farooque, Performance restoration and autopsy of NF membranes used in seawater pretreatment. *Desalination,* **178,** 2005, 261–271.
27. Hassan, A.M., A.M. Farooque, A.T.M. Jamaluddin, A.S. Al-Amoudi, M.AK. Al-Sofi, A.F. Al-Rubaian, N.M. Kither, I.A.R. Al-Tisan, A. Rowaili, A demonstration plant based on the new NF-SWRO process. *Desalination,* **131,** 2000, 157–171.
28. Pontié, M., J.S. Derauw, S. Plantier, L. Edouard, L. Bailly, Seawater desalination: Nanofiltration-a substitute for reverse osmosis? *Desalination and Water Treatment,* **51,** 2013, 485–494.
29. Pontié, M., H. Dach, A. Lhassani, C. Kéba, C.K. Diawara, Water defluoridation using nanofiltration vs. reverse osmosis: the first world unit, Thiadiaye (Senegal). *Desalination and Water Treatment,* **51,** 2013, 164–168.

30. Ravindran, V., H.-H. Tsai, M.D. Williams, M. Pirbazari, Hybrid membrane bioreactor technology for small water treatment utilities: Process evaluation and primordial considerations. *Journal of Membrane Science*, **344,** 2009, 39–54.
31. Al Abdulgader, H., V. Kochkodan, N. Hilal, Hybrid ion exchange–Pressure driven membrane processes in water treatment: A review, *Separation and Purification Technology,* **116,** 2013, 253–264.
32. Han, S.C., K.H. Choo, S.J. Choi, M.M. Benjamin, Modeling manganese removal in chelating polymer-assisted membrane separation systems for water treatment, *Journal of Membrane Science,* **290,** 2007, 55–61.
33. Yao, P., K.H. Choo, M.H. Kim, A hybridized photocatalysis—-microfiltration system with iron oxide-coated membranes for the removal of natural organic matter in water treatment: Effects of iron oxide layers and colloids. *Water Research,* **43,** 2009, 4238–4248.
34. Leiknes, T., H. Ødegaard, H. Myklebust, Removal of natural organic matter (NOM) in drinking water treatment by coagulation–microfiltration using metal membranes. *Journal of Membrane Science,* **242,** 2004, 47–55.
35. Stoquart, C., P. Servais, P.R. Bérubé, B. Barbeau, Hybrid Membrane Processes using activated carbon treatment for drinking water: A review. *Journal of Membrane Science,* **411- 412,** 2012, 1–12.
36. Baudin, I., M.R. Chevalier, C. Anselme, S. Cornu, J.M. Laîné, L'Apié and Vigneux case studies: First months of operation. *Desalination,* **113,** 1997, 265–271.
37. Tomaszewska, M., S. Mozia, Removal of organic matter from water by PAC/UF system. *Water Research,* **36,** 2002, 4137–4143.
38. Mozia, S., M. Tomaszewska, A.W. Morawski, Studies on the effect of humic acids and phenol on adsorption—-ultrafiltration process performance. *Water Research,* **39,** 2005, 501–509.
39. Chang, Y.J., K.H. Choo, M.M. Benjamin, S. Reiber, Combined adsorption—-UF process increases TOC removal. *J. AWWA,* **90,** 1998, 90–102.
40. Lee, K.W., K.H. Choo, S.J. Choi, K. Yamamoto, Development of an integrated iron oxide adsorption/membrane separation system for water treatment, water science and technology. *Water Supply,* **2,** 2002, 293–300.
41. Kang, S.K., K.H. Choo, Why does a mineral oxide adsorbent control fouling better than powdered activated carbon in hybrid ultrafiltration water treatment? *Journal of Membrane Science,* **355,** 2010, 69–77.
42. Wang, S., C. Liu, Q. Li, Impact of polymer flocculants on coagulation microfiltration of surface water. *Water Research,* **4,** 2013, 4538–4546.
43. Le-Clech, P., E.-K. Lee, V. Chen, Hybrid photocatalysis/membrane treatment for surface waters containing low concentrations of natural organic matters. *Water Research,* **40,** 2006, 323–330.
44. Song, H., J. Shao, Y. He, B. Liu, X. Zhong, Natural organic matter removal and flux decline with PEG–TiO_2-doped PVDF membranes by integration of ultrafiltration with photocatalysis, *Journal of Membrane Science*, 405–406, 2012, 48–56.
45. Park, H., Y. Kim, B. An, H. Choi, Characterization of natural organic matter treated by iron oxide nanoparticle incorporated ceramic membrane ozonation process. *Water Research,* **46,** 2012, 5861–5870.

46. Ben-Sasson, M., Y. Zidon, R. Calvo, A. Adin, Enhanced removal of natural organic matter by hybrid process of electrocoagulation and dead-end microfiltration. *Chemical Engineering Journal,* **232,** 2013, 338–345.
47. Oehmen, A., R. Valerio, J. Llanos, J. Fradinho, S. Serra, M.A.M. Reis, J.G. Crespo, S. Velizarov, Arsenic removal from drinking water through a hybrid ion exchange membrane—Coagulation process. *Separation and Purification Technology,* **83,** 2011, 137–143.
48. Pal, P., S. Chakrabortty, L. Linnanen, A nanofiltration—Coagulation integrated system for separation and stabilization of arsenic from groundwater. *Science of the Total Environment,* **476–477,** 2014, 601–610.
49. Park, H., H. Choi, As(III) removal by hybrid reactive membrane process combined with ozonation. *Water Research,* **45,** 2011, 1933–1940.
50. Cui, M., M. Jang, S. Ibrahim, B. Park, E. Cho, J. Khim, Arsenite removal using a pilot system of ultrasound and ultraviolet followed by microfiltration. *Ultrasonics Sonochemistry,* **21,** 2014, 1527–1534.
51. Wickramasinghe, S.R., B. Han, J. Zimbron, Z. Shen, M.N. Karim, Arsenic removal by coagulation and filtration: comparison of groundwaters from the United States and Bangladesh. *Desalination,* **169,** 2004, 231–244.
52. Mólgora, C.C., A.M. Domínguez, E.M. Avila, P. Drogui, G. Buelna, Removal of arsenic from drinking water: A comparative study between electrocoagulation-microfiltration and chemical coagulation-microfiltration processes. *Separation and Purification Technology,* **118,** 2013, 645–651.
53. Sen, M., A. Manna, P. Pal, Removal of arsenic from contaminated groundwater by membrane-integrated hybrid treatment system. *Journal of Membrane Science,* **354,** 2010, 108–113.
54. Chaturvedi, S., P.N. Dave, Removal of iron for safe drinking water. *Desalination,* **303,** 2012, 1–11.
55. El Azher, N., B. Gourich, C. Vial, M. Belhaj Soulami, M. Ziyad, Study of ferrous iron oxidation in Morocco drinking water in an airlift reactor. *Chemical Engineering Processing,* **47,** 2008, 1877–1886.
56. Ellis, D., C. Bouchard, G. Lantagne, Removal of iron and manganese from groundwater by oxidation and microfiltration. *Desalination,* **130,** 2000, 255–264.
57. Setyadhi, L., J.C. Liu, Oxidation-microfiltration removal of Fe(II) from water. *Desalination and Water Treatment,* **51,** 2013, 374–383.
58. Tufekei, N., G. Demir, Catalytic effects of Fe(III) during oxygenation of Fe(II) in continuous flow iron removal systems. *Fresenius Environmental Bulletin,* **11,** 2002, 863–867.
59. Ghosh, D., M.K. Sinha, M.K. Purkait, A comparative analysis of low-cost ceramic membrane preparation for effective fluoride removal using hybrid technique. *Desalination,* **327,** 2013, 2–13.
60. Mohapatra, M., S. Anand, B.K. Mishra, D.E. Giles, P. Singh, Review of fluoride removal from drinking water. *Journal of Environmental Management,* **91,** 2009, 67–77.
61. Pontié, M., M. Rumeau, M. Ndiaye, C.M. Diop, Sur le problème de la fluorose au Sénégal: bilan des connaissances et présentation d'une nouvelle méthode de défluoridation des eaux de boisson. *Cahiers Santé,* **6,** 1996, 27–36.

62. Hu, K., J.M. Dickson, Nanofiltration membrane performance on fluoride removal from water. *Journal of Membrane Science,* **279,** 2006, 529–538.
63. Ben Nasr, A., K. Walha, C. Charcosset, R. Ben Amar, Defluoridation of water by nanofiltration. *Journal of Fluorine Chemistry,* **150,** 2013, 92–97.
64. Pontié, M., H. Dach, J. Leparc, M. Hafsi, A. Lhassani, Novel approach combining physico-chemical characterizations and mass transfer modelling of nanofiltration and low pressure reverse osmosis membranes for brackish water desalination intensification. *Desalination,* **221,** 2008, 174–191.
65. Ben Nasr, A., K. Walha, F. Puel, D. Mangin, R. Ben Amar, C. Charcosset, Precipitation and adsorption during fluoride removal from water by calcite in the presence of acetic acid. *Desalination and Water Treatment,* **52,** 2014, 2231–2240.

8

Wastewater Treatment Using Integrated Membrane Processes

Jinsong Zhang and Anthony G. Fane
Singapore Membrane Technology Centre, Nanyang Technological University, Singapore

8.1 Introduction

The integrated membrane system (IMS) discussed in this chapter refers to a system that adopts two or more membrane-based technologies. The IMS is used as separation unit in the chemical industry, pharmaceutical industry, and petroleum industry to a large extent. In the water industry, it has been well accepted for purifying a water stream to potable level, such as seawater desalination and drinking water production. However, the IMS is not widely applied in wastewater treatment as it is not regarded as cost-effective in view of secondary effluent disposal. Until recently, there has been a paradigm shift in wastewater management which has changed the fate of secondary effluent from disposal to recycle and reuse. This change is driven by the global water crisis where the traditional water resources are under constant threat due to pollution, population growth, climate change, and so on. As such, non-traditional sources, namely seawater and wastewater, have to be exploited (Fane, 2007; Bennett, 2009). In view of urban development, wastewater management deserves greater emphasis. The conventional wastewater management system is a linear treatment process based on disposal; it has to be adapted into a sustainable, closed-loop system for the purpose of conservation, recycle, and reuse. In this context, water reclamation, rather than wastewater treatment, has become a more common term in recent years, especially in water-scarce countries and some water-intensive industries where there is a need to reduce the "water footprint."

Integrated Membrane Systems and Processes, First Edition. Edited by Angelo Basile and Catherine Charcosset.

Membrane technology is a widely adopted technology for water reclamation. The most popular application is the membrane bioreactor (MBR) (Judd, 2008, 2011). MBR is a combination of the biological degradation process and membrane filtration, where the biological removal of carbon and nutrients happens concurrently with solid–liquid separation by microfiltration (MF) or ultrafiltration (UF) membranes in the same container. Compared with conventional wastewater treatment process, MBRs have many advantages including reduced footprint, lower waste sludge production, and improved product water quality. Despite of the improved water quality of MBR permeate, it is still far from fulfilling the expectations for reuse in many cases. It is known that a single membrane process may not be able to manipulate complex matters in wastewaters, which comprise suspend solids, dissolved organic matters, numerous pathogenic microorganisms, as well as nutrients and toxic compounds generated by human activities and/or industrial discharge. Therefore, the membrane-based processes for water reclamation need to be integrated to achieve desired permeate water quality, to reduce energy and chemical inputs, and to minimize wastes.

The design of IMS for water reclamation is usually on a case-by-case basis. The rule of thumb is to understand the sequence of complementary membrane processes (Table 8.1). Normally, loose membranes (MF or UF) are usually located prior to tight membranes (nanofiltration (NF) or reverse osmosis (RO)) to achieve different water quality requirements. Forward osmosis (FO) and membrane distillation (MD), as leading-edge membrane technologies, could also be integrated into the system under certain conditions for special purposes. In the following sections of this chapter, examples and case studies of IMS involving various combinations of MF, UF, NF, RO, MD, and FO will be introduced and discussed.

8.2 IMS Application for Wastewater Treatment: Current Status

8.2.1 IMS for Textile Industrial Wastewater: Target to Zero Discharge

A case study of textile wastewater treatment provides a good example to understand IMS applications (Van der Bruggen et al., 2004, 2013). This is because textile wastewater is a typical industrial wastewater with heavy load of organic and inorganic pollutants (Calabro et al., 1991; Rozzi et al., 2000; Vishnu et al., 2008; Lau and Ismail, 2009; Vergili et al., 2012), which is posing a great threat to the environment in many developing countries.

Figure 8.1 shows the general design concept of IMS in textile industrial wastewater treatment. This design is based on a systematic fractionation of particulate matters, dissolved organics (dyes, additives), and inorganic contaminants in the waste stream. In this system, a dead-end MF is used as the pretreatment unit to remove the suspended solids (SS). The permeate then passes through a loose NF membrane (NF-A) at high temperature, where the major portion of organics are rejected, while the salts passing through. The retentate stream of NF-A is a warm, concentrated organic solution that is suitable for MD, where the temperature difference provides the driving force. In the MD unit, distilled water can be extracted from the retentate of NF-A and recycled as fresh water, while the concentrate organics will be discharged and finally incinerated.

In the meantime, the permeate of NF-A containing an elevated concentration of salt is sent to a tight NF membrane (NF-B) (or RO), where the remaining organic compounds and salt can be removed. The permeate of NF-B is fit for reuse as process water. On the other

Table 8.1 *Membrane process for wastewater treatment.*

Membrane process	Process driving force	Rejections to wastewater contained matters							
		Water	Monovalent ions	Multivalent ions	Low molecular weight organics	High molecular weight organics	Viruses	Bacteria	Suspended solids
MF	Low hydraulic pressure	✗	✗	✗	✗	✗	✗	✓	✓
UF	Low hydraulic pressure	✗	✗	✗	✗	✓	✓	✓	✓
NF	High hydraulic pressure	✗	✗	✓	✓	✓	✓	✓	✓
RO	High hydraulic pressure	✗	✓	✓	✓	✓	✓	✓	✓
MD	Water vapor pressure (via temperature difference)	✗	✓	✓	✓	✓	✓	✓	✓
FO	Osmotic pressure	✗	✓	✓	✓	✓	✓	✓	✓

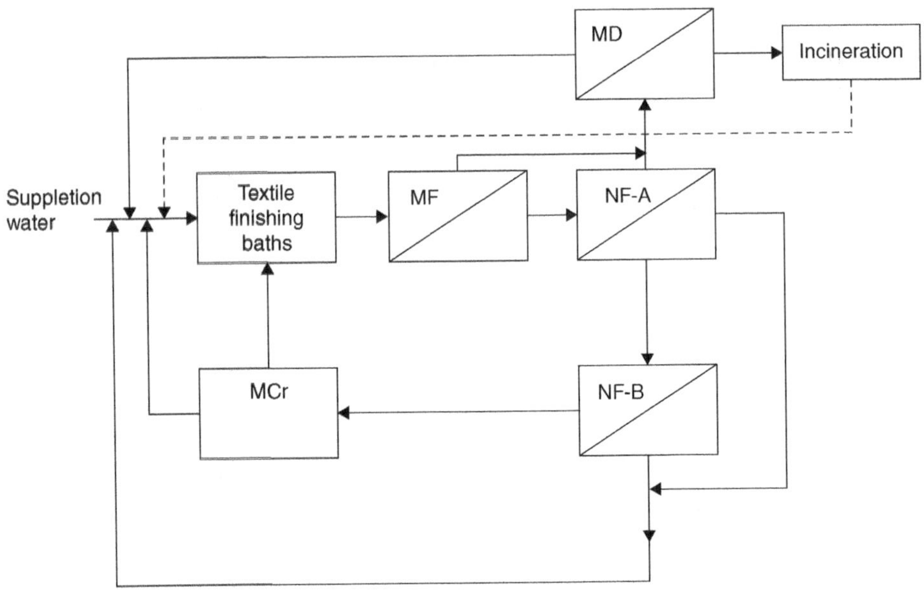

Figure 8.1 *Integrated membrane process for textile industrial treatment.* Source: *Reproduced from Van der Bruggen et al. (2004) with permission from Elsevier.*

hand, NF-B retentate contains a high concentration of salt that is comparable to the brine from desalination. The salts in the retentate can then be recovered as high-purity crystals in a membrane crystallizer (see Section 8.2.2), or being reused directly for new finishing baths with satisfactory purity.

With the application of integrated membrane processes, zero liquid discharge can be achieved for the treatment of textile dye bath waste streams (Vergili et al., 2012). Nevertheless, the success of the approach in practice would be largely determined by membrane fouling in the NF and the MD which could be a significant cost factor limiting the return on investment. In addition, membrane crystallization is also a challenge as a novel technology.

8.2.2 Integrated Pressure-Driven Membrane Process for Municipal Wastewater Reclamation

8.2.2.1 MF/UF-RO for Reclamation of Secondary Effluent of Municipal Wastewater

To tackle the challenges of water scarcity, the application of the dual membrane process to convert "used water" (secondary effluent) to potable water quality has seen dramatic growth in recent years. Since the development of the first asymmetric cellulose acetate RO membrane in the early 1960s by Loeb and Sourirajan, RO has been widely used to produce potable water from seawater, brackish water, and municipal wastewater. The IMS water reclamation process is a combination process, where secondary municipal effluent is pretreated by low pressure membranes (MF/UF) prior to RO. In the 1980s, this dual membrane approach was set up at Water Factory 21 in Orange County, and is now widely applied in many plants around the world (Fane and Leslie, 2004; Fane et al., 2011).

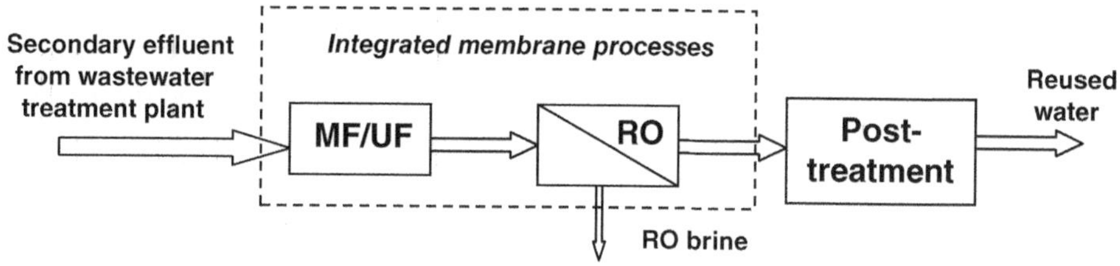

Figure 8.2 Schematic diagram of integrated MF/UF-RO membrane process for wastewater reclamation.

One of the challenges in integrated membrane processes is the need to produce high quality product water with high water recovery and sustainable operation at low energy consumption. Compared to 50% recovery of seawater desalination, the typical recovery of wastewater reclamation plants with IMS is 75%. The new wastewater reclamation plants designed in the United States target a recovery of 85% (Wetterau et al., 2011).

Figure 8.2 shows a typical schematic flow diagram of the membrane process that is commonly used in reclamation of municipal wastewater. As illustrated, the secondary effluent from an activated sludge treatment process passes through a series of steps involving MF/UF, RO, and post-treatment (typically ultraviolet). In this process, RO membrane plays an important role in producing high quality permeate water as it retains salts and other components.

8.2.2.2 MF/UF-RO Wastewater Reclamation Plant Performance Worldwide

According to a recent survey of global wastewater reclamation plants (Raffin et al., 2013), the United States, Australia, and Singapore are the top three countries for wastewater reclamation plants in the world. Among them, the United States and Singapore produce around 60% of the world's reclaimed water production. Table 8.2 shows some detailed information of nine plants with MF/UF-RO membrane processes. The design capacities of these nine plants are from 1.6 to 375 MLD, and plant locations are across the United

Table 8.2 The worldwide dual-membrane-based wastewater reclamation plants.

Site	Location	MF/UF supplier	RO supplier	Capacity (MLD)
Plant A	UK	Asahi Kasei/Pall	Koch	1.6
Plant B	Australia	Asahi Kasei/Pall	Toray	66
Plant C	USA	Asahi Kasei/Pall	Hydranautics	11.4
Plant D	Singapore	Asahi Kasei/Pall	Hydranautics	191
Plant E	Singapore	NI[a]	NI[a]	>200
Plant F	Spain	GE/Zenon	Dow	15
Plant G	Kuwait	Pentair/Norit	Toray	375
Plant H	Belgium	GE/Zenon	Dow	6.9
Plant I	USA	Siemens/Memcor	Hydranautics	265

Source: Reproduced from Raffin et al. (2013) with permission from Taylor & Francis.
[a] NI, not identified.

States, Europe, the Middle East, Southeast Asia, and Australia. The largest MF/UF-RO based wastewater reclamation plant is in Kuwait (Plant G) at a capacity of 375 MLD.

The product water from MF/UF-RO membrane reclamation plant is used for industrial process water, indirect potable reuse via a reservoir and groundwater recharge, and non-potable reuse. The reported overall specific energy demand (SED) of MF/UF-RO for wastewater reclamation (energy consumption per permeate volume) ranges from 0.8 to 2.3 kWh/m^3 which is about 25–70% of seawater desalination (Sauvet-Goichon, 2007; Raffin et al., 2013). Therefore, wastewater reclamation by the integrated membrane process is a sustainable option compared with seawater desalination, especially in terms of energy use and cost in urban areas.

To the integrated membrane wastewater reclamation plants, membrane fouling is a major challenge. Potential foulants for the IMS include turbidity (colloids), organics (more typical in wastewater), bacteria (leading to biofouling), and sparingly soluble salts (scale formers). The membrane fouling issue can be mitigated by applying effective pretreatment techniques. Pretreatment system includes some of the following technologies, for example, media filters, low pressure membranes, and scale inhibitors.

An effective pretreatment method should be able to

(i) improve water quality indices, such as Silt Density Index (SDI) or Modified Fouling Index (MFI);
(ii) improve water quality based on species removal (SS, total organic carbon (TOC), etc.);
(iii) have a demonstrable effect on fouling reduction and energy saving;
(iv) be relatively inexpensive, providing a clear cost benefit; and
(v) be flexible and responsive to varying loads.

MF/UF membranes are regarded as a pretreatment step for RO and are widely used in the integrated membrane process. All the large-scale water reclamation plants mentioned earlier use MF/UF modules, for example, hollow fiber (HF) modules from Pall/Asahi Kasei, Siemens/Memcor, and GE/Zenon and capillary tube (CT) modules from Pentair/Norit. The pretreatment membranes of Plants A to D (Table 8.3) are categorized as MF membranes (pore size at 0.1 μm and above), while the others are considered as "loose" UF membranes.

The design fluxes and permeabilities of MF/UF membranes are shown in Table 8.4. After temperature correction, the average flux of MF/UF ranges from 31 to 75 LMH. Significant difference was found in the corresponding permeabilities (flux/TMP ratio). Plants A, C, and D use similar Asahi Kasei/Pall HF membranes; however, the permeability of Plant A is only half of Plants C and D. Interestingly, similar phenomenon is also found in Plant I where same Siemens/Memcor HF membranes are applied. This observation indicates that the fouling propensity of the feed water is also a key factor which should be considered in the MF/UF pretreatment design. The feed water quality is an important factor that affects membrane fouling propensity, and consequently determines the operating flux of the membrane. In low pressure MF/RO, the flux-driven membrane fouling and the trade-off between capital (a function of flux) and operational cost (a function of energy and chemical usage) lead to considerations of threshold and sustainable flux.

MF/UF membranes used for pretreatment are usually in HF format to achieve high packing density. In addition, backwashing can be applied in HF and CT membrane system to mitigate the fouling (Table 8.5). Table 8.5 shows the cleaning parameters of MF/UF membrane in the nine wastewater reclamation plants. All plants apply water backwash,

Table 8.3 *The membrane applied in membrane-based wastewater reclamation plants.*

Site	Plant A	Plant B	Plant C	Plant D	Plant E	Plant F	Plant G	Plant H	Plant I
MICRO-/ULTRA-FILTRATION									
Membrane supplier	Asahi Kasei/Pall	Asahi Kasei/Pall	Asahi Kasei/Pall	Asahi Kasei/Pall	NI[b]	GE/Zenon	Pentair/Norit	GE/Zenon	Siemens/Memcor
Configuration	HF	HF	HF	HF	HF	HF	CT	HF	HF
Membrane type	Pressurized	Pressurized	Pressurized	Pressurized	Pressurized	Immersed	Pressurized	Immersed	Immersed
Pore size (μm)	0.1	0.2	0.1	0.1	0.04	0.02	0.03	0.02	0.04
Membrane area (m^2)	1700	85,400	3200	160,000	Approx. 250,000[a]	5016	304,640	15,600	730,000
REVERSE OSMOSIS									
Membrane supplier	Koch	Toray	Hydranautics	Hydranautics	NI[b]	Dow	Toray	Dow	Hydranautics
Membrane area (m^2)	1451	4000	43,200	427,000	Approx. 365,000[a]	2433	—	4002	580,000
Elements	6	7	7	7	7[a]	6	—	6	7
No. of stages	2	3	2	2	2	2	—	2	3

Source: Reproduced from Raffin et al. (2013) with permission from Taylor & Francis.
[a]Not in the reference, obtained from plant data.
[b]Not identified.

Table 8.4 *The MF/UF membrane flux, permeability in membrane-based wastewater reclamation plants.*

Site	Plant A	Plant B	Plant C	Plant D	Plant E	Plant F	Plant G	Plant H	Plant I	Avg.	SD
Operating flux (LMH)	35	65	60	44	47	28	65	40	33	**46**	**14.0**
Temperature (°C)	16	24	26	28	28	23	25	15	24	**23**	**4.7**
Temperature-corrected flux (LMH)	31	73	72	56	60	31	75	35	37	**52**	**19**
TMP (bar)	0.26	0.3	0.26	0.24	0.8	0.28	0.38	0.3	0.25	**0.3**	**0.2**
Permeability(LMH/bar)	120	244	276	232	74	109	198.3	115	149	**169**	**71**

Source: Reproduced from Raffin et al. (2013) with permission from Taylor & Francis.

Table 8.5 *The MF/UF membrane cleaning in membrane-based wastewater reclamation.*

Site	Plant A	Plant B	Plant C	Plant D	Plant E	Plant F	Plant G	Plant H	Plant I
MICRO-/ULTRA-FILTRATION									
Backwash frequency (min)	12	20	20 to 30	30	30	38	25	8 (winter) 10 (summer)	22
Backwash water flux (m^3/h)	200	250	250	250	300	4, 3	250	100	170
Backwash air (m^3/h)						91	N/A	–	
CIP frequency	1–2 per month	1 per month	6 per year	1 per year	1 per month	1 per month (1) 6 per year (2)	1 per year	1per month	21 days
CEB frequency per day or month	2 per month	1 per day (1) 1 per week (2)	–	1–3 per day	1 per day	–	1 per day	After 25–35 BW cycles	–
Chemical dosing for CIP	Alkaline NaOCl	NaOCl	Citric acid/ NaOH	Citric acid	NaOCl	NaOCl (1) Citric acid (2)	Oxalic acid	NaOCl	NaOH + Memclean/ citric acid
Chemical dosing for CEB	NaOCl	NaOCl (1) Citric acid (2)	–	NaOCl/ citric acid	Citric acid	–	Cl_2 + NaOH	NaOCl	–
REVERSE OSMOSIS									
Chemical dosing	Sulfuric	No	No	H_2SO_4	H_2SO_4	$NaHSO_3$	–	H_2SO_4	H_2SO_4
Anti-scalant dosing	Accepta 2651		Pretreat Plus100		–		–	–	AWC A-102 Plus
CIP frequency	6 months	6 months	6 per year	–	–	6 per month	–	4–5 per year	Every 6 months
Chemical dosing for CIP	Accepta 2068	HCl	Citric acid	Citric acid	–		–	NaOH	STPP

Source: Reproduced from Raffin et al. (2013) with permission from Taylor & Francis.

while seven of them also apply air to scour the feed side of the membrane during the water backwash. Backwash intervals vary between 8 and 38 minutes and depend on the feed water quality, membrane module design, as well as imposed flux in the different plants. However, backwashing can only partially remove reversible fouling; irreversible fouling has to be removed by chemically enhanced backwash (CEB) and thoroughly chemical cleaning in place (CIP). The frequency of CIP or CEB is normally once per 1 or 2 months and is strongly correlated to the feed water quality and operation flux. Sodium hypochlorite, NaOH, and citric acids are the chemicals extensively used for CIP and CEB. The frequent cleaning by corrosive and high pH solutions requires chemical resistant membrane materials, for example, polyvinylidene fluoride (PVDF). Both Asahi and Siemens membranes are made of PVDF.

As aforementioned, membrane fouling is a significant issue in the MF/UF-RO process. The lowest water permeability of UF occurred at Plant E, and was 74 LMH/bar, indicting serious fouling. A fouled membrane sample from Plant E was collected and autopsied. In order to determine the possible causes of membrane fouling, chemical cleanings were conducted and the results are shown in Figure 8.3.

Membrane surface cleaning by water flush and 500 mg/L sodium hypochlorite (NaClO) solution was applied to remove the fouling layer. The cross-flow velocity was controlled at 0.1 m/s. In this study, the permeability of the fouled membrane sample was 88 LMH/bar (Figure 8.3), which is slightly higher than 74 LMH/bar due to the sampling point and time. After 70 minutes of clean water flushing, the permeability recovery is 0. After cleaning with NaClO solution, the permeability of the membrane recovered to 119 LMH/bar within 40 minutes. With the increase of NaClO cleaning duration from 40 minutes to 14 hours, the permeability of the membrane increased from 119 to 184 LMH/bar. Another 4.5 hours of acid (pH 2.54) cleaning was carried 14 hours after NaClO cleaning. The results showed that the permeability of the fouled membrane further increased to 196 LMH/bar. The results shown in Figure 8.3 confirmed that the NaClO solution flushing was able to remove the organic foulants from membrane surface and pores effectively. In summary, the organic fouling contribution to the membrane permeability loss was about 96 LMH/bar, which

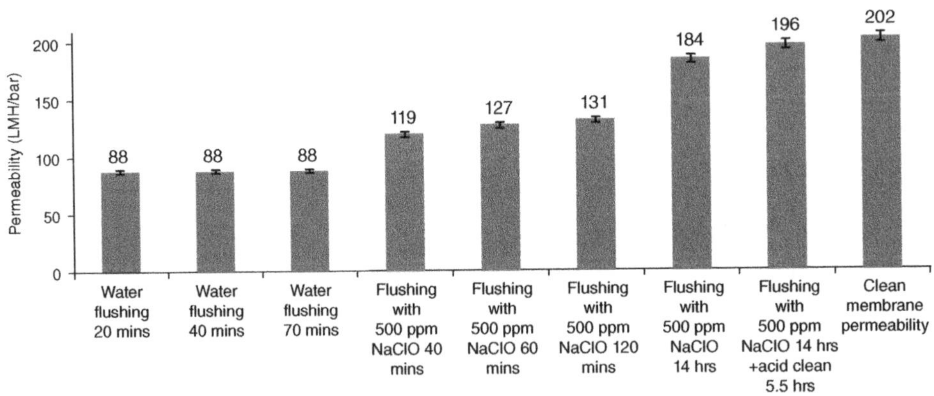

Figure 8.3 *Permeability of fouled membrane after sodium hypochlorite solution and acid clean.*

was about 48% of the clean water permeability. The inorganic fouling contribution to the permeability loss was around 14 LMH/bar, which was around 6% of the clean water permeability.

Table 8.6 illustrates the soluble organics contained in the feed and MF permeates of Plant E. It should be noted that the results are a one-off analysis for reference. The TOC concentrations in the feed and MF permeates were measured to be 21.9 mg/L and 16.8 mg/L, respectively, which suggested a 23% rejection. The concentration of the EPS, present as biopolymers at molecular weights above 20,000 Da, was around 14% of the total dissolved organic matters; the EPS rejection by the membrane was 58%, indicating a significant EPS removal by the MF membrane which was functioning as a UF membrane due to pore closure by fouling layer. In another words, the EPS was a key foulant for the MF filtration. As shown in Table 8.6, the concentration of humic substance at a molecular weight of 1000 Da was approximately 11% of the total dissolved organics, and the concentration of humic substances at a molecular weight of 350–500 Da and below 350 Da was 75% of the total dissolved organics. The average membrane rejection for the humic substance at a molecular weight of 1000 Da, 350–500 Da, and below 350 Da was 18%, which is lower than the rejection for the total dissolved organic carbon.

Figure 8.4 shows the SEM images of clean and fouled membrane surfaces obtained from Plant E. The surface of fouled membrane is found smoother and less porous than the clean membrane. This is because of the deposition of foulants on the membrane surface over the pores.

Figure 8.5 shows the results of inorganic constituent analysis of the membrane fouling layer (Plant E). Significant accumulations of iron, calcium, and aluminum in the membrane fouling layer are found at 3090 mg/m^2, 1063 mg/m^2, and 570 mg/m^2, respectively. No substantial accumulation of magnesium, calcium, manganese, and silicon is found. This is an interesting finding, as the pore size of the UF/MF membrane may not be able to effectively remove multivalent ions by size exclusion; the apparent rejection rates of aluminum, iron, and calcium were likely due to the formation of metal organic complexes with the humic acid-like or fulvic acid-like components in the feed water. It should also be noted that calcium–humate complexes could cause the highest flux decline due to their highly compactable floc-like structures).

8.2.2.2.1 RO Fouling and Wastewater Recovery

A significant portion of the dissolved inorganic and organic substances, such as silica, calcium, iron oxide, humic substances, and organic macromolecules, will pass through MF/UF pretreatment processes and can easily deposit onto the RO membrane surface causing membrane fouling. While most inorganic fouling can be well controlled by dosing anti-scalants in the feed stream, it is more difficult to control biological/organic fouling, which is probably the most prevalent fouling in the RO membranes for wastewater reclamation. For example, 150 RO membrane autopsy results from different plants showed that more than 50% of foulants were biological/organic materials (van Hoof et al., 2002). The formation of the fouling layer on the membrane surface will provide an additional hydraulic resistance to permeation. In addition, the solutes (e.g., sodium and chloride ions) have to diffuse through the tortuous path in the cake layer, so the effective back diffusion is greatly hindered. This phenomenon is known as the cake-enhanced osmotic pressure

Table 8.6 *LC-OCD analysis results for the feed and permeate of Plant E.*

	DOC	HOC	CDOC	>>20.000				approximate molecular weights (g/mol): ~1000 (see separate HS diagram)						300–500	<350	<350		
	Dissolved	Hydrophobic	Hydrophilic	BIOpolymers	DON (Norg)	N/C	% Proteins in BIOpol.*	Humic subst. (HS)	DON (Norg)	N/C	Aromaticity (SUVA-HS)	Mol-Weight ($\overline{Mn}$)	Position in HS diagram	Building blocks	LMW neutrals	LMW acids	Inorganic colloids SAC	SUVA (SAC/DOC)
	ppb-C	ppb-C	ppb-C	ppb-C	ppb-N	μg/μg	% BIOpol.	ppb-C	ppb-N	μg/μg	L/(mg m)	g/mol	--	ppb-C	ppb-C	ppb-C	(m^{-1})	L/(mg m)
	% DOC	% DOC	% DOC	% DOC	--	--	--	% DOC	--	--	--	--	--	% DOC	% DOC	% DOC	--	--
MF permeate	16,781	2705	14,076	1271	9	0.01	2	2389	144	0.06	2.92	414	A	1989	8313	114	1.10	1.25
	100%	16.1%	83.9%	7.6%	--	--	--	14.2%	--	--	--	--	--	11.9%	49.5%	0.7%	--	--
Wastewater feed	21,877	2889	18,988	3040	264	0.09	26	2518	140	0.06	2.83	422	B	2128	11206	96	1.84	1.07
	100%	13.2%	86.8%	13.9%	--	--	--	11.5%	--	--	--	--	--	9.7%	51.2%	0.4%	--	--

LMW = low-molecular weight

DON = dissolved organic nitrogen

n.q. = not quantifiable (< 1 ppb calculated)

n.m. = not measured

"Biopolymers" = polysaccharides, proteins, aminosugars

"Building Blocks" = breakdown products of humics

"Neutrals" = mono-oligosaccharides, alcohols, aldehydes, ketones

"Acids" = summaric value for monoprotic organic acids < 350 Da

*Under the presumption that all org. N in the BIOpolymer fraction originates from proteins.

HS diagram refers to separate humic substances plot by DOC Labor.

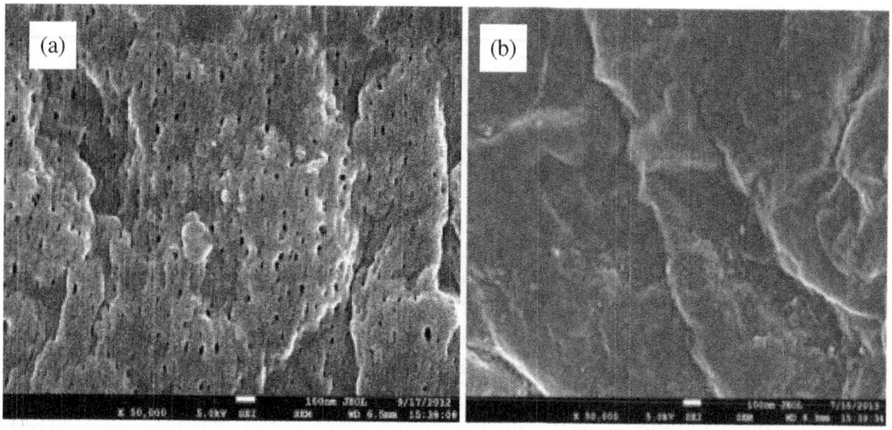

Figure 8.4 SEM micrograph of membrane surface. (a) Clean membrane. (b) Fouled membrane (Plant E).

(CEOP) effect (Hoek and Elimelech, 2003) and the increase in osmotic pressure means a loss in the effective driving force for filtration. Thus, the resistance of the membrane is determined by the interplay between the concentration polarization and the deposit layer. Strategies such as using biocides to eliminate the microorganisms and using cleaners to remove the biofilm are usually applied to control biofouling (McDonogh et al., 1994).

The wastewater recovery (portion of the feed recovered as permeate) is also a challenge to the current IMS in wastewater reclamation. Recovery is mainly limited by the potential fouling of the RO membranes, which includes inorganic scaling, organic fouling, and biofouling (Wetterau et al., 2011). High recovery can create conditions whereby the inorganic components are supersaturated. Preliminary analysis of the inorganic species over a range of recoveries suggests significant risks due to scalants (Ca^{2+}, Ba^{2+}, CO_3^{2-}, SO_4^{2-}, F^-,

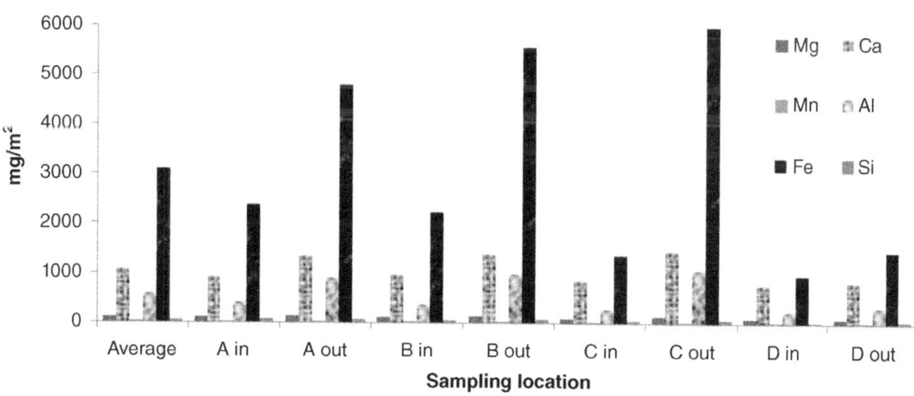

Figure 8.5 Accumulation of cations in the fouling layers of UF membrane from Plant E. (A, B, C, D indicate sampling points from different positions along the membrane fiber; subscripts "in" and "out" indicate inner and outer regions of the membrane fiber bundle in the module.) Cation bar charts: 1st Mg, 2nd Ca, 3rd Mn, 4th Al, 5th Fe, 6th Si.

Table 8.7 *Fouling control strategies of RO process in wastewater reclamation.*

Fouling challenges	Potential solutions
Scale control	Anti-scalants and water chemistry (pH, etc.) Chemical or seeded demineralization (ICD) Scale monitors
Organics/biofouling control	Membrane selection and surface treatment Biostats and biocides Organics and nutrient (TOC) limitation

etc.). High recovery can also lead to a concentrated TOC, which will increase the risk of organic fouling and biofouling (Zhao et al., 2010; Ridgway et al., 1984; Griebe and Flemming, 1998). Fouling control is the key to successful and sustainable operation of IMS at high recoveries for wastewater reclamation. Table 8.7 summarizes the control options.

8.2.2.2.2 *Organics/Biofouling Control*

TOC species with low molecular weight (<5kD), such as humics, neutrals, and organic acids, will pass through the MF/UF membranes and present in the permeate. A small proportion of these organics are likely to be biodegradable dissolved organic carbon (BDOC) which will contribute to the development of biofilm. The major portion of the organics is relatively refractory (non-BDOC?) and will contribute to organic fouling of the membranes. Effective strategies are desired to remove both BDOC and non-BDOC in the feed water.

For biofouling, it becomes a problem when the extent of biofilm growth exceeds an operationally defined threshold of interference. One of the attractive options to control biofilms is to limit nutrients in the feed water (Flemming et al., 1996, 1997; Marconnet et al., 2009). Biological activated carbon (BAC) is an effective method to remove most BDOC species; it is a low energy step (biological degradation and low bed pressure drop) (Kalkan et al., 2011) and has been used for biofouling control for RO processes (Griebe and Flemming, 1998).

8.2.2.2.3 *Scale Control—Intermediate Concentrate Demineralization and Monitors*

Recent studies on integrating concentrate demineralization (ICD) in brackish water desalination have shown promising results in enhancing water recovery (Gabelich et al., 2007, 2011; Sanciolo et al., 2008). The ICD treatment can be carried out with or without seeding to remove the mineral scale precursors. The treatment technique greatly depends on the feed stream water quality. Precipitation softening (without seeding) can be an effective ICD if the water contains adequate total alkalinity $CaCO_3$, allowing precipitation without high chemical dosing. The seeded precipitation technique is effective for removing calcium-bearing scalants from the concentrate which is supersaturated with respect to the mineral scalants (Rahardianto et al., 2010). Direct, non-destructive techniques for scale monitoring have been developed in recent years. These techniques allow direct observation of foulant accumulation and its dynamics on membranes. Uchymiak et al. (2007) developed a device called ex-situ scale observation detector (EXSOD), which can achieve direct visual observation and real-time monitoring of mineral surface scaling. That is, by using EXSOD, membrane surface imaging under RO operation can be obtained by high-resolution digital photography using an optical microscope. This can provide direct visualization of membrane fouling. Other techniques for real-time scale monitoring are also available, such as ultrasonic time domain reflectometry (UTDR) (Sim et al., 2012).

8.2.2.3 MBR/RO for Reclamation of Municipal Wastewater

The MBR is a combination of processes involving physical separation using membranes and the biological process that removes contaminants such as organics, nitrogen, and phosphorus. In order to achieve required permeate water quality for desired applications, RO process is usually used to further polish the MBR permeate. The combination of RO process with MBR is also considered as an IMS for wastewater treatment and reuse. The MBR/RO process has been validated in different wastewater treatments, for example, pulp paper wastewater, landfill leachate, petrochemical wastewater, coking wastewater, and textile wastewater (Ahn et al., 2002; Simens, 2008; Takht Ravanchi et al., 2009; Zhang et al., 2009; Hoinkis et al., 2012; Insel et al., 2013). However, the most common application is still for municipal water reclamation (Comerton et al., 2005; Côté et al., 2005; Qin et al., 2006; Tam et al., 2007; Dialynas and Diamadopoulos, 2009; Jacob et al., 2010; Joss et al., 2011; Kent et al., 2011; Sahar et al., 2011; Wu et al., 2013).

Figure 8.6 shows the working scheme of the MBR/RO integrated process for water reclamation. MBR/RO IMS has many advantages over the conventional wastewater treatment process, such as smaller footprint, higher loading rate, lower sludge production, and simplified process. A higher water recovery of up to 90% can be achieved, compared to the 75% of the UF/MF-RO process (Joss et al., 2011). A pilot testing of an MBR/RO system for reclamation of the domestic sewage was conducted at a capacity of 20 m^3 per day in Singapore in 2004. For comparison, an activated sludge process (ASP) plus MF/RO process was also operated in parallel. The new MBR/RO process demonstrated the capability of producing comparable or better (in terms of TOC) product water quality than that of the conventional ASP-MF-RO process in reclamation of domestic sewage. In addition, the RO membranes in the MBR-RO process could achieve 30% higher water flux than that in the ASP-MF/UF-RO process for wastewater reclamation (Qin et al., 2006). Nevertheless, the fouling propensities of RO membranes is strongly correlated to the MBR operation conditions, such as organic loading rate, which will significantly affect the soluble polysaccharide production in the RO feed (Wu et al., 2013).

Undesirable contaminants in the municipal wastewater such as bacteria, viruses, heavy metals, nitrate, chloride, sulfate, disinfection by-products, aromatic hydrocarbons, and

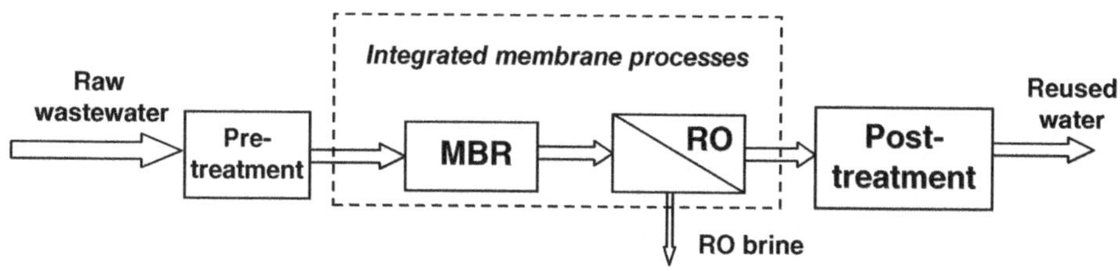

Figure 8.6 Schematics of MBR/RO process for waste water reclamation.

pesticides can be removed by integrated MBR/RO system and high quality product water can be achieved concurrently. In addition, the MBR/RO was found to be able to remove more than 90–99% of emerging trace organic contaminants and heavy metals (Alturki et al., 2010; Sahar et al., 2011; Dolar et al., 2012; Malamis et al., 2012; Cartagena et al., 2013; Nguyen et al., 2013). Another pilot study also found that the MBR/RO combination could achieve better removal of total estrogens than that of MF/RO (Tam et al., 2007).

8.2.3 Integrated Multiple Function Driven Membrane Process for Wastewater Reclamation

8.2.3.1 Membrane Distillation

MD is an emerging alternative technology for separation that is traditionally carried out by established separation processes such as distillation or RO (El-Bourawi et al., 2006). It is a thermally driven membrane separation process involving the transport of water vapor through microporous hydrophobic membranes. The main driving force for vapor transport across the membrane is the trans-membrane vapor pressure difference between the cold permeate and the hot feed (Lawson and Lloyd, 1997; El-Bourawi et al., 2006). Theoretically, the MD membrane is able to retain all non-volatiles including ions, macromolecules, colloids, and cells (Lawson and Lloyd, 1997). In addition, lower operating pressure is required when compared to the more prevalent pressure-driven methods (Lawson and Lloyd, 1997; El-Bourawi et al., 2006), and there is a lower energy requirement if waste heat or alternative heating sources like solar or geothermal heating are available (Lawson and Lloyd, 1997; El-Bourawi et al., 2006).

MD relies on vapor pressure difference across a microporous hydrophobic membranes to facilitate separation of volatile solvent (water from salt solution) or volatile solutes from impaired feed streams. The vapor pressure gradient is usually achieved by maintaining the temperature difference between a warm feed solution and a cold distillate that flow on the opposite side of the membrane. In MD, desalination and production of highly purified water can be achieved concurrently comparing to osmotically driven membrane processes (Section 8.2.3.2), at much lower temperatures compared to distillation processes, and at much lower pressures comparing to pressure-driven membrane processes. It can be most effectively and beneficially used when low-grade heat is readily available. Furthermore, compared to other membrane processes, the salinity of the feed stream minimally affects the driving force for mass transport through the membrane in MD. The salts, even at high concentrations, only slightly reduce the partial vapor pressure of such feed streams. Thus, MD can be beneficially used to high concentration salt feed desalination processes.

8.2.3.2 Fundamentals of MD

The main factors affecting flux in MD membranes include heat transfer through the boundary layers, the membrane thermal permeability, the partial pressure of air within the pores, and the feed composition (e.g., solute in the feed affects vapor pressure and film transfer coefficients (Schofield et al., 1990a, 1990b)). The heat and mass transfer mechanisms govern the permeation flux in MD and have been well researched with numerous models developed (Schofield et al., 1987, 1990a, 1990b; Martínez-Díez and Vázquez-González

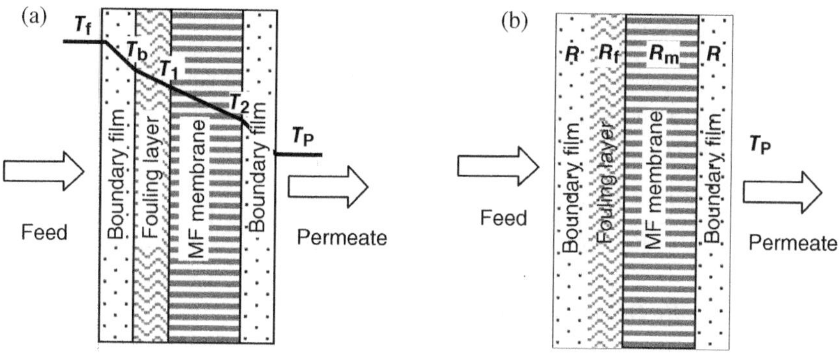

Figure 8.7 *Heat and mass transfer in MD.*

et al., 1999; Srisurichan et al., 2006; Gryta, 2008; Qtaishat et al., 2008). The approach is illustrated in Figure 8.7 for direct contact MD.

On both sides of the membrane surface, liquid boundary layers exist, namely the feed–membrane boundary layer and the permeate-membrane boundary layer. At the membrane surface on the feed side (Figure 8.7), temperature (denoted as T_1) will be lower than the temperature in the bulk feed (T_f) depending on the heat transfer coefficient (h_f) in the feed-boundary layer and the amount of heat loss during vaporization. Conversely, the temperature at the membrane surface on the permeate side (T_2) will be higher than the temperature in the permeate (T_p) due to the heat transfer coefficient (h_p) at the permeate-boundary layer and the heat gain from latent heat of condensation. In the case of a fouled membrane, T_1 is also influenced by the heat transfer coefficient (h_b) in the fouling layer and the heat balance across the MD membrane can be summarized by equation (8.1).

Heat balance equation:

$$h_f(T_f - T_b) = \left(\frac{k_b}{\delta_b}\right)(T_b - T_1) = \left(\frac{k_m}{\delta}\right)(T_1 - T_2) + J\Delta H_v = h_p(T_2 - T_p) \qquad (8.1)$$

Temperature polarization is corresponding to the net transmembrane temperature difference ($T_1 - T_2$) and overall temperature difference ($T_f - T_p$). The temperature polarization coefficient (TPC) can be expressed as follows:

$$\text{TPC} = \left(\frac{T_1 - T_2}{T_f - T_p}\right) \qquad (8.2)$$

TPC can be used to quantify the magnitude of the heat resistance in the boundary layers (and fouling layer in case of fouled membranes) over the total heat transfer resistance (Curcio and Drioli, 2005). It is thus an indication of the thermal efficiency of the system.

8.2.3.3 Mass Transfer in MD

The mass transfer through a fouled MD membrane is represented in Figure 8.7. The total mass transfer resistance includes resistance through the boundary layers (R_f and R_p), resistance due to fouling (R_b), and resistance in the membrane. The vapors transfer through the microporous, hydrophobic MF membrane via diffusion. The pore size of fouling layer is assumed to be smaller than the mean free path and Knudsen diffusion is assumed to be the dominant mass transfer mechanism through the fouling layer. The mass balance across the MD membrane can be summarized by the following equation (Schofield et al., 1987; Martínez-Díez and Vázquez-González, 1999; Srisurichan et al., 2006; Gryta, 2008; Qtaishat et al., 2008):

$$J = \left(\frac{P_f - P_b}{R_f}\right) = \left(\frac{P_b - P_1}{R_b}\right) = \left(\frac{P_1 - P_2}{R_m}\right) = \left(\frac{P_2 - P_p}{R_p}\right) \tag{8.3}$$

8.2.3.4 Osmotically Driven Membrane Processes

Osmotically driven membrane processes (ODMPs) are emerging membrane technologies that can be potentially utilized in desalination and water reuse. ODMP includes forward osmosis (FO) and pressure-retarded osmosis (PRO). ODMP is a natural process driven by the osmotic pressure difference across a semi-permeable membrane that retains solutes but allows water to transfer through the membrane. The asymmetric RO membrane, developed by Loeb and Sourirajan 1963), was initially used for ODMP. The asymmetric cellulose acetate membranes and the later developed thin-film composite membrane used for RO were also tried in FO or PRO tests (Kravath and Davis, 1975; Loeb et al., 1976; Mehta and Loeb, 1978). It was quickly observed that conventional RO membranes performed badly, due to internal polarization. Studies on ODMP have been comprehensive including the basic principles, applications, and challenges (Cath et al., 2006; Qin et al., 2012; Zhao et al., 2012).

Compared to pressure-driven membrane processes, FO operates at a very low hydraulic pressure for feed and draw solution (DS) recirculation, achieves a high rejection of salt and other contaminants like RO. FO is also expected to be a low fouling membrane process (Cath et al., 2006; Achilli et al., 2009; Cath, 2010). Furthermore, FO can be used in the condition of PRO to harvest osmotic power (Logan and Elimelech, 2012) (Section 8.2.3.4.1).

8.2.3.4.1 Fundamental Concept of ODMP

Osmosis is the spontaneous net movement of solvent molecules through a semipermeable membrane into a region of higher solute concentration, in the direction that tends to equalize the solute concentrations on the two sides. The osmotic pressure is defined to be the pressure required to maintain equilibrium, with no net movement of solvent. Osmosis in water and membrane processes will generate water flux across a semi-permeable membrane from a less concentrated solution to a more concentrated solution. The FO process utilizes osmotic pressure gradient ($\Delta\pi$) as a driving force instead of hydraulic pressure (ΔP), thus FO does not require input of energy. Theoretically, osmotic pressure (π) can be calculated using equation (8.4) (McNaught and Wilkinson, 1997).

$$\pi = i\phi CRT \tag{8.4}$$

where i is the dimensionless van't Hoff factor, ϕ is the osmotic coefficient which distinguishes the difference of solvent from ideal behavior referenced to Raoult's law, C is concentration of all solutes, $R = 0.08205746$ L atm/K/mol is the gas constant, and T is the thermodynamic (absolute) temperature.

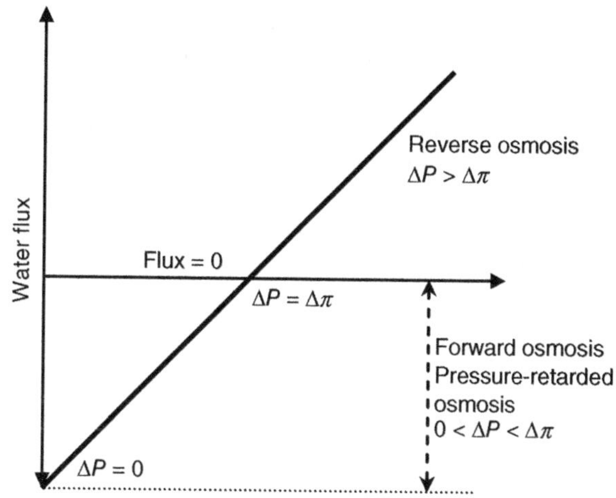

Figure 8.8 The relationship between FO, PRO, and RO processes.

Ideally, equation (8.5) describes the relationship between osmotic and hydraulic pressures and water flux (Figure 8.8):

$$J_W = A(\Delta\pi - \Delta P) \tag{8.5}$$

where J_W is the water flux across the membrane, A is the coefficient of water permeability, $\Delta\pi$ is the difference in osmotic pressures, and ΔP is the difference in hydraulic pressures (negative values of J_W indicating reverse osmotic flow). Figure 8.8 illustrates the relationship between FO, PRO, and RO. FO occurs when hydraulic pressure is zero ($\Delta P = 0$), and an osmotic pressure gradient ($\Delta\pi$) can drive water flow from a feed to a draw solution (DS). PRO applies a hydraulic pressure lower than the osmotic pressure to the draw solution side. The net water flux is in the direction toward the pressurized draw solution (same as FO) and this new volume of pressurized water can be used to generate power or for pumping. RO uses higher hydraulic pressure than the osmotic pressure ($\Delta P > \Delta\pi$), causing water flux from the concentrated solution to the permeate.

However, internal concentration polarization (ICP) can occur in FO and PRO due to the hindered diffusion of solute in the porous support layer. The ICP results in significant decline of water flux due to the reduction in $\Delta\pi$. An analytical model that incorporates the effect of ICP has been developed by Loeb et al. (1997). The effect of ICP depends on membrane orientation, because the membrane has an "active layer" (the thin skin layer) and a thicker support layer. The orientations are active layer facing draw solute (AL-DS) or active layer facing the feed water (AL-FW). According to this model, the FO permeate flux J_v can be determined by equation (8.6) for the AL-DS orientation and equation (8.7) for the AL-FW orientation:

$$J_w = K_m \ln\left(\frac{A\pi_D - J_w + B}{A\pi_F + B}\right) \text{(AL} - \text{DS)} \tag{8.6}$$

$$J_w = K_m \ln\left(\frac{A\pi_D + B}{A\pi_F + J_w + B}\right) \text{(AL} - \text{FW)} \tag{8.7}$$

where π_D and π_F are the osmotic pressures of the draw solution and feed solution, respectively; and A and B are the coefficients of water permeability and solute permeability of the rejection layer, respectively.

Equation (8.6) (AL-DS orientation) is preferred to be applied in the PRO mode (Gray et al., 2006). Equation (8.7) (AL-FW orientation) is employed in the FO mode (Tang et al., 2010; Jin et al., 2011).

The mass transfer coefficient K_m of the porous support layer is related to the solute diffusion coefficient D and the membrane structure parameter S,

$$K_m = \frac{D}{S} = \frac{\varepsilon \cdot D}{\tau \cdot \delta} \tag{8.8}$$

In equation (8.8), the structural parameter S is a length scale for the ICP in the porous support layer, and its value is given by membrane support layer thickness (δ) times its tortuosity (τ) over the porosity (ε).

In PRO, the AL-DS orientation is considered to be mechanically more stable, as the external hydraulic pressure is applied on the salty water side. In this case, concentrative ICP occurs in the porous substrate of the membrane, and water flux is given by equation (8.9) (Lee et al., 1981).

$$J_w = K_m \ln\left(\frac{A\pi_D - (J_w + A\Delta P) + B\left(1 + \frac{A\Delta P}{J_w}\right)}{A\pi_F + B\left(1 + \frac{A\Delta P}{J_w}\right)}\right) \tag{8.9}$$

8.2.3.5 *Novel Integrated Membrane Process for Wastewater Treatment*

The FO and MD processes have inherent limitations due to the potential need for the regeneration of the draw solution of FO, and the poor volatile substance rejection by MD. Thus, in some cases both FO and MD are difficult to use as a stand-alone process for wastewater treatment. Various hybrid membrane process concepts could overcome these difficulties. This is illustrated in the process flow charts in Figure 8.9.

The FO/RO system has been studied at laboratory and pilot scale for treating sewage, pharmaceutical contaminated water, dairy wastewater, and petrol industrial wastewater (Cath et al., 2005a, 2005b, 2010; Aydiner et al., 2013; Hancock et al., 2013). A schematic flow diagram of the FO/RO process is illustrated in Figure 8.9a.

In the FO membrane process, wastewater is continuously fed into the FO membrane element, tangentially to the active layer of the FO membrane. Permeate transfers across the membrane into the draw solution by the osmotic pressure gradient. The concentrated wastewater can be further treated by anaerobic digestion to produce methane or by an aerobic bioprocess to meet discharge standards. The diluted draw solution from the FO module can be concentrated by RO in a closed loop, where the draw solute can be synthetic NaCl solution, $MgSO_4$, NH_4HCO_3, or seawater. The permeate from the RO membrane should be high-quality product water.

A pilot study has evaluated the performance of FO/RO for the treatment of sewage to direct potable reuse water (Hancock et al., 2013). A commercial spiral wound FO membrane

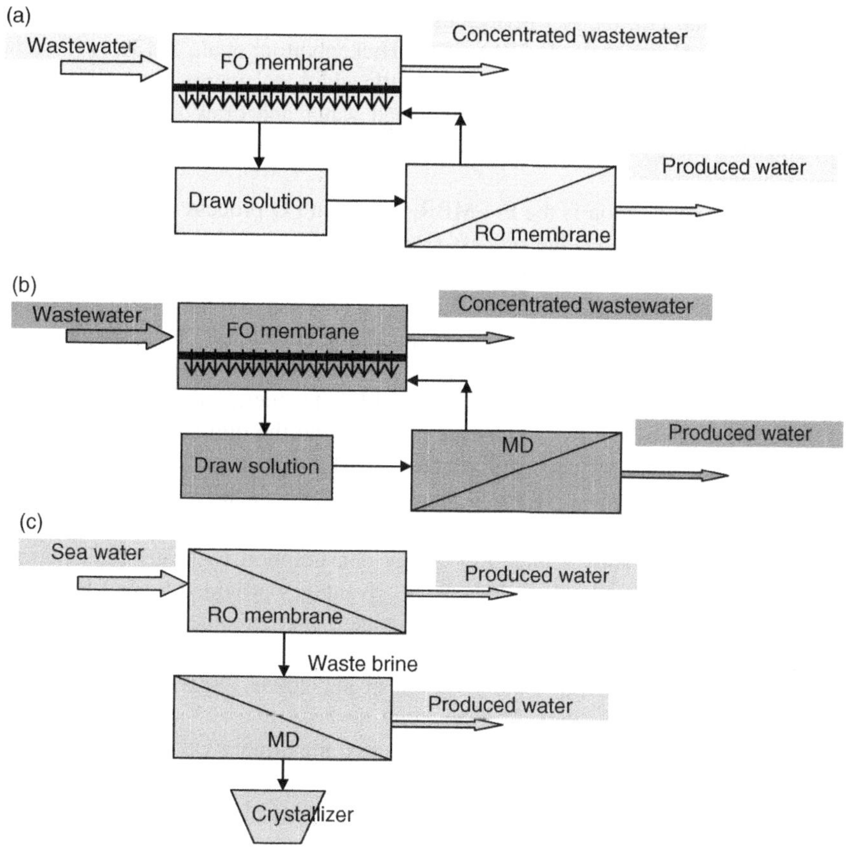

Figure 8.9 Schematics of novel integrated membrane processes for wastewater treatment. (a) FO/RO; (b) FO/MD; (c) RO/MD-crystallizer.

element from Hydration Technology Innovations (HTI) (Albany, OR) was installed and the membrane material was cellulose triacetate (CTA). Commercial seawater RO spiral wound membranes (SW30 2540, DOW Filmtec, Edina, MN) were used in the RO pilot system. The pilot demonstration was operated for approximately 1300 hours of continuous operation. Approximately 10,000 L of purified water was produced from 900,000 L of wastewater through the FO-RO process. The water flux was maintained at a relatively constant rate of 5.7±0.2 LMH. The dual-barrier FO-RO process achieved more than 99.9% removal of orthophosphate and dissolved organic carbon and more than 96% rejection of nitrate. The concentrations of all constituents in the wastewater were reduced by the FO/RO process to meet the US EPA drinking water standards. Thus, the FO/RO permeate water is drinkable and the FO/RO process is capable of treating wastewater into potable water in one step. In the study, membrane fouling occurred in the FO module; however, the fouling was mostly reversible and easily recovered.

Similar studies were conducted at lab and pilot scale. The results showed that if DS flows in a once-through pattern (seawater source available) to mitigate feed constituents accumulation in the DS, the RO permeates will be even better at the same rejection rate (Cath et al., 2010; Hancock et al., 2011, 2012; Hickenbottom et al., 2013).

FO is attractive as the process requires very little additional energy input (a small amount of electricity is required for delivering the feed water and draw solution) (Cath et al., 2006). Thus, the energy demand for FO separation can be much lower than other prevailing separation technologies.

Another niche application is the FO-MBR, where an FO process is used in conjunction with the aerobic activated sludge process (Lay et al., 2010, 2011; Xiao et al., 2011; Yap et al., 2012). In the aerobic FO-MBR, the non-porous FO membrane acts as a complete barrier for organic (COD) retention and biomass separation. Such a hybrid process enjoys the advantages of both high permeate water quality and potentially lower production cost.

A schematic diagram of the FO-MD process is shown in Figure 8.9b. Similar to the FO/RO process, the system also includes draw solute regeneration. The water is extracted from the wastewater by the FO membrane modules and the diluted DS is regenerated by the MD unit. This concept has been patented in 2007 (Cath and Martinetti, 2007). MD is a thermally driven process. If waste heat is available, such as that collocated with a wastewater treatment plant and a power plant, the integration of forward osmosis (FO) and MD would significantly reduce the energy consumption for water reclamation. The FO/MD process captures the complementary advantages provided by FO and MD. The MD membrane can provide complete retention and hence in-situ regeneration of the DS. The FO/MD process has been tested to treat dye wastewater, protein solutions, and synthetic sewage in lab-scale experiments (Cath et al., 2005a, 2005b; Wang et al., 2011; Ge et al., 2012).

In the FO/MD process, to drive the MD process, the diluted DS temperature has to be elevated. As a consequence, a temperature gradient across the FO membranes is developed, and substantially enhanced FO flux is observed (Cath et al., 2005a, 2005b; Koo et al., 2013). The increased water flux is due to the decrease in concentration polarization resistances in the boundary layer, supporting layers as well as the active layer. This is because mass diffusion kinetics are improved as the temperature increases (Zhao and Zou, 2011; Phuntsho et al., 2012; You et al., 2012). As such, there is an obvious synergy between the processes.

Another novel integrated wastewater treatment process is known as the MD bioreactor (MDBR), which is a thermophilic bioprocess combined with the MD process (Phattaranawik et al., 2008). MDBR systems typically follow the direct contact MD (DCMD) configuration in which the condensing liquid is in contact with the membrane (Phattaranawik et al., 2008, 2009; Khaing et al., 2010; Goh et al., 2013). Since the driving force for flux is vapor pressure gradient, the temperature difference between the feed and permeate should ideally be maintained at >30°C and the feed temperature preferably at >50°C. Due to the requirement for elevated temperature, special thermophiles (microbes that proliferate at an optimal temperature of 45–60°C) have to be used.

The application of the RO/MD-crystallizer is depicted in Figure 8.9c. The brine stream from the seawater desalination RO plant is typically 50% of the feed stream and double the concentration (~70 g/L salts) with an osmotic pressure of about 50 bar. Further recovery of water by RO becomes more difficult due to the need for increasingly high pressures. However, discharge of this brine concentrate to the ocean imposes some environmental concerns (Lattemann and Höpner, 2008; Liu et al., 2013; Zhou et al., 2013). This issue is especially significant for inland brackish water desalting RO.

In the RO/MD-crystallizer process, the RO brine is pumped into the MD process The highly concentrated MD brine goes directly to the crystallizer, the temperature of which is controlled by further heat input and cooling (a vertical gradient would be preferred). Additional MD modules could also be installed in the crystallizer, which could extract more water from the brine and bring the crystallizer to saturation.

The concept of membrane distillation crystallization (MDC) was introduced by Drioli and coworkers (Curcio and Drioli, 2005), who continue to work in this area (Macedonio et al., 2007; Al Obaidani et al., 2009; Di Profio et al., 2010). Tun et al. (2005) specifically addressed MDC applied to various inorganic salts at very high concentration illustrating the principle behind MD of salts with negative or positive temperature-solubility gradients. Three types of HF membranes, namely, single-layer PVDF, dual-layer hydrophobic–hydrophobic PVDF, and dual-layer hydrophobic–hydrophilic PVDF/PAN membranes, have been tested in the MD-crystallizer process. The single-layer PVDF (SL-PVDF) membrane was found to have superior wetting resistance compared to other types of dual-layer membranes (Ji et al., 2010; Edwie and Chung, 2012).

8.3 Strategic Co-location Concept for Integrated Process Involving RO, PRO, and Wastewater Treatment

Recoveries for seawater desalination are typically in the range of 35–55% (Desalination Committee, 2011), which implies that a significant amount of pre-treated seawater feed and the chemicals used are discarded wastefully during brine disposal. Waste is a misplaced resource. A study by Sim et al. (2012) has identified possible synergies of seawater desalination and water reuse processes using the concept of strategic co-location, and harnessing the inherent salinity difference of the processes to produce clean water more efficiently and cost-effectively. Minimal disruption of conventional processes and equipment was considered by Sim et al. when calculating the potential reductions in energy consumption and capital costs from the synergistic coupling of the processes. Focus was placed on the design depicted in Figure 8.10 that simultaneously dilutes the feed solution with impaired water and produces renewable energy via options of PRO, FO, or direct mixing. A second PRO process is then employed to harness the energy from salinity gradients between the brine streams of impaired water and seawater brine that are otherwise perceived as waste and require proper disposal. In summary, a higher recovery results relative to the independent operations of the water reuse and seawater desalination plants. Such integrated Hybrid Processes of RO and FO have gained attraction recently (Cath, 2010; Bamaga et al., 2011; Shaffer et al., 2012), but no attempts have been made to quantitatively calculate the potential savings.

Through the use of an impaired water source, it was observed that the Hybrid Process increased the recovery of water from the assumed 66.7% for conventional seawater desalination and water reuse processes to a potential of 80% (Victor et al., 2013). This was further complemented by a projected reduction of up to 23% in energy consumption and 8.7%–20% potential capital cost savings when compared to conventional seawater desalination plants. Insights gained in several areas of the study include the following: (1) the use of

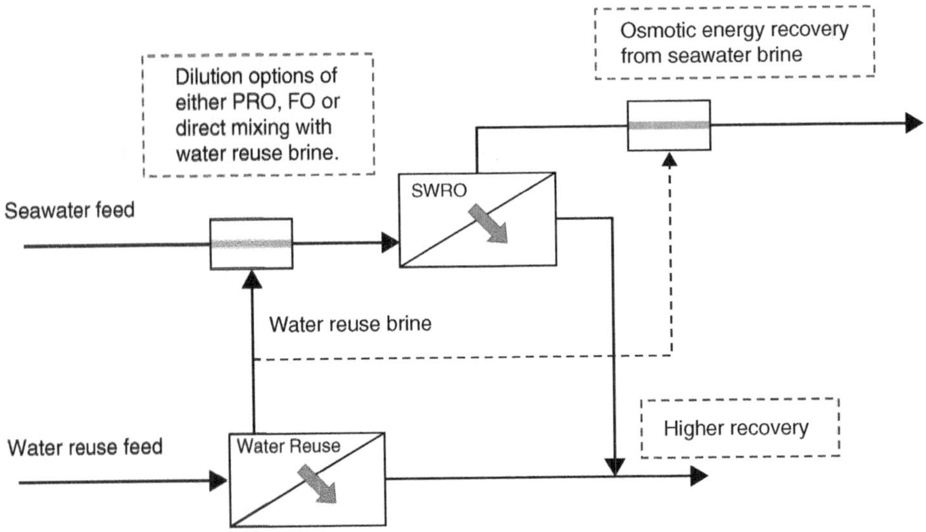

Figure 8.10 *Proposed configuration of the Hybrid Process.* Source: *Reprinted from Sim et al. (2012) with permission from Elsevier.*

osmotic pressure via different forms of energy recovery devices (ERDs); (2) the advantages of energy production versus other forms of energy reduction resulting from the reduction in seawater feed flow rate and reduced requirements of brine disposal; (3) the counter-intuitive reduction in capital costs even when taking into account the increased number of ERDs and membrane area. However, it is noted that this is a conservative method since a hypothetical scenario based on average process values was assumed as outlined in the design considerations, which can be potentially affected by other considerations such as the type of feed source for both the water reuse and SWRO plants. The feed source could affect the operating pressure, fouling and scaling behavior, pre-treatment, and brine disposal methods.

A key concern is that the principal savings come from the seawater desalination process; as such, the implementation of the Hybrid Process is critically dependent on the amount of impaired water available for the dilution of the seawater feed. The possible reduction in seawater feed pre-treatment and brine disposal is what makes the Hybrid Process a potentially successful technology. The availability of sufficient impaired water could be an issue since water stressed areas usually rely on seawater desalination to make up for the shortfall in the water supply. As such, for these areas there is a real probability of not having sufficient impaired water sources. The main concern in this design analysis is for the water reuse plant to have a capacity that is three times that of the SWRO plant. In response to this concern, a sensitivity analysis on the capacity of the water reuse plant has been done whereby it was reduced to one-third and two-thirds of the original capacity. As this might be anticipated in the absence of a sufficient amount of impaired water, thus, both the specific energy consumption and capital costs increase significantly. Nevertheless this IMS concept has potential benefits in water recovery and cost savings.

8.4 Conclusions

IMS is a treatment process that contains two or more membrane-based processes. It can be used to treat wastewaters for water reclamation and solids recovery, and potentially to achieve zero liquid discharge. Dual membrane processes using low-pressure MF/UF as pretreatment prior to RO are now commonly applied in water reclamation, and increasingly being adopted for seawater desalination. The novel processes MD and FO are also attractive for IMS applications, and can be combined with other processes (such as RO) to increase recovery and improve energy efficiency.

Nomenclature

A	Water permeability coefficient (L/m^2·h·atm)
A_m	Effective membrane surface area (m^2)
B	Salt permeability coefficient (L/m^2·h; m/s)
D	Solute diffusion coefficient (m^2/s)
h_b	Heat transfer coefficient in biofilm or fouling layer (W/m^2 K)
h_f	Heat transfer coefficient in feed boundary (W/m^2 K)
h_m	Heat transfer coefficient in membrane (W/m^2 K)
h_p	Heat transfer coefficient in permeate boundary (W/m^2 K)
i	Dimensionless van't Hoff factor
J	MD Permeation flux (kg/m^2 s)
J_w	FO permeation flux (kg/m^2 s)
k	Thermal conductivity of fluid (J/s m K)
k_b	Thermal conductivity of fouling layer (W/m K)
k_m	Thermal conductivity of membrane (W/m K)
K_m	Mass transfer coefficient (m/s)
P_1	Vapor pressure at membrane surface on the feed side (Pa)
P_2	Vapor pressure at membrane surface on the permeate side (Pa)
P_b	Vapor pressure at fouling layer (Pa)
P_f	Vapor pressure at bulk feed (Pa)
P_p	Vapor pressure at bulk permeate (Pa)
R	Gas constant (8.314 N m/K mol)
R_b	Resistance in biofilm or fouling layer (Pa m^2 s/kg)
R_e	Reynolds number ($d_h \rho v/\mu$)
R_f	Resistance in Feed boundary (Pa m^2 s/kg)
R_m	Resistance in membrane (Pa m^2 s/kg)
R_p	Resistance in permeate boundary (Pa m^2 s/kg)
R_t	Total resistance (Pa m^2 s/kg)
S	membrane structural parameter (m)
T_1	Temperature at membrane surface on the feed side (K)
T_2	Temperature at membrane surface on the permeate side (K)
T_b	Temperature at on fouling layer surface (K)
T_f	Temperature in bulk feed (K)

T_p Temperature in bulk permeate (K)
ΔH_v Heat of vaporization (J/kg)

List of Greek letters

ε	Membrane porosity
η	Viscosity of water (Pa s)
τ	Tortuosity
π_{draw}, π_{feed}	Osmotic pressures (atm)
$\Delta\pi$	Osmotic pressure difference across the membrane (atm)
σ	Geometric standard deviation (nm)
δ	Membrane thickness (m)
δ_b	Fouling layer thickness (m)

References

Achilli, A., Cath, T. Y., Marchand, E. A. and Childress, A. E. 2009. "The forward osmosis membrane bioreactor: A low fouling alternative to MBR processes." *Desalination* **239**, 1–3: 10–21.

Ahn, W.-Y., Kang, M.-S., Yim, S.-K. and Choi, K.-H. 2002. "Advanced landfill leachate treatment using an integrated membrane process." *Desalination* **149**, 1–3: 109–114.

Al Obaidani, S., Curcio, E., Di Profio, G. and Drioli, E. 2009. "The role of membrane distillation/crystallization technologies in the integrated membrane system for seawater desalination." *Desalination and Water Treatment* **10**, 1–3: 210–219.

Alturki, A. A., Tadkaew, N., McDonald, J. A., Khan, S. J., Price, W. E. and Nghiem, L. D. 2010. "Combining MBR and NF/RO membrane filtration for the removal of trace organics in indirect potable water reuse applications." *Journal of Membrane Science* **365**, 1–2: 206–215.

Aydiner, C., Topcu, S., Tortop, C., Kuvvet, F., Ekinci, D., Dizge, N. and Keskinler, B. 2013. "A novel implementation of water recovery from whey: "forward-reverse osmosis" integrated membrane system." *Desalination and Water Treatment* **51**, 4–6: 786–799.

Bamaga, O. A., Yokochi, A., Zabara, B. and Babaqi, A. S. 2011. "Hybrid FO/RO desalination system: Preliminary assessment of osmotic energy recovery and designs of new FO membrane module configurations." *Desalination* **268**, 1–3: 163–169.

Bennett, A. 2009. "Legislation and market forces: Regulators and wastewater reuse." *Filtration & Separation* **46**, 5: 16–21.

Calabro, V., Drioli, E. and Matera, F. 1991. Membrane Distillation in the Textile Waste-Water Treatment. In *12th International Symposium on Desalination and Water Re-Use*, Malta, Elsevier Science Bv.

Cartagena, P., El Kaddouri, M., Cases, V., Trapote, A. and Prats, D. 2013. "Reduction of emerging micropollutants, organic matter, nutrients and salinity from real wastewater by combined MBR–NF/RO treatment." *Separation and Purification Technology* **110**: 132–143.

Cath, T. Y. 2010. "Osmotically and thermally driven membrane processes for enhancement of water recovery in desalination processes." *Desalination and Water Treatment* **15**, 1–3: 279–286.

Cath, T. Y. and Martinetti, C. R. 2007. Combined membrane-distillation-forward-osmosis systems and methods of use. U.S. Patent.

Cath, T. Y., Adams, D. and Childress, A. E. 2005. "Membrane contactor processes for wastewater reclamation in space: Part II. Combined direct osmosis, osmotic distillation, and membrane distillation for treatment of metabolic wastewater." *Journal of Membrane Science* **257**, 1–2: 111–119.

Cath, T. Y., Gormly, S., Beaudry, E. G., Flynn, M. T., Adams, V. D. and Childress, A. E. 2005b. "Membrane contactor processes for wastewater reclamation in space: Part I. Direct osmotic concentration as pretreatment for reverse osmosis." *Journal of Membrane Science* **257**, 1–2: 85–98.

Cath, T. Y., Childress, A. E. and Elimelech, M. 2006. "Forward osmosis: Principles, applications, and recent developments." *Journal of Membrane Science* **281**, 1–2: 70–87.

Cath, T. Y., Hancock, N. T., Lundin, C. D., Hoppe-Jones, C. and Drewes, J. E. 2010. "A multi-barrier osmotic dilution process for simultaneous desalination and purification of impaired water." *Journal of Membrane Science* **362**, 1–2: 417–426.

Comerton, A. M., Andrews, R. C. and Bagley, D. M. 2005. "Evaluation of an MBR–RO system to produce high quality reuse water: Microbial control, DBP formation and nitrate." *Water Research* **39**, 16: 3982–3990.

Côté, P., Siverns, S. and Monti, S. 2005. "Comparison of membrane-based solutions for water reclamation and desalination." *Desalination* **182**, 1–3: 251–257.

Curcio, E. and Drioli, E. 2005. "Membrane distillation and related operations—A review." *Separation and Purification Reviews* **34**, 1: 35–86.

Di Profio, G., Curcio, E. and Drioli, E. 2010. "Supersaturation control and heterogeneous nucleation in membrane crystallizers: Facts and perspectives." *Industrial & Engineering Chemistry Research* **49**, 23: 11878–11889.

Dialynas, E. and Diamadopoulos, E. 2009. "Integration of a membrane bioreactor coupled with reverse osmosis for advanced treatment of municipal wastewater." *Desalination* **238**, 1–3: 302–311.

Desalination Committee. 2011. *Seawater Desalination Power Consumption*, White Paper. Desalination Committee, WaterReuse Association Alexandra, VA, USA.

Dolar, D., Gros, M., Rodriguez-Mozaz, S., Moreno, J., Comas, J., Rodriguez-Roda, I. and Barceló, D. 2012. "Removal of emerging contaminants from municipal wastewater with an integrated membrane system, MBR–RO." *Journal of Hazardous Materials* **239–240**: 64–69.

Edwie, F. and Chung, T.-S. 2012. "Development of hollow fiber membranes for water and salt recovery from highly concentrated brine via direct contact membrane distillation and crystallization." *Journal of Membrane Science* **421–422**: 111–123.

El-Bourawi, M. S., Ding, Z., Ma, R. and Khayet, M. 2006. "A framework for better understanding membrane distillation separation process." *Journal of Membrane Science* **285**, 1–2: 4–29.

Fane, A. G. 2007. "Sustainability and membrane processing of wastewater for reuse." *Desalination* **202**, 1–3: 53–58.

Fane, A. G. and Leslie, G. L. 2004. Membrane Technology—The Key to Water Reuse. In *Proceedings of the IWA Conference on Water Environment-Membrane Technology*. International Water Association. 875–883.

Fane, A. G., Tang, C. Y. and Wang, R. 2011. Membrane technology for water: Microfiltration, ultrafiltration, nanofiltration and reverse osmosis. In *Treatise on Water Science*. Wilderer, P., ed. Oxford, Academic Press. 4: 301–355.

Flemming, H.-C., Griebe, T. and Schaule, G. 1996. "Antifouling strategies in technical systems—A short review." *Water Science and Technology* **34**, 5–6: 517–524.

Flemming, H. C., Schaule, G., Griebe, T., Schmitt, J. and Tamachkiarowa, A. 1997. "Biofouling—The Achilles heel of membrane processes." *Desalination* **113**, 2–3: 215–225.

Gabelich, C. J., Williams, M. D., Rahardianto, A., Franklin, J. C. and Cohen, Y. 2007. "High-recovery reverse osmosis desalination using intermediate chemical demineralization." *Journal of Membrane Science* **301**: 131–141.

Gabelich, C. J., Rahardianto, A., Northrup, C. R., Yun, T. I. and Cohen, Y. 2011. "Process evaluation of intermediate chemical demineralization for water recovery enhancement in production-scale brackish water desalting." *Desalination* **272**: 36–45.

Ge, Q. C., Wang, P., Wan, C. F. and Chung, T. S. 2012. "Polyelectrolyte-promoted forward osmosis-membrane distillation (FO-MD) hybrid process for dye wastewater treatment." *Environmental Science & Technology* **46**, 11: 6236–6243.

Goh, S., Zhang, J., Liu, Y. and Fane, A. G. 2013. "Fouling and wetting in membrane distillation (MD) and MD-bioreactor (MDBR) for wastewater reclamation." *Desalination* **323**: 39–47.

Gray, G. T., McCutcheon, J. R. and Elimelech, M. 2006. "Internal concentration polarization in forward osmosis: role of membrane orientation." *Desalination* **197**, 1–3: 1–8.

Griebe, T. and Flemming, H. C. 1998. "Biocide-free antifouling strategy to protect RO membranes from biofouling." *Desalination* **118**, 1–3: 153–156A.

Gryta, M. 2008. "Fouling in direct contact membrane distillation process." *Journal of Membrane Science* **325**, 1: 383–394.

Hancock, N. T., Xu, P., Heil, D. M., Bellona, C. and Cath, T. Y. 2011. "Comprehensive bench- and pilot-scale investigation of trace organic compounds rejection by forward osmosis." *Environmental Science & Technology* **45**, 19: 8483–8490.

Hancock, N. T., Black, N. D. and Cath, T. Y. 2012. "A comparative life cycle assessment of hybrid osmotic dilution desalination and established seawater desalination and wastewater reclamation processes." *Water Research* **46**, 4: 1145–1154.

Hancock, N. T., Xu, P., Roby, M. J., Gomez, J. D. and Cath, T. Y. 2013. "Towards direct potable reuse with forward osmosis: Technical assessment of long-term process performance at the pilot scale." *Journal of Membrane Science* **445**: 34–46.

Hickenbottom, K. L., Hancock, N. T., Hutchings, N. R., Appleton, E. W., Beaudry, E. G., Xu, P. and Cath, T. Y. 2013. "Forward osmosis treatment of drilling mud and fracturing wastewater from oil and gas operations." *Desalination* **312**: 60–66.

Hoek, E. M. V. and Elimelech, M. 2003. "Cake enhanced concentration polarization: A new fouling mechanism for salt-rejecting membranes." *Environmental Science & Technology* **37**: 5581–5588.

Hoinkis, J., Deowan, S. A., Panten, V., Figoli, A., Huang, R. R. and Drioli, E. 2012. "Membrane bioreactor (MBR) technology—A promising approach for industrial water reuse." *Procedia Engineering* **33**: 234–241.

Insel, G., Dagdar, M., Dogruel, S., Dizge, N., Ubay Cokgor, E. and Keskinler, B. 2013. "Biodegradation characteristics and size fractionation of landfill leachate for integrated membrane treatment." *Journal of Hazardous Materials* **260**: 825–832.

Jacob, M., Guigui, C., Cabassud, C., Darras, H., Lavison, G. and Moulin, L. 2010. "Performances of RO and NF processes for wastewater reuse: Tertiary treatment after a conventional activated sludge or a membrane bioreactor." *Desalination* **250**, 2: 833–839.

Ji, X., Curcio, E., Al Obaidani, S., Di Profio, G., Fontananova, E. and Drioli, E. 2010. "Membrane distillation-crystallization of seawater reverse osmosis brines." *Separation and Purification Technology* **71**, 1: 76–82.

Jin, X., Tang, C. Y., Gu, Y., She, Q. and Qi, S. 2011. "Boric acid permeation in forward osmosis membrane processes: Modeling, experiments, and implications." *Environmental Science & Technology* **45**, 6: 2323–2330.

Joss, A., Baenninger, C., Foa, P., Koepke, S., Krauss, M., McArdell, C. S., Rottermann, K., Wei, Y., Zapata, A. and Siegrist, H. 2011. "Water reuse: >90% water yield in MBR/RO through concentrate recycling and CO2 addition as scaling control." *Water Research* **45**, 18: 6141–6151.

Judd, S. 2008. "The status of membrane bioreactor technology." *Trends in Biotechnology* **26**, 2: 109–116.

Judd, S. 2011. *The MBR Book,* 2nd ed. Oxford, Butterworth-Heinemann.

Kalkan, Ç., Yapsakli, K., Mertoglu, B., Tufan, D. and Saatci, A. 2011. "Evaluation of Biological Activated Carbon (BAC) process in wastewater treatment secondary effluent for reclamation purposes." *Desalination* **265**, 1–3: 266–273.

Kent, F. C., Farahbakhsh, K., Mahendran, B., Jaklewicz, M., Liss, S. N. and Zhou, H. 2011. "Water reclamation using reverse osmosis: Analysis of fouling propagation given tertiary membrane filtration and MBR pretreatments." *Journal of Membrane Science* **382**, 1–2: 328–338.

Khaing, T.-H., Li, J., Li, Y., Wai, N. and Wong, F.-S. 2010. "Feasibility study on petrochemical wastewater treatment and reuse using a novel submerged membrane distillation bioreactor." *Separation and Purification Technology* **74**, 1: 138–143.

Koo, J. W., Han, J. H., Yun, T., Lee, S. and Choi, J. S. 2013. "Integration of forward osmosis with membrane distillation: Effect of operating conditions." *Desalination and Water Treatment* **51**, 25–27: 5355–5361.

Kravath, R. E. and Davis, J. A. 1975. "Desalination of sea water by direct osmosis." *Desalination* **16**, 2: 151–155.

Lattemann, S. and Höpner, T. 2008. "Environmental impact and impact assessment of seawater desalination." *Desalination* **220**, 1–3: 1–15.

Lau, W.-J. and Ismail, A. F. 2009. "Polymeric nanofiltration membranes for textile dye wastewater treatment: Preparation, performance evaluation, transport modelling, and fouling control—A review." *Desalination* **245**, 1–3: 321–348.

Lawson, K. W. and Lloyd, D. R. 1997. "Membrane distillation (Review)." *Journal of Membrane Science* **124**: 1–25.

Lay, W. C. L., Liu, Y. and Fane, A. G. 2010. "Impacts of salinity on the performance of high retention membrane bioreactors for water reclamation: A review." *Water Research* **44**, 1: 21–40.

Lay, W. C. L., Zhang, Q., Zhang, J., McDougald, D., Tang, C., Wang, R., Liu, Y. and Fane, A. G. 2011. "Study of integration of forward osmosis and biological process: Membrane performance under elevated salt environment." *Desalination*, doi:10.1016/j.desal.2011.01.036.

Lee, K. L., Baker, R. W. and Lonsdale, H. K. 1981. "Membranes for power generation by pressure-retarded osmosis." *Journal of Membrane Science* **8**, 2: 141–171.

Liu, T.-K., Sheu, H.-Y. and Tseng, C.-N. 2013. "Environmental impact assessment of seawater desalination plant under the framework of integrated coastal management." *Desalination* **326**: 10–18.

Loeb, S. and Sourirajan, S. 1963. "Sea water demineralization by means of an osmotic membrane. Saline water conversion II. *American Chemical Society* **38**: 117–132.

Loeb, S., Van Hessen, F. and Shahaf, D. 1976. "Production of energy from concentrated brines by pressure-retarded osmosis: II. Experimental results and projected energy costs." *Journal of Membrane Science* **1**: 249–269.

Loeb, S., Titelman, L., Korngold, E. and Freiman, J. 1997. "Effect of porous support fabric on osmosis through a Loeb-Sourirajan type asymmetric membrane." *Journal of Membrane Science* **129**, 2: 243–249.

Logan, B. E. and Elimelech, M. 2012. "Membrane-based processes for sustainable power generation using water." *Nature* **488**, 7411: 313–319.

Macedonio, F., Curcio, E. and Drioli, E. 2007. "Integrated membrane systems for seawater desalination: Energetic and exergetic analysis, economic evaluation, experimental study." *Desalination* **203**, 1–3: 260–276.

Malamis, S., Katsou, E., Takopoulos, K., Demetriou, P. and Loizidou, M. 2012. "Assessment of metal removal, biomass activity and RO concentrate treatment in an MBR–RO system." *Journal of Hazardous Materials* **209–210**: 1–8.

Marconnet, C., Houari, A., Galas, L., Vaudry, H., Heim, V. and Di Martino, P. 2009. "Biodegradable dissolved organic carbon concentration of feed water and NF membrane biofouling: a pilot train study." *Desalination* **242**, 1–3: 228–235.

Martínez-Díez, L. and Vázquez-González, M. I. 1999. "Temperature and concentration polarization in membrane distillation of aqueous salt solutions." *Journal of Membrane Science* **156**, 2: 265–273.

McDonogh, R., Schaule, G. and Flemming, H.-C. 1994. "The permeability of biofouling layers on membranes." *Journal of Membrane Science* **87**, 1–2: 199–217.

McNaught, A. D. and Wilkinson, A. 1997. *IUPAC Compendium of Chemical Terminology*. Oxford, Blackwell Scientific Publications.

Mehta, G. D. and Loeb, S. 1978. "Performance of permasep B-9 and B-10 membranes in various osmotic regions and at high osmotic pressures." *Journal of Membrane Science* **4**: 335–349.

Nguyen, L. N., Hai, F. I., Kang, J., Price, W. E. and Nghiem, L. D. 2013. "Removal of emerging trace organic contaminants by MBR-based hybrid treatment processes." *International Biodeterioration & Biodegradation* **85**: 474–482.

Phattaranawik, J., Fane, A. G., Pasquier, A. C. S. and Bing, W. 2008. "A novel membrane bioreactor based on membrane distillation." *Desalination* **223**, 1–3: 386–395.

Phattaranawik, J., Fane, A. G., Pasquier, A. C. S., Wu, B. and Wong, F. S. 2009. "Experimental study and design of a submerged membrane distillation bioreactor." *Chemical Engineering and Technology* **32**, 1: 38–44.

Phuntsho, S., Vigneswaran, S., Kandasamy, J., Hong, S., Lee, S. and Shon, H. K. 2012. "Influence of temperature and temperature difference in the performance of forward osmosis desalination process." *Journal of Membrane Science* **415–416**: 734–744.

Qin, J.-J., Kekre, K. A., Tao, G., Oo, M. H., Wai, M. N., Lee, T. C., Viswanath, B. and Seah, H. 2006. "New option of MBR-RO process for production of NEWater from domestic sewage." *Journal of Membrane Science* **272**, 1–2: 70–77.

Qin, J.-J., Lay, W. C. L. and Kekre, K. A. 2012. "Recent developments and future challenges of forward osmosis for desalination: A review." *Desalination and Water Treatment* **39** 1-3: 123–136.

Qtaishat, M., Matsuura, T., Kruczek, B. and Khayet, A. 2008. "Heat and mass transfer analysis in direct contact membrane distillation." *Desalination* **219**, 1–3: 272–292.

Raffin, M., Germain, E. and Judd, S. 2013. "Wastewater polishing using membrane technology: A review of existing installations." *Environmental Technology* **34**, 5: 617-627.

Rahardianto, A., McCool, B. C. and Cohen, Y. 2010. "Accelerated desupersaturation of reverse osmosis concentrate by chemically-enhanced seeded precipitation." *Desalination* **264**: 256–267.

Ridgway, H. F., Justice, C. A., Whittaker, C., Argo, D. and Olson, B. H. 1984. "Biofilm fouling of RO membranes—Its nature and effect on treatment of water for reuse." *Journal AWWA* **76**, 6: 94–102.

Rozzi, A., Malpei, F., Bianchi, R. and Mattioli, D. 2000. "Pilot-scale membrane bioreactor and reverse osmosis studies for direct reuse of secondary textile effluents." *Water Science and Technology* **41**, 10–11: 189–195.

Sahar, E., David, I., Gelman, Y., Chikurel, H., Aharoni, A., Messalem, R. and Brenner, A. 2011. "The use of RO to remove emerging micropollutants following CAS/UF or MBR treatment of municipal wastewater." *Desalination* **273**, 1: 142–147.

Sanciolo, P., Zou, L., Gray, S., Leslie, G. and Stevens, D. 2008. "Accelerated seeded precipitation pre-treatment of municipal wastewater to reduce scaling." *Chemosphere* **72**: 243–249.

Sauvet-Goichon, B. 2007. "Ashkelon desalination plant—A successful challenge." *Desalination* **203**, 1–3: 75–81.

Schofield, R. W., Fane, A. G. and Fell, C. J. D. 1987. "Heat and mass transfer in membrane distillation." *Journal of Membrane Science* **33**, 3: 299–313.

Schofield, R. W., Fane, A. G. and Fell, C. J. D. 1990. "Gas and vapour transport through microporous membranes. II. Membrane distillation." *Journal of Membrane Science* **53**, 1–2: 173–185.

Schofield, R. W., Fane, A. G., Fell, C. J. D. and Macoun, R. 1990. "Factors affecting flux in membrane distillation." *Desalination* **77**: 279–294.

Shaffer, D. L., Yip, N. Y., Gilron, J. and Elimelech, M. 2012. "Seawater desalination for agriculture by integrated forward and reverse osmosis: Improved product water quality for potentially less energy." *Journal of Membrane Science* **415–416**: 1–8.

Sim, S. T. V., Chong, T. H., Krantz, W. B. and Fane, A. G. 2012. "Monitoring of colloidal fouling and its associated metastability using Ultrasonic Time Domain Reflectometry." *Journal of Membrane Science* **401–402**: 241–253.

Sim, V. S. T., She, Q., Chong, T. H., Tang, C. Y., Fane, A. G. and Krantz, W. B. 2013. "Strategic co-location in a hybrid process involving desalination and pressure retarded osmosis (PRO) " *Membranes* **3**, 3: 98–125.

Simens. 2008. "First use of an MBR with RO in the pulp and paper industry." *Membrane Technology* **5**: 10–11.

Srisurichan, S., Jiraratananon, R. and Fane, A. G. 2006. "Mass transfer mechanisms and transport resistances in direct contact membrane distillation process." *Journal of Membrane Science* **277**, 1–2: 186–194.

Takht Ravanchi, M., Kaghazchi, T. and Kargari, A. 2009. "Application of membrane separation processes in petrochemical industry: A review." *Desalination* **235**, 1–3: 199–244.

Tam, L. S., Tang, T. W., Lau, G. N., Sharma, K. R. and Chen, G. H. 2007. "A pilot study for wastewater reclamation and reuse with MBR/RO and MF/RO systems." *Desalination* **202**, 1–3: 106–113.

Tang, C. Y., She, Q., Lay, W. C. L., Wang, R. and Fane, A. G. 2010. "Coupled effects of internal concentration polarization and fouling on flux behavior of forward osmosis membranes during humic acid filtration." *Journal of Membrane Science* **354**, 1–2: 123–133.

Tun, C. M., Fane, A. G., Matheickal, J. T. and Sheikholeslami, R. 2005. "Membrane distillation crystallization of concentrated salts—flux and crystal formation." *Journal of Membrane Science* **257**, 1–2: 144–155.

Uchymiak, M., Rahardianto, A., Lyster, E., Glater, J. and Cohen, Y. 2007. "A novel RO ex situ observation detector (EXSOD) for mineral scale characterization and early detection." *Journal of Membrane Science* **291**: 86–95.

Van der Bruggen, B. 2013. "Integrated membrane separation processes for recycling of valuable wastewater streams: Nanofiltration, membrane distillation, and membrane crystallizers revisited." *Industrial & Engineering Chemistry Research* **52**, 31: 10335–10341.

Van der Bruggen, B., Curcio, E. and Drioli, E. 2004. "Process intensification in the textile industry: The role of membrane technology." *Journal of Environmental Management* **73**, 3: 267–274.

van Hoof, S. C. J. M., Minnery, J. G. and Mack, B. 2002. "Performing a membrane autopsy." *The International Desalination & Water Reuse* 11/4 Feb/March: 40–46.

Vergili, I., Kaya, Y., Sen, U., Gönder, Z. B. and Aydiner, C. 2012. "Techno-economic analysis of textile dye bath wastewater treatment by integrated membrane processes under the zero liquid discharge approach." *Resources, Conservation and Recycling* **58**: 25–35.

Vishnu, G., Palanisamy, S. and Joseph, K. 2008. "Assessment of fieldscale zero liquid discharge treatment systems for recovery of water and salt from textile effluents." *Journal of Cleaner Production* **16**, 10: 1081–1089.

Wang, K. Y., Teoh, M. M., Nugroho, A. and Chung, T. S. 2011. "Integrated forward osmosis-membrane distillation (FO-MD) hybrid system for the concentration of protein solutions." *Chemical Engineering Science* **66**, 11: 2421–2430.

Wetterau, G., Liu, P., Chalmers, B., Richardson, T. and VanMeter, H. B. 2011. "Optimizing RO design criteria for indirect potable reuse." *IDA Journal Fourth Quarter* 40–45.

Wu, B., Kitade, T., Chong, T. H., Uemura, T. and Fane, A. G. 2013. "Impact of membrane bioreactor operating conditions on fouling behavior of reverse osmosis membranes in MBR–RO processes." *Desalination* **311**: 37–45.

Xiao, D., Tang, C. Y., Zhang, J., Lay, W. C. L., Wang, R. and Fane, A. G. 2011. "Modeling salt accumulation in osmotic membrane bioreactors: Implications for FO membrane selection and system operation." *Journal of Membrane Science* **366**: 314–324.

Yap, W. J., Zhang, J., Lay, W. C. L., Cao, B., Fane, A. G. and Liu, Y. 2012. "State of the art of osmotic membrane bioreactors for water reclamation." *Bioresource Technology* **122**: 217–222.

You, S.-J., Wang, X.-H., Zhong, M., Zhong, Y.-J., Yu, C. and Ren, N.-Q. 2012. "Temperature as a factor affecting transmembrane water flux in forward osmosis: Steady-state modeling and experimental validation." *Chemical Engineering Journal* **198–199**: 52–60.

Zhang, Y., Ma, C., Ye, F., Kong, Y. and Li, H. 2009. "The treatment of wastewater of paper mill with integrated membrane process." *Desalination* **236**, 1–3: 349–356.

Zhao, S. and Zou, L. 2011. "Effects of working temperature on separation performance, membrane scaling and cleaning in forward osmosis desalination." *Desalination* **278**, 1–3: 157–164.

Zhao, S., Zou, L., Tang, C. Y. and Mulcahy, D. 2012. "Recent developments in forward osmosis: Opportunities and challenges." *Journal of Membrane Science* **396** 0: 1–21.

Zhao, Y., Song, L. and Ong, S. L. 2010. "Fouling behaviour and foulant characteristics of RO membranes for treated secondary effluent reclamation." *Journal of Membrane Science* **349**: 65–74.

Zhou, J., Chang, V. W. C. and Fane, A. G. 2013. "An improved life cycle impact assessment (LCIA) approach for assessing aquatic eco-toxic impact of brine disposal from seawater desalination plants." *Desalination* **308**: 233–241.

9

Membrane Reactor: An Integrated "Membrane + Reaction" System

Angelo Basile, Adolfo Iulianelli, and Simona Liguori
Institute on Membrane Technology – Italian National Research Council (ITM-CNR), University of Calabria, Cosenza, Italy

9.1 Introduction

Membrane reactor (MR) technology combines two distinct sciences such as membranes and catalysis to constitute an integrated "membrane + reaction" system. In many scientific studies published in the specialized literature, often there has been a difficult evaluation on which of them has more relevance in MR development. However, it is reasonable to affirm that catalyst performance is certainly an important aspect in the design of improved MRs, particularly in the field of hydrogen production units, for example, fulfilling the requirements for the integration of fuel processors with proton exchange membrane (PEM) fuel cells. One of the most important issues affecting the MR concerns the poisoning/contaminant effects due to the inhibition caused by some reaction products or various impurities contained in the feeds, constituting an important obstacle for their performance.

Particular attention should also be given to the feedstock used for hydrogen production through reforming reactions, because they could affect the composition of the reaction products and the presence of contaminants in such streams.

Reforming of natural gas is the dominant process for hydrogen generation at the industrial scale (Iulianelli et al., 2012; Moriarty and Honnery, 2007) and it is essentially used for stationary applications. Nevertheless, having in mind to solve the issues related to the formation of greenhouse gases (GHGs) and to limit the exploitation of fossil fuels, it is expected that alcohols such as ethanol, methanol, glycerol, or other bio-fuels as acetic acid,

Integrated Membrane Systems and Processes, First Edition. Edited by Angelo Basile and Catherine Charcosset.

biogas, etc. could have a major impact in future applications. Furthermore, the adoption of new technologies for producing hydrogen as the MRs, alternative to the conventional reactors (CRs) could constitute an important goal in the viewpoint of intensification of the whole process (Brunetti et al., 2014).

Therefore, the scope of this chapter is to provide an overview on the recent advances on MRs used to perform several reaction processes and particularly, for hydrogen generation by exploiting renewable sources.

9.2 Hydrogen Economy

Today, the world economy is still oriented toward fossil fuels exploitation. However, the rise of environmental pollution caused by GHGs and other harmful emissions is driving to the development of new technologies and the exploitation of renewable sources. In the last decades, hydrogen has been seen as an alternative and new energy vector, but unfortunately although it is the most abundant compound in the Universe; it is not naturally present under molecular form on the Earth. The benefits due to hydrogen utilization can be resumed in the wide availability of a variety of feedstock and various hydrogen generation technologies.

Nowadays, the main feedstock for producing hydrogen are the hydrocarbons, water, and derived of biomass and therefore, the extraction/generation of hydrogen involves various chemical processes and consequently, energy requirements. By taking into account the nature of chemical process and/or the energy requirements, hydrogen production can be shortly subdivided into three major fields: (1) thermochemical, (2) electrochemical, and (3) biological methods (Turner, 2004).

Furthermore, hydrogen is today mainly consumed in different fields such as (Chaubeyet et al., 2013):

1. Ammonia production (54%)
2. Chemical industry/refineries (35%)
3. Electronic industry (6%)
4. Metal–glass industry (3%)
5. Food industry (2%).

Owing to the increase in global energy demand, the depletion of fossil fuels has increased, speeding up the search for alternative resources. As a consequence, for mitigating this phenomenon it seems imperative to expand the hydrogen economy by increasing the capacity of producing hydrogen in order to meet the corresponding increased hydrogen demand (Nielsen, 2005).

Nevertheless, various scientists have expressed some doubts whether hydrogen can constitute the right solution for such key aspects as energy world requirements, environmental security, and sustainability issues since the current technologies for hydrogen generation as an energy vector are too costly and waste of energy (Shinnar, 2004).

9.2.1 Why Membrane Reactors?

Hydrogen generation can be conventionally pursued via several processes including natural gas steam reforming, partial oxidation, autothermal reforming, steam-iron, plasma

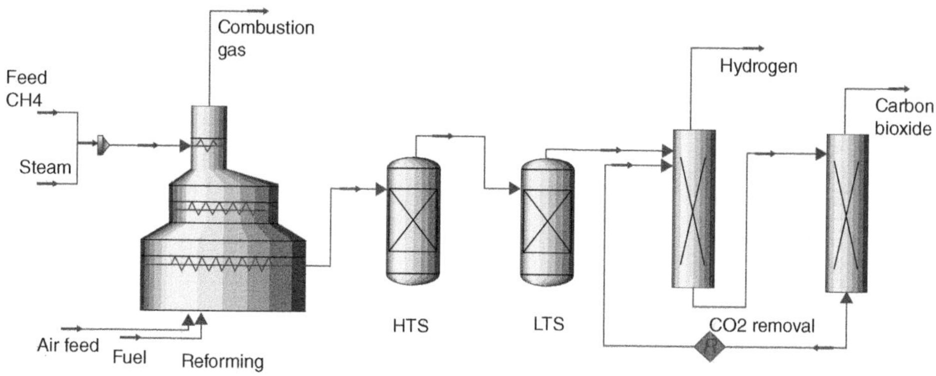

Figure 9.1 Conceptual scheme of a conventional system for high-grade hydrogen generation.

reforming, thermochemical water splitting, and biological processes (Chaubeyet et al., 2013). Up until now, natural gas reforming is the most used industrial process and its efficiency can reach ~85%, meeting up to 50% of the total world hydrogen consumption. Therefore, continuous research is oriented to enhance both the process performance and economics. In particular, in the contest of high-grade hydrogen generation (e.g., for PEM fuel cells supply), the industrial process is developed in a multi-stage system (Figure 9.1) consisting of a CR for carrying out natural gas steam reforming reaction, followed by two stages of water gas shift reactors (high and low temperature), useful to decrease the CO content of the hydrogen rich-stream coming out from the reformer, a preferential oxidation reactor, and a pressure swing adsorption stage to separate the CO_2 and further by-products.

Alternatively, inorganic MRs and housing hydrogen permselective membranes make possible the intensification of the aforementioned process through the integration of the reaction stage with the simultaneous hydrogen separation stage in only one device (Iulianelli et al., 2012; Lu et al., 2007; Basile et al., 2013) (Figure 9.2).

The concept of MRs has been introduced since the 1950s, even though a growing interest has been registered only with the development of innovative inorganic membrane materials operating at high temperature (Lin, 2001; Westermann and Melin, 2009). At industrial scale, various heterogeneous gas–solid catalytic processes, conventionally performed in

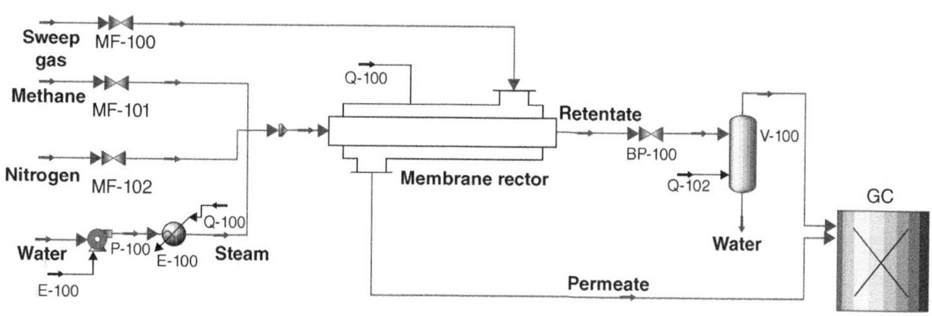

Figure 9.2 Conceptual scheme of a membrane reactor for high-grade hydrogen generation.

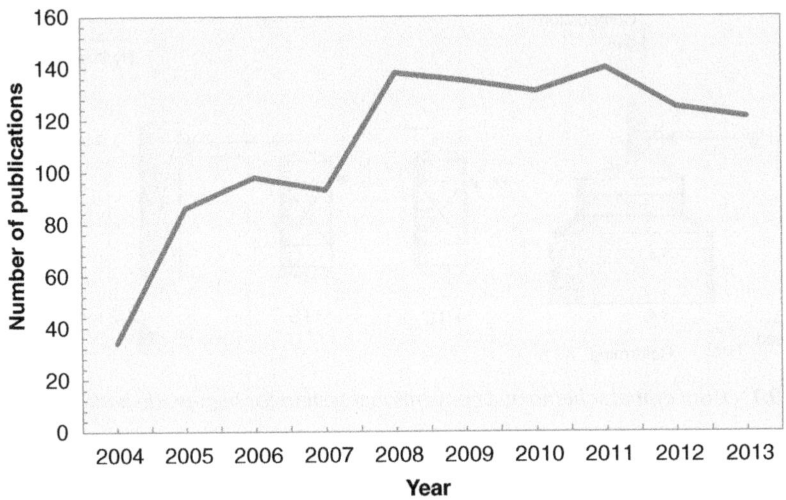

Figure 9.3 *Number of scientific publications about hydrogen production in MRs versus year. Database: www.scopus.com, searching keywords: "hydrogen production" AND "membrane reactor".*

various typologies of CRs such as fixed, fluidized, or trickle-bed reactors are performed as a combination of operations at high temperatures and in chemically harsh ambient.

Nevertheless, only few industrial scale applications can be noticed about MR utilization for hydrogen generation. Among them, the most representative is the Tokyo Gas Company Ltd. (web [1]), even if an extensive literature (Figure 9.3), is dedicated to scientific studies on MRs to propose them as a feasible solution at a larger scale.

As a general comment, MRs are applied not only in the field of hydrogen production and this depends also on the typology of membranes allocated inside the MR. Indeed, numerous scientists have proposed the application of membranes in combination with chemical and biochemical reactions with the aim of intensifying the whole process. As a general subdivision, it is possible to summarize the different classes of MRs depending on which kind of process is carried out and the membrane housed inside:

1. Dense and porous inorganic MRs (Mallada and Menéndez, 2008; Westermann and Melin, 2009; Saracco et al., 1999; Coronas and Santamaria, 1999; Lin, 2001; Smart et al., 2013; Wieland et al., 2002);
2. Polymeric MRs (Scholes et al., 2010);
3. Zeolite MRs (McLeary et al., 2006; Fong et al., 2008);
4. Photo-catalytic MRs (Mozia, 2010);
5. Enzyme MRs (Andrić et al., 2010);
6. Membrane bio-reactors using cells (Reij et al., 1998);
7. Electrochemical MRs (fuel cells, electrolytic cells, etc.) (Chatenet et al., 2010, Iulianelli and Basile, 2012);
8. Carbon MRs (Briceño et al., 2012a; Sznejer and Sheintuch, 2004; Itoh and Haraya, 2000).

As reflected in an extensive literature on the argument, among inorganic MRs, the gas permeation characteristics as well as the hydrogen permselectivity of dense and composite Pd-based membranes have been particularly studied in MR applications (Adhikari and Fernand, 2006; Arstad et al., 2006; Brunetti et al., 2014; Dittmeyer et al., 2001; Dolan et al., 2006; Soumitra et al., 2012; Basile et al., 2011; Basile, 2008; Tong et al., 2005; Lin and Rei, 2000). This is because there are several benefits due to the utilization of Pd-based MRs in reforming processes over the CRs. They can be shortly summarized as reported below:

1. Chemical reaction and hydrogen purification performed in only one stage;
2. Higher conversions than CRs (exercised at the same MRs operating conditions) or, vice versa, the same conversion of CRs but operating at milder conditions;
3. Production of high-grade hydrogen as a permeated stream;
4. Improvement of both hydrogen yield and hydrogen selectivity.

9.3 Membrane Reactors

As a general classification, an MR can operate in flat or tubular geometry. In the latter configuration, the density of the packed bed could be improved using multichannel tubular monoliths and depositing the catalyst inside the pores. However, in both cases, MRs can make possible: (a) the removal of such a product for a thermodynamically limited reaction; (b) the removal of a favorite product; and (c) the controlled addition of reactants.

Another kind of classification subdivides the MRs into: (a) catalytic membrane reactors (CMRs); (b) packed-bed membrane reactors (PBMRs); (c) catalytic non-permselective membrane reactors (CNMRs); (d) non-permselective membrane reactors (NMRs); (e) reactant-selective packed-bed reactors (RSPBRs); and (f) fluidized-bed membrane reactors (FBMRs). The most studied and applied typology of MR is the PBMR, which can be operated in both tubular and plane configuration.

In particular, in the PBMRs the catalyst is packed in a fixed region and generally, it is in contact with the membrane surface (Figure 9.4). In tubular configuration, the catalyst

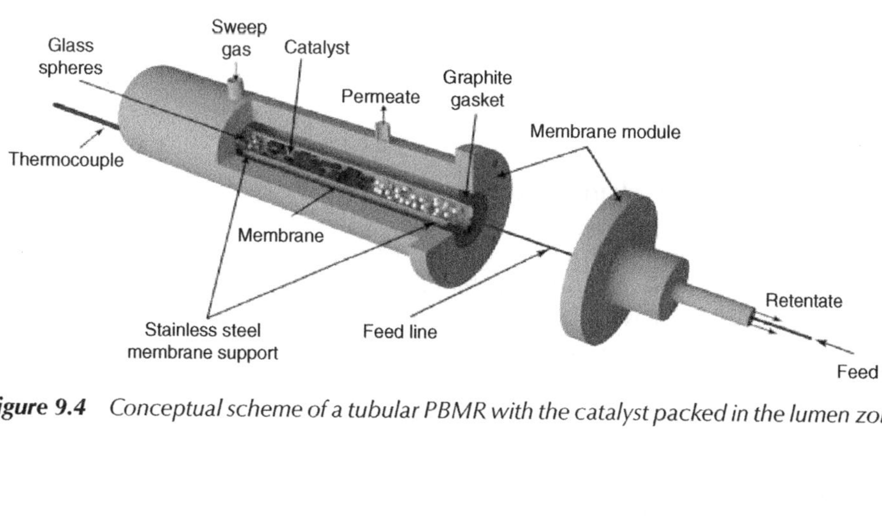

Figure 9.4 *Conceptual scheme of a tubular PBMR with the catalyst packed in the lumen zone.*

can be packed in the membrane lumen but also in the annulus, depending on which kind of inorganic membrane is allocated inside the MR. The particle dimension of the catalyst can affect a lot the performance of a PBMR because, if very small, they can increase the pressure drop along the reactor; otherwise, big particles can limit the internal mass transfer. As a further drawback, mass transfer limitation from bed to wall can take place in PBMRs, negatively affecting the hydrogen permeation owing to a variable temperature profile along the reactor, consequently provoking a detrimental effect on both catalyst and membrane performance (Tsotsis et al., 1992).

As reported in the recent literature, the FBMR represents an emerging technology and a typical configuration consists of a membrane immersed in a catalytic bed exercised in bubbling or turbulent regime (Adris and Grace, 1997). In particular, an FBMR makes possible a higher catalyst effectiveness and an enhanced mass and heat transfer rates with respect to a PBMR, favoring a more uniform temperature profile. On the contrary, the main FBMR drawback is the minimum fluidization velocity, which has to be guaranteed for keeping the catalyst bed in suspension (Deshmukh et al., 2007).

9.3.1 Membrane Reactors Utilization

Most of the research on MRs deals with their utilization to perform reactions limited by the equilibrium conversion. Indeed, the principal benefit of such MR is the separation capability of a membrane, used to enhance the performance of a generic catalytic system. Currently, there are two main generic approaches: the selective product separation, defined as the extractor modality, and selective reactant addition, distributor modality. The first configuration makes possible the *in-situ* removal of one of the products. For instance, in the field of hydrogen production from reforming reactions in CRs, the hydrogen yield and the conversion are limited by thermodynamics. Therefore, the utilization of an MR in extractor modality induces the selective hydrogen removal from the reaction to the permeate side, overcoming the thermodynamic equilibrium restrictions of an equivalent CR. This characteristic is the well-known shift effect, which allows the improvement of a generic reaction system operating also at milder conditions than a CR and favoring, consequently saving energy considerably (Iulianelli et al., 2012).

The second approach used in various MRs allows the control of the contact within reactants. As an example, permselective membranes can be housed in an MR to feed distributively one of the reactants. For instance, the oxygen distributive addition in preferential oxidation reaction can affect product selectivity and conversion (Miachon et al., 2003; Dalmon et al., 2007).

9.4 Membranes for Membrane Reactors

Among organic and inorganic membranes, the latter although more expensive than the organic ones, seems to be more adequate for their utilization in MRs because they possess high resistance toward solvents, well-defined stable pore structure (in the case of porous inorganic membranes), high mechanical stability and, in particular, elevated resistance at high operating temperatures. Inorganic membranes can be constituted by different materials such as metals (palladium, silver, niobium, titanium, etc. and their alloys), zeolite, ceramic, carbon, silica, and oxides (alumina, titania, zirconia). Commonly, they are stable in the temperature range of 300–800°C and in particular cases (ceramic membranes), up to 1000°C (Iulianelli and Basile, 2011a). Furthermore, they also show high resistance to chemical degradation if compared to organic membranes.

The general classification of inorganic membranes subdivides them into two classes: non-porous (dense) and porous membranes. Dense membranes such as Pd-based and/or perovskites make possible the selective permeation of certain gases (hydrogen and oxygen) via solution–diffusion mechanism or solid-state ionic conduction. Commonly, non-porous membranes exhibit extremely high permselectivity but limited permeability with high cost of investment due to the use of precious metals and/or extreme synthesis and operating conditions (Edlund and McCarthy, 1995).

Then, in the following sub-paragraphs, a small resume of the most representative inorganic membrane materials is reported and discussed.

9.4.1 Ceramic Membranes

Ceramic membranes are commonly constituted by alumina, titanium, or silica oxides and are chemically inert and stable at high temperatures (Smart et al., 2013). These characteristics make possible the utilization of ceramic membranes for microfiltration and ultrafiltration processes (Kim and Van der Bruggen, 2010; Fan et al. 2014), for food (Zhong et al., 2013; Alpatova et al., 2014), biotechnology, and pharmaceutical applications (Alventosa-deLara et al., 2012; Samei et al., 2013; Matos et al., 2013; Skouteris et al., 2012; Blanpain-Avet et al., 2011; Muvdi Nova et al., 2008), gas separation (Hashim et al., 2010; Adewole et al., 2013; Zhang et al., 2013; Hu et al., 2011; Lee et al., 2006) besides the applications in MRs (Lin et al., 2013; Wu et al., 2010; Volkov et al., 2012; Jiang et al., 2013; Wei et al., 2013; Iulianelli and Basile, 2011a).

Most of the drawbacks about ceramic membranes consists of the difficulties in their proper sealing in modules when exercised at high temperature as well as the high sensitivity to temperature gradient as a main cause of membrane cracking.

9.4.2 Zeolite Membranes

As a new class of inorganic membranes and unlike the most representative microporous metal oxides such as silica, alumina, and titania, zeolites can be described as microporous crystalline alumina-silicate with uniform pore size (Caro et al., 2000). They are relatively stable at high temperatures, acidic or basic in nature, and can show hydrophilic or organophilic characteristics. The molecular sieving abilities of zeolite membranes combined to selective sorption properties and catalytic activity, in addition to their thermal and chemical stability, make them attractive for gas separation (Bastani et al., 2013; Wang et al., 2014), pervaporation (Bowen et al., 2004), and several MR applications (Zhang et al., 2012; Fong et al., 2008; McLeary et al., 2006).

One of the main drawbacks related to these membranes is represented by their relatively low gas fluxes compared to other inorganic membranes (ceramic). Furthermore, the zeolites suffer the thermal effect; indeed, at high temperature, the zeolite layer can show negative thermal expansion (shrinkage), while the support continuously expands, involving in thermal stress at the attachment within the zeolite layer and the support as well as for the connection of the individual microcrystals within the zeolite layer.

9.4.3 Carbon Membranes

Carbon membranes are currently categorized on the basis of the pore size in carbon molecular sieve (CMS) and selective surface flow (SSF) membranes. Furthermore, carbon membranes can be subdivided into two categories: supported and unsupported.

CMS membranes can permit the transport of small molecules through the pores, blocking the passage of larger molecules. This characteristic (small pore size) makes CMS membranes highly permselective for separation of gas mixtures containing small gas species (Centeno et al., 2004; Briceño et al., 2012b). On the contrary, SSF membranes possess larger pores than the dimensions of the molecules passing through and, as a consequence, they are based on the preferential adsorption of some components in the gas mixture, which is followed by surface diffusion in the carbon matrix (Seo et al., 2002). In particular, CMS membranes have been identified as very interesting candidates for gas separation, both in terms of separation properties and stability. CMS membranes can be obtained by pyrolysis of many thermosetting polymers such as polyvinylidene chloride (PVDC), polyfurfuryl alcohol (PFA), cellulose triacetate, polyacrylonitrile (PAN), and phenol formaldehyde. Today, carbon membranes are also studied in applications in MRs for hydrogen production from methanol steam reforming (Sá et al., 2009, 2011; Briceño et al., 2012a) or, more generally, in dehydrogenation reaction process (Sznejer and Sheintuch, 2004; Hirota et al., 2013).

9.4.4 Metal Membranes

Commonly, dense metal membranes based on Pd, Ti, Ta, V, Nb, etc. are applied in the field of hydrogen separation from gas mixtures and/or in MR applications. Due to the characteristics of hydrogen permselectivity with respect to all other gases, palladium and its alloys play the role of the dominant material in this field (Amandusson et al., 2001; Yun and Oyama, 2011; Hatlevik et al., 2010; Dolan et al., 2006; Basile, 2008; Lin and Buxbaum, 2001; Rezac, 2000). Nevertheless, the presence of palladium in nature is scarce and, therefore, this involves a high cost for manufacturing Pd-based membranes (Criscuoli et al., 2001).

However, palladium is not the unique material possessing high hydrogen permselective characteristics and, in the last two decades, new hydrogen permselective membranes have been developed on the basis of such alternative materials as tantalum, vanadium, nickel, niobium, titanium, which are considered valid and—in some cases—less expensive than palladium (Dolan et al., 2012; Song et al., 2011; Ko et al., 2012; Dolan, 2010) or in combination with palladium (Tereschenko et al., 2007; Basile et al., 2008a; Ahmad et al., 2011; Nozaki and Hatano, 2013).

As a general consideration, a problem often associated to metal membranes is surface poisoning, which can have a major impact on thin metal membranes. In particular, H_2S or CO can seriously poison Pd-based membranes. Indeed, CO molecules can be adsorbed on the palladium surface, blocking available dissociation sites for hydrogen (Goldstein and George, 2011). Otherwise, H_2S can deactivate the membrane by absorption and can react with the surface for producing corrosion products. However, the detrimental effect caused by H_2S can be minimized by operating at higher temperature or by using particular alloys (Gade et al. 2011). On the contrary, CO poisoning can be recovered for desorption by operating at temperatures above 300°C (Catalano et al., 2010). As a further drawback, if pure dense Pd-based membranes are exposed to hydrogen permeation below 300°C and 20 bar, the "embrittlement phenomenon" can take place, causing the mechanical resistance loss as well as the depletion of the hydrogen permselectivity characteristics. This problem affecting pure palladium membranes can be overcome by alloying palladium with other metals such as silver, copper, gold. (Basile et al., 2008b).

A part from the poisoning effect of hydrogen sulfide and CO, other contaminants such as SO_2, Hg vapor, thiophene, arsenic, unsaturated hydrocarbons, or chlorine carbon from organic materials make possible the irreversible poisoning of palladium membranes.

9.4.5 Composite Membranes

In the last decade, special attention has been paid to composite Pd-based membranes with the scope of reducing the amount of palladium and consequently the whole membrane cost. Indeed, as addressed in an extensive literature, to pursue the scope of Pd-based membrane cost reduction, recently there has been a strong interest toward the development of membranes constituted of a thin Pd-layer coating (ranging from submicron to few microns) on porous supports (ceramic, Vycor, or porous stainless steel (PSS)) (David amd Kopac, 2011; Moparthi et al., 2010; Itoh et al., 2014; Pinacci and Drago, 2012; Chen et al., 2010; Wei et al., 2012; Liguori et al., 2014). On this route, the benefits of composite Pd-based membrane utilization take into account a reduced material cost, an enhanced resistance to mechanical strength, and higher permeating flux, even if the hydrogen permselectivity of these membranes show a finite value.

Commonly, composite membranes are constituted by a thin layer of palladium or its alloy deposited onto porous supports such as porous Vycor glass (silica gel), SiO_2, Al_2O_3 and B_2O_3 or PSS. High mechanical resistance and easy gas sealing are allowed by supports having a thermal expansion coefficient close to palladium like the PSS. However, PSS support shows as a main issue, the intermetallic diffusion of palladium at relatively high temperatures, inducing a depletion of the hydrogen permselectivity.

9.5 Mass Transport Mechanisms for Inorganic Membranes

The mass transport through inorganic membranes takes place by following different regimes. In porous membranes, molecular transport depends on the membrane properties. For example, such macroporous materials as α-alumina provide no separating function and are mainly useful for supporting mesoporous separation layer.

Knudsen diffusion regime occurs in mesoporous membranes such as Vycor glass or γ-alumina, while microporous membranes such as CMSs, silica, and zeolites show higher permselectivity performance owing to their molecular sieving effect.

In case of porous membranes, the Poiseuille (viscous flow) regime takes place if the average pore diameter is bigger than the average free path of fluid molecules. Indeed, the collisions within the different molecules are more frequent than molecules and the porous wall, making not possible any separation (Iulianelli and Basile, 2011a, Iulianelli et al., 2012). Otherwise, if the average pore diameter of the porous membrane is similar to the average free path of fluid molecules, the Knudsen regime takes place. Under this mass transport regime, the collisions of the molecules with the porous wall frequently occur and the flux of the component permeating through the membrane is calculated by means of equation (9.1) as reported below (Petropoulos and Papadokostaki, 2012):

$$J_i = \frac{G}{\sqrt{2 \cdot M_i \cdot R \cdot T}} \cdot \frac{\Delta p_i}{\delta} \tag{9.1}$$

where J_i is the flux of the i-species across the membrane, G the geometrical factor, M_i the molecular weight of the i-species, R the universal gas constant, T the absolute temperature, Δp_i the pressure difference of i-species, and δ the membrane thickness.

However, when permeating molecules are adsorbed on the pore wall of the membrane due to the active sites, then surface diffusion mechanism occurs. This transport mechanism can deplete the effective pore dimensions, limiting the transfer of different molecular species. Nevertheless, this diffusion regime can be also present in the case of a Knudsen transport. It is less significant by lowering the temperature owing to the progressive decrease of the bond strength between molecules and surface. The regime of capillary condensation takes place if one of the components condenses within the pores owing to capillary forces, which are sufficiently strong only at low temperature and in the presence of small pores. In conditions of small and homogeneous pore dimensions, uniformly distributed on the membrane, capillary condensation can show high permselectivity (Uchytil et al., 2003). Multi-layer diffusion describes the strong interactions within molecules and the membrane surface. This mechanism takes place as an intermediate flow regime between surface diffusion and capillary condensation. Molecular sieve regime occurs if the pore diameters are very small, making possible only the permeation of smaller molecules.

In case of dense membranes, the transport takes place via solution-diffusion mechanism, in which the dissociated molecules on the gas/membrane interface are adsorbed at the atomic level on the membrane surface. Then, the atoms diffuse through the membrane lattice, re-combine to form molecules at the gas/membrane interface and then, desorb.

Among the dense metal membranes, special attention should be paid to the pure and alloyed palladium membranes owing to their full hydrogen permselectivity with respect to all other gases. In detail, the hydrogen molecular transport in palladium membranes is governed by the solution-diffusion mechanism, which involves six different activated steps such as: (i) dissociation of molecular hydrogen at the gas/metal interface; (ii) adsorption of the atomic hydrogen on the membrane surface; (iii) dissolution of atomic hydrogen into the palladium matrix; (iv) diffusion of atomic hydrogen through the membrane; (v) re-combination of atomic hydrogen to form hydrogen molecules at the gas/metal interface; and (vi) desorption of hydrogen molecules (Basile et al., 2011). Commonly, the hydrogen permeating flux through a membrane can be described by the following general equation:

$$J_{H_2} = Pe_{H_2}\left(p^n_{H_{2,\text{retentate}}} - p^n_{H_{2,\text{permeate}}}\right)/\delta \tag{9.2}$$

where J_{H_2} is the hydrogen flux permeating through the membrane, Pe_{H_2} the hydrogen permeability, δ the membrane thickness, $p_{H_{2,\text{retentate}}}$ and $p_{H_{2,\text{permeate}}}$ the hydrogen partial pressures in the retentate (reaction side) and permeate (where the hydrogen permeated through the membrane is recovered) sides, respectively, n (variable in the range 0.5–1) the dependence factor of the hydrogen permeating flux on the hydrogen partial pressure.

For dense palladium membranes fully selective to hydrogen permeation, equation (9.2) becomes the Sieverts–Fick law with the "n" factor equal to 0.5:

$$J_{H2,\text{Sieverts-Fick}} = Pe_{H_2} \cdot \left(p^{0.5}_{H_2,\text{retentate}} - p^{0.5}_{H_2,\text{permeate}}\right)/\delta \tag{9.3}$$

Nevertheless, at high pressures in the palladium bulk, the hydrogen–hydrogen interactions become negligible, thus the "n" factor is equal to 1:

$$J_{H_2} = Pe_{H_2} \cdot (p_{H_2,\text{retentate}} - p_{H2,\text{permeate}})/\delta \tag{9.4}$$

$$J_{H_2} = Pe^0_{H_2}\left[\exp\left(-E_a/RT\right)\right] \cdot \left(p^{0.5}_{H_2,\text{retentate}} - p^{0.5}_{H_2,\text{permeate}}\right)/\delta \tag{9.5}$$

By expressing the hydrogen permeability as an Arrhenius-like equation, the Sieverts–Fick law (9.3) becomes Richardson's equation (9.5).

9.6 Applications of Inorganic Membrane Reactors

Many chemical reaction processes can be performed in inorganic MRs and Table 9.1 reports a small summary where the most recent experimental investigations on a number of reactions for producing hydrogen or as typical reaction processes in industry are shown.

In detail, this table shows the typology and some characteristics of the membrane housed inside the MRs beside the authors and the institution of the corresponding author. In particular, a variety of MRs are applied and studied for performing various chemical reactions and—as shown—ceramic, metallic, zeolite, and composite membranes are useful to be allocated inside the MRs. In brief, what proposed in Table 9.1 demonstrates that inorganic MRs constitute a mature technology and many studies are still ongoing to improve the performance of each chemical process with the scope to propose them as a realistic and valid alternative to the CRs.

9.6.1 Recent Advances on Hydrogen Production in MRs from Steam Reforming of Renewable Sources

As previously stated, in the last years, hydrogen has received considerable attention as the most viable energy vector for numerous applications. Recently, the scientific community involved in this field has been devoted to study hydrogen generation in more environmental friendly processes, based on the exploitation of bio-sources and, particularly, the bio-alcohols. Among them, ethanol and methanol as bio-derived and renewable sources constitute attracting candidates for producing hydrogen through reforming reactions in both CRs and MRs. In particular, ethanol emerged as a more competitive source for hydrogen production because it is renewably producible from biomass (Bej et al., 2014; Bichon et al., 2008; Chiou et al., 2014; Iulianelli et al., 2010a; Iulianelli et al., 2010b), constituting an important bio-source that seems to be more suitable than other renewable bio-feedstock such as methanol, glycerol, acetic acid, diethyl ether, etc., owing to its very low toxicity (Iulianelli and Basile, 2011b). Nevertheless, methanol exploitation in reforming reactions shows various benefits as a hydrogen carrier for fuel cell applications (Xiu and Shahbazi,

Table 9.1 *Some of the chemical reaction processes performed in MRs.*

Kind of reaction	Membrane	Material	Reference (Institution of corresponding author)
Steam reforming of methane	Composite	Supported Pd-based	Dittmar et al. 2013 (Linde AG, Engineering Division, Germany)
Steam reforming of methane	Metal	Vanadium-based	Matsuka et al. 2013 (Applied Chemistry Dpt., Faculty of Engineering, Kyushu University, Japan)
Partial oxidation of methane	Metal	Palladium	Chibane and Djellouli 2012 (Laboratoire de Génie des Procédés Chimiques (LGPC), Université Ferhat Abbas, Algery)
Dry reforming of methane	Composite	Supported Pd-based	García-García et al. 2013 (Department of Chemical Engineering, Imperial College London, UK)
Steam reforming of ethanol	Metal	Pd-Ag	Koch et al. 2013 (Institut de Te'cniques Energe'tiques, Universitat Polite'cnica de Catalunya, Spain)
Water–gas shift reaction	Composite	Supported Pd-Au	Hwang et al. 2013 (Korea Institute of Energy Research, Republic of Korea)
Autothermal steam reforming of glycerol	Metal	Pd-Ag	Lin et al. 2013 (Dept. of Chemical Engineering, Feng Chia University, Taiwan)
Steam reforming of methanol	Composite	Supported Pd-based	Liguori et al. 2014 (Institute on Membrane Technology of the National Research Council, Italy)
Direct cracking of methane	Ceramic	Pd-alloy	Han et al. 2013 (Icheon Branch, Korea Institute of Ceramic Engineering and Technology, Republic of Korea)
Acetic acid steam reforming	Metal	Pd-Ag	Iulianelli et al. 2008 (Institute on Membrane Technology of the National Research Council, Italy)
Dimethyl ether steam reforming	Composite	Silica/PSS	Park et al. 2008 (Nat. Res. Lab. for Funct. Membranes, Env. and En. Res. Center, Korea Research Institute of Chemical Technology, South Korea)

Table 9.1 *(Continued)*

Kind of reaction	Membrane	Material	Reference (Institution of Corresponding author)
Dehydrogenation of cyclohexane to benzene	Zeolite	FAU/Al_2O_3	Jeong et al. 2003 (Applied Chemistry Dpt., Kyushu University, Japan)
Decomposition of ammonia	Composite	Silica/Ru/γ-Al_2O_3/α-Al_2O_3	Li et al. 2012 (Chemical Engineering Dpt., Hiroshima University, Japan)
Dehydrogenation of ethane to ethylene	Ceramic	Composite alumina	Chalakov et al. 2009 (Max Planck Inst., Germany)
Dehydrogenation of isobutane to isobutene	Composite	Pd-Ag/PSS	Liang and Hughes 2005 (Institute of Materials Research, University of Salford, UK)
Dehydrogenation of propane	Composite	Pd/Al_2O_3	Gbenedio et al. 2010 (Chem. Eng. and Chem. Techn. Dpt., Imperial College London, UK)
Dehydrogenation of butane to butadiene	Ceramic	V/MgO-impregnated Al_2O_3	Alfonso et al. 2002 (Chemical Eng. Dpt., University of Zaragoza, Spain)
Dehydrogenation of ethane to ethylene	Ceramic	BYS/α-Al_2O_3	Akin and Lin 2002 (Chem. Eng. Dpt., University of Cincinnati, USA)
Dehydrogenation of ethylbenzene to styrene	Metal	Pd	Abo-Ghander et al. 2008 (Chemical and Biochemical Eng. Dpt., University of British Columbia, Canada)

2012). Indeed, it is conventionally performed at relatively low temperatures in the range between 240°C and 260°C (Chein et al., 2012) that, compared to methane steam reforming performed at 800–1000°C, seems to be really attractive for hydrogen production. Figure 9.5 shows an overview of the most significant results in the last decade about methanol steam reforming in both MRs and CRs in terms of methanol conversion at different temperatures. As shown, although a direct comparison is not possible because the various experimental results have been obtained at different experimental conditions, it is possible to state that qualitatively the MRs allow to reach better conversion than the CRs with the further advantage to recover concentrated hydrogen as a permeated stream with the purity depending on which kind of hydrogen selective membrane is allocated inside the MR. The same aspect can be observed for ethanol steam reforming reaction (Liguori et al., 2014; Lim et al. 2010; Lin et al., 2013; Lin et al., 2008; Mironova et al., 2014; Sekine et al., 2014; Yu et al., 2009; Yun et al., 2012) (Figure 9.6). In this case, the further benefit of MRs utilization is constituted by the possibility of exercise the MRs at temperature lower than the CRs favoring possible energy conservation with respect to the CRs.

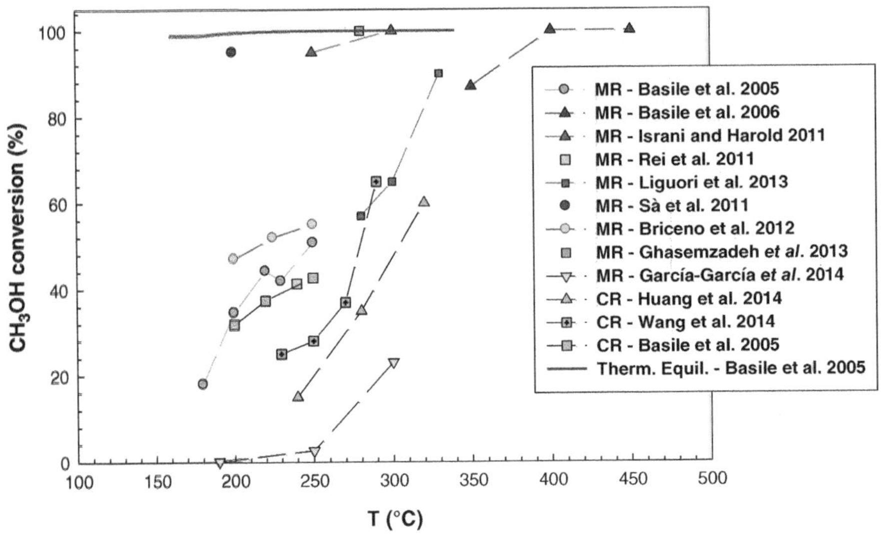

Figure 9.5 Methanol conversion versus temperature during methanol steam reforming reaction in MRs and CRs.

9.7 Conclusions

This chapter highlighted the importance of the MR application as a feasible and alternative technology to the conventional systems. Furthermore, MR technology has been introduced and discussed particularly in the field of inorganic membranes utilization to perform reaction processes for the generation of hydrogen, viewed as a new carrier for power production and

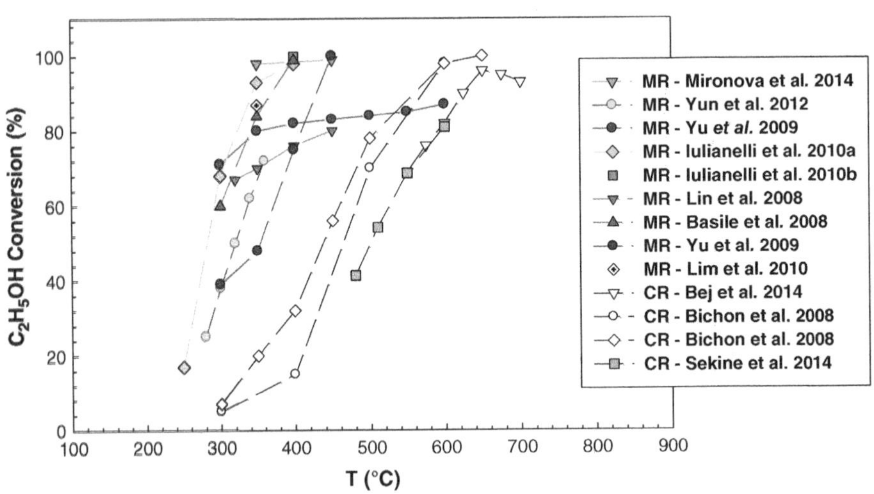

Figure 9.6 Ethanol conversion versus temperature during methanol steam reforming reaction in MRs and CRs.

alternative to the exploitation of derived fossil fuels. Furthermore, the various typologies of MRs have been described as well as the field of application beside the production of hydrogen. More in detail, this chapter has given a small overview on the various inorganic membranes useful in MR applications as well as their transport mechanisms. As a special case, a summary about the recent findings on hydrogen generation from reforming reactions of renewables such as ethanol and methanol in MRs has been given, making comparisons with recent experimental results obtained also in CRs.

List of Symbols

δ	Thickness
J_{H_2}	Hydrogen permeating flux
J_i	Permeating flux of i-species
ΔP_i	Transmembrane pressure of i-species
M_i	Molecular weight of i-species
R	Universal gas constant
T	Temperature
G	Geometrical factor
Pe_{H_2}	Hydrogen permeability
n	Dependence factor of the hydrogen permeating flux on the hydrogen partial pressure
$p_{H_2,\mathrm{retentate}}$	Hydrogen partial pressure on the retentate side
$p_{H_2,\mathrm{permeate}}$	Hydrogen partial pressure on the permeate side
E_a	Activation energy

List of Abbreviations

MR	Membrane reactor
PEM	Proton exchange membrane
GHG	Greenhouse gas
CR	Conventional reactor
CMR	Catalytic membrane reactor
CNMR	Catalytic non-permselective membrane reactor
NMR	Non-permselective membrane reactor
RSPBR	Reactant-selective packed-bed reactor
FBMR	Fluidized-bed membrane reactor
PBMR	Packed-bed membrane reactor
SSF	Selective surface flow
CMS	Carbon molecular sieve
PVDC	Polyvinylidene chloride
PFA	Polyfurfuryl alcohol
PAN	Polyacrylonitrile
PSS	Porous stainless steel

References

Abo-Ghander, N.S., Grace, J.R., Elnashaie, S.S.E.H., Lim, C.J., Modeling of a novel membrane reactor to integrate dehydrogenation of ethylbenzene to styrene with hydrogenation of nitrobenzene to aniline, *Chem. Eng. Sci.*, 2008; **63**:1817–1826.

Adewole, J.K., Ahmad, A.L., Ismail, S., Leo, C.P., Current challenges in membrane separation of CO_2 from natural gas: A review, *Int. J. Greenhouse Gas Contr.*, 2013; **17**: 46–65.

Adhikari, S., Fernand, S., Hydrogen membrane separation techniques, *Ind. Eng. Chem. Res.*, 2006; **45**:875–881.

Adris, A.M., Grace, J.R., Characteristics of fluidized-bed membrane reactors: scale-up and practical issues, *Ind. Eng. Chem. Res.*, 1997; **36**:4549–4556.

Ahmad, A.L., Jaya, M.A.T., Derek, C.J.C., Ahmad, M.A., Synthesis and characterization of TiO_2 membrane with palladium impregnation for hydrogen separation, *J. Membr. Sci.*, 2011; **366**:166–175.

Akin, F.T., Lin, Y.S., Selective oxidation of ethane to ethylene in a dense tubular membrane reactor, *J. Membr. Sci.*, 2002; **209**:457–467.

Alfonso, M.J., Menéndez, M., Santamaría, J., Oxidative dehydrogenation of butane on V/MgO catalytic membranes, *Chem. Eng. J.*, 2002; **90**:131–138.

Alpatova, A., Kim, E.-S., Dong, S., Sun, N., Chelme-Ayala, P., El-Din, M.G., Treatment of oil sands process-affected water with ceramic ultrafiltration membrane: Effects of operating conditions on membrane performance, *Sep. Purif. Techn.*, 2014; **122**:170–182.

Alventosa-deLara, E., Barredo-Damas, S., Alcaina-Miranda, M.I., Iborra-Clar, M.I., Ultrafiltration technology with a ceramic membrane for reactive dye removal: Optimization of membrane performance, *J. Hazardous Mat.*, 2012; **209–210**:492–500.

Amandusson, H., Ekedahl, L.-G., Dannetun, H., Hydrogen permeation through surface modified Pd and PdAg membranes, *J. Membrane Sci.*, 2001; **193**:35–47.

Andrić, P., Meyer, A.S., Jensen, P.A., Johansen, K.D., Reactor design for minimizing product inhibition during enzymatic lignocellulose hydrolysis II. Quantification of inhibition and suitability of membrane reactors, *Biotechnol. Adv.*, 2010; **28**:407–425.

Arstad, B., Venvik, H., Klette, H., Walmsley, J.C., Tucho, W.M., Holmestad, R., Holmen, A., Bredesen, R., Studies of self-supported 1.6 μm Pd/23 wt.% Ag membranes during and after hydrogen production in a catalytic membrane reactor, *Catal. Today*, 2006; **118**:63–72.

Basile, A., Hydrogen production using Pd-based membrane reactors for fuel cells, *Top. Catal.*, 2008; **51**:107–122.

Basile, A., Chakraborty, S., Iulianelli, A., Membrane and membrane reactors for sustainable growth, in *Membrane Processes for Sustainable Growth*, A. Basile and A. Cassano (Eds.), Nova Science Publishers, Inc., New York, 2013; pp. 121–151, ISBN: 978-1-62618-446-6.

Basile, A., Gallucci, F., Iulianelli, A., Tereschenko, G.F., Ermilova, M.M., Orekhova, N.V., Ti-Ni-Pd dense membranes – The effect of the gas mixtures on the hydrogen permeation, *J. Membr. Sci.*, 2008a; **310**:44–50.

Basile, A., Gallucci, F., Tosti, S., Synthesis, characterization, and applications of palladium membranes, *Membr. Sci. Technol.*, 2008b; **13**:255–323.

Basile, A., Iulianelli, A., Longo, T., Liguori, S., De Falco, M., Pd-based selective membrane state-of-the-art, in *Membrane Reactors for Hydrogen Production Processes*, L. Marrelli,

M. De Falco and G. Iaquaniello (Eds.), Springer, London, Dordrecht, Heidelberg, New York, 2011; pp. 21–55, ISBN: 978-0-85729-150-9, DOI: 10.1007/978-0-85729-151-6.

Bastani, D., Esmaeili, N., Asadollahi, M., Polymeric mixed matrix membranes containing zeolites as a filler for gas separation applications: A review, *J. Ind. Eng. Chem.*, 2013; **19**:375–393.

Bej, B., Pradhan, N.C., Neogi, S., Production of hydrogen by steam reforming of ethanol over alumina supported nano-NiO/SiO_2 catalyst, *Catal. Today*, 2014; 237: 80–88. http://dx.doi.org/10.1016/j.cattod.2014.01.019

Bichon, P., Haugom, G., Venvik, H.J., Holmen, A., Blekkan, E.A., Steam reforming of ethanol over supported Co and Ni catalysts, *Top. Catal.*, 2008; **49**:38–45.

Blanpain-Avet, P., Faille, C., Delaplace, G., Bénézech, T., Cell adhesion and related fouling mechanism on a tubular ceramic microfiltration membrane using *Bacillus cereus* spores, *J. Membr. Sci.*, 2011; **385–386**:200–216.

Bowen, T.C., Noble, R.D., Falconer, J.L., Fundamentals and applications of pervaporation through zeolite membranes, *J. Membr. Sci.*, 2004; **245**:1–33.

Briceño, K., Iulianelli, A., Montané, D., Garcia-Valls, R., Basile, A., Carbon molecular sieve membranes supported on non-modified ceramic tubes for hydrogen separation in membrane reactors, *Int. J. Hydrogen Energ.*, 2012a; **37**:13536–13544.

Briceño, K., Montané, D., Garcia-Valls, R., Iulianelli, A., Basile, A., Fabrication variables affecting the structure and properties of supported carbon molecular sieve membranes for hydrogen separation, *J. Membr. Sci.*, 2012b; **415–416**:288–297.

Brunetti, A., Drioli, E., Barbieri, P., Energy and mass intensities in hydrogen upgrading by a membrane reactor, *Fuel Proc. Technol.*, 2014; **118**:278–286.

Caro, J., Noack, M., Kölsch, P., Schäfer, R., Zeolite membranes – State of their development and perspective, *Microp. Mesop. Mater.*, 2000; **38**:3–24.

Catalano, J., Baschetti, M.G., Sarti, G.C., Hydrogen permeation in palladium-based membranes in the presence of carbon monoxide, *J. Membr. Sci.*, 2010; **362**:221–233.

Centeno, T.A., Vilas, J.L., Fuertes, A.B., Effects of phenolic resin pyrolysis conditions on carbon membrane performance for gas separation, *J. Membr. Sci.*, 2004; **228**:45–54.

Chalakov, L., Rihko-Struckmann, L.K., Munder, B., Sundmacher, K., Oxidative dehydrogenation of ethane in an electrochemical packed-bed membrane reactor: Model and experimental validation, *Chem. Eng. J.*, 2009; **145**:385–392.

Chatenet, M., Dubau, L., Job, N., Maillard, F., The (electro)catalyst membrane interface in the proton exchange membrane fuel cell: Similarities and differences with non-electrochemical Catalytic Membrane Reactors, *Catal. Today*, 2010; **156**:76–86.

Chaubeyet, R., Sahu, S., James, O.O., Maity, S., A review on development of industrial processes and emerging techniques for production of hydrogen from renewable and sustainable sources, *Ren. Sustain. En. Rev.*, 2013; **23**:443–462.

Chein, R.Y., Chen, Y.C., Lin, Y.S., Chung, J.N., Hydrogen production using integrated methanol-steam reforming reactor with various reformer designs for PEM fuel cells, *Int. J. Hydrogen Energ.*, 2012; **36**:466–476.

Chen, W., Hu, X., Wang, R., Huang, Y., On the assembling of Pd/ceramic composite membranes for hydrogen separation, *Sep. Purif. Technol.*, 2010; **72**:92–97.

Chibane, L., Djellouli, B., Role of periodic input composition and sweeping gas for improvement of hydrogen production in a palladium membrane reactor by partial oxidation of methane, *Chinese J. Chem. Eng.*, 2012; **20**:577–588.

Chiou, J.Y.Z., Lee, C.L., Ho, K.F., Huang, H.H., Yu, S.W., Wang, C.B., Catalytic performance of Pt-promoted cobalt-based catalysts for the steam reforming of ethanol, *Int. J. Hydrogen Energ.*, 2014; **39** (11): 5653–5670 http://dx.doi.org/10.1016/j.ijhydene.2014.01.156

Coronas, J., Santamaria, J., Catalytic reactors based on porous ceramic membranes, *Catal. Today*, 1999; **51**:377–389.

Criscuoli, A., Basile, A., Drioli, E., Loiacono, O., An economic feasibility study for water gas shift membrane reactor, *J. Membr. Sci.*, 2001; **181**:21–27.

Dalmon, J.-A., Cruz-López, A., Farrusseng, D., Guilhaume, N., Iojoiu, E., Jalibert, J.-C., Miachon, S., Mirodatos, C., Pantazidis, A., Rebeilleau-Dassonneville, M., Schuurman, Y., van Veen, A.C., Oxidation in catalytic membrane reactors, *Appl. Catal. A Gen.*, 2007; **325**:198–204.

David, E., Kopac, J., Development of palladium/ceramic membranes for hydrogen separation, *Int. J. Hydrogen Energ.*, 2011; **36**:4498–4506.

Deshmukh, S.A.R.K., Heinrich, S., Mörl, L., van Sint Annaland, M., Kuipers, J.A.M., Membrane assisted fluidized bed reactors: Potentials and hurdles, *Chem. Eng. Sci.*, 2007; **62**:416–436.

Dittmar, B., Behrens, A., Schödel, N., Rüttinger, M., Franco, Th., Straczewski, G., Dittmeyer, R., Methane steam reforming operation and thermal stability of new porous metal supported tubular palladium composite membranes, *Int. J. Hydrogen Energ.*, 2013; **38**: 8759–8771

Dittmeyer, R., Höllein, V., Daub, K., Membrane reactors for hydrogenation and dehydrogenation processes based on supported palladium, *J. Mol. Catal. A Chem.*, 2001; **173**:135–184.

Dolan, M.D., Non-Pd BCC alloy membranes for industrial hydrogen separation, *J. Membr. Sci.*, 2010; **362**:12–28.

Dolan, M.D., Dave, N.C., Ilyushechkin, A.Y., Morpeth, L.D., McLennan, K.G., Composition and operation of hydrogen-selective amorphous alloy membranes, *J. Membr. Sci.*, 2006; **285**:30–55.

Dolan, M.D., Song, G., McLennan, K.G., Kellam, M.E., Liang, D., The effect of Ti on the microstructure, hydrogen absorption and diffusivity of V–Ni alloy membranes, *J. Membr. Sci.*, 2012; **415–416**:320–327.

Edlund, D.J. and McCarthy, J., The relationship between intermetallic diffusion and flux decline in composite-metal membranes: Implications for achieving long membrane lifetime, *J. Membr. Sci.*, 1995; **107**:147–153.

Fan, X., Tao, Y., Wang, L., Zhang, X., Lei, Y., Wang, Z., Noguchi, H., Performance of an integrated process combining ozonation with ceramic membrane ultra-filtration for advanced treatment of drinking water, *Desalination*, 2014; **335**:47–54.

Fong, Y.Y., Abdullah, A.Z., Ahmad, A.L., Bhatia, S., Development of functionalized zeolite membrane and its potential role as reactor combined separator for para-xylene production from xylene isomers, *Chem. Eng. J.*, 2008; **139**:172–193.

Gade, S.K., DeVoss, S.J., Coulter, K.E., Paglieri, S.N., Alptekin, G.O., Way, J.D., Palladium–gold membranes in mixed gas streams with hydrogen sulfide: Effect of alloy content and fabrication technique, *J. Membr. Sci.*, 2011; **378**:35–41.

García-García, F.R., Soria, M.A., Mateos-Pedrero, C., Guerrero-Ruiz, A., Rodríguez-Ramos, I., Li, K., Dry reforming of methane using Pd-based membrane reactors fabricated from different substrates, *J. Membr. Sci.*, 2013; **435**:218–225.

Gbenedio, E., Wu, Z., Hatim, I., Kingsbury, B.F.K., Li, K., A multifunctional Pd/alumina hollow fibre membrane reactor for propane dehydrogenation, *Catal. Today*, 2010; **156**:93–99.

Goldstein, D.N., George, S.M., Surface poisoning in the nucleation and growth of palladium atomic layer deposition with Pd(hfac)2 and formalin, *Thin Solid Films*, 2011; **519**:5339–5347.

Han, K.-S., Kim, J.H., Kim, H.K., Hwang, K.T., Direct methane cracking using a mixed conducting ceramic membrane for production of hydrogen and carbon, *Int. J. Hydrogen Energ.*, 2013; **38**:16133–16139.

Hashim, S.M., Mohamed, A.R., Bhatia, S., Current status of ceramic-based membranes for oxygen separation from air, *Adv. Colloid Interf. Sci.*, 2010; **160**:88–100.

Hatlevik Ø., Gade, S.K., Keeling, M.K., Thoen, P.M., Davidson, A.P., Way, J.D., Palladium and palladium alloy membranes for hydrogen separation and production: History, fabrication strategies, and current performance, *Sep. Purif. Technol.*, 2010; **73**: 59–64.

Hirota, Y., Ishikado, A., Uchida, Y., Egashira, Y., Nishiyama, N., Pore size control of microporous carbon membranes by post-synthesis activation and their use in a membrane reactor for dehydrogenation of methylcyclohexane, *J. Membr. Sci.*, 2013; **440**: 134–139.

Hu, X., Yu, J., Song, J., Wang, X., Huang, Y., Toward low-cost Pd/ceramic composite membranes for hydrogen separation: A case study on reuse of the recycled porous Al_2O_3 substrates in membrane fabrication, *Int. J. Hydrogen Energ.*, 2011; **36**:15794–15802.

Hwang, K.R., Lee, S.K., Ryi, S.K., Kim, D.K., Kim, T.H., Park, J.S., Water-gas shift reaction in a plate-type Pd-membrane reactor over a nickel metal catalyst, *Fuel Proc. Technol.*, 2013; **106**:133–140.

Itoh, N., Haraya, K., A carbon membrane reactor, *Catal. Today*, 2000; **56**:103–111.

Itoh, N., Suga, E., Sato, T., Composite palladium membrane prepared by introducing metallic glue and its high durability below the critical temperature, *Sep. Purif. Technol.*, 2014; **121**:46–53.

Iulianelli, A., Basile, A., An experimental study on bio-ethanol steam reforming in a catalytic membrane reactor. Part I: Temperature and sweep-gas flow configuration effects, *Int. J. Hydrogen Energ.*, 2010a; **35**:3170–3177.

Iulianelli, A., Basile, A., Ceramic membranes: Preparation techniques, characterization procedures and applications in membrane reactors, *Int. J. Green Energ., Energ. & Environ.*, 2011a; **2**(3):1–17.

Iulianelli, A., Basile, A., Hydrogen production from ethanol via inorganic membrane reactors technology: A review, *Catal. Sci. Technol.*, 2011b; **1**:366–379.

Iulianelli, A., Basile, A., Sulfonated PEEK-based polymers in PEMFC and DMFC applications: A review, *Int. J. Hydrogen Energ.*, 2012; **37**:15241–1525.

Iulianelli, A., Liguori, S., Longo, T., Basile, A., Inorganic membrane and membrane reactor technologies for hydrogen production, in *Hydrogen Production: Prospects and Processes*, D.R. Honery and P. Moriarty (Eds.), Nova Science Publishers, Victoria, Australia, Series: Energy Science, Engineering and Technology, 2012; pp. 377–398, ISBN: 978-1-62100-246-8.

Iulianelli, A., Liguori, S., Longo, T., Tosti, S., Pinacci, P., Basile, A., An experimental study on bio-ethanol steam reforming in a catalytic membrane reactor. Part II: Reaction pressure, sweep factor and WHSV effects, *Int. J. Hydrogen Energ.*, 2010b; **35**:3159–3164.

Iulianelli, A., Longo, T., Basile, A., CO-free hydrogen production by steam reforming of acetic acid carried out in a Pd–Ag membrane reactor: The effect of co-current and counter-current mode, *Int. J. Hydrogen Energ.*, 2008; **33**:4091–4096.

Jeong, B.-H., Sotowa, K.-I., Kusakabe, K., Catalytic dehydrogenation of cyclohexane in an FAU-type zeolite membrane reactor, *J. Membr. Sci.*, 2003; **224**:151–158.

Jiang, H., Meng, L., Chen, R., Jin, W., Xing, W., Xu, N., Progress on porous ceramic membrane reactors for heterogeneous catalysis over ultrafine and nano-sized catalysts, *Chinese J. Chem. Eng.*, 2013; **21**:205–215.

Kim, J., Van der Bruggen, B., The use of nanoparticles in polymeric and ceramic membrane structures: Review of manufacturing procedures and performance improvement for water treatment, *Environ. Pollut.*, 2010; **158**:2335–2349.

Ko, W.-S., Jeon, J.-B., Shim, J.-H., Lee, B.-J., Origin of hydrogen embrittlement in vanadium-based hydrogen separation membranes, *Int. J. Hydrogen Energ.*, 2012; **37**:13583–13593.

Koch, R., López, E., Divins, N.J., Allué, M., Jossen, A., Riera, J, Llorca, J., Ethanol catalytic membrane reformer for direct PEM FC feeding, *Int. J. Hydrogen Energ.*, 2013; **38**: 5605–5615.

Lee, H.-J., Yamauchi, H., Suda, H., Haraya, K., Influence of adsorption on the gas permeation performances in the mesoporous alumina ceramic membrane, *Sep. Pur. Technol.*, 2006; **49**:49–55.

Li, G., Kanezashi, M., Lee, H.R., Maeda, M., Yoshioka, T., Tsuru, T., Preparation of a novel bimodal catalytic membrane reactor and its application to ammonia decomposition for COx-free hydrogen production, *Int. J. Hydrogen Energ.*, 2012; **37**:12105–12113.

Liang, W., Hughes, R., The catalytic dehydrogenation of isobutane to isobutene in a palladium/silver composite membrane reactor, *Catal. Today*, 2005; **104**:238–243.

Liguori, S., Iulianelli, A., Dalena, F., Piemonte, V., Huang, Y., Basile, A., Methanol steam reforming in an Al_2O_3 supported thin Pd-layer membrane reactor over $Cu/ZnO/Al_2O_3$ catalyst, *Int. J. Hydrogen Energ.*, 2014; **39**:18702–18710. doi: http://dx.doi.org/10.1016/j.ijhydene.2013.11.113.

Liguori, S., Iulianelli, A., Dalena, F., Pinacci, P., Drago, F., Broglia, M., Huang, Y., Basile, A., Performance and long-term stability of Pd/PSS and Pd/Al_2O_3 for hydrogen separation, *Membranes*, 2014; **4**(1):143–162. doi: 10.3390/membranes/40x000x.

Lim, H., Gu, Y., Oyama, S.T., Reaction of primary and secondary products in a membrane reactor: Studies of ethanol steam reforming with a silica–alumina composite membrane, *J. Membr. Sci.*, 2010; **351**:149–159.

Lin, H., Peng, W., Zhang, M., Chen, J., Hong, H., Zhang, Y., A review on anaerobic membrane bioreactors: Applications, membrane fouling and future perspectives, *Desalination*, 2013; **314**:169–188.

Lin, K.H., Chang, A.C.-C., Lin, W.H., Chen, S.H., Chang, C.Y., Chang, H.F., Autothermal steam reforming of glycerol for hydrogen production over packed-bed and Pd/Ag alloy membrane reactors, *Int. J. Hydrogen Energ.*, 2013; **38**:12946–12952.

Lin, W.H., Hsiao, C.S., Chang, H-F., Effect of oxygen addition on the hydrogen production from ethanol steam reforming in a Pd–Ag membrane reactor, *J. Membr. Sci.*, 2008; **322**:360–367.

Lin, Y.M., Rei, M.H., Process development for generating high purity hydrogen by using supported palladium membrane reactor as steam reformer, *Int. J. Hydrogen Energ.*, 2000; **25**:211–219.

Lin, Y.S., Microporous and dense inorganic membranes: current status and prospective, *Sep. Purif. Technol.*, 2001; **25**:39–55.

Lin, Y.S., Buxbaum, R.E., Metal membranes: Membrane separations, Reference Module in Chemistry, Molecular Sciences and Chemical Engineering, from Encyclopedia of Separation Science, 2000, 3365–3372.

Lu, G.Q., Diniz de Costa, J.C., Duke, V, Giessler, S., Socolow, R., Williams, R.H., Kreutz, T., Inorganic membranes for hydrogen production and purification: A critical review and perspective, *J. Coll. Interface Sci.*, 2007; **314**:589–603.

Mallada, R., Menéndez, M., Inorganic Membranes: Synthesis, Characterization and Applications, Elsevier, 2008; pp. 1–460.

Matos, M., Suárez, M.A., Gutiérrez, G., Coca, J., Pazos, C., Emulsification with microfiltration ceramic membranes: A different approach to droplet formation mechanism, *J. Membr. Sci.*, 2013; **444**:345–358.

Matsuka, M., Higashi, M., Ishihara, T., Hydrogen production from methane using vanadium-based catalytic membrane reactors, *Int. J. Hydrogen Energ.*, 2013; **38**:6673–6680.

McLeary, E.E., Jansen, J.C., Kapteijn, F., Zeolite based films, membranes and membrane reactors: Progress and prospects, *Microp. Mesop. Mat.*, 2006; **90**:198–220.

Miachon, S., Perez, V., Crehan, G., Torp, E., Ræder, H., Bredesen, R., Dalmon, J.-A., Comparison of a contactor catalytic membrane reactor with a conventional reactor: Example of wet air oxidation, *Catal. Today*, 2003; **82**:75–81.

Mironova, E.Y., Ermilova, M.M., Orekhova, N.V., Muraviev, D.N., Yaroslavtsev, A.B., Production of high purity hydrogen by ethanol steam reforming in membrane reactor, *Catal. Today*, 2014; **236**: 64–69. http://dx.doi.org/10.1016/j.cattod.2014.01.014

Moparthi, A., Uppaluri, R., Gill, B.S., Economic feasibility of silica and palladium composite membranes for industrial dehydrogenation reactions, *Chem. Eng. Res. Des.*, 2010; **88**:1088–1101.

Moriarty, P., Honnery, D., Intermittent renewable energy: The only future source of hydrogen?, *Int. J. Hydrogen Energ.*, 2007; **32**:1616–1624.

Mozia, S., Photocatalytic membrane reactors (PMRs) in water and wastewater treatment. A review, *Sep. Purif. Technol.*, 2010; **73**:71–91.

Muvdi Nova, C.J., Paolucci-Jeanjean, D., Belleville, M.-P., Barboiu, M., Rivallin, M., Rios, G., Elaboration, characterization and study of a new hybrid chitosan/ceramic membrane for affinity membrane chromatography, *J. Membr. Sci.*, 2008; **1**:81–89

Nielsen, T.R., Manufacture of hydrogen, *Catal. Today*, 2005; **106**:293–296.

Nozaki, T., Hatano, Y., Hydrogen permeation through a Pd/Ta composite membrane with a HfN intermediate layer, *Int. J. Hydrogen Energ.*, 2013; **38**:11983–11987.

Park, S.-J., Lee, D.-W., Yu, C.-Y., Lee, K.-Y., Lee, K.-H., Hydrogen production from a DME reforming-membrane reactor using stainless steel-supported Knudsen membranes with high permeability, *J. Membr. Sci.*, 2008; **318**:123–128.

Petropoulos, J.H., Papadokostaki, K.G., May the Knudsen equation be legitimately, or at least usefully, applied to dilute adsorbable gas flow in mesoporous media? *Chem. Eng. Sci.*, 2012; **68**:392–400.

Pinacci, P., Drago, F., Influence of the support on permeation of palladium composite membranes in presence of sweep gas, *Catal. Today*, 2012; **193**:186–193.

Reij, M.W., Keurentjes, J.T.F., Hartmans, S., Review—Membrane bioreactors for waste gas treatment, *J. Biotechnol.*, 1998; **59**:155–167.

Rezac, M.E., Membrane Separations: Catalytic Membrane Reactors, Reference Module in Chemistry, Molecular Sciences and Chemical Engineering, from Encyclopedia of Separation Science, 2000, 1676–1682.

Sá, S., Silva, H., Sousa, J.M., Mendes, A., Hydrogen production by methanol steam reforming in a membrane reactor: Palladium versus carbon molecular sieve membranes, *J. Membr. Sci.*, 2009; **339**:167–170

Sá, S., Silva, H., Sousa, J.M., Mendes, A., Steam reforming of methanol over a $CuO/ZnO/Al_2O_3$ catalyst part II: A carbon membrane reactor, *Chem. Eng. Sci.*, 2011; **66**:5523–5530.

Samei, M., Mohammadi, T., Asadi, A.A., Tubular composite PVA ceramic supported membrane for bio-ethanol production, *Chem. Eng. Res. Des.*, 2013; **91**:2703–2712.

Saracco, G., Neomagus, H.W.J.P., Versteeg, G.F., van Swaaij, W.P.M., High-temperature membrane reactors: potential and problems, *Chem. Eng. Sci.*, 1999; **54**:1997–2017.

Scholes, C.A., Smith, K.H., Kentish, S.E., Stevens, G.W., Review—CO_2 capture from pre-combustion processes – Strategies for membrane gas separation, *Int. J. Greenhouse Gas Con.*, 2010; **4**:739–755.

Sekine, Y., Nakazawa, Y., Oyama, K., Shimizu, T., Ogo, S., Effect of small amount of Fe addition on ethanol steam reforming over Co/Al_2O_3 catalyst, *Appl. Catal. A: Gen.*, 2014; **472**:113–122

Seo, Y.G., Kum, G.H., Seaton, N.A., Monte Carlo simulation of transport diffusion in nanoporous carbon membranes, *J. Membr. Sci.*, 2002; **195**:65–73.

Shinnar, R., Demystifying the hydrogen myth, *Chem. Eng. Prog.*, 2004; November: 5–6.

Skouteris, G, Hermosilla, D., López, P., Negro, C., Blanco, A., Anaerobic membrane bioreactors for wastewater treatment: A review, *Chem. Eng. J.*, 2012; **198–199**:138–148.

Smart, S., Liu, S., Serra, J.M., Diniz Da Costa, J.C., Iulianelli, A., Basile, A., Porous ceramic membranes for membrane reactors, in *Handbook of Membrane Reactors—Fundamental Materials Science, Design and Optimisation*, A. Basile Ed., 2013; pp. 298–336, Woodhead Publishing Ltd Series in Energy, Cornwall. ISBN: 0857094149, DOI: 0.1533/9780857097330.2.298.

Song, G., Dolan, M.D., Kellam, M.E., Liang, D., Zambelli, S., V–Ni–Ti multi-phase alloy membranes for hydrogen purification, *J. Alloys Comp.*, 2011; **509**:9322–9328.

Soumitra, K., Bindal, R.C., Prabhakar, S., Tewari, P.K., The application of membrane reactor technology in hydrogen production using S–I thermochemical process: A roadmap, *Int. J. Hydrogen Energ.*, 2012; **37**:3612–3620.

Sznejer, G., Sheintuch, M., Application of a carbon membrane reactor for dehydrogenation reactions, *Chem. Eng. Sci.*, 2004; **59**:2013–2021.

Tereschenko, G.F., Ermilova, M.M., Mordovin, V.P., Orekhova, N.V., Gryaznov, V.M., Iulianelli, A., Gallucci, F., Basile, A., New Ti-Ni dense membranes with low palladium content, *Int. J. Hydrogen Energ.*, 2007; **32**:4016–4022.

Tong, J., Shirai, R., Kashima, Y., Matsumura, Y., Preparation of a pinhole-free Pd-Ag membrane on a porous metal support for pure hydrogen separation, *J. Membr. Sci.*, 2005; **260**:84–89.

Tsotsis, T.T., Champagnie, A.M., Vasileiadis, S.P., Ziaka, Z.D., Minet, R.G., Packed bed catalytic membrane reactors, *Chem. Eng. Sci.*, 1992; **47**:2903–2908.

Turner, J.A., Sustainable hydrogen production, *Science*, 2004; **305**:972–974.

Uchytil, P., Petrickovic, R., Thomas, S., Seidel-Morgenstern, A., Influence of capillary condensation effects on mass transport through porous membranes, *Sep. Purif. Technol.*, 2003; **33**:273–281.

Volkov, V.V., Novitskii, E.G., Dibrov, G.A., Samokhin, P.V., Kipnis, M.A., Yaroslavtsev, A.B., Catalytic conversion of methanol to dimethyl ether on polymer/ceramic composite membranes, *Catal. Today*, 2012; **193**:31–36.

Wang, H., Dong, X., Lin, Y.S., Highly stable bilayer MFI zeolite membranes for high temperature hydrogen separation, *J. Membr. Sci.*, 2014; **450**:425–432.

Web [1]. http://www.tokyo-gas.co.jp/index_e.html, February 2014.

Wei, L., Yu, J., Hu, X., Huang, Y., Fabrication of H_2-permeable palladium membranes based on pencil-coated porous stainless steel substrate, *Int. J. Hydrogen Energ.*, 2012; **37**:13007–13012.

Wei, Y., Yang, W., Caro, J., Wang, H., Dense ceramic oxygen permeable membranes and catalytic membrane reactors, *Chem. Eng. J.*, 2013; **220**:185–203.

Westermann, T., Melin, T., Review—Flow-through catalytic membrane reactors—Principles and applications, *Chem. Eng. Proc.*, 2009; **48**:17–28.

Wieland, S., Melin, T., Lamm, A., Membrane reactors for hydrogen production, *Chem. Eng. Sci.*, 2002 **57**:1571–1576.

Wu, Z., Wang, B., Li, K., A novel dual-layer ceramic hollow fibre membrane reactor for methane conversion, *J. Membr. Sci.*, 2010; **352**:63–70.

Xiu, S., Shahbazi, A., Bio-oil production and upgrading research: A review, *Int. J. Hydrogen Energ.*, 2012; **16**:4406–4414.

Yu, C.Y., Lee, D.W., Park, S.J., Lee, K.Y., Lee, K.H., Study on a catalytic membrane reactor for hydrogen production from ethanol steam reforming, *Int. J. Hydrogen Energ.*, 2009; **34**:2947–2954.

Yun, S., Lim, H., Oyama, S.T., Experimental and kinetic studies of the ethanol steam reforming reaction equipped with ultrathin Pd and Pd–Cu membranes for improved conversion and hydrogen yield, *J. Membr. Sci.*, 2012; **409–410**:222–231

Yun, S., Oyama, S.T., Correlations in palladium membranes for hydrogen separation: A review, *J. Membr. Sci.*, 2011; **375**:28–45.

Zhang, Y., Sunarso, Y., Liu, S., Wang, R., Current status and development of membranes for CO_2/CH_4 separation: A review, *Int. J. Greenhouse Gas Control*, 2013; **12**:84–107.

Zhang, Y., Wu, Z., Hong, Z., Gu, X, Xu, N., Hydrogen-selective zeolite membrane reactor for low temperature water gas shift reaction, *Chem. Eng. J.*, 2012; **197**:314–321.

Zhong, Z., Xing, W., Zhang, B., Fabrication of ceramic membranes with controllable surface roughness and their applications in oil/water separation, *Ceramics Int.*, 2013; **39**:4355–4361.

10

Membranes for IGCC Power Plants

Kamran Ghasemzadeh,[1] *Angelo Basile,*[2] *and Seyyed Mohammad Sadati Tilebon*[1]
[1]*Chemical Engineering Department, Urmia University of Technology, Urmia, Iran*
[2]*Institute on Membrane Technology – Italian National Research Council (ITM-CNR), University of Calabria, Cosenza, Italy*

10.1 Introduction

Gasification is one of the favorite methods of producing syngas from heavy fossil feedstocks such as petcoke, coal, and refinery residues. In the integrated gasification combined cycle (IGCC) power plants, the feedstock, an oxidizer and a temperature moderator (usually steam) react inside a gasifier to produce a syngas stream that is rich in CO and H_2, which is ultimately fed into a gas turbine-based combined cycle for power generation. However, research on the IGCC power plant technology started in the 1970s, but coal-based IGCC power plants are still not fully commercial.

On the other hand, it is estimated that the total global power generation in 2030 will be about 1.6 times more than that of 2010, and it is also expected that the use of coal-fired thermal power generation will continue to grow as it remains the main source of electrical power. Coal, due to its stable and cheap price, large amounts of minable reserves, a good geographical distribution of coal mines and, depending on their location, found in politically stable regions, will remain an important primary energy source for the future (Nagasaki et al., 2013).

According to scientific reports, over 50% of the electricity demand in the EU is provided by burning fossil fuels, especially by burning coal and natural gas. It should be noted that combustion of these fuels is accompanied by emissions of large amounts of pollutants into the atmosphere, namely, CO_x, SO_x, NO_x and dust, and some of these pollutants, such as CO_2, CH_4, NO_x and H_2O, are called greenhouse gases, which cause the greenhouse effect (Kotowicz et al., 2010).

Integrated Membrane Systems and Processes, First Edition. Edited by Angelo Basile and Catherine Charcosset.

However, the amount of CO_2 emitted per unit of energy produced from fossil fuels such as natural gas or oil is low compared to that produced by coal. This problem has created a prospective future for clean coal technologies. There is also a tendency towards introducing regulations on the per-unit emissions of CO_2 from thermal power plants, and this creates the potential for requirements that are difficult to achieve in practice, such as higher efficiency and the use of biomass in multi-fuel combustion (Nagasaki et al., 2013).

The IGCC power plants are being developed as an economical method of producing electricity, while simultaneously capturing CO_2 for sequestration. In these kinds of power plants, conventional cold absorption processes are the common technology for CO_2 separation from gasified coal syngas. Separation and sequestration of the CO_2 by these procedures increase the levelized cost of the electricity (LCOE) produced by about 30%.

Therefore, membrane technology can be suggested as a suitable option for use in the IGCC plant environments. On the other hand, membranes are compact and the modular design is well suited to being scaled up. The high pressure of the syngas stream in IGCC plants works well with the driving forces of the species transport across the membrane. Depending on the kind of membrane material used (ceramic, metallic, or polymeric), different locations throughout the IGCC process can be appropriate for the membrane integration. To the best of the authors' knowledge, polymeric membranes are commercially available and can be integrated at lower temperatures, i.e., below 150°C. However, the metallic and ceramic membranes require a higher temperature. According to the literature, using membranes in IGCC plants can cause an efficiency loss of the order of 4–10%. Hence, in recent years, research has focused on improving membrane performances in IGCC plants to compensate for the efficiency loss (Franz et al., 2014).

This chapter deals with the application of membranes for oxygen, hydrogen, and carbon dioxide separation in an IGCC plant for power generation. Hence, an extensive description of how to use the various membranes to improve IGCC plant efficiency will be presented.

10.2 IGCC Technology for Power Generation

As shown in Figure 10.1, the IGCC process flow scheme is significantly complex. Typically, an air separation unit is first of all applied to produce oxygen, which with water is then used to gasify coal at a high temperature and pressure. Generally, the syngas produced (CO and H_2) also includes other components, such as CO_2, N_2, CH_4, Ar, H_2S, particulates and tars. The gas stream is quenched and scrubbed to remove tar and particulates. If CO_2 capture or H_2 production is desired in the IGCC plant, with regard to the water gas shift (WGS) reaction, high and low temperature shift reactors are used to convert CO and H_2O to H_2 and CO_2. Sulfur compounds, and optionally CO_2, are then eliminated by a low temperature absorption process. The high pressure syngas stream is then burned with air and the hot high-pressure gas product stream is used to drive a gas turbine and generate electricity. The hot turbine exhaust stream is applied to produce steam that provides additional electricity in a steam turbine. The cooled gas stream is vented into the atmosphere. In general, an IGCC plant has an overall heat-to-electrical power efficiency of about 45%, significantly better than the 35% efficiency of a conventional subcritical pulverized coal (PC) power plant. However, this advantage is more than outweighed by the higher capital cost of an IGCC plant (Merkel et al., 2012). As far as the authors are aware, without CO_2 capture, the

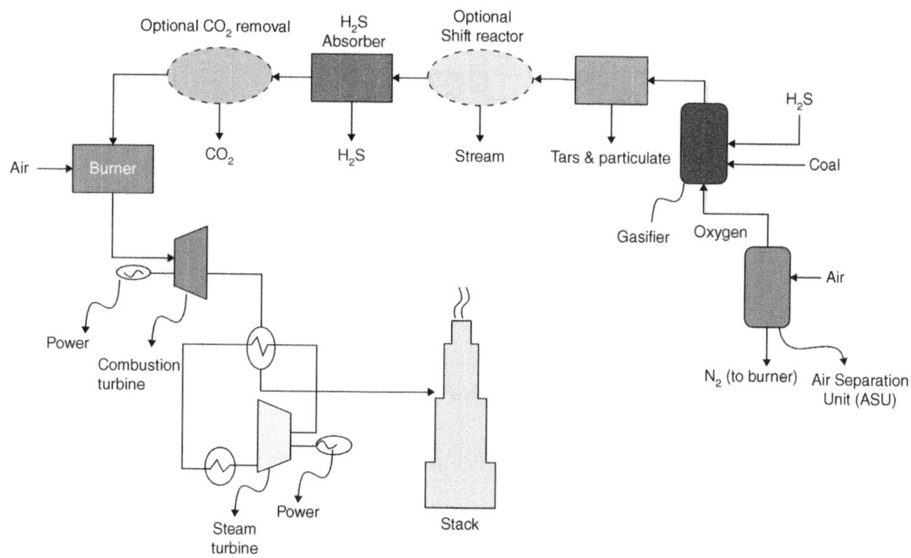

Figure 10.1 *Process flow diagram for an IGCC power plant*

electricity produced in an IGCC plant is estimated to be 25% more expensive than electrical power produced in the subcritical PC power plants (Merkel et al., 2012).

Therefore, two kinds of configurations can be considered for IGCC power plants. Indeed, when considering the CO_2 capture stage in the IGCC design, two additional process blocks are needed (besides the compression of CO_2 for transportation) (Maurstad et al., 2005):

- a shift WGS reactor in which the CO reacts with H_2O to H_2 and CO_2;
- an absorption or membrane process for the CO_2 capture or hydrogen production.

Hence, the block flow diagrams (BFD) for both processes are presented in Figures 10.2 and 10.3. It should be noted that the real benefit of IGCC technology can be bold if a cost is placed on CO_2 emissions. This is because the CO_2 removal from high-pressure, high-concentration gasification streams will be significantly less costly than CO_2 removal from conventional PC power plant flue gas. This preference can be intensified if membrane technology is used in the IGCC plant. Therefore, in the following, we focus on one of the IGCC plant configurations, in which CO_2 capture and hydrogen production are considered by using membrane technology (Carbo et al., 2006).

10.3 Application of Membranes in an IGCC Power Plants

According to the literature, Figure 10.4 shows areas where membranes might be effectively integrated into the IGCC process. As indicated in Figure 10.4, each area has potential benefits and disadvantages that should be considered for any proposed CO_2 removal technology. To the best of our knowledge, currently, the best existing technology is a two-stage physical absorption technology employing the solvent Selexol^{TM}. H_2S is removed in the first stage

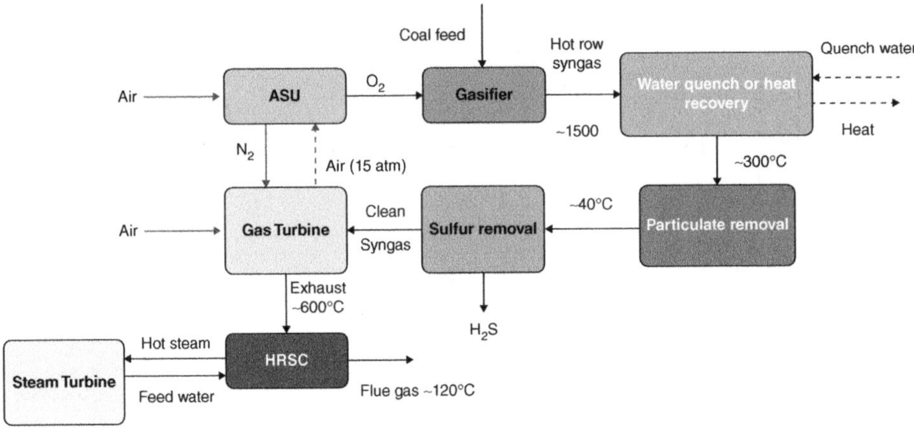

Figure 10.2 The process scheme of an IGCC power plant without CO_2 capture processing

and CO_2 in the second. On the other hand, unlike other gas separation technologies (e.g., adsorption and absorption), membranes are compact and modular, and could be placed at more than one location to separate H_2 from CO_2 and other gas components.

It should be noted that process conditions, gas composition, temperature, and pressure are different at the various locations recognized in Figure 10.4. Each location will have its own unique set of advantages and disadvantages in respect to hydrogen separation and recovery. Moreover, other technologies under development, such as warm or hot gas clean-up processes (e.g., H_2S or acid-gas removal (AGR)), may affect the selection of a H_2/CO_2 separation process, and may or may not complement the membrane separation. Indeed, the operating conditions for any membrane technology presented should match the conditions

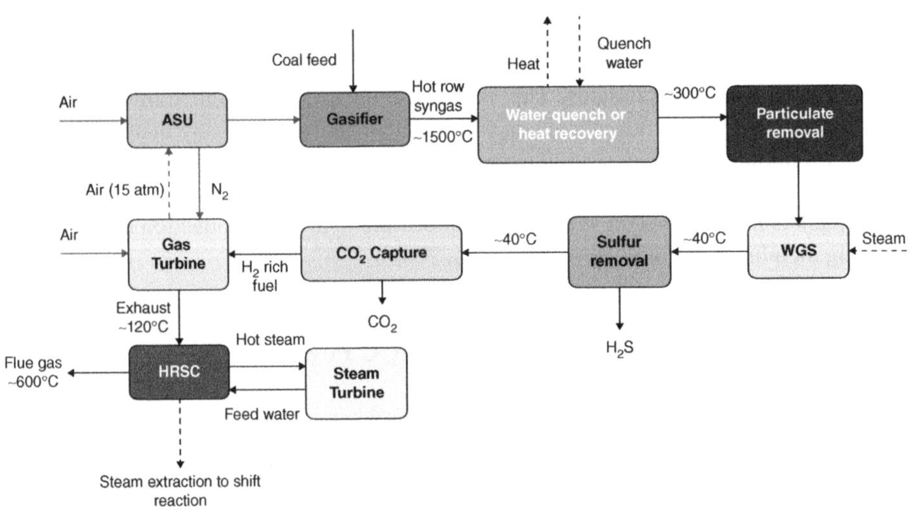

Figure 10.3 The process scheme of an IGCC power plant with CO_2 capture processing

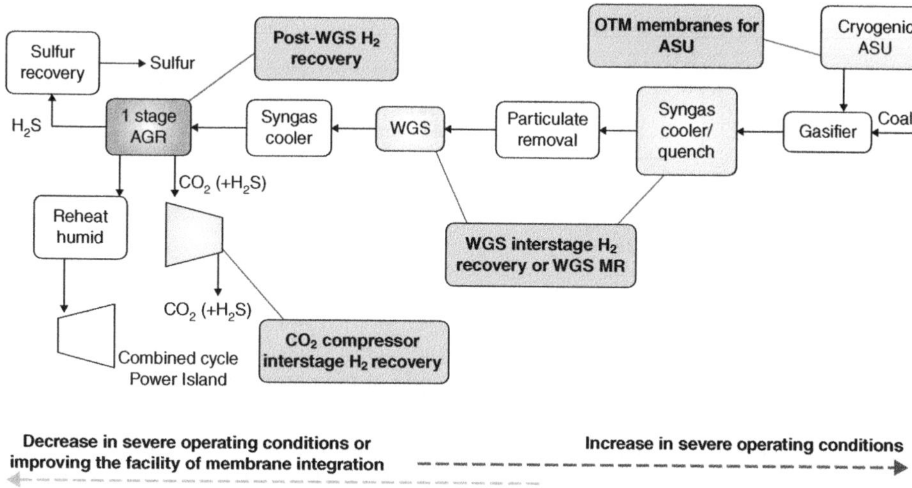

Figure 10.4 *Identification of membrane unit locations in IGCC plants*

where it is located in the process. If it is possible, it is not necessary to compress/expand or heat/cool the product and feed streams. In fact, these additional operations can only cause a decrease in the overall efficiency and will increase the overall cost of an IGCC plant. The appropriate location of a membrane unit in the process design is critical, and it appears unlikely that a single membrane material could adequately apply in all the feasible placements in the process. Therefore, the challenge is to take advantage of the exclusive characteristics of the individual membrane technologies, while modifying any inadequacies. Indeed, the five situations identified in Figure 10.4 are described below (Manzolini et al., 2013):

Post-WGS H_2 recovery – This operation is the current location for CO_2 and H_2S absorption set up in the IGCC process. At this point, the syngas stream has been cleaned of all impurities that might have a damaging effect on the membrane materials. Current technologies operate at "cold gas" temperatures that require the syngas leaving the WGS reactors to be cooled before entering the absorption process. Syngas stream cooling condenses the water in the syngas stream and can cause lower plant efficiencies, since the fuel gas must be reheated prior to firing in the gas turbine and condensation reduces the mass flow to the turbine. "Warm gas" H_2S removal units are being developed to increase efficiency; however, they will be useless if the syngas stream must be cooled anyway to remove CO_2.

The CO_2 is produced at relatively low pressures (3–10 bar) by the current two-stage absorption processes, and also the compression required to deliver the CO_2 for sequestration is maximized. The suitable gas separation membranes located here would replace the second stage of the absorption process, and also the first stage, if H_2S and CO_2 can be co-sequestered. The CO_2 is transported to the compression train at high pressure; however, the recovered H_2 must be re-compressed unless a diluent such as N_2 is used as a sweep gas to lower the H_2 partial pressure on the membrane permeate side. The diluent or sweep gas will also increase the mass flow to the gas turbine. If warm gas H_2S removal is employed, the membrane must be permeable to H_2O to avoid later condensation of

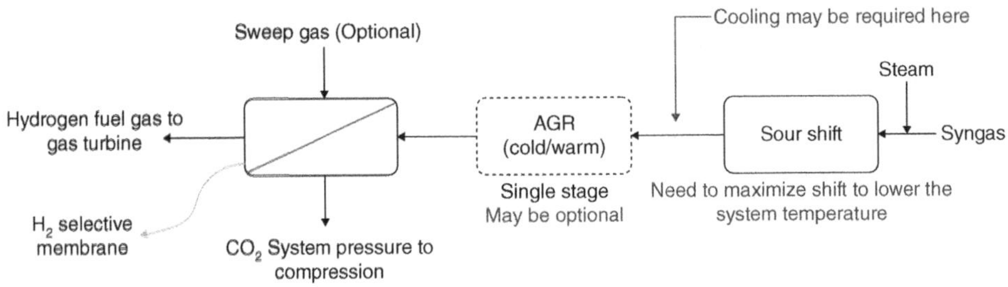

Figure 10.5 *Membrane integration after the WGS reaction*

this water in the CO_2 compression train. This style of membrane integration is illustrated in Figure 10.5.

CO_2 Compressor interstage H_2 recovery – The membranes' location, as shown in Figure 10.4, has the benefit that the high feed gas pressures will increase the driving force of H_2 permeance, thus, maximizing the H_2 recovery or minimizing the membrane area requirements. Application of a sweep gas stream is still beneficial and theoretical H_2 recoveries as high as 98% are possible. In this option, CO_2 compression is minimized and H_2 can be delivered at the required service pressure without re-compression. However, some H_2 will be over-compressed as it passes through the CO_2 compression train before recovery. If multiple membrane units are located prior to compression, between stages and post compression, the process can be optimized. As with the post-WGS H_2 recovery, the feed gas stream to the membrane has been cleaned to remove impurities. This type of membrane integration is illustrated in Figure 10.6.

WGS interstage H_2 recovery – It is clear that removing H_2 between reactor stages drives the WGS reaction towards completion by shifting the balance in favor of H_2 production. However, the membrane must be adequately impermeable to water for this strategy to be useful. Moreover, interstage H_2 recovery causes the catalyst volume to be minimized, and possibly reduces the number of required shift reactors. It also facilitates the excess steam co-fed to the reactor and the volume of interstage cooling to be decreased, improving the efficiency of the process. Normally, a sulfur-tolerant shift catalyst is used with H_2S removed

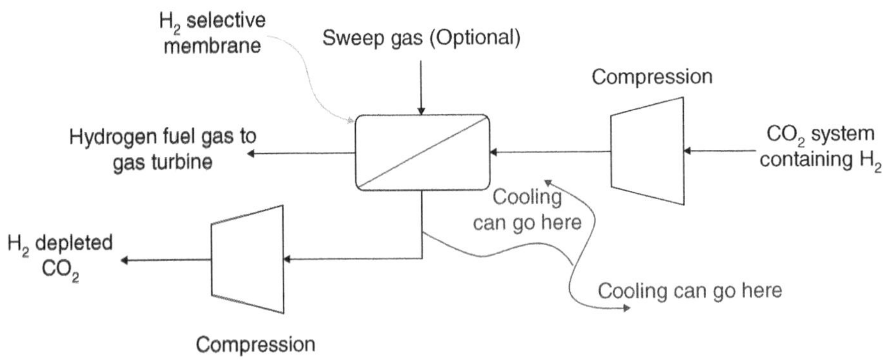

Figure 10.6 *Membrane integration between compressors*

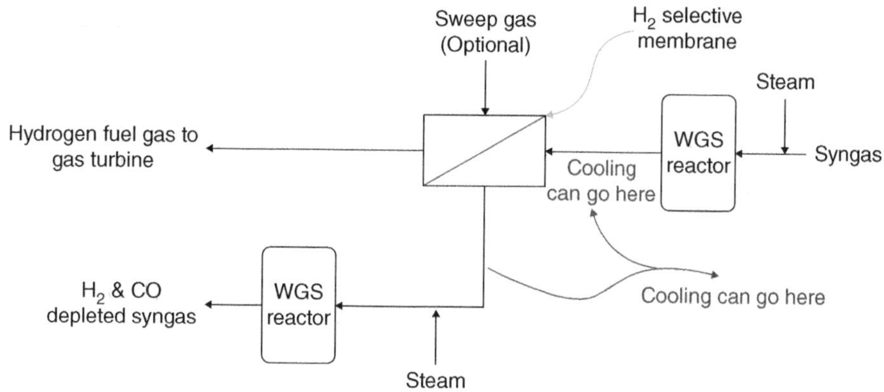

Figure 10.7 *Membrane integration between WGS reactors*

downstream of the WGS via absorption-based, cold gas AGR. As mentioned before, warm gas removal technologies are still being developed. It should be noted that a membrane integrated in this section would need to be resistant to sulfur compounds. Alternatively, hot gas H_2S removal could be performed upstream of the WGS reactor. In this case, iron and cobalt-based shift catalysts would be applied (as indicated in Figure 10.7).

The homogeneous WGS reaction will occur at high temperatures, if the membrane unit is integrated within the syngas cooling step. This location is the most difficult one, and any membrane placed here would need to have a high tolerance for a wide range of contaminants, including sulfur and nitrogen compounds, particulates, and trace metals present in the raw syngas leaving the gasifier. Moreover, as shown for the membrane integration in Figure 10.8, use of a sweep gas stream is even more beneficial when coupled with the WGS reaction, since it will improve H_2 removal and change the WGS balance further in favor of H_2 production.

WGS Membrane Reactor (MR) – By using the membrane with the WGS shift reaction, the benefits described above can be maximized. This can be accomplished by packing the retentate flow space with a WGS catalyst or by integrating a membrane with a surface

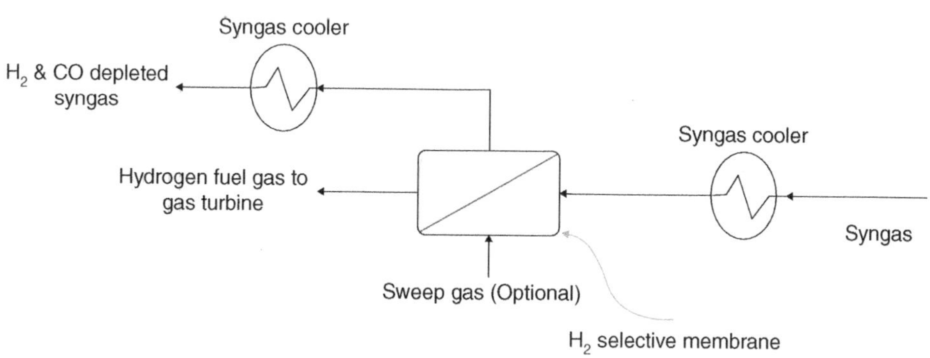

Figure 10.8 *Membrane integration between syngas coolers*

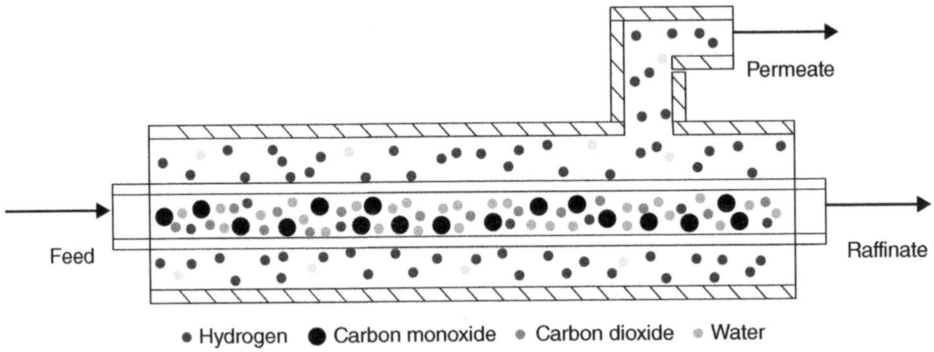

Figure 10.9 A schematic of the WGS-MR

that is catalytically active for the WGS reaction. However, this presents a very challenging operating environment for the membrane. As well as the sulfur tolerance discussed above, the membrane would need to be resistant to a number of other compounds, such as high molecular weight hydrocarbons, methanol, and coke (carbon deposition), produced as side-products of the catalyzed WGS reaction. The membrane would also be subject to temperature gradients resulting from the exothermic reaction heat, and some form of internal cooling might be needed, complicating the design and fabrication of the MR. A schematic of a WGS MR is shown in Figure 10.9.

Membrane set up for oxygen separation – Oxygen transport membranes (OTM) are possible alternatives to a cryogenic air separation unit (ASU) to produce oxygen in the IGCC plants. The presence of a gas turbine cycle producing high pressure and a high temperature O_2-rich stream makes it possible to use an OTM in the plant, as illustrated in Figure 10.10 (Leo et al., 2009). The compressed air exits the main compressor at 400–450°C and 18–20 bar (the common conditions for large-scale, heavy-duty combustion turbines), is optionally compressed more with a boost compressor, and is heated up to the typical OTM operating (temperature around 800–900°C).

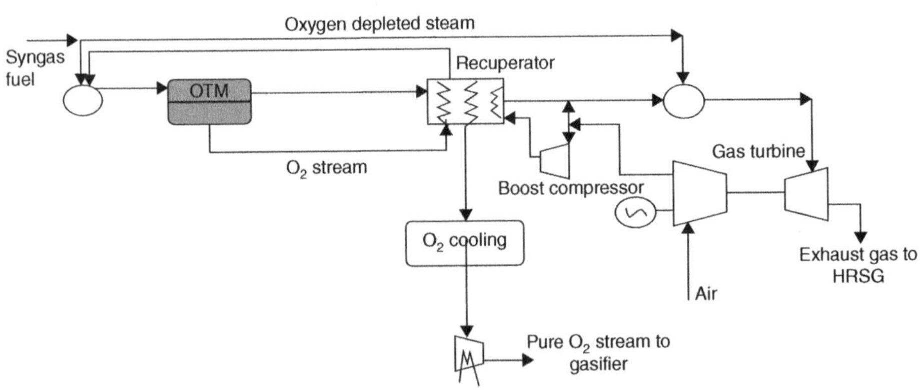

Figure 10.10 Integration of an OTM permeator in a gas turbine of an IGCC plant

Table 10.1 *Preferred membrane temperature condition for IGCC plants*

Location	Temperature range, °C		
	Interstage syngas cooling		
Water quench/convective cooler		200–425	
Radiant cooling/gas quench		700–900	
	Inlet		Outlet
WGS reactor – High temperature shift	300–350		400–500
WGS reactor – Low temperature shift	200–350		240–320
Post-cold-gas AGR		40–450	
Post warm-gas AGR		100–450	
Post hot-gas AGR		>450	
Compressor	Inlet		Outlet
CO_2 compressor interstage	~40		65–200

The air heating operation is carried out by combusting part of the syngas stream and using a recovering heat exchanger. The optimal temperature of the air exiting the heat exchanger is determined by choosing economic optimization. On one hand, high temperatures require expensive equipment to heat the air; the cost of the heat exchanger exceeds that of the combustor. On the other hand, higher heat exchanger outlet temperatures decrease the flow rate of the syngas burned upstream of the membrane, which enhances the difference in the O_2 partial pressure across the membrane, thus reducing its area and cost. Oxygen-depleted (or "vitiated") air from OTM is cooled in the recovering heat exchanger and sent to the main combustor, where the remaining syngas is burned and the design turbine inlet temperature is raised prior to the expansion operation.

Table 10.1 indicates the operating temperatures for the locations in the IGCC plant discussed in Figure 10.1. The low and high ranges presented for interstage cooling are based on whether the gasifier employs a syngas quench or a radiant cooler, respectively. Similar to the WGS set-up, the low temperature range corresponds to conditions after the interstage coolers and the high temperature range to the outlet of the reactors. The AGR technology under development may operate in a number of different temperature regimes. For interstage compression, the low temperature value comes after cooling, upstream of the next stage of compression, and the high temperature range corresponds to the possible compressor outlet.

The IGCC plant operating pressures are summarized in Table 10.2. The pressure ranges given show two common styles of gasifier operation. In general, the lower pressures are employed in systems using amine-based AGR, while the higher pressures are more illustrative of a system employing Selexol™ technology. Moreover, the high pressure mode is more desirable for a membrane-based gas unit. For the CO_2 compression inlet, the ranges presented match possible pressures exiting any upstream H_2S/CO_2 AGR process. The compression outlet pressure matches the CO_2 transport pipeline. The desired permeate pressure is set by the gas turbine design used in the IGCC topping-cycle.

As illustrated in Table 10.2, the reported low pressure range is typical of IGCC plant applications. It should be noted that it is the H_2 partial pressure difference that is the driving force for membrane separation, not the total pressure differential across the membrane.

Table 10.2 *Pressure and H2 concentration of preferred membranes for IGCC plants*

	Pressure range, bar Gasifier			
Location	Low		High	H_2 Concentration (wet)
Syngas cooling	~28 ↓		~70 ↓	16–29%
WGS reactor				~42%
Gas clean-up	~21 Cold-gas 3.4–10		~45 Warm/hot gas 21–45	~54%
Outlet of CO_2 compression		150		<2%
H_2 in permeate stream	For power 24–31		For chemicals >31	For power 44–80%

However, the total differential does have a direct bearing on the membrane structural integrity. Moreover, approximate molar H_2 concentrations are also listed in Table 10.2.

10.3.1 Hydrogen Selective Membranes

As mentioned before, the application of hydrogen selective membranes is limited to power plants containing CO_2 capture or those that deliver H_2 as the final product. Among the usable membranes technologies, low temperature polymeric membranes, high temperature palladium-based dense membranes, silica membranes, zeolite membranes, and mixed conducting membranes can be considered for integration in the IGCC plants. However, high temperature mixed conducting membranes are not suitable for use in the IGCC plants, because their high operating temperature (>700°C) would either enforce the operation on the raw syngas from the gasifier or require a major syngas reheating after cleaning.

The integration of hydrogen selective membranes into the IGCC plants with a CO_2 capturing operation and/or hydrogen production offers the potential for improved performance efficiency. However, optimal integration and operation of membranes consistently include a trade-off between capital and operating costs that cannot be evaluated by considering the membrane alone, but its effect on the entire power plant should be carefully investigated.

The following general aspects should be considered when the plant configuration of an IGCC integrating hydrogen selective membranes is defined:

- The main goal of minimizing the total IGGC plant cost is achieved by limiting the membrane surface area, which presents the operation at high driving force of H_2 permeance across the membrane.
- In order to minimize the membrane surface area and improve the overall IGCC plant performance, catalysts can be integrated to promote the WGS reaction on the feed side of the membrane. The WGS reaction increases the H_2 partial pressure and reduces the required membrane surface area.
- A sweep gas stream can also be employed on the permeate side, when possible, to decrease the partial pressures of the permeate gas and thus enhance its permeance across the membrane.

- It is necessary to operate the hydrogen selective membrane at favorable temperatures, by considering both the effects on membrane permeance and on the equilibrium and kinetics of the WGS reaction.
- The severe conditions favoring membrane degradation should be avoided. When defining a membrane-based IGCC plant layout, care should be taken to avoid conditions leading to thermal and mechanical stresses, corrosion, fouling, and degradation, and also poisoning of the membrane.
- In the case of the sulfur-tolerant membranes, the behavior of H_2S should also be carefully investigated when selecting the process integration strategy. It should be noted that, if the H_2/H_2S selectivity is not high enough, H_2S concentration in the gas turbine fuel leading to unacceptable levels of SO_2 emissions might result.

However, according to the literature, typically five types of hydrogen selective membranes can be considered, and their properties are presented in Table 10.3.

Therefore, the following sections (Sections 10.3.1.1, 10.3.1.2, and 10.3.1.3) discuss the most promising options for integrating hydrogen separation membranes into the IGCC plants from the experimental and modeling point of views. As mentioned above, of these five types of hydrogen selective membranes, the metallic, micro-porous ceramic and polymeric membranes have more potential than the other membranes for use in the IGCC plants.

10.3.1.1 Integration of Micro-porous Ceramic Membranes in IGCC Power Plants

Among the different ceramic membranes, silica and zeolite membranes can offer a good performance for hydrogen separation in the IGGC plants. According to the literature, several studies have presented interesting results on various membrane materials for possible application in the IGCC process.

As a first approach, by Bracht et al. in 1997, the application of silica membranes was studied in the WGS reaction for CO_2 recovery in the IGCC process. In this study, the net efficiency of the IGCC process was 42.8%, obtained by using the WGS-silica MR, while the value for the conventional process was 40.5%. Moreover, by economic analysis, Bracht et al. indicated that a promising investment and operational cost can be achieved by using the WGS-silica MR.

Costa et al. (2009) examined the integration of a cobalt doped silica membrane for the separation of H_2, in the WGS reaction for the IGCC plant. As presented in Figure 10.11, they proposed an advanced scheme for the IGCC plant. According to their results, application of a higher temperature silica membrane allows further simplification of the gas cooling, conditioning and separation step, but this scheme needs a high temperature membrane tolerant of the syngas stream.

Franz et al., in 2010 and 2011, presented a modeling study to investigate the potential of CO_2 capture from IGCC plants by hydrogen selective ceramic membranes. Detailed models of a reference power plant and three different CO_2 capture concepts were arranged with Aspen Plus. The parameter variations were arranged to investigate the influence of the membrane characteristics and the IGCC plant-specific boundary conditions on the performance of the CO_2 capture concepts. In their study, for ceramic membranes with a selectivity of H_2/N_2 and H_2/CO_2 around 500, the results showed that for a sour CO shift and a sweet CO shift, efficiency losses of 9.07 and 9.43 percentage points are feasible,

Table 10.3 *The properties of five types of hydrogen selective membranes*

	Dense polymer	Microporous ceramic	Dense ceramic	Porous carbon	Dense metallic
Temperature range (°C)	<100	200–600	600–900	500–900	300–600
H_2 selectivity	Low	Moderate	Very high	Low	Very high
H_2 permeance	Low	High	Moderate	Moderate	High
Poisoning problems	HCl, Sox, CO_2		H_2S	Strong vapors, organics	H_2S, HCl, CO
Examples of membrane material	Polymers	Silica, alumina, zirconia, titania, zeolites	$SrCeO_{3-\delta}$, $BaCeO_{3-\delta}$	Carbon	Palladium alloys, Pd-Cu. Pd-Au
Transport mechanism	Solution-diffusion	Molecular sieving	Solution-diffusion	Surface diffusion, molecular sieving	Solution/diffusion

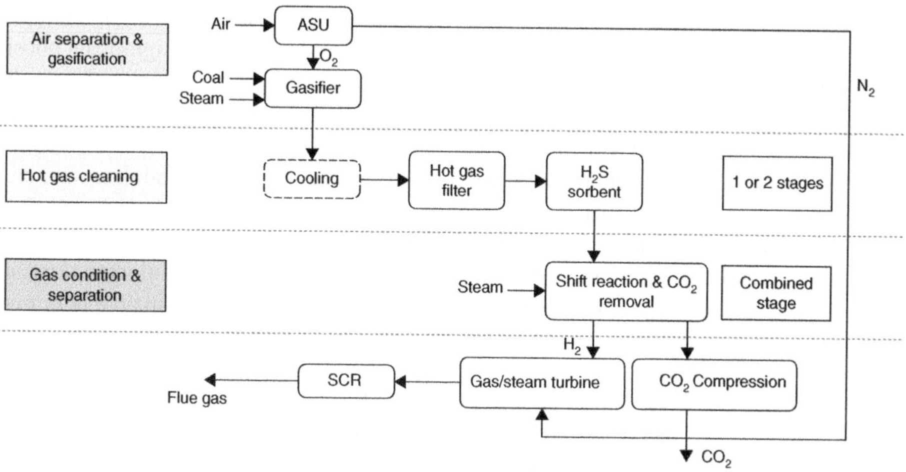

Figure 10.11 Advanced scheme of the IGCC plant presented by Costa et al., in 2009

respectively, while the separating factor was about 97% of the CO_2 with 95% purity. One of the promising concepts for CO_2 capture in the Franz et al. studies was a ceramic MR concept with simultaneous CO_2 separation and CO shift. By using this concept, the separation factor of 96.6% was achieved and purities above 95% with an efficiency loss of 6.7 percentage points.

In 2011, Ku et al. carried out a theoretical study to analyze the hydrogen selective membrane performance for CO_2 recovery in the IGCC process. As illustrated in Figure 10.12,

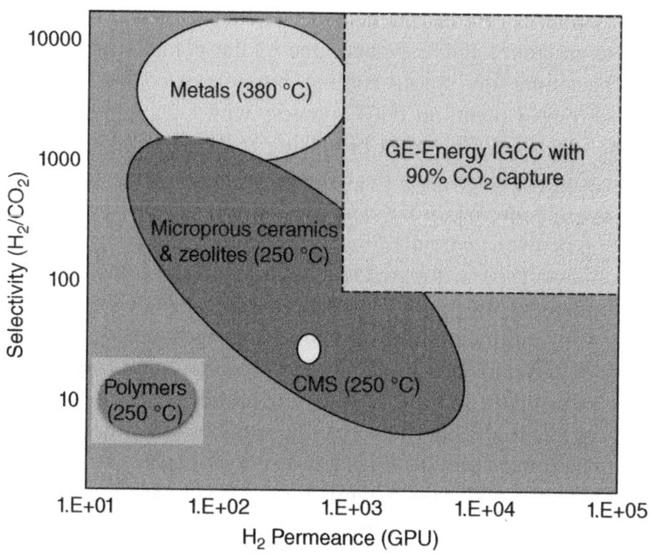

Figure 10.12 Trade-off diagram for H_2 selective membranes

their results showed that high temperature membranes can separate CO_2 at high temperature and pressure and therefore these kinds of membranes can provide a significant boost in performance when properly integrated into an IGCC power plant. In contrast to H_2 production systems where the high H_2/CO_2 selectivity favors ultra-high selectivity membranes, the requirements for the IGCC process indicate that a wider range of materials, including silica and zeolite-based membranes, should be considered. However, candidate membrane materials must satisfy a number of performance requirements, including H_2/CO_2 and H_2/N_2 selectivity targets, in addition to exhibiting promising permeance, manufacturability, and stability. As candidate membranes transfer from the lab scale to field demonstrations, and ultimately commercialization, it is important that development efforts involve both material improvements and systems design components to ensure the optimal profits of the technology are realized.

In 2007 and 2012, Verweij et al. and Zhang et al. used zeolite membranes for a WGS reaction in the IGCC plants. But the results are not promising with respect to other ceramic membranes such as silica membranes.

In 2013, Koutsonikolas et al. evaluated the integration of membrane technology in the IGCC process, in order to decrease the CO emissions of power plants. In their investigation, three different types of membrane were studied (polymeric, ceramic, and metallic) taking into account the latest advances in membrane development. The simulation of the membranes' performance was conducted in a Visual Fortran code and this was incorporated in the Aspen Plus software for the overall performance assessment. The energy analysis of the alternative cases illustrates that CO_2 capture in the hybrid IGCC process is technically feasible but with an accompanying energy disadvantage in the power plant's output. Taking into account both technical and economic issues, the most promising scenario appears to be the integration of a two-stage ceramic membrane process for H_2 separation and CO_2 capture.

Franz et al. (2014) presented an economic evaluation of pre-combustion CO_2 capture in the IGCC plant by porous ceramic membranes. The analysis was based on a comparison with two different reference IGCC plants. One of the plants was a conventional IGCC plant without CO_2 capture (dry syngas cooling, bituminous coal, efficiency of 47.4%). As a second reference power plant, an IGCC process with CO_2 removal based on standard Selexol scrubbing was used. The most promising technology for CO_2 capture by membranes in the IGCC applications was the combination of a WGS reactor and a hydrogen selective membrane into one WGS-MR. For the case of WGS-MR, efficiency losses can be restricted to about 6 percentage points for a CO_2 separation factor of 90%. This is a severe reduction in the efficiency loss compared to Selexol (10.3 percentage points) or IGCC–CM (8.6 percentage points). It should be noted that parameters such as membrane costs and lifetime, costs of CO_2 certificates and annual operating hours were taken into account. The main aim of these evaluations was to identify the minimum cost of electricity for the different capture cases under varying boundary conditions. Setting 90% CO_2 separation, the analysis clearly identifies that the economic minimum for the cost of the electricity and the maximum thermodynamic efficiencies do not coincide. The electricity cost for the second reference case was 67 €/MW h, while for the WGS-MR integration with 90% CO_2 separation was 57 €/MW h, of which certificate costs of 30 €/t_{CO2}, membrane costs of 300 €/m^2, and 8000 operating hours/year were considered.

10.3.1.2 Integration of Metallic Membranes in IGCC Power Plants

Of the different types of metallic membranes, Pd-based membranes are surely the most investigated technology for membrane integration in IGCC plants, thanks to their high selectivity and relatively high H_2 permeability. However, low tolerance to sulfur is still a drawback of Pd membranes. Only a sulfur level in the range of ppm or even lower is usually needed, depending on the Pd alloy adopted. Sometimes sulfur-tolerant membranes can resist an H_2S concentration as high as 1–20 ppm while sensible alloys needs H_2S to be lower than 1 ppm. In general, the higher sulfur tolerance is traded off by a lower permeance. Typically, research has focused on the addition of other metals such as Cu and Co to Pd to enhance the tolerance to sulfur.

In 2007, Amelio et al. studied the capture and storage of CO_2 from fossil fuels in the IGCC plant. The main purpose of their evaluation was to analyze the thermodynamic performances and the technological aspects of two IGCC plants, as well as give a prediction of the investment costs for the plants and the resulting energy consumption. In the above-mentioned work, a one-dimensional computational model of the Pd-based MR was proposed, to simulate and evaluate the capability of the IGCC plant to capture CO_2. After an analysis of the energetic performances of these plants, the results present a rather moderate performance disadvantage of an IGCC plant (with WGS-Pd-MR) with respect to the conventional IGCC (3.46% penalty on efficiency and 3.48% penalty on net power), basically owing to the higher electrical power consumption for the CO_2 compression and to produce the steam applied as the sweep gas stream for the MR. On the contrary, from an economic point of view, the provision of the investment and the cost of the electricity production (COE) of the IGCC plant with Pd-MR is more competitive compared to the conventional IGCC plant in the medium term (investment cost and COE of the IGCC plant with Pd-MR are, respectively, 4% and 1.4% lower) because of the lower costs of MRs and because of the absence of the CO_2 absorption system.

In 2009, Rezvani et al. presented a comparative assessment of coal-fired IGCC systems with CO_2 capture, using three different procedures, namely, physical absorption, WGS-Pd-MRs, and chemical looping. According to their results, despite the lowest efficiency loss among the investigated systems, the economic performance of the double stage chemical looping combustion (CLC) cycle was outperformed by systems that integrated the physical absorption and WGS-MRs.

Moreover, in 2010, Anderson et al. proposed a scaled-up study of Pd-based membranes for the IGCC plant. In their study, tubular membrane styles up to 2ft long were tested under a range of conditions. Membrane tubes were tested under simulated WGS gas compositions at temperatures between 300°C and 400°C and at feed stream pressures up to 700 psig. According to the results achieved, data was collected for a membrane tested up to 600 psig feed pressure in a 100% H_2 feed stream. Flux rates up to 175 $SCFH/ft^2$ were observed at these high hydrogen partial pressures. They have completed the design and construction of a 12 lb_m/day MR.

As a first attempt, Badwal et al. in 2013 undertook a comprehensive study of the application of Pd-based MR during the WGS reaction in the IGCC plant. In this work, the design, construction, and ordering of a reliable MR and an adaptable test station for the evaluation of hydrogen permeation membrane (metals, ceramics or cermets) at high temperatures and high differential pressures were described. The MR system has been designed to operate at high temperatures up to 800°C and pressure gradients across the membrane up to 10 bar. The system has multiple levels of built-in safety redundancy, which comprise a range of controls and monitors for both operator and system safety. A number of Pd and Pd-Ag alloys of nominal thicknesses in the 20 and 140 μm range were sourced and an Al-based porous ceramic support structure was fabricated to evaluate the Pd-based membranes. The test station has been validated with Pd and Pd-Ag alloys of different thicknesses. Indeed, a flexible MR system was described which allows fast screening and evaluation of composite Pd-based membranes for hydrogen permeation flux measurements at high temperatures up to 700–800°C and with a high pressure gradient across the membrane to 10 bar.

In 2014, Manzolini et al. discussed the thermodynamic performance of an IGCC plant, in which the Pd-based membranes were used in a configuration involving two WGS-MR and two membrane permeators placed in series. They reported the analysis results by comparing the advanced IGCC plant (with the membrane unit) with the conventional IGCC plant (without the CO_2 capture operation), as follows:

- The thermal power input to the gasifier is enhanced due to keeping the same gas turbine size. Indeed, this is owing to the higher cold gas efficiency of the reference IGCC plant case than the ones achieved in the advanced IGCC plant.
- In the plant without post-firing, the steam cycle power productivity losses are due to the WGS steam demand, while in the post-firing case, the steam cycle power output rises because about 10% of the permeated H_2 is used for the post-firing case.
- The overall plant efficiency range was obtained between 7 and 8 percentage points below the reference IGCC case without CO_2 capture. It means that the membrane application reduces the efficiency drawback by about 30% compared to the IGCC, with CO_2 capture using commercially ready technology.
- CO_2 avoidance by membrane separation is about 85%, a value somewhat lower than one achievable by conventional pre-combustion CO_2 capture with solvents.

Their results indicate that, with regard to membrane technology, the adoption of the Pd-Cu-Au alloy instead of Pd-Ag allows the use of the Selexol technology as a sulfur removal process, with around 2% efficiency improvement. However, the lower permeability of the Pd-based membrane almost doubles the membrane area, stopping the slight thermodynamic benefits.

10.3.1.3 Integration of Polymeric Membranes in IGCC Power Plants

In recent decades, polymeric membranes have become available commercially, have a lower cost, and are easily made into large and defect-free modules compared to other types of membranes. The perm-selective polymeric membranes can be classified into two basic categories: glassy and rubbery. Glassy polymers show low chain intra-segmental mobility and long relaxation times, while rubbery polymers exhibit the opposite characteristics, namely, high intra-segmental mobility and short relaxation times. At this present moment, more than 90% of industrial perm-selective polymeric membrane processes for gas separation use the glassy polymeric membranes, due to their high gas selectivity and acceptable mechanical properties.

The characteristics of polymeric membranes, which mostly affect their integration into IGCC plants and sometimes can limit their application, are the moderate operating temperature, the relatively low selectivity in respect of CO_2 and the tolerance of H_2S. Indeed, the low operating temperature means that the polymeric membrane can only operate on the cooled syngas. Thus, in the case of an IGCC with CO_2 capture, a hydrogen membrane should follow the low temperature WGS reactor (operating at about 200°C) and could not be integrated into a WGS-MR. Therefore, several reports about the integration of polymeric membrane technology are being proposed in the following.

In 2004, Kaldis et al. presented a comparison study on the hydrogen selective low-temperature polymer and high-temperature ceramic membranes for pre-combustion capture and achieved an IGCC plant efficiency 8–14% lower than the one obtained for the IGCC plant without the CO_2 capture. In particular, up to 57% of the CO_2 could be recovered at 88% purity with three polymer membrane stages. However, the study did not include the significant pressure drops in equipment between the gasifier and the membrane system, which would lower the available separation driving force, and did not provide the energy required to compress the captured CO_2 to delivery pressure.

Furthermore, in 2009, Krishnan et al. reported that the hydrogen selective Polybenzimidazole (PBI) membranes could cover the economics of CO_2 capture in an IGCC process, compared to a base case Selexol process. An Aspen Plus-based process model has shown that a high temperature membrane system can compare with the Selexol-based process. The advantage of the PBI membrane system is the maintenance of the CO_2 stream at higher pressures decreasing the energy required for compression to pipeline pressures. It should be noted that if the H_2S remains with the CO_2 stream and can be sequestered, at that time the cost of electricity appears to be lower than Selexol-based separation systems.

Franz et al. (2010) investigated the potential of hydrogen selective polymeric membranes to separate CO_2 from the coal-based ones of IGCC power plants. In their study, as a reference power plant, a modified case of the Puertollano gasification plant was considered. A detailed Aspen model for the reference case was developed. In this study, a four-end membrane configuration integrated into a complete IGCC plant was assessed. According to the results, the retentate stream containing the CO not converted in the WGS section and the H_2 not separated by the membrane is burned with O_2. In this way, the heating value of the syngas stream is fully exploited and the CO_2 purity is improved. By considering a membrane operating at nearly ambient temperature with a H_2/CO_2 selectivity around of 50 and a target CO_2 separation factor around of 85%, an efficiency drop of ~9% was estimated with respect to the reference IGCC without CO_2 capture in their model.

Moreover, Merkel et al. (2012) showed that the hybrid membrane-condensation processes using today's polymer membranes in an IGCC plant can compete with the conventional CO_2 capture technology. Currently, Selexol absorption technology is the preferred approach to CO_2 capture. These conventional processes will increase the LCOE by 30% at 90% CO_2 capture. In particular, the application of a nitrogen sweep on a hydrogen selective membrane can decrease the compression requirements, and is a low energy configuration. According to their results, the best process designs applying hydrogen selective countercurrent/sweep membranes that operate for bulk hydrogen recovery, combined with low-temperature CO_2 selective membranes, which help to improve the CO_2 purification and liquefaction, have the potential to decrease the elevated LCOE from 90% CO_2 capture to around 15%. In addition, Merkel et al. even proposed a larger cost saving might be possible if higher permeance, and especially, higher H_2/CO_2 selectivity membranes can be developed.

10.3.2 Oxygen Selective Membranes

Currently, there are several technologies available to produce oxygen. The selection of technology used is highly dependent on the type and scale of the application, the final purity of the oxygen and the energy cost. According to their technology readiness level (TRL), the oxygen production technologies can be categorized, grouped here into (1) mature; (2) DDandD (development, demonstration and deployment); and (3) embryonic categories for simplicity. Indeed, mature technologies offer very small scope for further improvements as they have already undergone the main technical advances for at least the past two decades. However, they offer the optimum performance and risk minimization profiles, and are therefore the most attractive ones for deployment in the first generation of IGCC plants. Hence, Table 10.4 provides more detailed information relating to each technology, including typical energy demands and the capital costs of the most important discussed air separation technologies. It should be noted that the data presented was chosen for the ASU technology, where this technology was integrated into an oxyfuel coal combustion or IGCC power plant (Costa et al., 2013).

It is obvious that the ion transport membrane (ITM) or the mixed conducting membrane (MCM) has high potential with respect to other technologies or especially with respect to oxygen selective membranes such as polymeric membranes.

As previously mentioned, oxygen transport membranes (OTMs) offer a promising technology for applications in oxy-fuel and pre-combustion CO_2 capture processes, for gas and coal power plants. Indeed, OTMs are dense ceramic membranes which display mixed conductivity of oxygen ions and electrons. Therefore, in the following, different research studies on the potential of application OTMs in IGCC plant are investigated.

In 2000, Dyer et al., as a first approach, applied the ITM technology for oxygen separation from air in the IGCC plant. They indicated that the ITM oxygen process has the potential to produce significant benefits when integrated with power generation cycles. The ITM syngas process, by combining air separation and high-temperature syngas generation processes into a single compact ceramic MR, has the potential to noticeably decrease the capital investment of gas-to-liquid (GTL) plants and of distributed hydrogen.

Kaldis et al. (2004) proposed a short-term challenge would be the integration of membranes as ASU plant in coal power plants, in particular in oxy-fuel or coal gasification processes. As the ITM membranes operate at high temperatures (700–1000°C), there will be a large energy disadvantage associated with heating a high air volume, of which the maximum theoretical recovery rate is the concentration of oxygen (20%) in air. Their energy and cost modeling analysis directed towards a hybrid coal IGCC system, consisting of a gasifier, a shift reactor and an oxygen selective membrane, showed that CO_2 removal is technically feasible. As a consequence, the calculated energy penalty of incorporating ceramic oxygen separation unit varies between 8 and 14%.

In 2011, Kunze et al. proposed a novel IGCC plant integrating oxygen separation from air by MCM and post-combustion CO_2 capture by a twin fluidized bed carbonation–calcinations loop. An Aspen Plus simulation of this conceptualized plant presented a noteworthy enhancement to the plant efficiency compared to an IGCC plant with

Table 10.4 *Comparison of different technologies for ASUS from the viewpoints of energy consumption and total cost*

Technology	Cryogenic air separation	Pressure swing adsorption	Polymeric membrane	Ionic transport membrane
Current status	Mature	Mature	Mature	Demonstration
O_2 purity limit (% vol) (Remaining impurities)	99+ (Ar)	95 (Ar)	40 (N_2, CO_2, H_2O)	100
O_2 flow rate in largest commercial installations (tO_2d^{-1})	>3000[4]	<350[6]	<20 for oxygen-enriched combustion applications	5 (soon to grow to 100 with installation of new plant)
Suitability for CCS	high	moderate	Alone – low Hybrid – moderate[5]	high
Time frame to CCS commercialization	Immediate	Short	NA	Medium/long
Installed capital cost in oxy-fuel or IGCC plant ($US2008kWe^{-1})	310–500*[2,6]	150–200 for standalone PSA system producing	95-160 for a 30 % O_2 stream[8]	260–295#[2,8]
Energy consumption in oxy-fuel or IGCC plant (kWh tO_2^{-1})	245–670*[2,8]	~450–700[9]	260 for 40% O_2 Stream 190^- –40& for hybrid membrane/cryogenic system	100–655[2,8]
Energy penalty for integration with CCS	~25%[2]	NA	NA	~25%[2]

The wide range in these value reflects the differing rates required and energy integration opportunities for oxy fuel and IGCC applications
#The largest includes a heating value of NG to generate the required operating temperatures for ITM, whereas the smallest value represents only the electrical energy input
$Defined as $1-\eta_{CCS}/\eta_{Ref}$
^For plants producing > 1000tO_2d^{-1}
&For plants producing > 100tO_2d^{-1}

conventional pre-combustion CO_2 capture. To obtain a relative improvement, a comparative analysis of the present concept with a conventional IGCC plant with integrated ASU and pre-combustion CO_2 capture for similar operating conditions has been discussed. In particular, the new IGCC concept provided a net efficiency of around of 43.2%, based on low heating value (LHV) for hard coal. Furthermore, the concept allows CO_2 recovery of almost 98%, resulting in substantially lower specific CO_2 emissions of 22.1 g/kWh.

Anantharaman et al. (2011) proposed some system considerations to integrate an OTM-based ASU in the IGCC plant with CO_2 capture. An IGCC process using an OTM without a sweep gas stream and 100% air side integration with GT was presented. The OTM-integrated IGCC indicates an efficiency improvement of 0.7% over the reference IGCC plant with cryogenic ASU. However, the trade-off is the tighter integration and reduced flexibility of the plant. Also, in this case, the number of plant components is increased. Indeed, a cryogenic ASU for N_2 production is added to the OTM-based ASU for O_2 production.

Furthermore, in 2012, Kunze et al. presented an assessment of alternative process configurations for the IGCC plants with regard to important long-term efficiency potential. In their study, as shown in Figures 10.13, 10.14 and 10.15, three different future hard coal IGCC concepts were evaluated, applying emerging technologies and different CO_2 capture approaches. The proposed pre-combustion capture concept (IGCC-MR) exhibited an efficiency potential of 45.1% and is predominantly based on mature technology. However, the MR is not state of the art, and detailed experimental investigations are required in order to realize the assumed performance. Their investigation clearly demonstrates the noteworthy influence of the membrane unit on plant efficiency and therefore justifies further development.

In the post-combustion capture concept (IGCC-CL), a favorable efficiency potential (45.87%) can only be obtained with an integrated OTM process. The high dependency on both the carbonate looping and the OTM assumptions indicates higher uncertainty about the simulation results and possible future realization. Furthermore, the high level of integration might cause reduced flexibility and decreased availability of the IGCC plant.

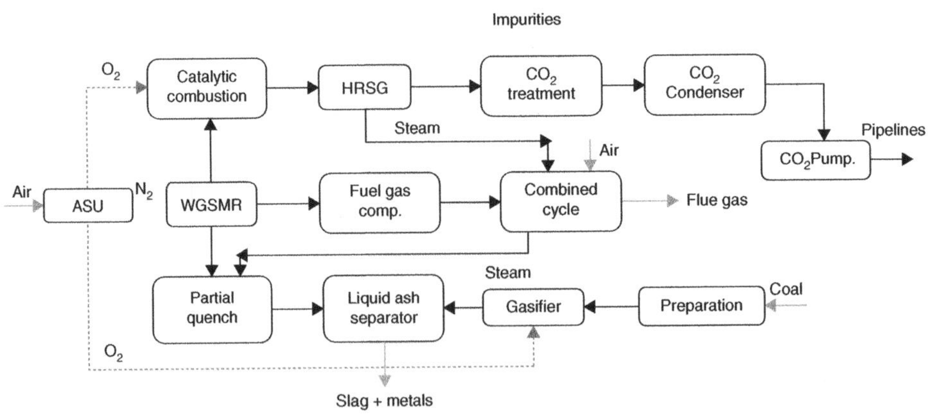

Figure 10.13 A simplified process scheme for the IGCC-MR concept

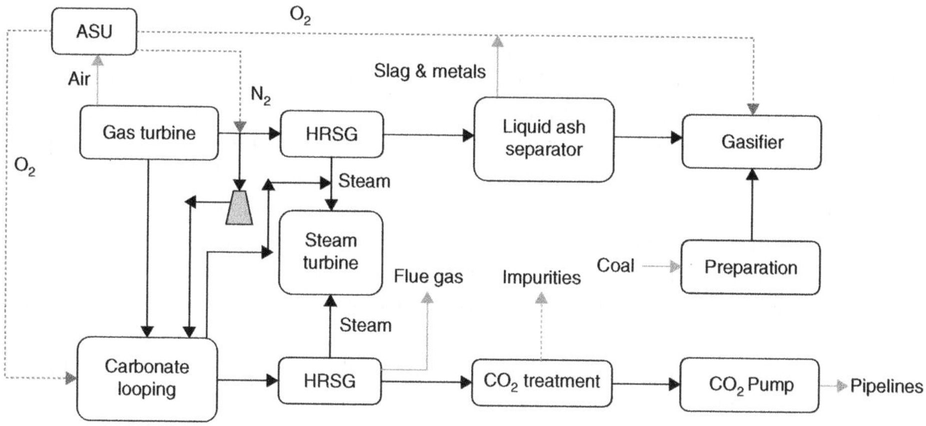

Figure 10.14 *A simplified process scheme for the IGCC-CL concept*

The IGCC concept based on the oxy-fuel process (IGCC-Oxy) also revealed a considerable long-term potential of 45.74%. However, the concept indicates considerable modifications in the combined cycle section as well as entirely new equipment. A new gas turbine and condenser are required, owing to the oxygen combustion and different working fluid. It seems highly uncertain whether the high financial and construction effort will be devoted in future despite the high thermodynamic potential of the concept.

10.3.3 CO_2 Selective Membranes

As mentioned before, there are two locations where it is possible to capture CO_2 from coal-fired power plants: (1) capturing CO_2 from syngas before the combustion operation

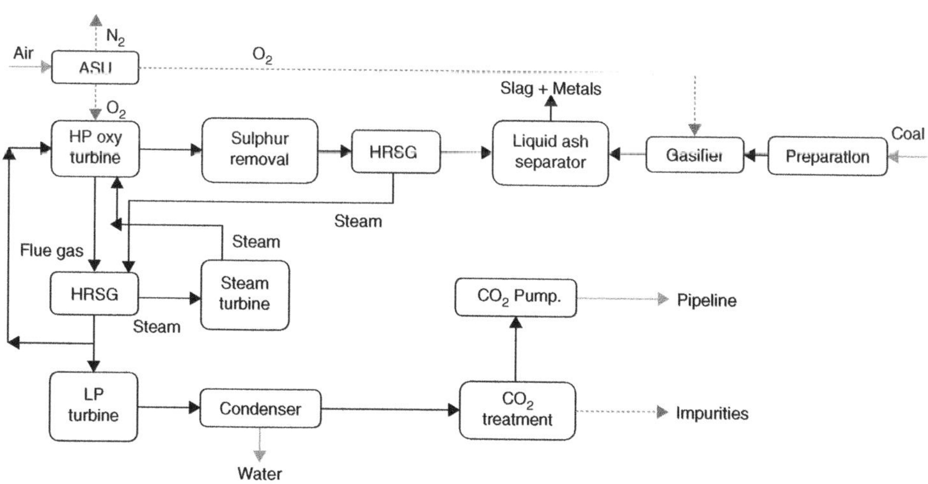

Figure 10.15 *A simplified process scheme for the IGCC-Oxy concept*

(i.e., pre-combustion capture); or (2) capturing CO_2 from the boiler tail-end flue gas stream (i.e., post-combustion capture). The IGCC plant integrates the pre-combustion capture technology due to the highest capability in capturing CO_2 under the conditions of a small amount of process gas under higher pressure; this strategy results in higher plant efficiency than using the post-combustion capture procedure and leads to higher economic aspects. However, currently, the main purpose of the ongoing research is the capture of CO_2 to provide a purity more than 90%. At the same time, such methods of CO_2 separation are sought, in which the enhancement of electricity production costs will not be higher than 10% compared to conventional systems without CO_2 capture (Kotowicz et al., 2010).

According to the literature, a broad variety of membranes have been extensively developed for separation and purification applications involving CO_2 in power generation. The membranes for CO_2 separation at high temperatures would be considered in pre-combustion CO_2 capture to keep H_2 at feed pressure and simultaneously to separate CO_2 for sequestration, free of other contaminations. As indicated in Table 10.5, similar to the hydrogen selective membranes, the different types of CO_2 membranes can be categorized such as zeolite, silica, polymeric, mixed matrix and carbon membranes.

According to the condition limitations of the IGCC plant, only some of the CO_2 selective membranes can possibly be applied in this process. The polymeric membranes present low selectivities and also need gas cooling before the separation operation. The facilitated transport membranes offer higher selectivities, and, therefore, opportunities to decrease the energy consumption by providing cooling requirements in a hybrid membrane-cryogenic separation process or by reducing the costs involved in permeate recompression and recycling. However, both types of membranes (polymeric and facilitated membranes) can suffer from compression-related stability problems at high pressures, and the mixed matrix membranes, as a suitable approach, can be creatively used to deal with these issues. Nevertheless,

Table 10.5 *Some of the best CO_2-selective membranes*

Membrane	CO_2-permeance ($mol.m^{-2}.s^{-1}.Pa^{-1}$)	CO_2/N_2	CO_2/CH_4	CO_2/H_2	T (°C)	References
Zeolite (ZSM-5)	$10^{-7} \times 3$	9.5	–	–	30	Shekhawat et al., 2003
Zeolite (Y-Type)	$10^{-7} \times 3$	11	8	–	130	Shekhawat et al., 2003
Zeolite (Sapo-34)	$10^{-7} \times 1.5$	16	36	14	27	Shekhawat et al., 2003
Carbon	$10^{-7} \times 4$	8	20	–	65	Shekhawat et al., 2003
Polymer (PN8)	250 (barrier)	70	18	–	30	Orme et al., 2005
Matrimid-5218	44	27	200	–	30	Shekhawat et al., 2003
Facilitated	100	110	–	–	40	Shekhawat et al., 2003
Silica	$10^{-6} \times 3$	80	149	0.0625	400	Pakizeh et al., 2007

Table 10.6 The results of optimization by exergy analysis

IGCC technology	Net output (MW_e)	η (%)	$\eta_{penalty}$ (% pt)	Carbon capture ratio (%)	Shift steam flow (kmol/kmol)	Inlet temp. WGS-MR (°C)	Sweep steam flow (t/h)
Reference IGCC	500.0	48.9	–	–	–	–	–
IGCC with H_2-selective WGS-MR	443.9	43.4	5.5	100	1.0	260[2]	–
IGCC with CO_2-selective WGS-MR	403.3	39.4	9.5	90	1.0	250	50
IGCC with HT/LT-WGS and Selexol CO_2 capture	413.5	40.4	8.5	90	1.1	–	–

for the pre-combustion process, ceramic membranes such as a silica membrane can be suggested as a good option. Therefore, in the following, different membranes that can present a high performance in IGCC plants are discussed.

In 2006, Carbo et al. presented a theoretical study of an efficiency comparison of IGCC plants with a CO_2 selective membrane, IGCC plants with an H_2 selective membrane, and conventional IGCC plants. The results of the initial system assessments that were directed in order to follow optimum operating conditions for advanced separation improved WGS-MRs. The application of H_2 and CO_2 selective WGS-MR in the IGCC plant was optimized with exergy analyses and compared with respect to efficiency. Hence, the optimization resulted in optimum operating conditions, which guides the selection of a designated material for the development of advanced CO_2 selective MRs. As reported in Table 10.6, the results of energy analysis are presented for three cases.

The application of steam as sweep flow within the CO_2 selective WGS-MR case results in an absolute efficiency loss of around of 3.0%. Decreasing the steam sweep flow rate by operating the WGS-MR at lower retentate pressures seems mandatory in order to simplify comparison with the H_2 selective WGS-MR. Furthermore, the CO_2 selective WGS-MR allows flexibility in the CO_2 ratio. This enables utility companies to speculate on CO_2 market fluctuations by changing the CO_2 ratio.

According to their results, the IGCC plant with CO_2 capture through CO_2 selective WGS-MR has one important drawback compared to the H_2 selective WGS-MR, i.e., the elevated efficiency penalty. However, the benefits of the CO_2 selective WGS-MR are the flexibility and CO_2 purity with respect to CO_2 separation and fuel conversion. The application of slurry-fed gasification will also be evaluated in future system assessments, possibly with primary sour WGS upstream of the AGR (Carbo et al., 2006).

Grainger et al. (2008) proposed a simulation study for the investigation of an IGCC plant with CO_2 capture, based on data taken from the ELCOGAS 315 MWe Puertollano plant. This process integrated a fixed site carrier polyvinylamine (PVA) membrane, developed at MEMFO at the Norwegian University of Science and Technology (NTNU), to separate the

CO_2 from a CO-shifted syngas stream. It seems that the modified process, using a sour shift catalyst to convert CO to CO_2 before the sulfur clean-up, could obtain 85% CO_2 recovery at a suitable purity for separation. The efficiency reduction for such a process would be approximately 10 percentage points, including CO_2 compression to 110 bar. The IGCC plant was considered to cost €2320/kW, producing electricity at a cost of €7.6 cents/kWh and with a CO_2 avoidance cost of €39/tCO_2. The advantages obtainable by using a fixed-site carrier (FSC) membrane unit are:

- Lower energy consumption than in an absorption unit.
- No solvent make-up or chemical inventory required.

The aim of this evaluation was to apply this new membrane technology using existing plant data to obtain as practical an investigation as possible. However, there are opportunities for improvement in the performance of an integrated membrane for the IGCC plant. Improvement in the membrane's CO_2 permeance at higher pressures would decrease the membrane area required and this is the subject of ongoing research. Commercially available gasifiers running at higher pressures would allow lower membrane areas, less hydrogen loss, and higher permeate pressures.

In 2010, a simulation analysis was presented by Franz et al., to evaluate an IGCC plant with a CO_2 selective membrane and with an H_2 selective membrane based on the Aspen Plus model. In this study, the potential of CO_2 selective and H_2 selective polymeric membranes to separate CO_2 from IGCC plants was investigated. As a reference power plant, a modified version of the Puertollano gasification plant was considered. For the CO_2 selective membranes, single membrane module concepts as well as cascade concepts were examined. The simulation results show that with H_2 selective and CO_2 selective membranes (CO_2/H_2 selectivity 15.5, H_2/CO_2 selectivity 5.91), the current requirements concerning CO_2 purity and the CO_2 separation factor cannot be satisfied. According to their results, a CO_2/H_2 selectivity of 150 for a CO_2 selective membrane would be required to attain power plant efficiency losses below 10 percentage points with a separation factor above 85%. However, for a cascade concept, the CO_2/H_2 selectivity needed would be in the order of 60 to achieve the same values. For H_2 selective membranes with an H_2/CO_2 selectivity of 50 separation degrees of 85%, efficiency losses less than 10% points can be achieved.

In 2010, Hussain et al. conducted a feasibility analysis using HYSYS software integrated with an in-house membrane program to investigate the influence of process parameters on energy demands and costs using a novel CO_2 selective FSC membrane. Their results specified that a membrane system using high performance FSC membranes was feasible for CO_2 capture, even at a low CO_2 concentration ($\sim$10%) in a flue gas, compared to amine absorption in terms of energy requirement, and it was also possible to obtain more than 90% CO_2 recovery and >95% CO_2 purity in a permeate stream.

Merkel et al. (2012) reported that the economic feasibility of using the IGCC technology as a way of producing electricity while simultaneously capturing CO_2 is connected to the cost of separating the CO_2 from high-pressure CO_2-hydrogen streams. Both cases of the IGCC plants with CO_2 selective membrane and with H_2 selective membrane were compared with the conventional Selexol absorption technology. As indicated in Table 10.7, the results of this study demonstrate that hybrid membrane-condensation processes using today's polymer membranes can be compared with conventional IGCC with CO_2 capture technology.

Table 10.7 *Comparison of conventional IGCC plant with two different types of IGCC plant combined with H_2 and CO_2 selective membrane*

System/cost parameter	Selexol + CO_2 compression
Compression power used (MW)	31
Refrigeration (or steam) used (MW)	19
Total power used (MW)	50
Membrane area used (m^2)	–
Membrane or Selexol equipment cost (million $)	74
Compression and/or refrigeration equipment cost (million $)	18
Bare erected plant cost (million $)	166
Engineering fees (million $)	16
Contingencies (million $)	69
Total plant cost (million $)	252
Estimated enhancement in LCOE (%)	31

As a general consequence of Merkel et al.'s work, the best design, using recently developed membranes, has 40% of the capital cost and consumes 50% of the energy of cold absorption.

The resulting enhancement in the LCOE to separate and sequester the CO_2 is then about 15%. However, if higher permeance, and especially more selective membranes can be developed, the cost of the technology described will decrease even further.

He et al. (2012) investigated the application of hollow fiber carbon membranes for CO_2 capture from flue gas. They reported a capital cost of 100 \$/tCO_2 avoided using the carbon membrane, which is still higher than traditional chemical absorption process of MEA (59 \$/tCO_2 avoided), but the referred carbon membranes had clear potential for further optimization.

In 2013, an experimental study for CO_2 removal from flue gas was presented by Sandru et al. The results obtained in the Nanoglowa EU project used polymeric polyvinylamine (PVA) FSC membranes developed at NTNU Norway. The pilot scale testing using real flue gas was carried out at power plant sites of EDP in Portugal. The aim of the work was CO_2 separation from the flue gas of coal-fired IGCC plants using membrane technology, and involved several aspects: membrane up-scaling, material durability, and pilot testing in a power plant. Gas permeation experiments and material analyses validated that the membrane material and separation performances were positively affected by exposure to synthetic and real flue gas impurities. A pilot scale module with a 1.5 m^2 of NTNU membrane installed was tested continuously for 6 months. The membranes displayed constant separation performances with a maximum purity of 75% CO_2 in permeate stream. The performances were kept constant despite several challenges related to the IGCC plant operation such as high levels of NO_x (600 mg/Nm^3) and 200 mg/Nm^3 SO_2.

In general, the CO_2 selective membranes offer considerable promise as alternatives to SOTA absorption processes for CO_2 capture and H_2 purification. The challenges offered by different scenarios vary greatly. Typically, in the post-combustion process, the operating conditions are slight (close to ambient pressure and temperature), but the driving force is limited. It is critical, therefore, to improve the membrane separation performance, both CO_2 permeance and CO_2/N_2 selectivity.

The pre-combustion process presents harsher operating conditions of high temperatures (100–200°C) and pressures (>50 bars).Another important challenge is to separate CO_2 from H_2, which is the second smallest gas molecule. Therefore, improving membrane selectivity, evaluating and increasing the operational stability will be critical to practical applications.

10.4 Conclusion and Future Trends

This chapter has presented the principles behind integrating membranes in IGCC power plants with hydrogen production, especially in systems implementing CO_2 capture. The discussion indicated that simplifications in IGCC plant design and efficiency improvements of a few percentage points can be achieved by doing the following:

- replacing the ASU with the ITM membrane system that provides the oxygen for gasification;
- integrating low temperature polymeric hydrogen membranes downstream in a WGS process;
- intensifying the CO_2 formation and separation by combining H_2 selective (Pd-based or silica) membranes with the WGS reaction process;
- integrating CO_2 selective membrane (silica and zeolite membranes) for pre-combustion process.

Also the potential economics of membrane separation in electricity production appear to be positive even though the membrane lifetime duration is still an open issue which may affect this aspect. Therefore, the main challenge to employing CO_2 capture strategies is the development of membrane materials. If higher permeance, and especially more selective, membranes can be developed, the cost of the technology described will decrease even further.

Moreover, the advantages of H_2 separation membranes in IGCC plants can be discovered if hydrogen succeeds as an extensive decarbonized energy carrier on the road to reducing CO_2 emissions in the fight against climate change. In this context their many potential benefits may accelerate their deployment compared to plants for simple power generation with CO_2 capture.

Abbreviations

AGR	Acid gas removal
ASU	Air separation unit
BFD	Block flow diagram
CLC	chemical looping combustion
COE	Cost of the electricity
DD&D	development, demonstration and deployment
EDP	Entrepreneurship Development Programs
EU	European Union
FSC	Fixed site carrier

GT	Gas turbine
GTL	Gas-to-liquid
IGCC	Integrated gasification combined cycle
ITM	Ion transport membrane
LCOE	Levelized cost of the electricity
LHV	Lower heating value
MCM	Mixed conducting membrane
MDEA	Methyl-di-ethanol amine
MEMFO	
MEA	Mono-ethanol amine
MIEM	Mixed ion and electron membrane
MR	Membrane reactor
NO_X	Nitrogen oxides
NTNU	The Norwegian University of Science and Technology
OTM	Oxygen transport membrane
PC	Pulverized coal
Pd	
PVA	Polyvinylamine
SOTA	State of the art
SO_X	Sulfur oxides
TRL	Technology readiness level
WGS	Water gas shift

References

Amelio, M., Morrone, P., Gallucci, F. and Basile, A. (2007) Integrated gasification gas combined cycle plant with membrane reactors: Technological and economical analysis. *Energy Conversion and Management*, **48**, 2680–2693.

Anantharaman, R. and Bolland, O. (2011) Integration of oxygen transport membranes in an IGCC power plant with CO_2 capture. *Chemical Engineering Transactions*, **2**, 25–30.

Anderson, D.H., Iv, C.R.E., Harkins, T.H. et al. (2009) Hydrogen separation using dense composite membranes Part 2: Process integration and scale-up for H_2 production and CO_2 sequestration, in *Inorganic Membranes for Energy and Environmental Applications*, Springer, Berlin, pp. 155–171.

Badwal, S.P.S. and Ciacchi, F.T. (2013) High temperature membrane reactor system for hydrogen permeation measurements and validation with Pd based membranes. *Journal of Membrane and Separation Technology*, **2**, 13–26.

Bracht, M., Alderliesten, P.T., Kloster, R. et al. (1997) Water gas shift membrane reactor for CO_2 control in IGCC systems: Techno-economic feasibility study. *Energy Conversion and Management*, **38**, 164.

Carbo, M.C., Jansen, D. and Verkooijen, A.H.M. (2006) Advanced membrane reactors in IGCC : H_2 or CO_2 separation. Paper presented at Fifth Annual Conference on Carbon Capture and Sequestration, Alexandria, VA, pp. 8–11.

Costa, J. Da., Smart, S., Motuzas, J., Lui, S. and Zange, D. (2013) *State of the Art (SOTA) Report on Dense Ceramic Membranes for Oxygen Separation from Air*. University of Queensland Press, Brisbane.

Dyer, P.N., Richards, R.E., Russek, S.L. and Taylor, D.M. (2000) Ion transport membrane technology for oxygen separation and syngas production. *Solid State Ionics*, **134**, 21–33.

Franz, J., Maas, P. and Scherer, V. (2014) Economic evaluation of pre-combustion CO_2-capture in IGCC power plants by porous ceramic membranes. *Applied Energy*, **130**, 532–542.

Franz, J. and Scherer, V. (2010) An evaluation of CO_2 and H_2 selective polymeric membranes for CO_2 separation in IGCC processes. *Journal of Membrane Science*, **359**, 173–183.

Franz, J. and Scherer, V. (2011) Impact of ceramic membranes for CO_2 separation on IGCC power plant performance. *Energy Procedia*, **4**, 645–652.

Grainger D. and Hägg M.B. (2008) Techno-economic evaluation of a PVA CO_2-selective membrane in an IGCC power plant with CO_2 capture. *Fuel*, **87**, 14–24.

He, X. and Hägg, M.B. (2012) Membranes for environmentally friendly energy processes. *Membranes*, **2**, 706–726.

Hussain, A. and Hägg, M.-B. (2010) A feasibility study of CO_2 capture from flue gas by a facilitated transport membrane. *Journal of Membrane Science*, **359**, 140–148.

Kaldis, S.P., Skodras, G., and Sakellaropoulos, G.P. (2004) Energy and capital cost analysis of CO_2 capture in coal IGCC processes via gas separation membranes. *Fuel Processing Technology*, **85**, 337–346.

Kotowicz, J. and Skorek-osikowska, A. (2010) Membrane separation of carbon dioxide in the integrated gasification combined cycle systems: Introduction. *Archives of Thermodynamics*, **31**, 145–164.

Koutsonikolas, D.E., Kaldis, S.P., Pantoleontos, G.T., Zaspalis, V.T. and Sakellaropoulos, G.P. (2013) Techno-economic assessment of polymeric, ceramic and metallic membranes integration in an advanced IGCC process for H_2 production and CO_2 capture. *Chemical Engineering Transactions*, **35**, 715–720.

Krishnan, G., Steele, D., Brien, K.O., Callahan, R., Berchtold, K. and Figueroa, J. (2009) Simulation of a process to capture CO_2 from IGCC syngas using a high temperature PBI membrane. *Energy Procedia*, **1**. 4079–4088.

Ku, A.Y., Kulkarni, P., Shisler, R. and Wei, W. (2011) Membrane performance requirements for carbon dioxide capture using hydrogen-selective membranes in integrated gasification combined cycle (IGCC) power plants. *Journal of Membrane Science*, **367**, 233–239.

Kunze, C., De, S., and Spliethoff, H. (2011) A novel IGCC plant with membrane oxygen separation and carbon capture by carbonation-calcinations loop, *International Journal of Greenhouse Gas Control*, **5**, 1176–1183.

Kunze, C. and Spliethoff, H. (2012) Assessment of oxy-fuel, pre- and post-combustion-based carbon capture for future IGCC plants, *Applied Energy*, **94**, 109–116.

Leo, A., Liu, S., and Costa, Jo. C.D. da. (2009) Development of mixed conducting membranes for clean coal energy delivery. *International Journal of Greenhouse Gas Control*, **3**, 357–367.

Manzolini, G., Gazzani, M., Turi, D.M. and Macchi, E. (2013) Application of hydrogen selective membranes to IGCC. *Energy Procedia*, **37**, 2274–2283.

Maurstad, O. (2005) *An Overview of Coal based Integrated Gasification Combined Cycle (IGCC): Report*, Laboratory for Energy and the Environment, Cambridge, MA.

Merkel, T.C., Zhou, M. and Baker, R.W. (2012) Carbon dioxide capture with membranes at an IGCC power plant. *Journal of Membrane Science*, **389**, 441–450.

Nagasaki, N., Sasaki, K., Suzuki, T., Dodo, S. and Nagaremori, F. (2013) Near-zero-emission IGCC power plant technology. *Hitachi Review*, **62**, 39–47.

Orme, J. and Stewart, F. (2005) Mixed gas hydrogen sulfide permeability and separation using supported polyphosphazene membranes, *Journal of Membrane Science*, **253**, 243–249.

Pakizeh, M., Omidkhah, M.R. and Zarringhalam, A. (2007) Synthesis and characterization of new silica membranes using template–sol–gel technology. *International Journal of Hydrogen Energy*, **32**, 1825–1836.

Rezvani, S., Huang, Y., Mcilveen-Wright, D., Hewitt, N. and Mondol, J.D. (2009) Comparative assessment of coal fired IGCC systems with CO_2 capture using physical absorption, membrane reactors and chemical looping. *Fuel*, **88**, 2463–2472.

Sandru, M., Kim, T., Capala, W., Huijbers, M. and Hägg, B. (2013) Pilot scale testing of polymeric membranes for CO_2 capture from coal-fired power plants. *Energy Procedia*, **37**, 6473–6480.

Shekhawat, D., Luebke, D.R., and Pennline, W.H. (2003) *A Review of Carbon Dioxide Selective Membranes: A Technical Report*, National Energy Technology Laboratory, USA.

Verweij, H., Lin, Y.S., and Dong, J. (2006) Microporous silica and zeolite membranes for hydrogen purification. *MRS Bulletin*, **31**, 756–764.

Zhang, Y., Wu, Z., Hong, Z., Gu, X. and Xu, N. (2012) Hydrogen-selective zeolite membrane reactor for low temperature water gas shift reaction. *Chemical Engineering Journal*, **197**, 314–321.

11

Integration of a Membrane Reactor with a Fuel Cell

Viktor Hacker, Merit Bodner, and Alexander Schenk
Institute of Chemical Engineering and Environmental Technology, Graz University of Technology, Graz, Austria

11.1 Introduction

Over the last few decades, environmental concerns have become a more and more important issue. The substitution of fossil fuels with renewable resources to reduce greenhouse gas emissions is the aim of many research groups around the world. Many novel methods of energy production have been proposed, yet so far only a few have reached the stage of commercialisation. As a downside, many of those systems – commercially available or not – are very limited in their possibilities for application. Whereas the fuel cell can replace many currently used forms of energy production.

Fuel cells have long been known. The first hydrogen fuel cell was simultaneously developed by Sir William Grove [1, 2] and Schönbein [3] in the 1800s and has ever since been subject to intense research. With the invention of polymer electrolytes [4], the fuel cell system has become lighter and easier to handle. This development rendered the fuel cell

Integrated Membrane Systems and Processes, First Edition. Edited by Angelo Basile and Catherine Charcosset.

to become a reasonable alternative to conventional internal combustion engines in automotive applications, but nevertheless, fuel cells are also suitable for the stationary use as a combined source of electricity and heat.

In all recent studies, hydrogen is considered one of the most important energy carriers when it comes to prognoses for a sustainable future. Even though hydrogen does not occur in nature in its molecular form, it is yet very abundant, since it is bound in water and hydrocarbons. Besides the conventional pathways, using fossil fuels and conventional energy sources, hydrogen deriving from sustainable processes using renewable resources and energy sources is gaining importance and recognition [5–8].

Using hydrogen from renewable feedstocks, fuel cells are an environmentally friendly and efficient source of electrical energy. They cause very low or no local emissions and no noise pollution. Another main advantage of the fuel cell in comparison with other energy conversion systems is its high energy density.

The major holdbacks of a fuel cell are the high costs and the limited lifetime. The costs mainly derive from the catalyst. Platinum is expensive, but via novel preparation methods, the metal content of the catalyst can be decreased significantly without loss of power output [9]. Lifetime limitation is mainly caused by impurities, which poison the catalyst or accelerate membrane degradation. Contaminations can have various sources and often show vast effects on the fuel cell. Catalyst poisoning causes a decrease in the performance of fuel cells, whereas membrane degradation reduces the lifetime by the formation of pinholes vastly. Most of the harmful impurities derive from fuel processing, using renewable or fossil fuels likewise. Pure hydrogen is therefore essential to successfully run a fuel cell and the combination of fuel cells with membrane reactors (MRs) is especially suitable.

11.2 Fuel Cell Basics

Fuel cells convert the chemical energy of molecules or chemical compounds into electrical energy by making the electrons, transferred in the combustion redox reaction, accessible for further usage. Thereby, oxidation and reduction take place at the anode and the cathode, respectively.

$$2H_2 \rightarrow 4H^+ + 4e^- \quad E_0 = \mathbf{0.000\ V}$$

$$O_2 + 4H^+ + 4e^- \rightarrow 2H_2O \quad E_0 = \mathbf{1.229\ V}$$

$$2H_2 + O_2 \rightarrow 2H_2O \quad E_0 = \mathbf{1.229\ V}$$

Both electrode compartments are separated by a membrane, which must be ion-conductive, must be electrically non-conductive, must not allow gas to pass through and has to be chemically, mechanically and thermally stable. The first fuel cell systems developed use liquid electrolytes, but modern fuel cells rather use polymer electrolytes (ionomers). Avoiding liquid electrolytes like phosphoric acid for acidic or potassium hydroxide for alkaline fuel cells (AFCs), respectively, simplifies the overall handling of a fuel cell system. Most commercially available acidic ionomers use a PTFE backbone, functionalised with acid groups. The membrane sandwiched between two electrodes is referred to as membrane electrode assembly (MEA).

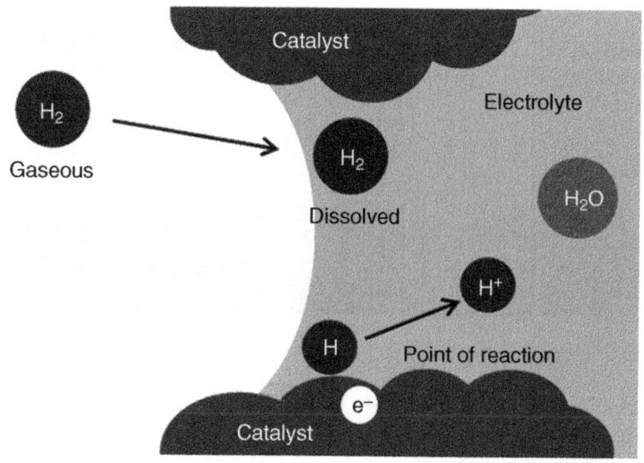

Figure 11.1 Three-phase boundary, active layer in fuel cell electrodes.

The electrodes consist of a backing layer which is coated with the catalyst layer. Up to operating temperature of 200°C, the electrode backing layer consists of carbon paper or cloth; metal foams might also be used. A high porosity of the gas diffusion layer benefits the mass transport properties, ensures a high active surface area and thereby reduces the catalyst loading.

Onto the backing layer, the catalyst layer (active layer) is applied, which contains catalyst on carbon support and electrolyte. The backing layer as well as the carbon support of the catalyst provides mechanical strength and electrical conductivity to the electrode. By adding electrolyte to the catalyst layer, ionic species can travel from one electrode to the other. Fuel conversion is only possible at the three-phase boundary (as shown in Figure 11.1), which consists of electron conductive material (e.g. carbon), ion-conductive electrolyte and catalyst [10]. Therefore, the right ratio of all compounds is essential.

In low-temperature fuel cells, hydrophobicity is induced to avoid clogging of the pores by product water or water from humidification. Catalysts improve the reaction rate by reducing the activation energy without being consumed. Platinum is the most common catalyst in low-temperature acidic fuel cells both on the anode and the cathode side. When reformate gas is used instead of pure hydrogen, Pt alloys are more suitable anode catalysts.

Hydrogen and oxygen both enter the fuel cell through bipolar plates in which flow fields are carved. They determine the mass flow properties and the homogeneous distribution of reactants throughout the electrodes. Further, bipolar plates have to be electrically conductive. Flow field plates usually consist of polymer-enforced graphite, but sometimes they are also made out of metal or stainless steel.

11.2.1 Reaction Mechanisms

A deeper understanding of the fuel cell reaction mechanisms is essential in order to understand the fuel cell in its entirety. The overall reaction can be divided into anode and cathode partial reactions. At the anode of an acidic hydrogen-based fuel cell, the hydrogen

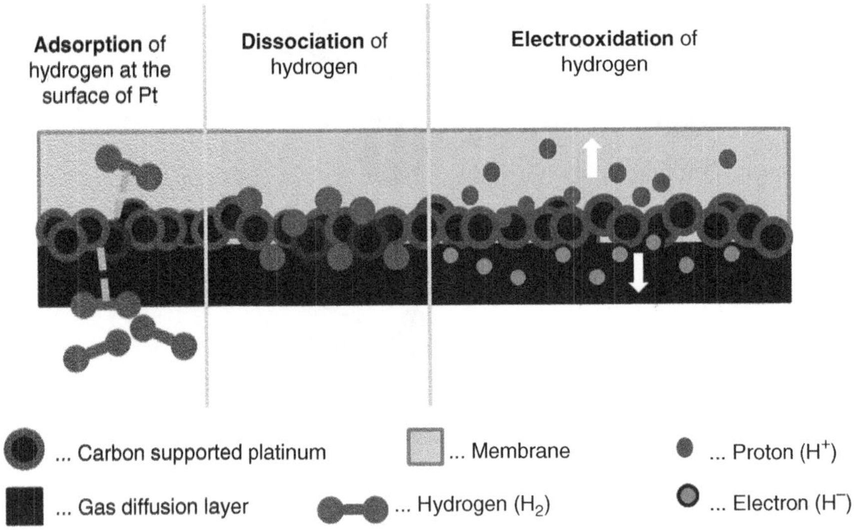

Figure 11.2 *Hydrogen oxidation reaction in three steps.*

oxidation reaction (HOR) takes place. Compared to the oxygen reduction reaction (ORR) at the cathode, the HOR follows a rather simple reaction mechanism.

11.2.1.1 The Hydrogen Oxidation Reaction

The Tafel–Volmer mechanism [11] describes the three essential steps of the HOR at the platinum catalyst (Figure 11.2).

First, the hydrogen molecule adsorbs at the platinum surface. At the surface, the H–H bond is broken. Two hydrogen atoms are adsorbed at the platinum. In a third step, the hydrogen atoms are then oxidised, relieving electrons into the electrode and protons into the electrolyte.

$$H_2 + 2Pt \leftrightarrow H_2 \cdots 2Pt$$

$$H_2 \cdots 2Pt \leftrightarrow 2(Pt - H)$$

$$2(Pt - H) \leftrightarrow 2Pt + 2H^+ + 2e^-$$

Fuel cells using other fuels than hydrogen exist as well. The oxidation reactions taking place in those cells are significantly more complex than the HOR and not fully understood yet.

11.2.1.2 The Oxygen Reduction Reaction

The reaction either follows the associative or the dissociative mechanism [12,13]. The order of steps of the ORR depends on the onset potential of the electrode.

Following the associative mechanism (as shown in Figure 11.3), oxygen adsorbs on the platinum surface in a first step. The adsorbed oxygen molecule then reacts with protons in

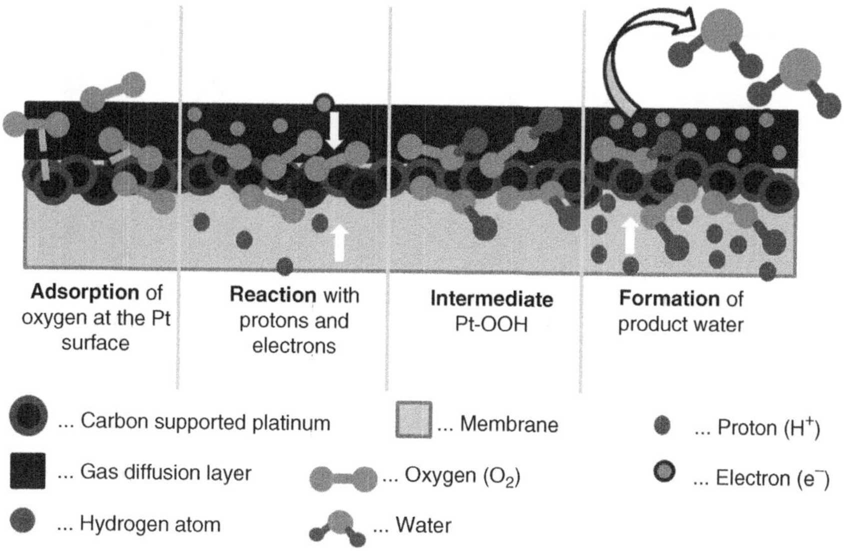

Figure 11.3 Associative mechanism for the oxygen reduction reaction.

multiple steps until fully reduced to water.

$$O_2 + Pt \rightarrow O_2 \cdots Pt$$

$$O_2 \cdots Pt \rightarrow O_2 - Pt$$

$$O_2 - Pt + H^+ + e^- \rightarrow Pt - OOH$$

$$Pt - OOH + 3H^+ + 3e^- \rightarrow Pt + 2H_2O$$

The other possibility is that the dissociative mechanism differs widely from the associative mechanism (Figure 11.4).

The first step of the ORR is the dissociative chemisorption of oxygen on the platinum surface. The O–O bond is broken immediately, and in the second step, each oxygen atom accepts one electron. The oxygen is protonated and water is released.

$$O_2 + 2Pt \rightarrow 2(Pt - O)$$

$$Pt - O + 2H^+ + 2e^- \rightarrow Pt + H_2O$$

11.2.2 Electrochemical Basics of the Fuel Cell

Fuel cells, like batteries, are galvanic cells and can be described by certain electrochemical parameters. One experimentally simple way to characterise a fuel cell is the current–voltage curve (I–V curve). During the measurement, the different current densities in dependence of voltages between the open circuit voltage (OCV; no current) and almost 0 V are recorded. By recording multiple pairs of values at equilibrium, the curve gives the power densities as well as mass transport limitations.

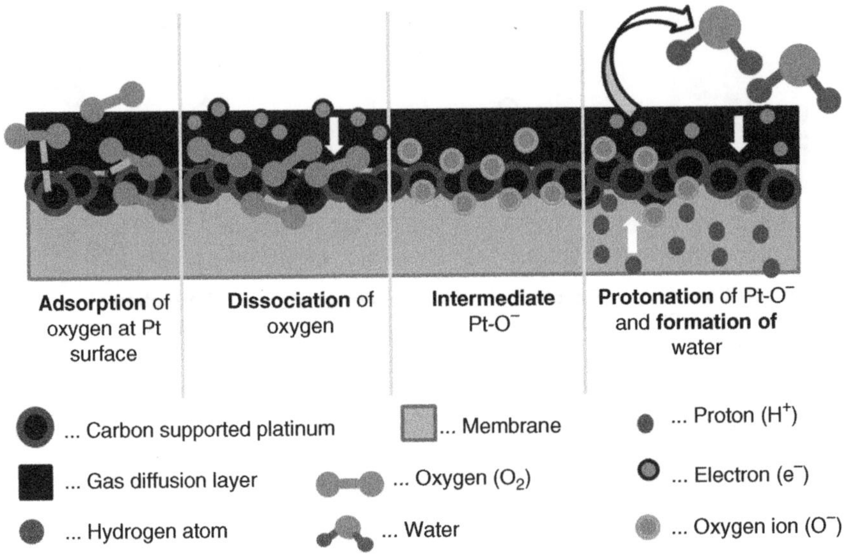

Figure 11.4 *Dissociative mechanism of the oxygen reduction reaction.*

11.2.2.1 Theoretical Current–Voltage Characteristics

The theoretical OCV (a.k.a electromotive force) of any galvanic cell is the sum of the standard potentials of both electrodes.

$$E_0 = E_{0,\text{Cathode}} - E_{0,\text{Anode}}$$

where E_0 is the open circuit voltage (V) and $E_{0,\text{Cathode/Anode}}$ is the cathode/anode standard potentials versus standard hydrogen electrode (SHE; V).

Voltage drops in dependence of the current density due to internal resistance (Figure 11.5, segment I).

$$E = E_0 - i^* R_i$$

$$E = i^*(R_i + R_e)$$

where E is the voltage at current i (V), i the current (A), R_i the internal resistance (Ω) and R_e the external resistance (Ω).

11.2.2.2 Real Current–Voltage Characteristics

The real current–voltage curve of a fuel cell is the sum of the anode and the cathode potential. At a given current, the anode potential increases and the cathode potential decreases; the measurable voltage again is the sum of both potentials.

The real I-V curve differs from the theoretical curve due to losses. The losses (over-potentials or polarisations shown in Figure 11.5) account for the difference between the theoretically calculated potential and the experimentally observable potential. The

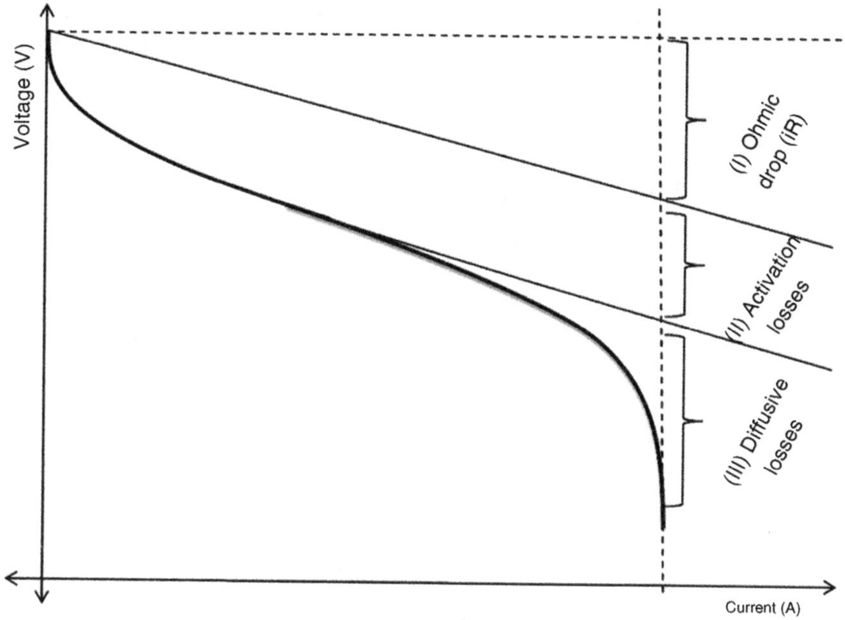

Figure 11.5 *Real current–voltage curve.*

overpotential (a.k.a polarisation) is the difference between a theoretically calculated potential and the experimentally observable potential.

$$\eta = E - E_0$$

where η is the overpotential (V).

Overpotentials result from the limitations (losses) in the reaction pathways. The most important types of losses for a fuel cell are the reaction rate loss (in the region of activation polarisation), the resistance loss (in the region of ohmic polarisation) and the gas transport loss (in the region of concentration polarisation).

- *Ohmic Voltage Drop* (Figure 11.5, *segment I*): The linear voltage drop is directly proportional to the overall resistance of electrolyte, electrodes and current collectors and wiring. The slope can be decreased by minimising the resistance.
- *Reaction Rate and Resistance Loss* (Figure 11.5, *segment II*): The limitation of the electrons travelling through the electrodes' double layer is called resistance loss. It is influenced by the catalyst, the concentration of the reaction gases and the electrolyte. Reaction rate loss derives from a limiting partial reaction. For example, this might be the slow dissociation of hydrogen on the platinum catalyst for a fuel cell. This loss results in a non-linear voltage drop in the low current region of the I–V curve.
- Furthermore, the reduction of the open circuit voltage also results from mixed potentials and internal currents. Reaction gases diffusing through the membrane result in mixed potentials. Together with internal currents, mixed potentials shift the I–V curve towards lower voltages.

- *Gas Transport Loss* (Figure 11.5, *segment III*): In high current regions, voltage drops rapidly due to diffusion limitation. The reaction rate is limited by the reagents' partial pressures at the electrodes.

11.3 Different Types of Fuel Cells

Besides the rather common acidic low-temperature polymer electrolyte fuel cell (PEFC), multiple other types of fuel cells exist (as shown in Figure 11.6) or are under development. Different types of electrolytes can be used to conduct not only protons but also other ionic species, changing the entire cell chemistry. Also, fuels more complex than hydrogen can be processed. The fuel can either be reformed in the anode before the oxidation or can adsorb on the catalyst, being directly oxidised. The different systems described require different operating conditions. Especially the temperature has a significant impact.

11.3.1 Methods of Classification

The different types of fuel cells are most commonly distinguished by either their electrolyte, fuel or working temperature. All types of fuel cells use oxygen as oxidising agent.

11.3.1.1 By Temperature

The operation conditions have a major influence on the fuel cell systems. Especially the temperatures have a vast range from room temperature up to 1000°C, depending on the

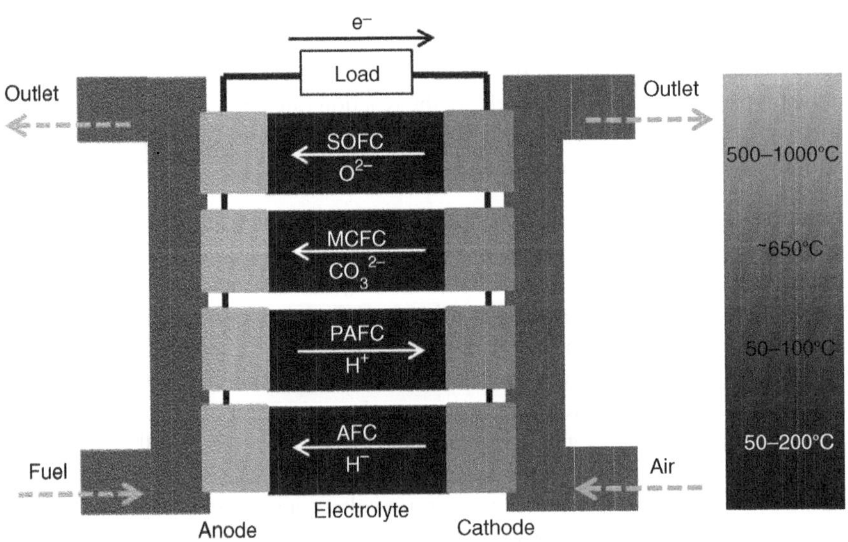

Figure 11.6 *Different types of fuel cells. SOFC, solid oxide fuel cell; MCFC, molten carbonate fuel cell; PAFC, phosphoric acid fuel cell; AFC, alkaline fuel cell.*

type of fuel cell. If classified by temperature, three different ranges of fuel cell temperatures are set.

- *Low-Temperature Fuel Cells:* The operating temperature lies below 100°C. Therefore, aqueous electrolytes can be used. These are either aqueous acidic or alkaline solutions or acidic or alkaline polymer electrolytes, which require water in order to become ion conductors. For example, polymer electrolyte membrane fuel cells.
- *Mid-Temperature Fuel Cells:* The operating temperature lies between 100°C and 250°C. Water at these temperatures is no longer liquid. Therefore, acids or molten salts are used as electrolyte either in a liquid, solid or matrix form. Neither requires water to conduct ions. For example, phosphoric acid fuel cells.
- *High-Temperature Fuel Cells:* The operating temperature exceeds 400°C. Solid oxides or molten salts are used as electrolyte. For example, molten carbonate fuel cells or solid oxide fuel cells.

11.3.1.2 By Electrolyte

Electrolytes can be either classified by their state of matter (liquid or solid) or by the ions they conduct. Their chemical composition has a significant effect on the entire system.

- *Acidic Fuel Cells:* Protons serve as ionic species. The electrolyte can be either a liquid acid such as phosphoric acid or an acidic polymer such as Nafion. Acidic fuel cells are usually operated at low to intermediate temperatures.
- *Alkaline Fuel Cells:* Hydroxide ions serve as ionic species. Again, the electrolyte can be either liquid (e.g. an aqueous potassium hydroxide solution) or an alkaline polymer. AFCs are usually operated at low temperatures.
- *Molten Carbonate Fuel Cells:* Carbonate anions serve as ionic species and molten carbonate serves as liquid electrolyte. Molten carbonate fuel cells (MCFCs) require high operation temperatures.
- *Solid Oxide Fuel Cells:* Oxygen anions serve as ionic species. The electrolyte consists of solid oxides. Solid oxides are ion-conductive at high temperatures and are therefore operated at temperatures above those of MCFCs.

11.3.1.3 By Fuel

Fuel cells can process different types of fuels. If fuels other than hydrogen are used, they are often reformed internally. Especially in molten carbonate and solid oxide fuel cells (SOFCs), water-gas shift reaction takes place. But it is also possible to directly oxidise the fuel at the anode catalyst surface. This requires different types of catalysts.

- *Hydrogen*: Both acidic and alkaline cells using hydrogen as fuel exist. In both cases, water is the only reaction product. Hydrogen is mainly produced via reforming of natural gas.
- *Alcohol*: It is possible to either reformate the alcohol or to use it directly in the fuel cell. For the direct alcohol fuel cells (DAFCs), primarily methanol is used, but other alcohols such as ethanol or glycol can also be converted. Direct fuel cells mostly use acidic electrolytes, but alkaline DAFCs exist as well. When fully oxidised, the reaction

products of any alcohol are water and carbon dioxide. Most alcohols can be produced by fermentation of sugars or starchy compounds.

- *Reformate gas*: It is primarily used in molten carbonate and SOFCs, since these two types are operated at high temperatures and are therefore resistant towards poisoning of the catalyst by carbon monoxide. Reaction products are water and carbon dioxide. Reformate gas is mostly produced from natural gas.

11.3.2 Fuel Cell Types

Following the classifications above, different types of fuel cells can be described. Some of the most important are as follows.

11.3.2.1 Polymer Electrolyte Fuel Cell

The most common type of fuel cell is the acidic hydrogen fuel cell [10, 14]. PEFCs are usually operated between room temperature and 80°C. Acidic polymers (ionomers such as Nafion) are used as electrolyte. At the anode side, hydrogen is provided; the cathode is supplied with air at low pressures. Hydrogen adsorbs on the anode at the platinum catalyst and is oxidised, producing proton species. The H^+ ions diffuse through the proton-conductive electrolyte towards the cathode. At the cathode, oxygen adsorbs on the catalyst and is reduced forming H_2O. For each molecule of hydrogen, two electrons travel from the anode to the cathode and can be utilised.

11.3.2.2 Alkaline Fuel Cell

AFCs are operated using hydrogen at temperatures up to 80°C [10, 14, 15]. The electrolyte is an aqueous potassium hydroxide solution with a concentration between 7 and 9 mol L^{-1}, being the peak conductivity. The AFC has some advantages compared to the PEFC. KOH as electrolyte is inexpensive and the cathode reaction kinetics are significantly improved at high pH values. Furthermore, a large variety of metals is stable in alkaline media. Therefore, other catalyst systems than the conventional platinum containing ones can be utilised. For example, silver as cathode and nickel as anode catalyst.

The AFC has a few disadvantages as well. To avoid the formation of potassium carbonate from potassium hydroxide and carbon dioxide, pure oxygen has to be used at the cathode. Further, the KOH concentration in the electrolyte decreases with the formation of product water. Therefore, it is necessary to circulate the electrolyte.

Alternatives are hydroxide-conducting alkaline polymer membranes, but their molar conductivities do not reach those of proton exchange membranes.

11.3.2.3 Direct Alcohol Fuel Cell

DAFCs range from room temperature up to 80°C [14, 16]. The electrolyte usually consists of acidic polymers analogous to the ones used in PEFCs, but alkaline polymers and acidic and alkaline solutions are also possible. Alcohols are directly oxidised at the anode in a complex reaction, forming water and carbon dioxide. Platinum and platinum group metals are common catalysts, but again, the usage of alkaline electrolytes allow the usage of non-noble metals such as nickel.

11.3.2.4 High-Temperature Polymer Electrolyte Fuel Cell

With operation temperatures between 120°C and 250°C, high-temperature polymer electrolyte fuel cells (HT-PEFCs) are considered mid-temperature fuel cells [14, 17]. The HT-PEFC works analogously to the PEFC, but due to the increased temperatures, only electrolytes independent of humidification can be used. These fuel cells are less sensitive towards poisoning of the catalyst.

11.3.2.5 Phosphoric Acid Fuel Cell

Phosphoric acid fuel cells (PAFCs) are operated at about 200°C, using hydrogen as fuel and platinum as catalyst [10, 14]. Phosphoric acid in a neutral polymer/SiC matrix is used as electrolyte. Since phosphoric acid is highly corrosive, the number of materials that can be used is limited.

11.3.2.6 Molten Carbonate Fuel Cell

MCFCs are operated at about 650°C, using reformate gas as fuel and molten carbonate salts as electrolyte [10, 14]. Catalyst poisoning is no longer an issue and carbon monoxide can even be used as fuel. Therefore, and in combination with the improved reaction kinetics, non-noble metals such as nickel can be used as catalysts.

11.3.2.7 Solid Oxide Fuel Cell

Temperatures range from 650°C up to 1000°C [10, 14]. The SOFC uses reformate gas as fuel and ion-conductive solid oxides such as zirconium dioxide as electrolyte. Like the MCFC, the SOFC can use non-noble metals as catalysts.

11.4 Contaminations of the PEFC

A vast variety of contaminations can affect, especially PEFCs, negatively in means of power output and lifetime. Contaminations often derive from fuel processing, but also from the oxidant, especially if ambient air is used, but also from system compounds such as the bipolar plates. Different contaminations can harm the fuel cell via three different mechanisms.

Via *poisoning of the catalyst*: Contaminations such as carbon monoxide or hydrogen sulphide inhibit the catalysis of the HOR at the anode and the ORR at the cathode, respectively.

The membrane resistance is increased via *membrane degradation* mechanisms. A change in membrane structure disturbs the water management of the electrolyte and inhibits the proton transportation.

Via *structural alterations of the active layer*, mass transport properties change fundamentally.

11.4.1 Anode Gas Stream

PEFCs demand pure hydrogen in order to avoid shortening of the lifetime by degradation. Hydrogen is conventionally produced by the reforming of hydrocarbons, such as methane.

Most common methods are steam reforming, autothermal reforming and partial oxidation. The resulting reformate gas already possesses a high content of hydrogen, but also high amounts of carbon monoxide, carbon dioxide and traces of inert gases, vaporous nitrogen compounds and sulphuric compounds.

Steam reforming consists of two reaction steps. First, long chain hydrocarbons are broken into smaller molecules and part of the methane is converted in a pre-reforming step at high temperatures. The product gas of the first step contains hydrogen, carbon monoxide, carbon dioxide and methane. In a second step, methane is further converted to carbon dioxide and hydrogen at lower temperatures. Also, the not fully oxidised carbon monoxide is converted by the water gas shift reaction, increasing the hydrogen yield. The reformate gas yet still has to be purified in further steps [18].

For the partial oxidation, natural gas, coal or oil are oxidised at sub-stoichiometric conditions, mainly producing hydrogen and carbon monoxide [19].

Autothermal reforming is the combination of both steam reforming and partial oxidation, increasing the conversion efficiency. Air and water content, as well as temperature, are optimised in order to obtain the highest yield [20].

Many other hydrogen-producing processes are known, but have not reached the level of commercialisation yet.

When hydrogen is produced with conventional methods, contaminations with compounds such as carbon monoxide, carbon dioxide, hydrogen sulphide, ammonia and multiple different organic sulphur compounds are inevitable.

11.4.1.1 Carbon Monoxide (CO)

The poisoning of the platinum catalyst by carbon monoxide is one of the most well-documented effects of a contamination on the fuel cell. Important factors for the poisoning are not only the content of carbon monoxide in the anode stream, but also the duration of exposure, temperature and composition of the catalyst. Even small amounts of CO can result in a significant voltage loss.

$$2CO + 2(Pt - H_{ads}) \rightarrow 2(Pt - CO_{ads}) + H_2$$

Even though the catalyst poisoning is such a well-known effect, the exact mechanism is not fully understood yet.

When adsorbed, carbon monoxide is strongly bound to the platinum surface, blocking it [21]. CO can even replace already adsorbed hydrogen molecules [22]. The HOR rate is therefore reduced due to the loss of active catalyst surface area.

11.4.1.2 Carbon Dioxide (CO_2)

Hydrogen produced via steam reforming contains high amounts of carbon dioxide. Especially at high current densities, the negative influences of CO_2 are observable. A reversion of the water gas shift reaction occurs, catalysed by platinum. Carbon dioxide is reduced by hydrogen to carbon monoxide, which then poisons the catalyst [23].

$$CO_2 + 2(Pt - H_{ads}) \rightarrow Pt - CO_{ads} + H_2O + Pt$$

For this reason, even small amounts of carbon dioxide can have a significant negative effect on the fuel cell performance.

11.4.1.3 Hydrogen Sulphide (H_2S)

Reformate gas further contains traces of hydrogen sulphide. H_2S as well as other sulphuric compounds bind strongly at the platinum surface, blocking it and thereby inhibiting the HOR. Due to the multiplicity of electrochemical potentials, sulphuric compounds, once adsorbed, have a high number of possibilities to further react at the catalyst surface.

Mathieu and Primet proposed a dissociative interaction between hydrogen sulphide and platinum [24].

$$H_2S + Pt \rightarrow Pt - S + H_2$$

$$H_2 + 2Pt \rightarrow 2PtH$$

This mechanism was confirmed for the interaction between H_2S contaminations and the Pt catalyst in PEFCs by Lopes et al. [25].

11.4.1.4 Ammonia (NH_3)

The reformate gas itself can contain traces of ammonia, but also if the anode stream contains residues of nitrogen-containing hydrocarbons, ammonia can be formed within the fuel cell [26].

Furthermore, the formation of traces of ammonia from nitrogen and hydrogen at high temperatures during fuel processing has been reported [27].

Unlike other contaminations, ammonia does not poison the catalyst, but adsorbs at the polymer electrolyte membrane, reacting with protons, forming the ammonium cation. Ammonium is stabilised by the acidic membrane and blocks proton pathways, increasing the membrane resistance and thereby reducing the cell performance.

$$NH_3(g) \rightarrow NH_3(\text{Membrane})$$

$$NH_3(\text{Membrane}) + H^+ \rightarrow NH_4^+$$

The concentration of ammonia and the duration of the exposure influence the degradation. In the early stages of the contamination, the negative effect of the contamination can be partially reversed by the usage of ultrapure hydrogen. After a long time exposure though, the effect becomes irreversible [27].

11.4.2 Cathode Gas Stream

Most fuel cells are using ambient air as a source of oxygen. Depending on the environmental situation, the air pollution is more or less severe. Common contaminations are nitrogen, sulphur and carbon oxide, but also volatile organic compounds deriving from transportation and industries.

11.4.2.1 Nitrogen Oxides (NO_X)

Nitrogen oxides are, in the presence of oxygen, oxidised to nitrogen dioxide. NO_2 does not poison the catalyst, but influences the ionomer and the ionomer-catalyst boundaries. Also, nitrogen dioxide can be reduced at the catalyst surface area, decreasing the oxygen reduction rate [28].

In the presence of water, NO_2 disproportionates into nitric acid (HNO_3) and nitrous acid (HNO_2); nitrous acid being further oxidised by oxygen to form nitric acid. HNO_3 is a strong inorganic acid and increases the proton concentration. The excess protons change the cathode stoichiometry causing a lack of oxygen and increasing the cathode overpotential [29].

Degradation caused by short-term exposure of the fuel cell to nitrogen dioxide is reversible by the usage of pure air, but becomes irreversible after long-term exposure.

11.4.2.2 Sulphur Oxides (SO_X)

Sulphur oxides are, just like hydrogen sulphide in the anode gas, common impurities in air. Sulphur dioxide and sulphur trioxide poison the catalyst by strongly binding at the catalyst surface. Even small amounts of sulphur in the cathode gas stream decrease the catalysts activity towards the ORR significantly. If the coverage of the catalyst's surface with SO_x increases, sulphur can change the kinetics of the ORR, switching from the four-electron pathway to a two-electron pathway [30].

Especially traces of sulphur dioxide are known to show a severe effect on the cell performance. The voltage loss is proportional to the amount of SO_2 in the cathode gas stream. Increasing the humidification can reduce this effect [31].

11.4.2.3 Carbon Oxides (CO_X)

Contaminations of the cathode gas stream with carbon monoxide and carbon dioxide show the same effect on the platinum catalyst as at the anode. By reducing the catalyst activity, the cell performance is decreased significantly. Furthermore, CO contaminations of the anode gas stream can harm the cathode catalyst as well by diffusing through the membrane or passing through pinholes [32].

11.4.3 Contaminations of Components

Because of the corrosive conditions in PEFCs, metal ions from cell and stack components can be dissolved. Bipolar plates, seals, current collectors, gas inlets and humidification components all can be sources of a variety of contaminations.

Especially corrosion of the fuel cell components releases a large variety of alkali metals, alkaline earth metals and transition metals, as well as rare earth metals in the form of metal cations into the MEA. Metal ions usually show no influence on the catalysts, but they do affect the polymer electrolyte membrane. They inactivate the acid groups of the membrane, reducing the proton transportation. Furthermore, metal ions catalyse the membrane degradation by increasing radical formation [33]. The ion conductivity of the ionomer is decreased and the water management is influenced negatively, resulting in an overall reduction of the fuel cell performance.

11.5 Methods to Avoid Poisoning

Traces of certain compounds can have a significantly negative effect on the entire fuel cell performance and lifetime. Therefore, it is important to identify harmful contaminations and their sources.

Hydrogen-rich reformate gas is most commonly used as fuel in PEFCs, but it does contain considerable amounts of carbon monoxide and carbon dioxide. Therefore, either an optimisation of the reforming process, the purification of the reaction gas or the optimisation of fuel cell materials and operation conditions is necessary.

11.5.1 Increasing the Fuel Cell Tolerance towards Contaminations

The approach is to alter the affected material or the operation conditions, so the system becomes more resistant towards harmful compounds.

11.5.1.1 Contamination-Resistant Electrocatalysts

The poisoning of the catalyst within the fuel cell results from the strong bond between the contaminating compound (i.e. carbon monoxide) and platinum at low temperatures. CO tolerant catalysts weaken this bond and usually consist of alloys containing platinum and different noble and transition metals.

When noble metals are added to the platinum catalyst, the bifunctional mechanism is favoured. Water is electrooxidised and thereby oxidises the adsorbed CO, releasing carbon dioxide [34].

$$M + H_2O \rightarrow M - OH + H^+ + e^-$$

$$Pt - CO + M - OH \rightarrow Pt + M + CO_2 + H^+ + e^-$$

The combination of platinum and ruthenium has shown to be the most tolerant towards carbon monoxide contaminations, but other binary and ternary combinations are possible as well [35].

Catalysts such as platinum cobalt on carbon support are known to be resistant towards sulphur contamination and can be used in fuel cells. A tolerant catalyst has to bind impurities weaker than pure platinum, but the hydrogen-catalyst or oxygen-catalyst binding has to be sufficiently strong [36].

The catalyst surface must resist the conditions inside a fuel cell and must not be deactivated thereof. Catalyst metals must not form stable oxides, since oxides might segregate in the alloy.

11.5.1.2 Operating the PEFC at Increased Temperatures

Operating a PEFC at higher temperatures holds many advantages and prevents the cell from being harmed by contaminations. The adsorption of carbon monoxide is highly depending on the operation temperature of the fuel cell. At increased temperatures, the bond between CO and platinum is weakened, thus the effect impurities have on performance and lifetime is reduced. The higher the temperature, the higher is the CO tolerance. Also, the reaction kinetics is improved, but the temperature is limited by the boiling point of water and the material. Most materials commonly used in low-temperature PEFCs do not withstand temperatures exceeding 100°C [37].

11.5.2 Avoiding Contaminations

Contaminations in the anode gas stream can be avoided either by optimising the reformation step or by purifying the gas. For the cathode gas, the usage of filters is common.

11.5.2.1 Pre-treatment of the Reformate Gas

The reformate gas is treated before entering the fuel cell. Preferential oxidation or selective oxidation are the methods of choice [38]. Impurities of the reformate gas are removed in multiple steps like desulphurisation, steam reforming, autothermal reforming, water gas shift reaction, pressure-swing adsorption and methanation [8, 39]. All these reactions require highly selective catalysts to avoid the consumption of hydrogen [40], but also in order to gain high activities, duration times and high resistance against catalyst poison.

Methods to avoid Poisoning by Sulphur Compounds: While there are many methods to reduce the influence which carbon monoxide has on the fuel cell performance, there are only a few approaches to avoid poisoning by sulphur compounds.

To remove organic sulphur compounds, either chemical reaction techniques or adsorption technologies can be used. Hydrodesulphurisation (HDS) and alkylation are considered chemical approaches. For the HDS, the organo-sulphur is hydrogenated and sulphur released as H_2S. For the alkylation, the sulphur-containing compounds are selectively alkylated. Thereby the molecular weight and the boiling point are increased. The sulphuric compounds can then be removed by distillation [8].

Purging of the anode with ultrapure hydrogen can clean the catalyst from adsorbed sulphur, partially regenerating it. The degree of regeneration highly depends on how long the catalyst was exposed to impurities [41].

Also, short voltage impulses [42] or cyclic voltammetry [36] can force sulphur compounds to desorb and regenerate the fuel cell performance.

11.5.2.2 Optimisation of the Reforming Step

By optimising the reforming step, the formation of harmful compounds can either be avoided or the purification step can be included in the reforming process itself.

- *Metal-Steam Process:* Within metal-steam processes, such as the iron-steam process, metal oxides are reduced by reducing agents, such as hydrogen, carbon monoxide or methane. In a second step, hydrogen is released by the oxidation of the metal with water. Thereby, hydrogen of a high purity can be produced. The metal oxides can then be recycled. The reduced metals can furthermore be used for energy storage [43].
- *Catalytic Decomposition of Ammonia:* Ammonia can be liquefied at room temperature and 8 bar pressure. Therefore, ammonia can also be used for hydrogen storage. The cracking of ammonia into hydrogen and nitrogen is a mildly endothermic process, resulting in no other by-products than nitrogen and traces of ammonia and water. Yet for the usage in acid fuel cells, all traces of ammonia have to be removed before entering the fuel cell. This can be done by hydrogen stream scrubbing [44].
- *Electrolysis:* Water electrolysis is a relatively simple process, producing hydrogen of high purity and results in no local pollution. For the acidic water electrolysis, protons are reduced at the cathode and O^{2-} is oxidised at the anode, giving hydrogen and oxygen,

respectively. The sustainability of the process highly depends on the primary energy source though [45].

- *Membrane Reactors:* Another method of generating ultrapure hydrogen is the use of MRs. MRs, using dense metal membranes, are able to produce hydrogen with purities of up to 100% and promise weight and space saving [46].

Furthermore, using an MR can increase the conversion efficiency significantly and an operation at moderate temperatures is possible. Therefore, the water gas shift (WGS) reaction is favoured and the amount of CO in the exhaust gas is reduced. By removing the produced hydrogen, the equilibrium is shifted towards the product side, enhancing hydrogen production in MRs. High pressures show a negative effect on the equilibrium conversion, but simultaneously increase the driving force for hydrogen removal. Therefore, high pressures are counterbalanced by the effect of hydrogen removal and MRs achieve higher conversion rates than conventional reactors at all pressures [47].

Hydrogen generation can either be independent of the fuel cell operation (external reforming) or in combination (internal reforming) with it. The combined system utilises the thermal energy of the exothermic electrochemical reaction of the fuel cell for the endothermic hydrogen production reaction in the membrane reactor. Therefore, heat exchange between fuel cell and reactor can increase the system efficiency. Yet due to the temperature gap, a PEFC being operated at up to 80°C and a typical MR being operated at about 250°C; the internal reforming is especially suitable for HT-PEFCs [46].

11.5.2.3 Methods to Avoid Contaminations of the Air

One of the most practical methods to reduce the contamination of the fuel cell by impurities of the cathode gas flow is the usage of air filters. By choosing a material with proper physical properties, contaminations such as nitrogen oxide, sulphur oxide and carbon oxide can be removed from the cathode gas stream [48–50].

11.5.2.4 Methods to Avoid Contaminations at the Membrane

Due to the severe effect which membrane degradation has on lifetime and performance of a PEFC, any contamination accelerating the degradation has either to be avoided, or the membrane itself has to be stabilised towards degradation. As a major part of the membrane degradation process is caused by the formation of free radicals, cerium(III) ions and manganese(II) ions can be added to the electrolyte, since they have been shown to quench radicals [51].

11.5.2.5 Oxygen Bleeding

Oxygen bleeding is a very simple and effective method to avoid damage by carbon monoxide in a fuel cell [52]. Small amounts of oxygen are added to the fuel, oxidising carbon monoxide via the water gas shift reaction. Since this includes CO that is already bound on the catalyst surface, an almost full regeneration can be achieved. Yet excessive oxygen in the anode stream bears danger and reduces the cell performance [53].

The oxygen or air bleed technique has also been shown to have a positive effect on contaminations with H_2S, yet not as significant as with CO contaminations [25].

11.6 Conclusion

Among the large variety of different fuel cell types, the PEFCs are an especially promising and highly efficient system for future energy production. Even though they have been and are subject to intense research, lifetime is still a limitation. This is mostly caused by impurities in the fuel and in the oxidant. Therefore, an effective approach to either purify the gases or to make the fuel cell more resistant against contaminations is necessary.

Impurities, such as sulphuric components or CO_2, can be removed by the usage of filters or by gas scrubbing. This rather simple approach is not possible for all types of contaminations. Especially the concentration of the component with the most severe effects on the fuel cells efficiency, carbon monoxide, has to be reduced to very low levels, even when using improved catalysts. This is barely achievable with simple methods, such as filters.

Since most alterations of the fuel cell system itself cause increasing costs, the favoured approach is to remove contaminations during hydrogen production. Among various approaches, the integration of a MR is the most suitable. It is a very promising attempt to achieve high hydrogen purities and simultaneously improve the conversion rate of the reforming step.

List of Abbreviations

AFC	Alkaline fuel cell
DAFC	Direct alcohol fuel cell
HDS	Hydrodesulphurisation
HOR	Hydrogen oxidation reaction
HT-PEFC	High temperature polymer electrolyte fuel cell
I	Current
MCFC	Molten carbonate fuel cell
MEA	Membrane electrode assembly
MR	Membrane reactor
OCV	Open circuit voltage
ORR	Oxygen reduction reaction
PAFC	Phosphoric acid fuel cell
PEFC	Polymer electrolyte fuel cell
PTFE	Polytetrafluoroethylene
SOFC	Solid oxide fuel cell
V	Voltage
WGS	Water gas shift

List of Symbols

E	Voltage at current i (V)
E_0	Open circuit voltage (V)
$E_{0,\text{Cathode/Anode}}$	Cathode/anode standard potentials versus SHE (V)

i	Current (A)
R_e	External resistance (Ω)
R_i	Internal resistance (Ω)
η	Overpotential (V)

References

1. Grove WR, "On a new voltaic combination," *London Edinburgh Philos. Mag. J. Sci.*, vol. 13, pp. 430–431, 1838.
2. Grove WR, "On voltaic series and the combination of gases by platinum," *London Edinburgh Philos. Mag. J. Sci.*, vol. 14, pp. 127–129, 1839.
3. Schönbein CF, "On the voltaic polarization of certain solid and fluid substances," *London Edinburgh Philos. Mag. J. Sci.*, vol. 14, pp. 43–45, 1839.
4. Peckham TJ, Yang Y and Holdcroft S, "Proton exchange membranes," in *Proton Exchange Membrane Fuel Cells Material Properties and Performance*, D. P. Wilkinson, J. Zhang, R. Hui, J. Fergus and X. Li, Eds. Boca Raton, FL: CRC Press, 2010, pp. 107–191.
5. Levin DB and Chahine R, "Challenges for renewable hydrogen production from biomass," *Int. J. Hydrogen Energy*, vol. 35, no. 10, pp. 4962–4969, May 2010.
6. Dincer I, "Green methods for hydrogen production," *Int. J. Hydrogen Energy*, vol. 37, no. 2, pp. 1954–1971, Jan. 2012.
7. Cipriani G, Di Dio V, Genduso F, La Cascia D, Liga R, Miceli R and Ricco Galluzzo G, "Perspective on hydrogen energy carrier and its automotive applications," *Int. J. Hydrogen Energy*, vol. 39, no. 16, pp. 1–13, May 2014.
8. Holladay JD, Hu J, King DL and Wang Y, "An overview of hydrogen production technologies," *Catal. Today*, vol. 139, no. 4, pp. 244–260, Jan. 2009.
9. Gasteiger HA, Kocha SS, Sompalli B and Wagner FT, "Activity benchmarks and requirements for Pt, Pt-alloy, and non-Pt oxygen reduction catalysts for PEMFCs," *Appl. Catal. B Environ.*, vol. 56, no. 1–2, pp. 9–35, Mar. 2005.
10. Hamann CH and Vielstich W *Elektrochemie*, 4, vollst. Weinheim: Wiley-VCH Verlag GmbH & Co. KGaA, 2005, p. 662.
11. Mann RF and Thurgood CP, "Evaluation of Tafel–Volmer kinetic parameters for the hydrogen oxidation reaction on Pt(110) electrodes," *J. Power Sources*, vol. 196, no. 10, pp. 4705–4713, May 2011.
12. Nørskov JK., Rossmeisl J, Logadottir A, Lindqvist L, Kitchin JR, Bligaard T and Jónsson H, "Origin of the overpotential for oxygen reduction at a fuel-cell cathode," *J. Phys. Chem. B*, vol. 108, no. 46, pp. 17886–17892, Nov. 2004.
13. Karlberg GS, Rossmeisl J and Nørskov JK, "Estimations of electric field effects on the oxygen reduction reaction based on the density functional theory.," *Phys. Chem. Chem. Phys.*, vol. 9, no. 37, pp. 5158–5161, Oct. 2007.
14. Sharaf OZ and Orhan MF, "An overview of fuel cell technology: Fundamentals and applications," *Renew. Sustain. Energy Rev.*, vol. 32, pp. 810–853, Apr. 2014.
15. Merle G, Wessling M and Nijmeijer K, "Anion exchange membranes for alkaline fuel cells: A review," *J. Memb. Sci.*, vol. 377, no. 1–2, pp. 1–35, Jul. 2011.

16. Zhao TS, Liang ZX and Xu JB, "Fuel cells—Direct alcohol fuel cells overview," in *Encyclopedia of Electrochemical Power Sources*, Volume 2, J. Garche, Ed. Amsterdam: Elsevier B.V., 2009, pp. 362–369.
17. Heinzel A, Bandlamudi G and Lehnert W, "Fuel cells—Proton-exchange membrane fuel cells high temperature PEMFCs," in *Encyclopedia of Electrochemical Power Sources*, Volume 2, J. Garche, Ed. Amsterdam: Elsevier B.V., 2009, pp. 951–957.
18. Izquierdo U, Barrio VL, Cambra JF, Requies J, Güemez MB, Arias PL, Kolb G, Zapf R, Gutiérrez AM and Arraibi JR, "Hydrogen production from methane and natural gas steam reforming in conventional and microreactor reaction systems," *Int. J. Hydrogen Energy*, vol. 37, no. 8, pp. 7026–7033, Apr. 2012.
19. Liu S, Li W, Wang Y and Xu H, "Catalytic partial oxidation of methane to syngas in a fixed-bed reactor with an O2-distributor: The axial temperature profile and species profile study," *Fuel Process. Technol.*, vol. 89, no. 12, pp. 1345–1350, Dec. 2008.
20. Yoon H and Erickson P "Hydrogen from coal-derived methanol via autothermal reforming processes," *Int. J. Hydrogen Energy*, vol. 33, no. 1, pp. 57–63, Jan. 2008.
21. Kohlmayr G and Stonehart P, "Adsorption kinetics for carbon monoxide on platinum in hot phosphoric acid," *Electrochim. Acta*, vol. 18, no. 10, p. 765, Oct. 1973.
22. Stonehart P and Kohlmayr G, "Effect of poisons on kinetic parameters for platinum electrocatalyst sites," *Electrochim. Acta*, vol. 17, no. April 1971, pp. 369–382, 1972.
23. Janssen GJM, "Modelling study of CO_2 poisoning on PEMFC anodes," *J. Power Sources*, vol. 136, no. 1, pp. 45–54, Sep. 2004.
24. Mathieu M and Primet M, "Sulfurization and regeneration of platinum," *Appl. Catal.*, vol. 9, pp. 361–370, 1984.
25. Lopes T, Paganin VA and Gonzalez ER, "The effects of hydrogen sulfide on the polymer electrolyte membrane fuel cell anode catalyst: H_2S–Pt/C interaction products," *J. Power Sources*, vol. 196, no. 15, pp. 6256–6263, Aug. 2011.
26. Borup RL, Inbody MA, Tafoya JI, Semelsberger T and Perry L, "Durability Studies: Gasoline/Reformate Durability 2002 National Laboratory R&D Meeting Program Technical Objectives: Examine Fuel Effects on FP/Stack Durability," 2002.
27. Uribe FA, Gottesfeld S and Zawodzinski TA, "Effect of ammonia as potential fuel impurity on proton exchange membrane fuel cell performance," *J. Electrochem. Soc.*, vol. 149, no. 3, p. A293, 2002.
28. Mohtadi R, Lee WK and Van Zee JW, "Assessing durability of cathodes exposed to common air impurities," *J. Power Sources*, vol. 138, no. 1–2, pp. 216–225, Nov. 2004.
29. Yang D, Ma J, Xu L, Wu M and Wang H, "The effect of nitrogen oxides in air on the performance of proton exchange membrane fuel cell," *Electrochim. Acta*, vol. 51, no. 19, pp. 4039–4044, May 2006.
30. Garsany Y, Baturina OA and Swider-Lyons KE, "Impact of sulfur dioxide on the oxygen reduction reaction at Pt/Vulcan carbon electrocatalysts," *J. Electrochem. Soc.*, vol. 154, no. 7, p. B670, 2007.
31. Uribe FA, Brosha E, Garzon F and Mikkola M "VII. I. 4 effect of fuel and air impurities on PEM fuel cell performance," DOE Hydrogen Program, pp. 1046–1051, 2005.
32. Qi Z, He C and Kaufman A, "Poisoning of proton exchange membrane fuel cell cathode by CO in the anode fuel," *Electrochem. Solid-State Lett.*, vol. 4, no. 12, p. A204, 2001.

33. Cheng X, Shi Z, Glass N, Zhang L, Zhang J, Song D, Liu Z-S, Wang H and Shen J, "A review of PEM hydrogen fuel cell contamination: Impacts, mechanisms and mitigation," *J. Power Sources*, vol. 165, no. 2, pp. 739–756, Mar. 2007.
34. Roth C, Benker N, Mazurek M, Scheiba F and Fuess H, "Pt-Ru fuel cell catalysts subjected to H_2, CO, N_2 and air atmosphere: An X-ray absorption study," *Appl. Catal. A Gen.*, vol. 319, pp. 81–90, Mar. 2007.
35. Watanabe M and Motoo S, "Electrocatalysis by ad-atoms: Part III. Enhancement of the oxidation of carbon monoxide on platinum by ruthenium ad-atoms," *J. Electroanal. Chem.*, vol. 60, pp. 275–283, 1975.
36. Garsany Y, Baturina OA and Swider-Lyons KE, "Oxygen reduction reaction kinetics of SO_2-Contaminated Pt_3Co and Pt/Vulcan carbon electrocatalysts," *J. Electrochem. Soc.*, vol. 156, no. 7, p. B848, 2009.
37. Zhang J, Xie Z, Zhang J, Tang Y, Song C, Navessin T, Shi Z, Song D, Wang H, Wilkinson DP, Liu Z-S and Holdcroft S, "High temperature PEM fuel cells," *J. Power Sources*, vol. 160, no. 2, pp. 872–891, Oct. 2006.
38. Ghenciu AF, "Review of fuel processing catalysts for hydrogen production in PEM fuel cell systems," *Curr. Opin. Solid State Mater. Sci.*, vol. 6, no. 5, pp. 389–399, Oct. 2002.
39. Ledjeff-Hey K, Roes J and Wolters R, "CO2-scrubbing and methanation as purification system for PEFC," *J. Power Sources*, vol. 86, no. 1–2, pp. 556–561, Mar. 2000.
40. Korotkikh O and Farrauto R "Selective catalytic oxidation of CO in H_2: fuel cell applications," *Catal. Today*, vol. 62, no. 2–3, pp. 249–254, Nov. 2000.
41. Mohtadi R, Lee W-K, Cowan S, Van Zee JW and Murthy M, "Effects of hydrogen sulfide on the performance of a PEMFC," *Electrochem. Solid-State Lett.*, vol. 6, no. 12, p. A272, 2003.
42. Rockward T and Uribe FA, "Cleaning PEMFC Pt-electrodes with adsorbed S-species," *Meet. Abstr.*, vol. MA2005–02, p. 2014, 2014.
43. Thaler M and Hacker V, "Storage and separation of hydrogen with the metal steam process," *Int. J. Hydrogen Energy*, vol. 37, no. 3, pp. 2800–2806, Feb. 2012.
44. Choudhary T and Goodman D, "CO-free fuel processing for fuel cell applications," *Catal. Today*, vol. 77, pp. 65–78, 2002.
45. Wang M, Wang Z, Gong X and Guo Z, "The intensification technologies to water electrolysis for hydrogen production—A review," *Renew. Sustain. Energy Rev.*, vol. 29, pp. 573–588, Jan. 2014.
46. Iulianelli A, Ribeirinha P, Mendes A and Basile A, "Methanol steam reforming for hydrogen generation via conventional and membrane reactors: A review," *Renew. Sustain. Energy Rev.*, vol. 29, pp. 355–368, Jan. 2014.
47. Kyriakides A-S, Rodríguez-García L, Voutetakis S, Ipsakis D, Seferlis P and Papadopoulou S, "Enhancement of pure hydrogen production through the use of a membrane reactor," *Int. J. Hydrogen Energy*, vol. 39, no. 9, pp. 4749–4760, Mar. 2014.
48. Kennedy DM, Cahela DR, Zhu WH, Westrom KC, Nelms RM and Tatarchuk BJ, "Fuel cell cathode air filters: Methodologies for design and optimization," *J. Power Sources*, vol. 168, no. 2, pp. 391–399, Jun. 2007.
49. Ma X, Yang D, Zhou W, Zhang C, Pan X, Xu L, Wu M and Ma J, "Evaluation of activated carbon adsorbent for fuel cell cathode air filtration," *J. Power Sources*, vol. 175, no. 1, pp. 383–389, Jan. 2008.

50. Yang H, Lu Y and Tatarchuk BJ, "Glass fiber entrapped sorbent for reformates desulfurization for logistic PEM fuel cell power systems," *J. Power Sources*, vol. 174, no. 1, pp. 302–311, Nov. 2007.
51. Coms FD, Liu H and Owejan JE, "Mitigation of perfluorosulfonic acid membrane chemical degradation using cerium and manganese ions," *ECS Trans.*, 2008, vol. 16, pp. 1735–1747.
52. Gottesfeld S and Pafford J, "A new approach to the problem of carbon monoxide poisoning in fuel cells operating at low temperatures," *J. Electrochem. Soc.*, pp. 2651–2652, 1988.
53. Zamel N and Li X, "Effect of contaminants on polymer electrolyte membrane fuel cells," *Prog. Energy Combust. Sci.*, vol. 37, no. 3, pp. 292–329, Jun. 2011.

12

Solar Membrane Reactor

Kamran Ghasemzadeh,[1] *Angelo Basile,*[2] *and Abbas Aghaeinejad-Meybodi*[3]

[1]*Chemical Engineering Department, Urmia University of Technology, Urmia, Iran*

[2]*Institute on Membrane Technology – Italian National Research Council (ITM-CNR), University of Calabria, Cosenza, Italy*

[3]*Chemical Engineering Department, Sahand University of Technology, Tabriz, Iran*

12.1 Introduction

Membrane reactor (MR) is a chemical plant that combines a reaction chamber with a membrane to add reactants or remove products of the reaction. Chemical reactors making use of membranes are usually known as MRs (Koc et al., 2012). The membrane can be used for different applications as follows:

- Separation
  - Selective extraction of reactants
  - Retention of the catalyst
- Distribution of a reactant
- Catalyst support (often combined with distribution of reactants).

In general, the conversion and selectivity of a chemical reaction can be improved by carrying out reactions in the MR. Solar MR is one of the newest MR technologies which can be used for some applications such as hydrogen production and wastewater treatment (Koc et al., 2012). In recent years, many studies have focused on MR technologies (Basile et al., 2005, Basile, et al., 2006, 2008; Harold et al., 2003; Nair and Harold, 2006; Wieland et al., 2002; Iulianelli et al., 2008a, 2008b; Gallucci and Basile, 2006, 2007, 2008; Ghasemzadeh et al., 2013a, 2013b, 2013c, 2015) in which the feasibility of achieving higher performance of MRs with respect to traditional reactors (TRs) was demonstrated.

Integrated Membrane Systems and Processes, First Edition. Edited by Angelo Basile and Catherine Charcosset.

In recent years, the combination of two emerging technologies like solar collectors for solar energy conversion (production of heat and/or electrical power) and membranes to enhance separation of chemicals is one of the novel attractive fields. Therefore, solar MRs technology attracted attention of researchers to this new field.

Indeed, the growing interest toward the exploitation of renewable energy sources (RES) has supported the development and optimization of efficient and cost-effective technologies to capture and convert solar energy; solar photovoltaic (PV) cells and concentrating solar power (CSP) plants are today widespread and mature technologies, adaptable for power production both at the small (decentralized) and large (centralized) scales. However, researchers are ongoing to make these energy conversion systems more efficient, cost-effective, and competitive with the more traditional fossil-fuel-based energy economy (De Falco et al., 2008).

Similarly, membrane technology was widely applied for several industrial applications (e.g., polymeric membranes for reverse osmosis, electrolysis, electrodialysis) but R&D efforts are focused on the development of new cost-effective methods for gas separation at small and large scales.

In principle, the TRs can be replaced by the MRs technology and new MR applications can be identified, especially when the separation of products is an issue in terms of investment, management, and power cost. This is the case for hydrogen production plants, where chemical separation is a key object at different stages of the process, also including the hydrogen purification by hydrogen selective membranes and hydrogen production in MRs.

Therefore, this chapter specifically shows the application of solar MR technology in hydrogen production. The extensive description of the solar MR process in other applications such as water treatment is beyond the scope of one chapter. Hence, after introducing various configurations of solar MRs in fields mentioned, a literature review is exclusively presented about different applications of solar MRs for hydrogen production from experimental and modeling point of views.

12.2 Configurations of Solar MR Systems

As a power source, the solar PV cell is a well-known mature technology to directly convert the solar radiation to electrical power, from 1 kW up to several tens of MW electricity capacity. Otherwise, the CSP technology is based on several mirrors setting on a "solar field," which track the sun and concentrate the solar radiation several hundred times over a linear tube or spot called the "solar receiver" (Winter et al., 1991). The solar receiver converts the concentrated solar radiation to heat at temperatures ranging from 300°C to more than 1400°C, depending on the CSP technology and the solar radiation "concentration factor." In general, a heat transfer fluid (HTF) is used to remove and transfer the absorbed heat from the receiver to the process via different types of heat exchangers; in other cases, the concentrated solar radiation directly powers chemical reactors without the use of HTFs (Chueh et al., 2010). However, there are different available CSP technologies (Figure 12.1) differing on the type of receiver, the mirror's shape, tracking system, etc. (Chueh et al., 2010): solar towers, solar dishes, solar troughs (or linear parabolic collectors) and linear Fresnel collectors. Although these systems have originally been considered for solar–electrical power production, the same concept can be applied to partially or totally replace

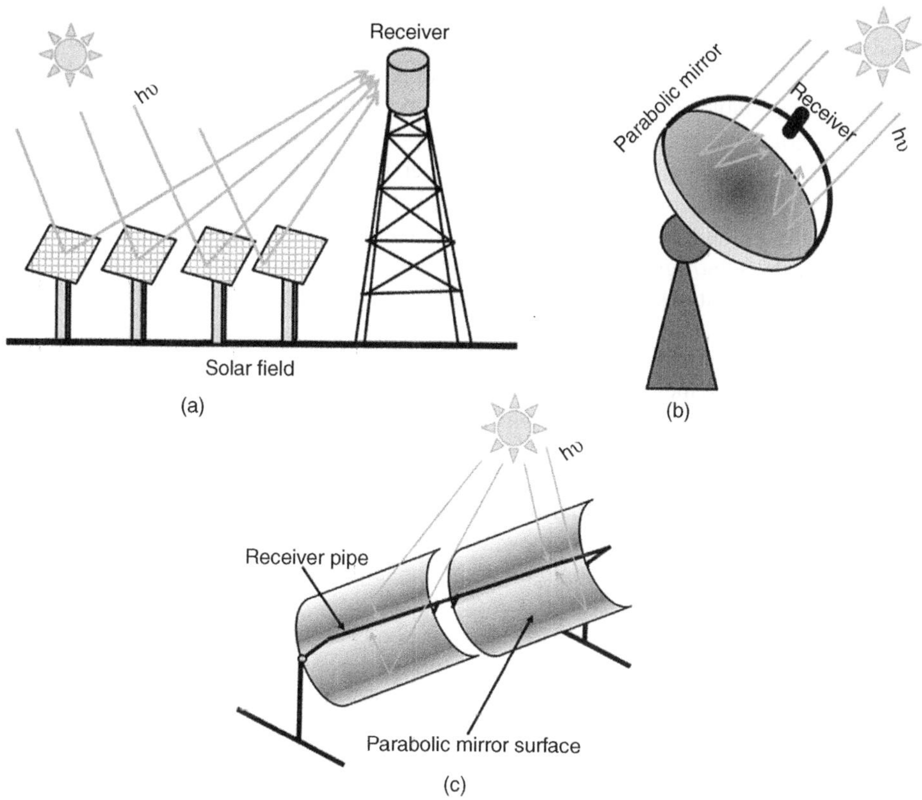

Figure 12.1 Types of CSP technologies. (a) Solar towers, (b) solar dishes, and (c) solar troughs.

fossil fuels as the heat supplier in thermochemical conversion. Therefore, the technologies mentioned can also be used as a heat supplier for MR systems nominated to solar MR technology.

According to the authors' best knowledge, solar MR systems for two conventional processes, namely, water treatment and hydrogen production were used. However, as mentioned in the introduction section, the main focus of this chapter is based on hydrogen production process.

12.2.1 Solar MRs for Water and Wastewater Treatment

In this process, combining of water/wastewater treatment reaction and separation step can be considered, simultaneously. According to the literature, the solar MRs can be divided generally into two main cases (Mozia, 2010):

- MRs with photocatalyst suspended in feed solution (case I)
- MRs with photocatalyst supported in/on the membrane (case II)

In case I, membrane filtration may be used as a single step for the complete recovery of catalyst particles from the solution (Augugliaro et al., 2006). The major disadvantages of such configuration are deterioration of the permeate flux and membrane fouling, which are especially observable in case of pressure-driven membrane techniques.

In both configurations, utilizing photocatalyst in suspension or immobilized on/in a membrane, the membrane might act as a barrier for degraded molecules, products and byproducts of their decomposition. However, this role strongly depends on the separation characteristics of the membrane used and the membrane process applied.

In case I, the solar/light source is positioned above the membrane, whereas in case II, three main configurations can be distinguished:

(a) Irradiation of the feed tank
(b) Irradiation of the membrane module
(c) Irradiation of an additional reservoir (photo-reactor) located between the feed tank and the membrane module.

In some configurations, the solar/light source can be positioned above both, membrane module and feed tank (or inside them, when the immersed UV lamps are used). Schematic diagrams of the most common configurations of solar MRs in this process are presented in Figures (12.2–12.5), (Mozia, 2010).

It should be noted that a solar MR design depends on the process mode (dead end or cross flow, batch or continuous flow, etc.), membrane technique applied (MF, UF, NF, etc.), membrane modules used (flat sheet, hollow fiber, submerged, etc.) or the type of the light source (Mozia, 2010).

However, most of the solar MRs are typical membrane installations equipped with an additional unit which is the photo-reactor with a solar/light source (Figure 12.4) or

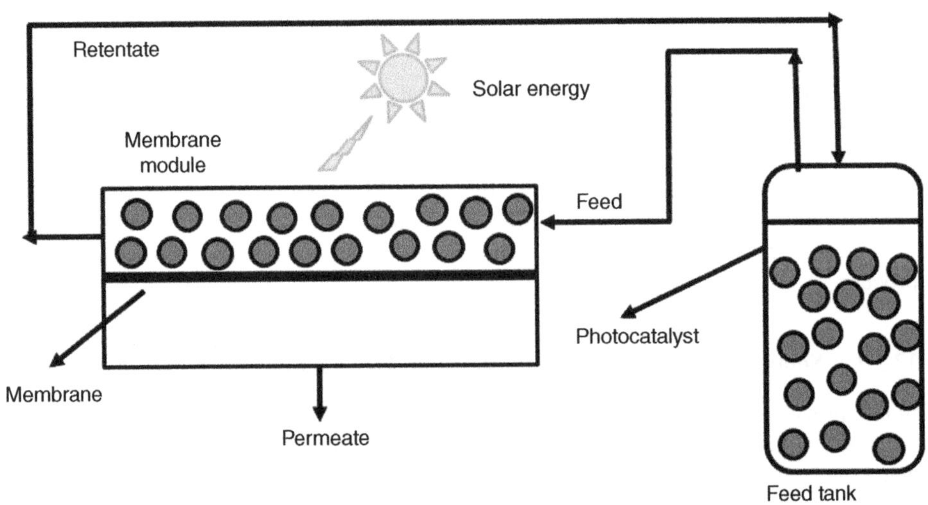

Figure 12.2 *Solar MR utilizing photocatalyst in suspension: irradiation of the membrane module.*

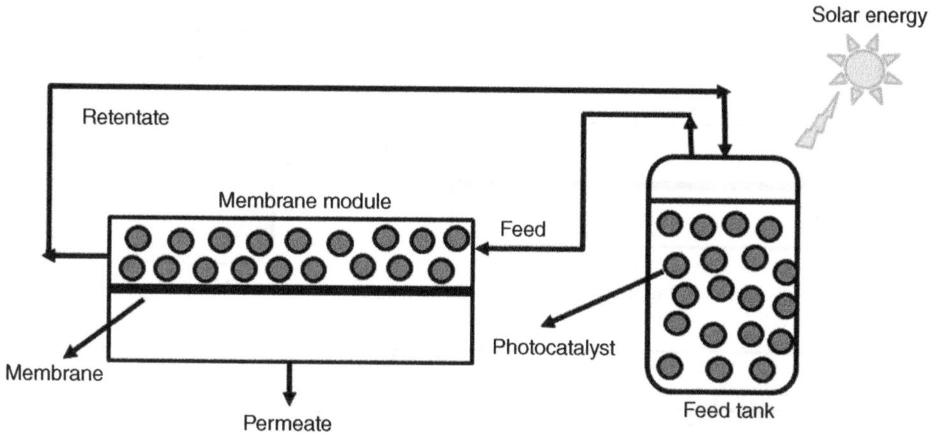

Figure 12.3 Solar MR utilizing photocatalyst in suspension: irradiation of the feed tank.

solar/light source only (Figures 12.2, 12.3, and 12.5). The construction of a membrane module is the most important difference between a membrane installation and solar MR in case of solar MRs with photocatalytic membranes (Mozia, 2010).

Moreover, Figure 12.6 presents two possible types of asymmetric photocatalytic membranes (Bosc et al., 2005). In the first case (Figure 12.6a), the photoactive layer, being also the separation layer (skin) is formed on a porous non-photoactive support, whereas in the second case (Figure 12.6b), the separation layer is non-photoactive deposited on a porous active support. The most applied configuration is the photocatalytic membrane with photoactive skin layer (Figure 12.6a).

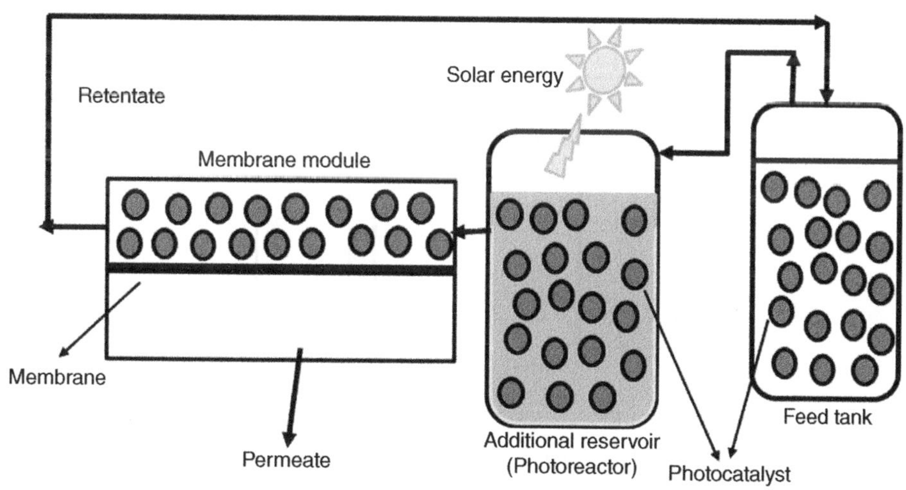

Figure 12.4 Solar MR utilizing photocatalyst in suspension: irradiation of the additional reservoir (photo-reactor) located between the feed tank and the membrane module.

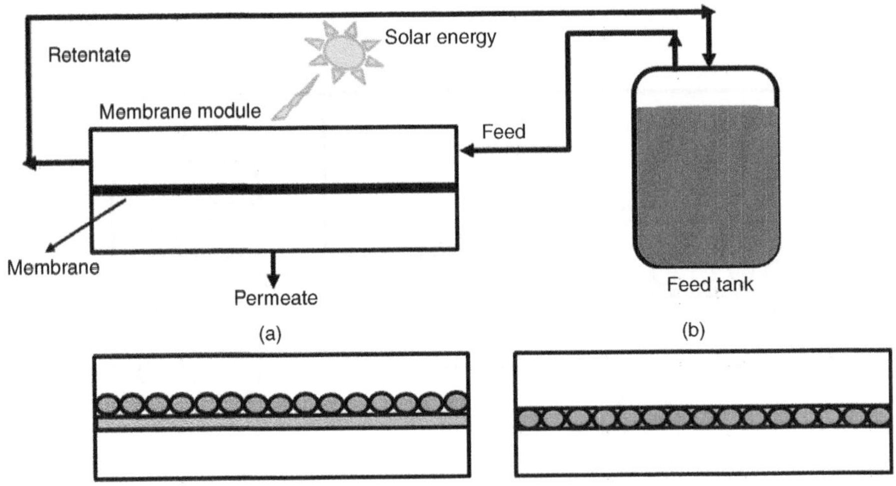

Figure 12.5 Solar MR utilizing photocatalyst immobilized (a) on a membrane and (b) within a membrane structure.

It should be noted that the description of water and wastewater treatment by solar MR systems is not presented in details. The most popular configurations of solar MRs for hydrogen production are discussed in the following sections.

12.2.2 Solar MRs for Hydrogen Production

Generally, there are three main different processes for hydrogen production: water decomposition, water electrolysis, and steam reforming, in which solar MRs can be used.

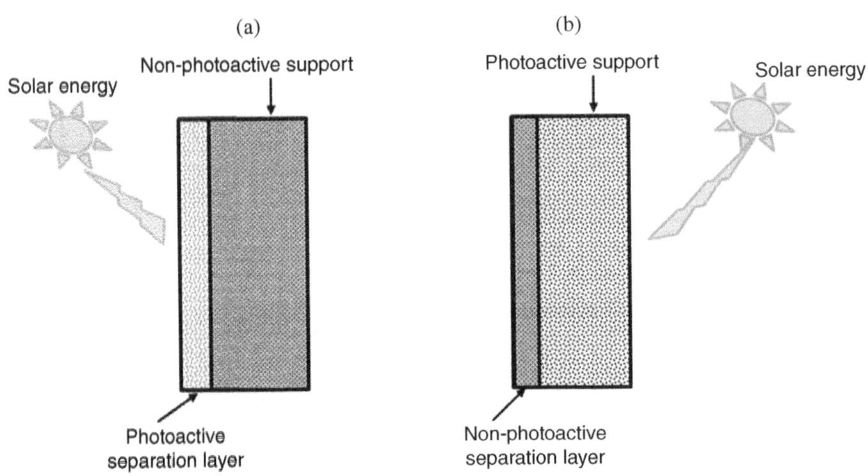

Figure 12.6 Asymmetric photocatalytic membranes. (a) Type 1: photoactive separation layer is deposited on a non-photoactive porous support; (b) type 2: non-photoactive separation layer is deposited on a photoactive porous support. Source: *Reprinted from Bosc et al. (2005) with permission from Elsevier.*

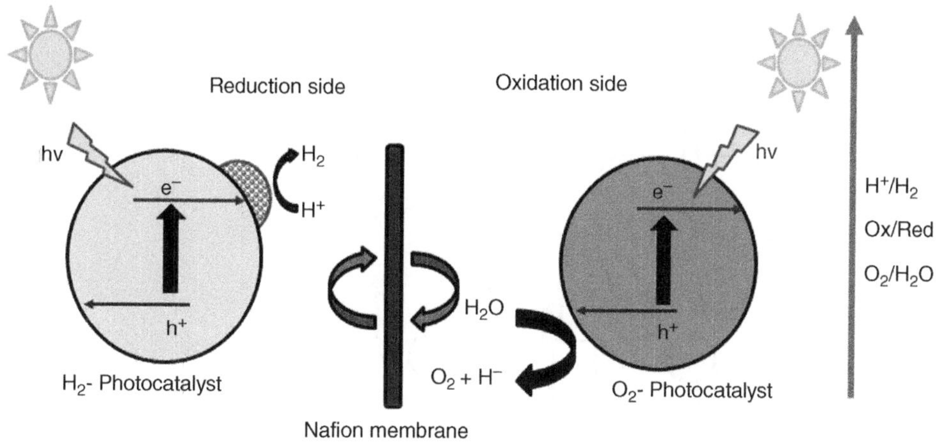

Figure 12.7 The schematic diagram of water decomposition by Z-scheme MR. Source: *Reprinted from Yu et al. (2011) with permission from Elsevier.*

12.2.2.1 Solar MRs in Water Decomposition

12.2.2.1.1 Z-Scheme MR

Most of the solar MRs applied for hydrogen production in water decomposition patterns from the Z-scheme mechanism are used by green plants for natural photosynthesis. The Z-scheme basically includes a dual photocatalyst system for water oxidation (which leads to O_2 formation) and water reduction (with H_2 generation). In addition, a redox couple is used to regenerate the catalyst transporting the charge between the two compartments of the reactor. A membrane is required to separate the aqueous suspensions containing the H_2- and the O_2-photocatalysts and to exchange electrons via the electron redox mediator (Figure 12.7). In fact, often a modified Nafion membrane is used to enable the ions transport (Molinari et al., 2014).

Lo et al. (2010) developed a novel two-compartments reactor in which the advantages of the Z-scheme and the H-type reactor were combined. In this system, powders of $Pt/SrTiO_3$: Rh (H_2-photocatalyst) and WO_3 (O_2-photocatalyst) are loaded separately in two compartments of a connected reactor divided by a modified Nafion ion-exchange membrane to perform the water decomposition reaction under visible-light irradiation (as shown in Figure 12.8).

The compartments of the twin reactor were filled with an aqueous solution containing Fe^{3+}/Fe^{2+} as the electron-transfer facilitator. By this way, hydrogen and oxygen can be generated separately.

12.2.2.1.2 Membrane Photocatalytic Systems

In membrane photocatalytic systems (MPS), the processes of oxidation and reduction are carried out over different sides of the membrane on the same photocatalyst. Therefore, the electrons must transfer from one side of the membrane to the other side. The basic idea of the developed MPS is shown in Figure 12.9.

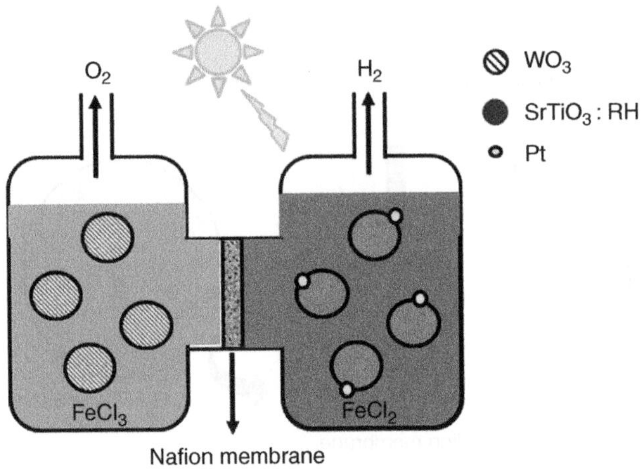

Figure 12.8 A novel schematic diagram of the Z-scheme MR for water decomposition. Source: Reprinted from Lo et al. (2010) with permission from Elsevier.

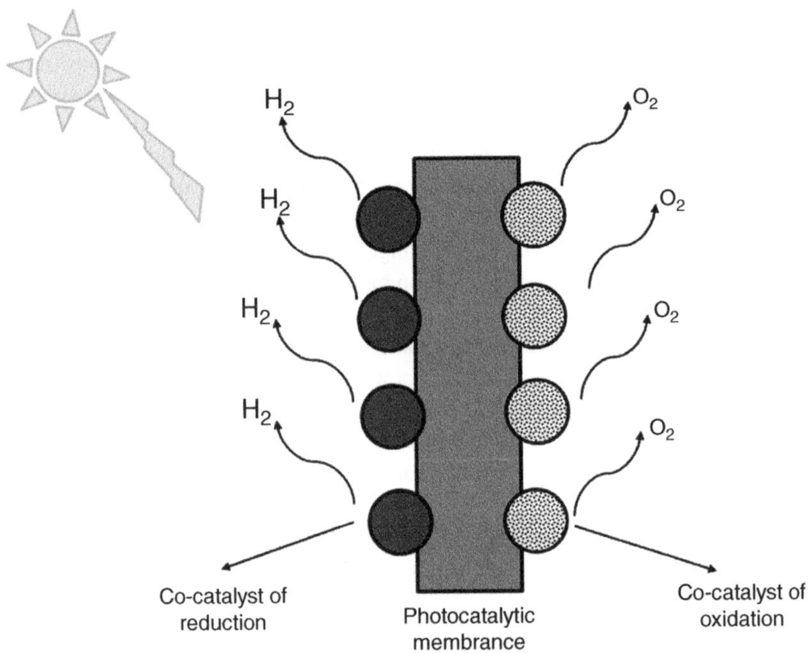

Figure 12.9 A schematic representation of the MPS to separate hydrogen and oxygen evolution. Source: Reprinted from Tsydenov et al. (2012) with permission from Elsevier.

In this case, the membrane is impregnated with TiO_2 photocatalyst and reduction as well as oxidation of co-catalysts Pt occurred on different sides of the membrane. Efficient separation of photo-generated charges and water decomposition products on the different sides of the membrane can be obtained by this way.

12.2.2.1.3 Thermochemical Processes

In thermochemical water decomposition, water splits into hydrogen and oxygen using a series of chemical reactions. All chemical intermediates are recycled internally within the process so that water is the only raw material and hydrogen and oxygen are only products (Rosen, 2010; Orhana et al., 2012).

Several studies of hydrogen production using thermochemical processes have been recently presented, including reports of several cycle statues such as the iodine–sulfur (S-I) cycle, the ISPRA Mark 9 cycle, the hybrid sulfur cycle, the Ca–Br cycle, the Cu–Cl cycle, and the adiabatic UT-3 cycle (Rosen, 2010). A majority of these cycles are driven by nuclear or solar energy sources. After detailed reviewing of all published literature about thermochemical cycles (more than 100 cycles from 800 references), the adiabatic UT-3 cycle and the S-I cycle were considered to have the greatest potential, and the S-I cycle was selected for further development and recommended as a suitable thermochemical process for water demonstration plant (Rosen, 2010).

The S-I process consists of the following three chemical reactions (Xinxin and Kaoru, 2007; Orhana et al., 2012):

Step 1: $I_2 + SO_2 + 2H_2O \rightarrow 2HI + H_2SO_4$
Step 2: $2HI \rightarrow H_2 + I_2$
Step 3: $H_2SO_4 \rightarrow H_2O + SO_2 + \frac{1}{2}O_2$

However, H_2 separation membranes can play a critical role in the gaseous HI decomposition reaction. The decomposition reaction of HI ($HI \leftrightarrow H_2 + I_2$) is an equilibrium reaction including conversion efficiency of around 22% at 450°C. The enhancement of decomposition rate by removing hydrogen from the reaction zone is possible. Therefore, selective membranes that can separate hydrogen from the reaction zone will result in a reduction of the HI gas amount that requires recycling through the reactor. In addition, an effective separation membrane can maintain the purity of the H_2 gas produced. The configurations of the S-I cycle-containing membrane, as the solar MR system, are shown in Figure 12.10.

The researcher's effort in the direction of development of different configurations of MR for thermochemical water decomposition is still new and a lot of opportunities as well as challenges are associated with each of the configurations studied till date.

12.2.2.2 Solar MRs in Water Electrolysis

Electrolysis of water is the decomposition of water into oxygen and hydrogen gas due to an electric current being passed through the water. External charge fields and additives can enhance water electrolysis or hydrogen production rate. The ohmic voltage drop, reaction, and potential are decreased due to more rapid bubble disconnection from the electrolytic cell and higher gas growing rate. However, thermodynamic decomposition voltage is not changed. Thus, it is a promising method by increasing the temperature or changing the

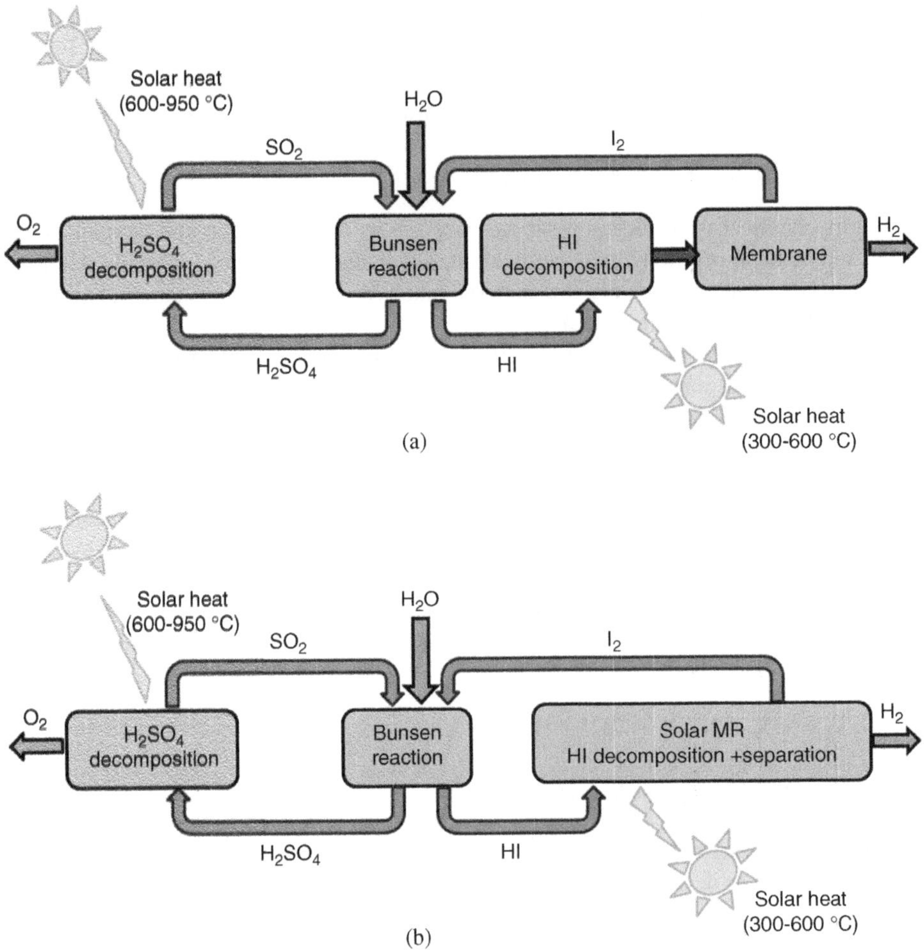

Figure 12.10 *Configurations of solar MR can be presented for thermochemical water decomposition.*

reaction route to reduce theoretical decomposition voltage of water from maximizing energy saving. Indeed, two different configurations of solar membrane technology were introduced for this process: solid oxide electrolyte cell (SOEC) and carbon-assisted water electrolysis (CAWE).

12.2.2.2.1 Solid Oxide Electrolyte Cell

The principle of SOEC is illustrated in Figure 12.11. The cathode for hydrogen evolution reaction and anode for oxygen evolution reaction are isolated by solid electrolyte. Actually at higher temperature (600–1000°C), water is decomposed to H_2 and O^{-2} on cathode. Then, O^{-2} is transported through solid electrolyte to anode and oxidized to O_2. In general, the conventional solid electrolyte is ZrO_2 doped withY_2O_3 (YSZ) which possesses excellent

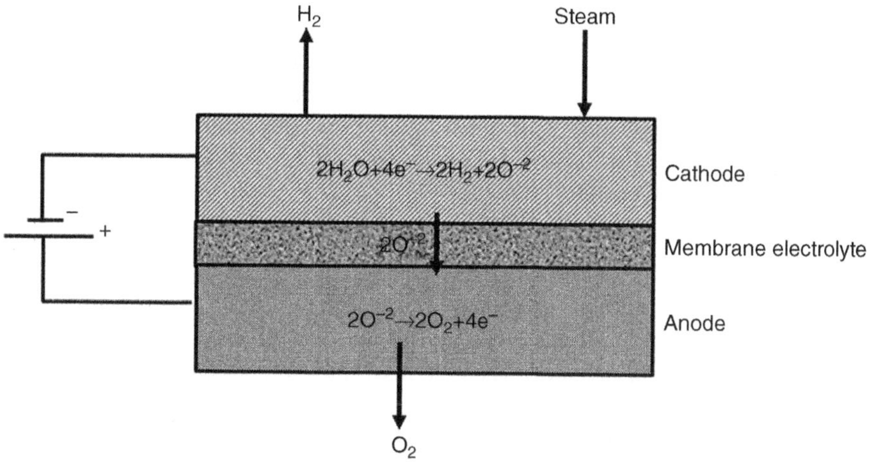

Figure 12.11 *The schematic of solid oxide electrolyte cells (SOECs).*

conductivity and stability (Jiang et al., 2007; Yu et al., 2012). In addition, scandia-doped ZrO_2 and ceria- and $LaGaO_3$-based materials are also used as solid electrolyte.

12.2.2.2.2 Carbon-assisted Water Electrolysis

The CAWE technology for hydrogen production was introduced by Coughlin and Farooque around 30 years ago (Coughlin and Farooque, 1979). In an acidic electrolyte, hydrogen is still produced on cathode, as shown in Figure 12.11, while, the reaction for anode is replaced by Equation (12.1)

$$C + 2H_2O(l) \rightarrow CO_2\,(g) + 4H^+ + 4e^- \quad E^0 = 0.21\ V \tag{12.1}$$

In contrast to the conventional water electrolysis, carbon is oxidized to CO_2 on anode instead of oxygen evolution reaction. For conventional water electrolysis, theoretical decomposition voltage of water is 1.23 V and energy consumption for hydrogen production is around 237.2 kJ mol^{-1}, whereas, for the CAWE technology, theoretical decomposition voltage and energy consumption are only 0.21 V and 40.2 kJ mol^{-1}, respectively.

As far as the solar plant is concerned, the solar-to-electrical conversion efficiency in PV cells is usually within the range 10–20%. On the other hand, solar CSP plants are characterized by solar–thermal efficiency around 50–70%, followed by thermal–electrical conversion efficiency around 5–38%, resulting in overall solar-to-electrical conversion efficiency within the range of 13–27% (Coughlin and Farooque, 1979).

However, the two types of solar technologies should be compared in terms of yearly power production (e.g., GWh/year) or capacity factor, which strongly depends on the location of the plant. Thus, the two solar conversion technologies are comparable in terms of the solar-to-electrical conversion efficiency.

12.2.2.3 Solar MRs in Steam Reforming

Besides water decomposition processes (electrolysis and thermochemical cycles) by solar energy, membranes can also be applied to solar reforming. Nowadays, more than 75% of industrial hydrogen production is obtained by hydrocarbons reforming, typically natural gas and oil derivatives. The most widely employed process is steam reforming according to the following general reaction scheme.

$$C_nH_{2n+2} + 2n\ H_2O\ nCO_2 + (n+2)\ H_2$$

where C_nH_{2n+2} is the hydrocarbon feedstock which can even be replaced by an oxygenated chemical (e.g., ethanol or glycerol) when the fossil fuel is replaced by a biomass-derived fuel.

Since methane reforming in MRs is carried out at lower temperatures (<600°C) with respect to conventional systems (>900°C), the heat supply in MR is somehow simpler and other heat sources can be used (De Falco et al., 2010).

The general layout of the plant is shown in Figure 12.12. The reactant mixture, composed of methane and steam, is fed to the MR reaction zone, while a sweeping gas (steam) is fed to the permeation zone to keep the hydrogen partial pressure as low as possible and to increase the hydrogen flux through the Pd-based membrane. A molten salt stream is used as a heat carrier to provide the heat duty required by the steam-reforming reactions. The retentate gas mixture consists of unreacted methane and steam, hydrogen, carbon monoxide, and dioxide. After the steam recovery by condensation, the outlet gas mixture has a high net heat value, which can be used to generate a high pressure steam and consequently electrical energy, resulting in an enhancement in the overall energy efficiency (De Falco et al., 2010).

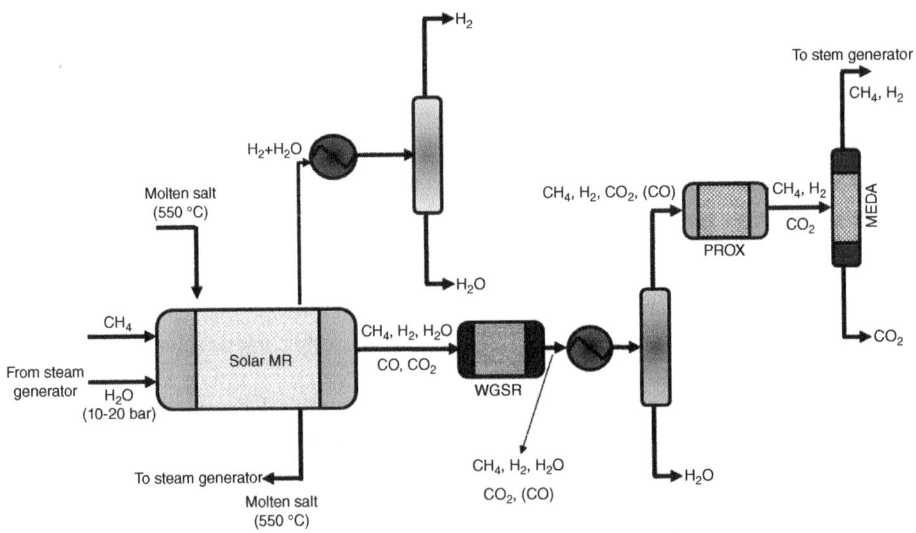

Figure 12.12 A general plant layout of the solar MR for steam reforming.

The outlet of permeation stream is a mixture of hydrogen and steam as hydrogen is recovered by condensation, compressed, and stored. Therefore, the proposed plant is co-generative as it is able to produce both hydrogen and electricity, and it is a hybrid plant considering the simultaneous utilization of solar source and natural gas.

12.3 Solar MRs Application from a Modeling Point of View

Numerical simulation models can be used in order to avoid high experimental costs and develop a better understanding of the several parameter effects to design and for the study of solar MR performance in different processes. Nevertheless, few works have been presented from the modeling point of view. For instance, there is no study about water electrolysis and photocatalytic water decomposition by solar MR technology from the modeling aspect. Therefore, in this section of chapter, a summarized literature review will be presented about the solar MR modeling aspect.

12.3.1 Water Decomposition Literature

12.3.1.1 Direct Thermal Decomposition

Direct solar thermal decomposition of water and separation of the products using ceramic membranes was modeled by Kogan et al. (2000a, 2000b). The reactor thermodynamic efficiency in this process increases by increasing the reaction temperature and by decreasing the pressure at the downstream side of the gas-separation membrane. Kogan et al. (2000a, 2000b) presented two studies to intensify the reactor efficiency. The first, by heat and mass recovery, using a steam injector in a two-stage reactor and the other, by lowering the reactor exit pressure of the hydrogen-enriched gas stream, using a multistage ejector. Computational fluid dynamics (CFD) analysis showed that it is possible to obtain a significant improvement of reactor efficiency in solar thermal water decomposition by using a multiple ejector, to lower the pressure of the hydrogen-rich stream.

Hydrogen production in water thermal decomposition via solar MR was investigated by a plug flow and isothermal mathematical model based on the equilibrium reaction above 2500 K (Ohya et al., 2002). In this modeling study, the characteristics of zirconia–silica membrane such as the Knudsen diffusion mechanism were considered. Numerical integration of ordinary differential equations (ODE) carried out using the fourth-order Runge–Kutta method. Ohaya et al. 2002) concluded that hydrogen production by PFR-type MR will be about 1.5 times larger than those obtained by CSTR-type MR.

12.3.1.2 Thermochemical Decomposition

In a HYTHEC study (Le Duigou et al., 2007), effective potential for massive hydrogen production of the S-I thermochemical cycle was investigated and compared with the hybrid Westinghouse (WH) cycle. Applying a liquid phase membrane separation unit to the HI_x processing section has two potential advantages, an increase in efficiency and less extreme operating conditions. The membrane separation unit is considered in the following system positions; on the column feed, at the column reboiler and on a column side-stream. However, for selecting the membrane unit, three key factors should be regarded:

- The solution is a highly corrosive, strongly acidic, aqueous system.
- The process temperature is 398 K.
- The membrane must be selective to either HI or H_2O.

In order to investigate the effect of dewatering, simulations were carried out using a modified version of the process flowsheeting code ProSimPlus, which includes the membrane separation unit. The membrane is modeled as a splitter with zero enthalpy drop. This means that the retentate exits at a lower temperature than it would in practice. This is accounted for the heat integration and so will not significantly affect the calculated efficiencies. Therefore, the simulation results show a little benefit in applying membrane unit for this process.

12.3.2 Steam Reforming Literature

The application of concentrated solar energy as an external heat source for methane steam reforming has been investigated by Giaconia et al. (2008). As reported by these authors, molten salts at temperatures up to 550°C can be used as solar heat carrier and storage system, and hydrogen selective membranes can be applied to drive reforming reaction at lower temperatures with respect to conventional systems (<550°C). Through the combination of new technologies, such as membranes and MRs, CSP systems, and molten salt heat carriers, a partial decarbonation of the fossil fuel as well as possibility to carry solar energy in the natural gas network is possible. Different plant configurations and operating conditions have been analyzed by using 1D mathematical model and Aspen Plus simulator for flowsheet development with mass and energy balances.

Solar MR has been modeled as double tube heat exchangers, in which gas mixture enters to the inner tube packed with catalyst bed at 500°C, and molten salts counter-current flows in the annular zone entering at 550°C and exiting at 530°C temperature. Moreover, in this 1D mathematical model, the mass, energy, and momentum balances have been used to obtain composition, pressure, and temperature axial profiles in the MRs, both in gas and molten salt sides. Indeed, the model was based on the following assumptions:

1. Steady-state conditions;
2. Plug flow behavior;
3. Pseudo-homogeneous catalyst bed;
4. Ideal gas behavior.

In this model, the Pd-based thin layer membranes supported on a ceramic or metallic material operating at 500°C was considered. Furthermore, 85% of the inlet hydrogen was recovered by the Pd-based membrane with 100% selectivity. As constant temperature and pressure conditions have been assumed along the permeate side, the adopted model only involves the mass balance across the membrane.

In the Aspen Plus simulations, membrane modules have been simulated as isothermal component separators with hydrogen split fraction of 0.85, and hydrogen permeate extraction pump simulated as a three-stage isentropic multistage compressor.

The modeling results demonstrate that the application of hydrogen permselective membranes allows to sensibly improve single pass methane conversion and heat value upgrading, either in multistage arrangements or by using a single solar MR with partial recycle of the non-permeated stream.

De Falco et al. (2008) investigated the reformer and membrane modules plant powered by a nuclear reactor or by a solar heated molten salts and evaluated the design parameters and production costs. The integration of reforming and membrane modules technology within a nuclear power station or coupled with a solar molten salts' system makes the production of hydrogen 10% less expensive than in the case of conventional steam reforming and reduces the overall CO_2 production of 28%.

In other research work, De Falco et al. (2010) investigated solar membrane natural gas steam-reforming process via a 2D model. In this study, the solar MR model was presented based on mass, energy, and momentum balances. The 2D nature of the model allows axial and radial concentrations and temperature profiles to be evaluated. Following assumptions are considered for simulation of solar MR:

- Steady-state conditions;
- Negligible axial dispersion and radial convective terms;
- Ideal gas behavior;
- Each double tube is representative of any other tube;
- Pseudo-homogenous condition inside the reaction zone;
- A single pseudo-effectiveness factor, independent of reaction and of local conditions;
- 100% Pd-based membrane hydrogen perm-selective.

The solar MR consists of a bundle of coaxial double tubes assembled in a shell. The annular section of each tube is the reaction zone in which Ni-based catalyst pellets are packed, whereas the inner tube is a dense Pd-based selective membrane that is able to remove hydrogen from the reaction zone (as shown in Figure 12.13). By combination of reaction and hydrogen separation, equilibrium constrains inside the MR are avoided and also high methane conversions at relatively low temperatures are achieved. The heat

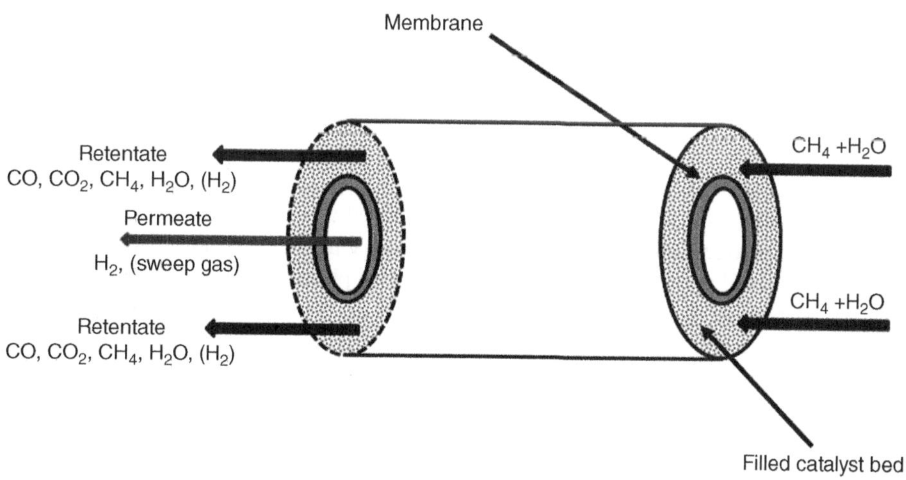

Figure 12.13 MR scheme applied to steam reforming reaction.

needed for the steam-reforming reaction at this low operating temperature can be provided by using a molten salt stream, heated up to 550°C by a parabolic mirror solar plant, as the heating fluid. The effects of some operating conditions on MR performance such as gas mixture residence time, reaction pressure and steam-to-carbon ratio are assessed. The results of this parameter study show an enhancement of methane conversion for solar MR with respect to the solar TR, in which methane conversion improvement is evaluated in the range 40.5–130.9% for the same operating conditions. Moreover, owing to the use of a solar source for chemical process heat duty requirements, the greenhouse gases (GHG) reduction is estimated to be in the range 33–67%.

12.4 Solar MRs Application from an Experimental Point of View

Many studies have focused on MR technologies for hydrogen production, but among them, few research studies are related to solar MR systems from an experimental point of view. Therefore, in this section, a summarized literature review will be presented.

Most of the solar or photocatalytic MRs studied for hydrogen production is based on steam reforming and water decomposition processes that mimic the Z-scheme mechanism used by green plants for the natural photosynthesis.

12.4.1 Water Decomposition Literature

According to the literature, water decomposition can be divided into three cases: photocatalytic, thermochemical, and direct water decomposition. Therefore, the processes review mentioned is performed separately.

12.4.1.1 Photocatalytic Water Decomposition

The Z-scheme basically consists of a dual photocatalyst system for water oxidation (which lead to O_2 formation) and water reduction (with H_2 generation). In addition, a redox couple is used to regenerate the catalyst transporting the charge between the two compartments of the reactor (Lo et al., 2010).

As mentioned before, the use of twin reactors shows several advantages in comparison to the Z-scheme in a batch reactor. The backward reaction of hydrogen and oxygen is minimized, the purification process allowing the use of pure generated hydrogen is not necessary, and the safety issue of H_2–O_2 explosion is overcame for a future industrial process (Yu et al., 2011a, 2011b).

As a first approach, Fujihara et al. (1998) proposed a method of water decomposition to produce hydrogen and oxygen in separated sides in the Z-scheme MR. Unlike the other conventional systems, this device combines two photocatalytic reactions using TiO_2 powders suspended in two different solutions and using a two-compartment cell equipped with Pt electrode and a proton exchange membrane (Figure 12.14).

By UV or solar light irradiation on both compartments, O_2 was produced on rutile TiO_2 particles suspended in a solution containing Fe(III)/Fe(II), while H_2 was produced on the TiO_2 anatase modified with Pt in a solution containing Br_2/Br^-. In the proposed system, Fe(II) ions formed in one compartment as the result of reduction of Fe(III) ions promoted by the photo-generated electrons were oxidized electrochemically through an external circuit

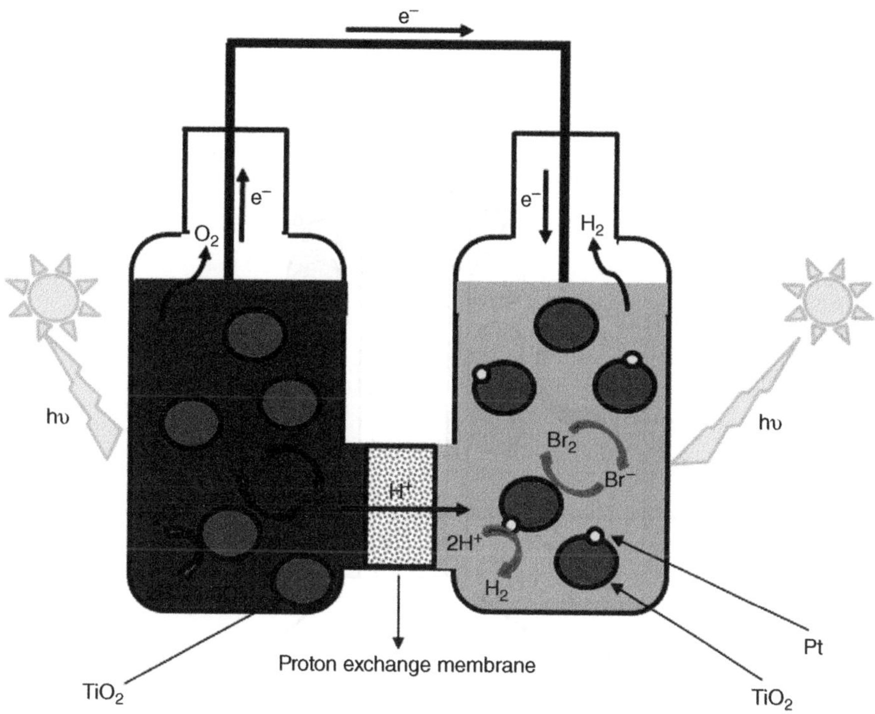

Figure 12.14 *Schematization of the photocatalytic MR used by Fujihara et al. (1998) for separated hydrogen and oxygen production.*

based on the release of Br ion. The electrical neutrality and the pH of the solution in each compartment were maintained constant through a proton exchange membrane allowing the proton transfer from one side to the other side of the system. Gases generation became stable values after 100 minutes (1.3 and 2.8 μmol h^{-1} for O_2 and H_2, respectively) and problems related to the deposition of iron compounds on the Pt electrode leading to its contamination were observed by using this system and achieved a reduction of gas generation rate during the time.

Seger and Kamat (2009) reported a TiO_2-Nafion-Pt assembly membrane (Figure 12.15) system using methanol as the sacrificial agent and UV light irradiation, without any external bias. The TiO_2 photocatalyst provided the methanol oxidation while, at the same time, H ions, driven to the Pt particle surface through the Nafion membrane, were reduced leading to a H_2 evolution rate of 69 μL h^{-1} cm^{-2}.

A novel two-compartment reactor including the advantages of the Z-scheme and the H-type MR was developed by Lo et al. (2010). In this system, powders of WO_3 and Pt/$SrTiO_3$: Rh were employed to produce oxygen and hydrogen, respectively. A cation exchange membrane was placed between the two semi-cells of the reactor, allowed the transfer of both protons and Fe(II)/Fe(III) ions in solution.

In another study by Yu et al. (2011a), hydrogen and oxygen were produced via Pt/$SrTiO_3$:Rh and $BiVO_4$ in a twin reactor system using a Nafion membrane (Figure 12.16).

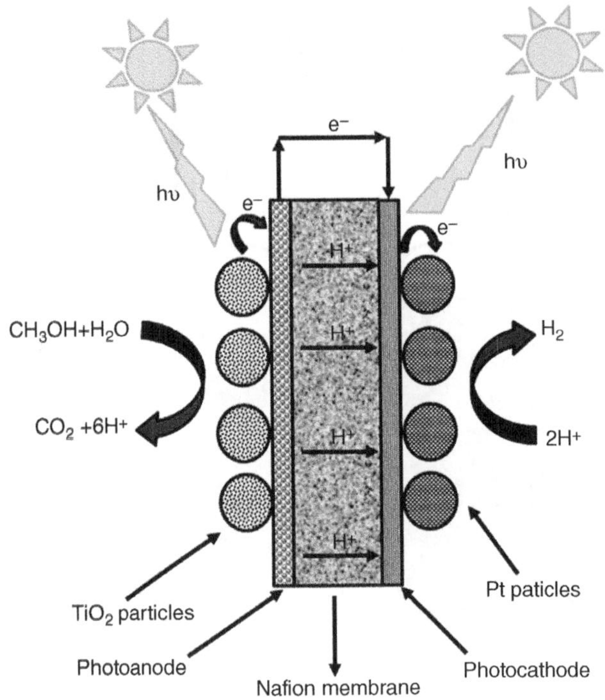

Figure 12.15 *Scheme of the proton exchange solar MR used by Seger and Kamat (2009).*

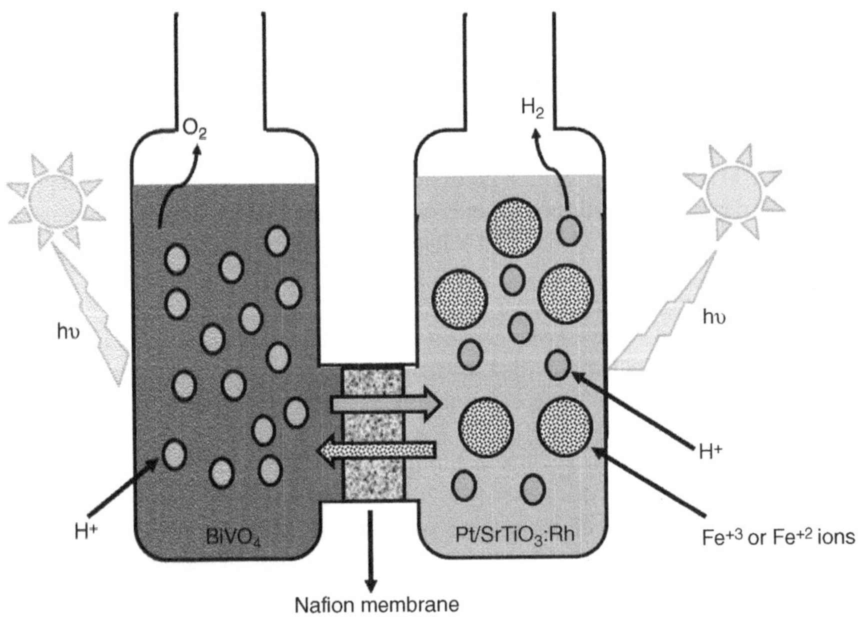

Figure 12.16 *Scheme of membrane twin reactor used by Yu et al. (2011b).*

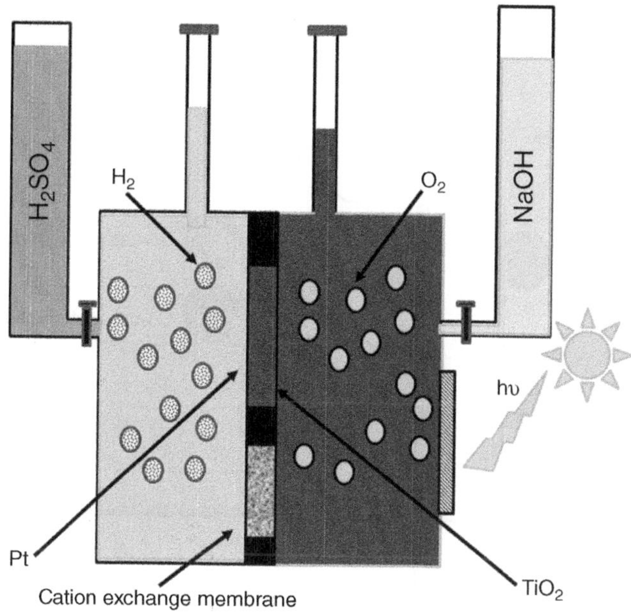

Figure 12.17 *Schematic of the photocatalytic MR used by Selli et al. (2007).*

The influence of the initial concentration of the iron species was also investigated and the obtained data exposed that hydrogen production increased with the enhancement of the iron concentration, although it was not proportional to the gas production.

In the work by Selli et al. (2007), the water decomposition reaction was studied in two Plexiglas compartments that caused separate production of hydrogen and oxygen on the two different sides of a photoactive Ti electrode, obtained using deposition of thin TiO_2 films on discs of pure Ti. One cell side was irradiated with a UV lamp or solar light through a window. The modified Ti disc was mounted between the two compartments of the cell, with the TiO_2 layer exposed to the irradiation source. A cation exchange membrane was attached over the Ti disc separating the two compartments of the system. Also in this case, in order to generate a small electrochemical current which ensured the transfer of electrons, two aqueous solutions of NaOH and H_2SO_4 were used (as shown in Figure 12.17).

The experimental results showed that oxygen was the main gas produced on the Ti-irradiated side, acting as the photoanode, while hydrogen was the main gas produced from water reduction, on the Pt/Ti side. However, the photocatalytic tests were carried out using Ti electrodes with different chemical–physical characteristics, demonstrating a promising efficiency for the electrode prepared with the magnetron sputtering deposition of Pt on TiO_2 disc operating at 600°C which maintained high durability continuously for over 180 hours. Under 300 nm wavelengths of light/solar irradiation, the H_2 and O_2 production rate was 100 and 26 mol min^{-1}, respectively. It is interesting to note that the production of the two gases increased when the normality of NaOH and H_2SO_4 was increased at values higher than 1.0 N, while their production was negligible when the two compartments were filled with pure water.

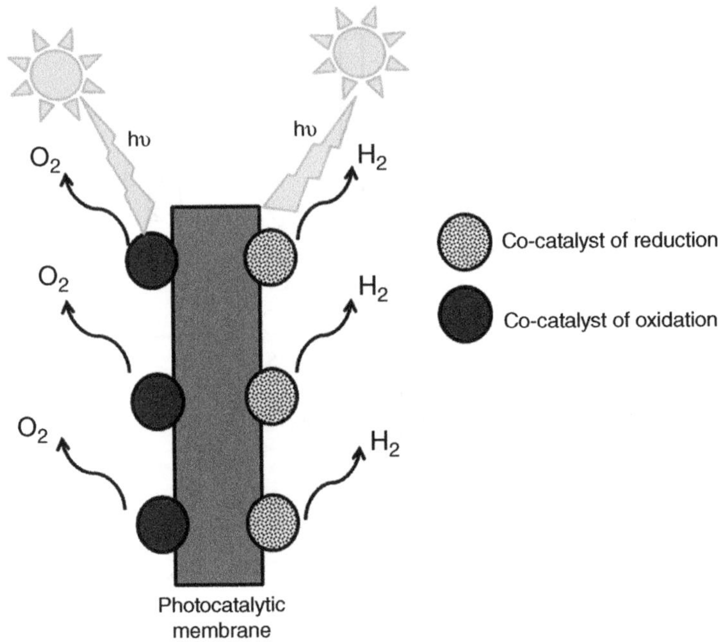

Figure 12.18 *Schematization of the process of photocatalytic hydrogen and oxygen production from water decomposition based on the MR system used by Tsydenov et al. (2012).*

Tsydenov et al. (2012) recently proposed an innovative membrane photosystem based on a porous polymeric membrane modified with Pt and Ti to produce hydrogen using ethanol dehydrogenation. A polytetrafluoroethylene filter modified with polypropylene, due to its high physical–chemical stability under light irradiation, was used as support (Figure 12.18).

The influence of several parameters such as the optimum value of Ti and metal loading, mechanical resistance, permeability, and morphology of the membrane were analyzed in order to improve the performance of the catalytic membrane system for hydrogen production. Therefore, photocatalytic tests were carried out using aqueous ethanol solutions under inert atmosphere to produce hydrogen. The reaction led to the formation of methane, ethane, carbon dioxide, and hydrogen as gaseous products and acetic acid and acetaldehyde as liquid by-products.

A significant efficiency of about 13% was reported using aUV irradiation source (l < 345 nm, irradiance 187 mW cm^{-2}) and presented a H_2 production rate of about 15 mL h^{-1}.

Moreover, a dual-layer photoelectrode preparation was investigated by Liao et al. (2012) for visible light water decomposition into hydrogen and oxygen. The photocatalyst, consisting of WO_3 and TiO_2 layers deposited on a Pt-coated foil, was tested in a H-type MR and in photo-voltammetry experiments, showing good catalytic properties in visible light adsorption. Photocatalytic runs were also carried out using UV and visible light, which showed a better performance compared to that of the Ti electrode.

Furthermore, Marschall et al. (2012) built an MR system (Figure 12.19) with two chambers for pure hydrogen and oxygen collection in which the membrane separated and at

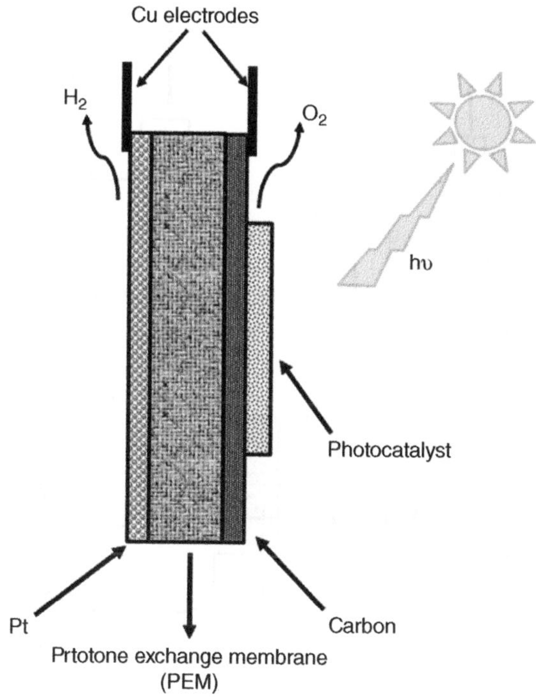

Figure 12.19 *Schematic representation of membrane electrode-photocatalyst assembly by Marschall et al. (2012).*

the same time supported a carbon coated Degussa TiO_2 P25 photoanode and a Pt cathode. No sacrificial agent was used in this system. Several membrane types such as commercial Nafion, FKE (Fumatech) membranes, handmade sulfonated polyethersulfone (SPES) flat sheet (Klaysom et al., 2010) and SPES/mesoporous-Si-MCM-41 nanoparticles (Marschall et al., 2009; Klaysom et al., 2010) were studied in the photocatalytic MR system.

The novel composite SPES-40 membrane containing 0.5 wt% sulfonated mesoporous Si-MCM-41 nanoparticles (SN) led to promising results for hydrogen production (82.4 mmol h^{-1} for Ti P25 in water and 104.5 mmol h^{-1} using diluted HCl) that, in addition to the absence of sacrificial agents, makes this system a promising MR system.

In recent years, Sasaki et al. (2013) investigated a visible-light-irradiated MR system using tris(2,2-bipyridyl) cobalt(II), ($[Co(bpy)_3]^{3/2}$) electron mediator. The used membrane having 10 nm pore size was used to separate two aqueous solutions containing $Ru/SrTiO_3$: Rh and $BiVO_4$ catalysts for hydrogen and oxygen evolution, respectively (Figure 12.20). After 3 hours of irradiation, a hydrogen production rate of 110 μL h^{-1} was obtained in the $Ru/SrTiO_3$: Rh-side, while in the compartment containing $BiVO_4$ both oxygen (43 μL h^{-1}) and hydrogen (4.3 μL h^{-1}) were produced. It is probably due to H_2 passage from the $Ru/SrTiO_3$:Rh-compartment through the membrane. The results achieved also showed that $[Co(bpy)_3]^{3/2}$ play a key role in gase formation as long as it can move from one side to the other of the system.

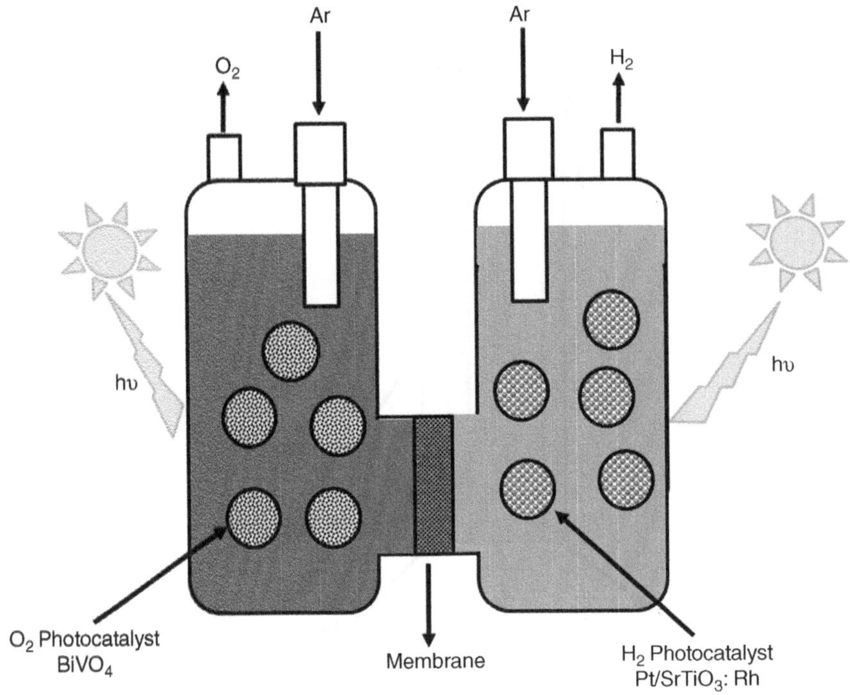

Figure 12.20 *Schematic of the solar or photocatalytic MR system used by Sasaki et al. (2013).*

The systems and experimental devices described in this section show a main role of a membrane for the photo-generation of separate streams of hydrogen and oxygen from water. However, the MRs (or hybrid processes using also membranes) in this field for photocatalytic hydrogen (and oxygen) production are still at an early stage of development. Too many efforts must be done to build a robust MR system able to work with high overall conversion efficiency.

12.4.1.2 Thermochemical Water Decomposition

As mentioned before, the S-I cycle shows an energy efficiency of 50%, which makes it as one of the most effective cycles among all water decomposition processes (Kar et al., 2012). In the gaseous HI decomposition reaction, H_2 separation membranes can play a critical role. The decomposition reaction $HI \rightarrow H_2 + I_2$ is an equilibrium reaction that has conversion efficiency around 22% at 450°C. The enhancement of decomposition rate by removing H_2 from the reaction zone is possible.

According to Onuki et al. (1997), there are no metallic materials to be stable under HI, H_2, and water vapor streams. For instance, Pd-based membranes offer the greatest (infinite) ideal H_2 selectivity for defect-free membranes. However, Pd-based membranes suffer from a variety of instability issues; Pd becomes increasingly brittle in the presence of superheated steam (Saracco et al., 1999). While silica membranes showing lower selectivity, have many advantages with respect to Pd-based membranes for applications mentioned. In a case study (Soumitra et al., 2011), a H_2-permselective MR had been analyzed to improve HI conversion and it was observed that the silica membrane was stable under H_2, HI, and steam at 723 K for 48 hours. However, the application effects of a H_2-permselective membrane on HI decomposition reaction are not entirely clear yet till date.

12.4.1.3 Direct Thermal Water Decomposition

Feasibility of experimental study for direct solar thermal decomposition of water and on-site separation of the products was carried out by Kogan (1998). In this process, developed at the Weizmann Institute of Science (WIS), water vapor is partially dissociated in a solar MR at temperatures approaching 2500 K. Hydrogen is separated from the hot mixture of water dissociation products by gas diffusion. Hence, part of the gas mixture is extracted from the reaction zone through a porous ceramic membrane. The permeate stream is enriched with hydrogen and the oxygen-enriched (retentate stream) gas mixture leaves the reaction zone bypassing the porous membrane.

12.4.2 Water Electrolysis Literature

As presented in Section 12.2, two novel solar MR systems were introduced for hydrogen production during water electrolysis that are well known to solid oxide electrolysis cells and CAWE cells. Therefore, in the following, a review study is presented for both systems from an experimental point of view.

12.4.2.1 Solid Oxide Electrolysis Cells

The thermodynamics of water electrolysis was calculated by the HSC software (Hino et al. (2004)). The values of electric energy demand (ΔG) and theoretical decomposition voltage (U_0) are 474 kJ mol^{-1} and 1.23 V at 25°C and 1 atm, respectively. In addition, decomposition process of water is an endothermic reaction. The electric energy demand and decomposition voltage decrease by the enhancement of temperature. For example, at 900°C, the electric energy demand and the decomposition voltage of water (i.e., steam) are 366 kJ/mol and 0.95 V, respectively. Moreover, theoretical energy saving is up to 23%, with respect to that at 25°C. Meanwhile, over potentials and ohmic voltage drop are also decreased obviously at high temperature. The facility for water steam electrolysis is called as SOEC which is the reverse process of solid oxide fuel cell (SOFC).

As shown in Figure 12.11 (Section 12.2), the cathode for hydrogen evolution reaction and the anode for oxygen evolution reaction are isolated by a solid electrolyte. At higher temperatures (600–1000°C), water is decomposed to H_2 and O_2 on the cathode electrode. Then O^{-2} is transported through the solid electrolyte to the anode and oxidized to O_2.

In general, the common SOEC stacks are tubular configurations with higher mechanical strength and planar configuration with better manufacturability. As a consequence, Hino et al. (2004) proposed that the planar SOEC stack is quite advantageous due to more uniform distribution of gas species and easier mass production.

Furthermore, Udagawa et al. (2007) studied steady state performances of SOEC at a current density of 7000 A m^{-2} and cathode inlet steam temperature of 850°C. They reported that cell voltage is only 1.30 V corresponding to an electricity consumption of about 3.1 kWh per normal m^3 of H_2. Joel et al. (2003) also found that cell voltage reduction by natural gas associate SOEC system was up to 1 V, compared to conventional water electrolysis. Consequently, the lower energy consumption and higher efficiency of SOEC system attract more attentions. However, energy source is noticeable issue that heats steam to operating temperature (600–1000°C). On the other hand, nuclear power produces heat and electricity simultaneously and is coupled with SOEC system (Yildiz et al., 2006). Therefore, this technology for hydrogen production was listed as one of the top ten progresses of science and technology by the Chinese Academy of Sciences (CAS) in 2004. Zhang et al. (2012) investigated the efficiencies of the SOEC system coupled with nuclear reactors. At operating temperature of 850°C, overall thermal-to-hydrogen conversion efficiencies are in the range 27–55%, by considering the outlet temperatures of steam from 250°C to 800°C. Moreover, Utgikar and Thiesen (2006) carried out the life cycle assessment of the SOEC system coupled with nuclear energy. The results indicated that the SOEC system exhibited higher energy efficiency with respect to conventional alkaline electrolysis due to lower cell voltage and electrical energy demand. The global warming potential (GWP) and acidification potential (AP) of this system are only one-sixth and one-third of those for the hydrogen production by methane steam reforming, respectively. The environmental impacts are comparable to hydrogen production from conventional water electrolysis powered by wind- or hydroelectricity. Also, alternative energy sources to heat steam is waste heat from chemical, metallurgical, and thermal power generation industries (Doenitz et al., 1980; Ni et al., 2007). This route recovers waste heat to preheat steam to the operating temperature and possesses better economy. In addition, solar energy and geothermal heat are also considered as heat sources for the SOEC system (Sigurvinsson et al., 2007).

12.4.2.2 Carbon-assisted Water Electrolysis Cells

In this process, compared to conventional water electrolysis, carbon is oxidized to CO_2 on anode instead of the oxygen evolution reaction. In general, carbon sources for anodic reaction are graphite, activated carbon, and coal. Coughlin and Farooque (1979, 1980a,1980b) found that the real operating voltage was 0.8–1.0 V for hydrogen production using coal as the carbon source.

Therefore, actual energy consumption of CAWE cells is only one-third to half of the energy for conventional water electrolysis and apparently, the cost for hydrogen production is reduced. However, the composition and structure of coal are very complicated. In addition to C element, S, O, N, and metal elements (M) also exist abundantly in various forms such as 3OH,—COOH,-S-,-NH-,MO_x, MS_y, and so on (Aho and Pirkonen, 1997; Yu et al., 1999; Zhao et al., 2008; Chen et al., 2012).

Coughlin and Farooque (1979) also concluded that the oxidation rate of coal increased with the enhancement of coal slurry concentration and the temperature in the H_2SO_4 solution. Moreover, Patil et al. (2006) investigated the electrocatalytic activities of various noble metal electrodes for coal oxidation. It was shown that Pt-Ir (80:20) electrode was the best anode material and the maximum faradic efficiency was 24% for CO_2 production. Also, energy consumption of hydrogen production was 21 Wh g^{-1}.

Yin et al. (2009) used Ti-supported metal oxides as anode materials. Their results indicated that Ti/IrO_2 with Fe^{3+} liquid catalyst revealed the best catalytic effect on coal oxidation and CO_2 and a spot of alkanes and olefins were main anode products. A suitable choice is partial electrochemical oxidation of coal on the anode to produce desirable small molecular compounds. By this way, the emission of CO_2, and GHG can be avoided. Meanwhile, better economy for coal-assisted water electrolysis is attained based on the hydrogen production on cathode and hydrocarbon organic products on anode. However, the selective control of coal oxidation to target products is the most challenging problem. By applying coal as the carbon source, some uncertainties may occur due to the complexity of the coal structure. Pure carbon materials such as activated carbon and graphite are better alternatives to coal even with higher cost.

Hence, Seehra et al. (2007) used a fine activated carbon (GX203) as the carbon source for CAWE system. The measurable H_2 was observed at cell voltage of 0.54 V, while hydrogen production rate (HR) was very low. In order to increase HR, various carbon materials were evaluated (Seehra and Bollineni, 2009).

As a general consequence, these new technologies show high potential and cost effectiveness for hydrogen production. It is clear that by using solar energy as an electricity power supplier or a heating system supplier can intensify the rate of commercializing these new electrolysis technologies.

12.4.3 Steam Reforming Literature

As mentioned before, besides solar-aided water decomposition processes (electrolysis and thermochemical cycles), membranes can also be applied in solar reforming. Clearly, the reforming process is highly heat-demanding for steam generation (an excess of steam is required to prevent reactor fouling and guarantee stable operation) and to drive the high temperature (>800°C) steam reforming reactor. Hence, the process heat is supplied by additional fuel, while its scan alternatively is supplied by an external carbon-free source like a CSP plant.

The CSP system is sensitively affected by weather condition; therefore it is better to utilize the thermal storage with molten salt (Figure 12.21). By controlling the flow rate of the molten salt, the heat supply to the steam reforming MR will be kept constant. Since the heat of concentrated solar thermal energy can be stored with molten salt, and also can be utilized for steam reforming at night, the facility for the steam reforming process can be scaled down, by comparing the solar reformer operated only during day time (Giaconia et al., 2008).

Several research projects (Berman et al., 2006; Moeller et al., 2006; Giaconia et al., 2008) have been carried out so far dealing with hydrocarbon conversion to hydrogen powered by concentrated solar energy, including methane steam reforming, dry reforming, and cracking. In particular, the solar methane steam reforming process was successfully demonstrated in a solar receiver reactor (volumetric reformer) operating at high temperatures (>850°C).

According to Watanuki et al. (2005), the steam reforming reaction of natural gas proceeds around 1000 K with high methane conversion and the decomposition of molten nitrate mixtures like $NaNO_3/KNO_3$ (60/40 w/w) occurs around 870 K (550–600°C). Therefore, the Pd-based MR has been investigated to lower the reaction temperature of methane steam reforming. In this study, a composite Pd membrane on alumina support was used. They found high performance of MR versus different feed molar ratio (CH_4/H_2O).

Clearly, the utilization of such molten salts can ensure constant rate of solar heat supply also for an energy-demanding industrial chemical process like steam reforming (Giaconia

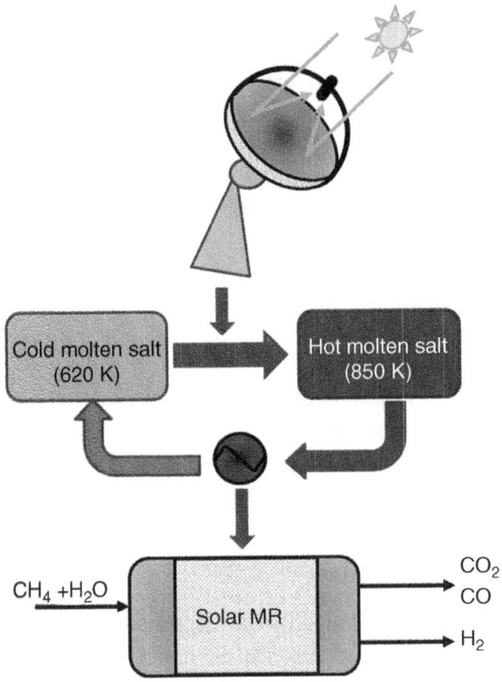

Figure 12.21 *Schematic outline of steam reforming of natural gas system using thermal storage with molten salt and solar concentrating system.*

et al., 2008). However, continuous solar energy supply to enhance the process management can avoid daily start-up and shut-down operations despite the intermittent primary source.

As another innovative project in the solar MR field for steam reforming process, a solar MR powered with molten nitrates is being developed in the framework of the European project CoMETHy (Compact Multifuel-Energy to Hydrogen converter).

In this case, the molten salts are applied to transfer the heat collected from the CSP plant and other possible back up heat sources to the steam reforming plant. This new technology imposes to operate the reformer at temperatures lower than 550°C, and selective membranes can be applied to recover high-purity hydrogen and increase conversion even with the thermodynamic limitations.

Indeed, by comparing a conventional steam reforming process, the solar MR technology for steam reforming operates at lower temperatures, consequently leading to significant improvement in material costs; so no special steel alloy for high-temperature operation is required. Therefore, the high-temperature furnace can be replaced by a flameless heat exchanger heated by a liquid molten salts stream, making the whole reactor envelope more compact. Moreover, by operating the process at lower temperatures, it is possible to combine steam reforming and water-gas shift (WGS) reactions in a single stage at 400–550°C, resulting in a low outlet CO content ($\ll$10% vol) and a saving of the reformer heat duty. Integration with membranes avoids dedicated hydrogen separation and purification units, to further improve the compactness and to increase the conversion regardless of the thermodynamic limitations of low-temperature reforming.

An additional advantage of using molten salts as HTF for steam reforming is presented when some stand-by periods of the plant are predicted. However, it is the case of small–medium-scale reformers for hydrogen refueling stations. In this case, the continuous molten salts recirculation facilitates the overall process management by maintaining all plant components (e.g., catalyst, membrane) at working temperature (400–550°C) also during stand-by periods; thus minimizing start-up periods and materials ageing resulting from thermal cycling.

On the other hand, considering the operational range (400–550°C, 1–10 bar), in general, composite membranes (with a few microns thickness of Pd-based layer deposited on a porous metal or ceramic support) or self-supported membranes (consisting of a rolled Pd-Ag foil with thickness >50 μm) can be applied for hydrogen removal from the molten salts heated reactor.

Giaconia et al. (2008) presented that via the incorporation of the catalyst with the membrane, two options can be considered: a multistage membrane reformer (MSMR), where the membrane is external to the reactor, or an integrated membrane reactor (IMR), where the membrane is incorporated with the catalyst and the heat exchanger.

The MSMR scheme consists of a number of reformers similar to that shown in Figure 12.22, where the thermodynamic equilibrium is approached, each one monitored by a

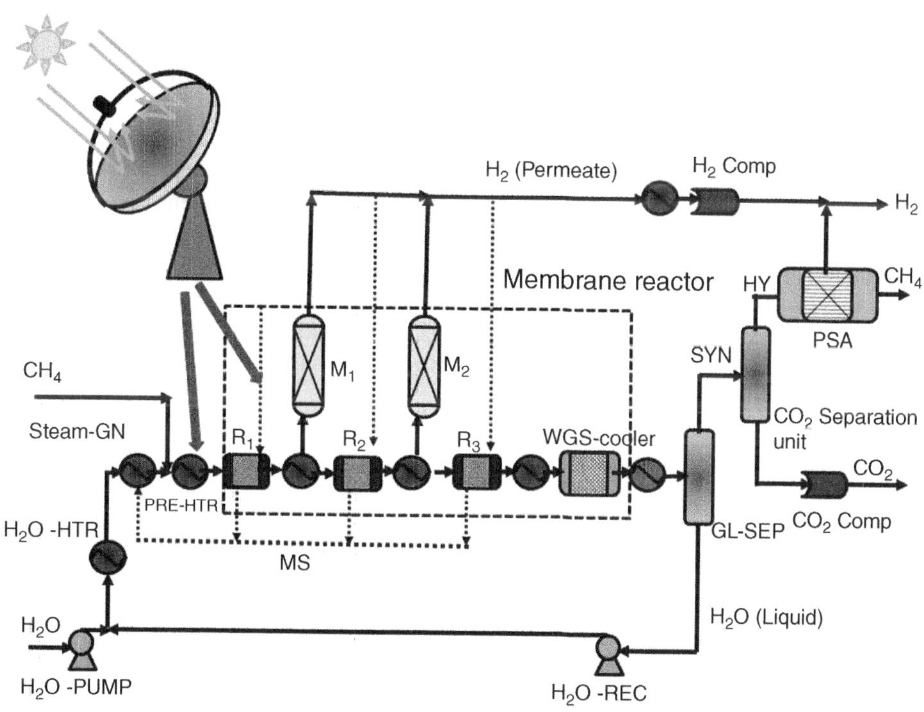

Figure 12.22 *Multistage membrane reactor (MSMR) scheme for solar steam reforming of methane using molten salts (MS), with three reactors in series (R1, R2, R3) and two intermediate membrane modules (M1, M2).*

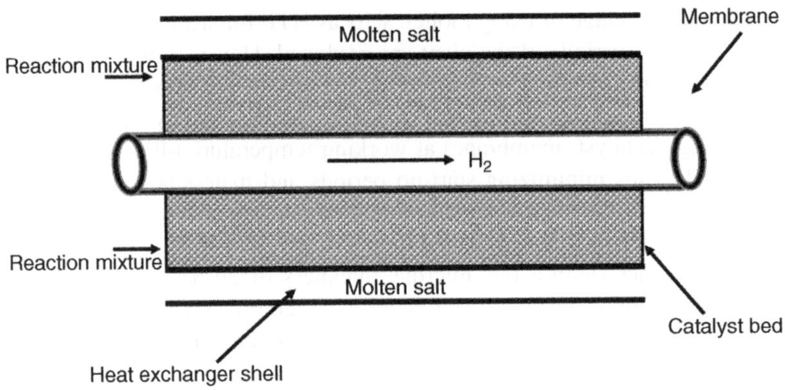

Figure 12.23 *Indicative sketch of a molten salts powered membrane reactor (IMR).*

membrane separation module (Giaconia et al., 2008). Thus, a continuous enhancement of methane conversion is achieved in the MR owing to hydrogen removal in the intermediate membrane units.

Therefore, the application of MRs in solar steam reforming increases the complexity of the chemical plant, while significantly reduces the size and costs of the power plant (heat supplier) by improving the overall thermal efficiency. This is a key point, since in solar-powered thermochemical processes, the CSP plant is always a major cost item (Nair et al., 2006; Giaconia et al., 2008), so that the membrane application will make the overall plant more competitive.

The above considerations are effective in the case of an IMR system too. The general scheme of a molten salt heated IMR is represented in Figure 12.23. Clearly, the IMR system with respect to the MSMR is more compact and leads to higher efficiency and feedstock conversion on a single pass. However, the development of this kind of reactor involves a notable engineering challenge with several design and mechanical issues dealing with the catalyst/membrane coupling. Therefore, the harmonization of the three key transport mechanisms (heat transfer from the molten salts to the catalyst bed, reaction kinetics, and hydrogen permeance through the membrane) is considered (Giaconia and Caputo, 2014).

12.5 The Main Challenges

It should be noted that many challenges are associated in exploiting the advantages of solar MRs for hydrogen production. However, most of these challenges are related to applying the MR technologies. Thereby, the main steps are presented as follows:

- Need for defect-free membranes with both high selectivity and permeability.
- Difficulty in reproducing laboratory-scale results on a larger scale suitable for commercialization.
- Ceramic and metallic membrane tubes have to be housed within a reactor assembly with feed/product lines and other reactor equipment.
- Necessity of progress in developing cheap and high temperature sealants.

- Need to develop technologies for heat supply/removal and temperature control for large- or medium-scale modules.
- Need to develop durable catalytic membranes that are less sensitive to poisoning.
- Need for a suitable method to impregnate the catalytic layer onto the membrane surface.
- Need for a design criteria for MR: optimal size, flow patterns, stages, and recycle streams.
- Need for durable membrane application in high temperatures (>800°C) for water decomposition processes.
- Complete corrosion studies on the various compartments of MR system, especially in water decomposition application.
- Need for using solar energy during continuous MR processes (more than 24 hours).

12.6 Conclusion and Future Trends

In this chapter, the major issues dealing with the coupling between a solar plant and a membrane-enhanced hydrogen production process have been overviewed considering three different processes: water decomposition, water electrolysis, and steam reforming.

Water electrolysis powered by PV or CSP plants is the most mature technology for the short-term centralized and decentralized CO_x-free hydrogen production.

Another solar-powered water decomposition route was represented by thermochemical cycles, photocatalytic, and direct thermal processes, which are considered as an interesting option for the longer-term centralized massive hydrogen production. The possibility to apply membranes in different process sections has also been considered: the case of the S-I process was discussed here, involving the potential exploitation of polymeric membranes (cation exchange membranes or hydrophobic microporous membranes) and/or ceramic membranes for high-temperature operations.

Finally, solar steam reforming using MRs has been suggested as an alternative low-temperature reforming route for fuel-flexible hydrogen production. In this case, Pd-based membranes for hydrogen separation were applied.

It is generally shown that the introduction of membranes in the chemical process can improve the whole solar-aided process in terms of efficiency and costs; also the management of the plant can be relieved. By comparing the different available process options, it is highlighted that major benefits can be obtained when the chemical plant such as MR is coupled with CSP plants equipped with suitable heat storage buffering systems. Therefore, applying solar MR technology according to advantages mentioned can show more potential in different applications especially for hydrogen production.

List of Abbreviations

AP	Acidification potential
CAS	Chinese Academy of Sciences
CAWE	Carbon-assisted water electrolysis
CFD	Computational fluid dynamics
CSP	Concentrating solar power

CSTR	Continuous stirred-tank reactor
GHG	Greenhouse gases
GWP	Global warming potential
HR	Hydrogen production rate
HTF	Heat transfer fluid
HYTHEC	Hydrogen thermochemical cycles
IMR	Integrated membrane reactor
MF	Microfiltration
MPS	Membrane photocatalytic systems
MR	Membrane reactor
MSMR	Multistage membrane reformer
NF	Nanofiltration
ODE	Ordinary differential equations
PFR	Plug flow reactor
PROX	Preferential oxidation
PV	Photovoltaic
R&D	Research and development
RES	Renewable energy sources
S-I	Sulfur–iodine
SOEC	Solid oxide electrolyte cell
SOFC	Solid oxide fuel cell
SPES	Sulfonated polyethersulfone
TR	Traditional reactor
UF	Ultrafiltration
WGSR	Water-gas-shift reactor
WH	Westinghouse
WIS	Weizmann Institute of Science

References

Aho, M.J., Pirkonen, P.M. (1997) Effects of pressure, gas temperature and CO_2 and O_2 partial pressures on the conversion of coal-nitrogen to NO, N_2O and NO_2. *Fuel*, **74**, 1677–1681.

Augugliaro, V., Litter, M., Palmisano, L., Soria, J. (2006) The combination of heterogeneous photocatalysis with chemical and physical operations: A tool for improving the photoprocess performance. *J Photochem Photobiol*, **7**, 127–144.

Basile, A., Gallucci, F., Paturzo, L. (2005) A dense Pd/Ag membrane reactor for methanol steam reforming: experimental study. *Catal Today*, **104**, 244–250.

Basile, A., Parmaliana, A., Tosti, S., Iulianelli, A., Gallucci, F., Espro, C., Spooren, J. (2008) Hydrogen production by methanol steam reforming carried out in membrane reactor on Cu/Zn/Mg-based catalyst. *Catal Today*, **137**, 17–22.

Basile, A., Tereschenko, G.F., Orekhova, N.V., Ermilova, M.M., Gallucci, F., Iulianelli, A. (2006) An experimental investigation on methanol steam reforming with oxygen addition in a flat Pd–Ag membrane reactor. *Int J Hydrogen Energ*, **31**, 1615–1622.

Berman, A., Karn, R.K., Epstein, M. (2006) A new catalyst system for high-temperature solar reforming of methane *Energy & Fuels*, **20**, 455–462.

Bosc, F., Ayral, A., Guizard, Ch. (2005) Mesoporous anatase coatings for coupling membrane separation and photocatalyzed reactions. *J Membr Sci*, **265**, 13–19.

Chen, Y., Mastalerz, M., Schimmelmann, A. (2012) Characterization of chemical functional groups in macerals across different coal ranks via micro-FTIR spectroscopy. *Int J Coal Geol*, **104**, 22–33.

Chueh, W.C., Falter, C., Abbott, M., Scipio, D., Furler, P., Haile, S.M., Steinfeld, A. (2010) High-flux solar-driven thermochemical dissociation of CO_2 and H_2O using nonstoichiometric ceria. *Science*, **330**, 1797–1801.

Coughlin, R.W., Farooque, M. (1979) Hydrogen production from coal, water and electrons. *Nature*, **279**, 301–303.

Coughlin, R.W., Farooque, M. (1980a) Consideration of electrodes and electrolytes for electrochemical gasification of coal by anodic oxidation. *J Appl Electrochem*, **10**, 729–740.

Coughlin, R.W., Farooque, M. (1980b) Electrochemical gasification of coal-simultaneous production of hydrogen and carbon dioxide by a single reaction involving coal, water and electrons. *Ind Eng Chem Process Des Dev*, **19**, 211–209.

De Falco, M.D., Barba, D., Cosenza, S., Iaquaniello, G., Marrelli, L. (2008) Reformer and membrane modules plant powered by a nuclear reactor or by a solar heated molten salts: Assessment of the design variables and production cost evaluation. *Int J Hydrogen Energ*, **33**, 5326–5334.

De Falco, M.D., Basile, A., Gallucci, F. (2010) Solar membrane natural gas steam reforming process: Evaluation of reactor performance. *Asia-Pacific J. Chem. Eng*, **5** (1), 179–190.

Doenitz, W., Schmidberger, R., Steinheil, E., Streicher, R. (1980) Hydrogen production by high temperature electrolysis of water vapour. *Int J Hydrogen Energ*, **5**, 55–63.

Fujihara, K., Ohno, T., Matsumara, M. (1998) Splitting of water by electrochemical combination of two photocatalytic reactions on TiO_2 particles. *J Chem Soc Faraday Trans*, **94** (24), 3705–3709.

Gallucci, F., Basile, A. (2006) Co-current and counter-current modes for methanol steam reforming membrane reactor. *Int J Hydrogen Energ*, **31**, 2243–2249.

Gallucci, F., Basile, A. (2008) Palladium membrane reactor for steam reforming reaction: A comparison between different fuels. *Int J Hydrogen Energ*, **33**, 1671–1687.

Gallucci, F., Basile, A., Tosti, S., Iulianelli, A., Drioli, E. (2007) Methanol and ethanol steam reforming in membrane reactors: An experimental study. *Int J Hydrogen Energ*, **32**, 1201–1210.

Ghasemzadeh, K., Liguori, S., Morrone, P., Iulianelli, A., Piemonte, V., Babaluo, A.A., Basile, A. (2013a) H_2 production by low pressure methanol steam reforming in a dense Pd–Ag membrane reactor in co-current flow configuration: Experimental and modeling analysis. *Int J Hydrogen Energ*, **38**, 16685–16697.

Ghasemzadeh, K., Morrone, P., Iulianelli, A., Liguori, S., Babaluo, A.A., Basile, A. (2013b) H2 production in silica membrane reactor via methanol steam reforming: Modeling and HAZOP analysis. *Int J Hydrogen Energ*, **38**, 10315–10326.

Ghasemzadeh, K., Morrone, P., Liguori, S., Babaluo, A. A., Basile, A. (2013c) Evaluation of silica membrane reactor performance for hydrogen production via methanol steam reforming: Modeling study. *Int J Hydrogen Energ*, **38**, 16698–16709.

Ghasemzadeh, K., Morrone, P., Babalou, A. A. and Basile, A. (2015) A simulation study on methanol steam reforming in the silica membrane reactor for hydrogen production. *Int J Hydrogen Energ*, **40** (10), 3909–3918.

Giaconia, A., Caputo, G. (2014) Membrane technologies for solar-hydrogen production, Woodhead Publishing Limited, 325–346.

Giaconia, A., De Falco, M.D., Caputo, G., Grena, R., Tarquini, P., Marrelli, L. (2008) Solar steam reforming of natural gas for hydrogen production using molten salt heat carriers. *AIChE J*, **54** (7), 1932–1944.

Harold, M.P., Nair, B. Kolios, G. (2003) Hydrogen generation in a Pd membrane fuel processor: Assessment of methanol-based reaction systems. *Chem Eng Sci*, **58**, 2551–2571.

Hino, R., Haga, K., Aita, H., Sekita, K. (2004) R&D on hydrogen production by high-temperature electrolysis of steam. *Nucl Eng Des*, **233**, 363–375.

Iulianelli, A., Longo, T., Basile, A. (2008a) Methanol steam reforming in a dense Pd–Ag membrane reactor: The pressure and WHSV effects on CO-free H_2 production. *J Membrane Sci*, **323**, 235–240.

Iulianelli, A., Longo, T., Basile, A. (2008b) Methanol steam reforming reaction in a Pd–Ag membrane reactor for CO-free hydrogen production. *Int J Hydrogen Energ*, **33**, 5583–5588.

Jiang, Y.Z., Gao, J.F., Liu, M.F., Wang, Y.Y., Meng, G.Y. (2007) Fabrication and characterization of Y_2O_3 stabilized ZrO_2 films deposited with aerosol-assisted MOCVD. *Solid State Ionics*, **177**, 3405–3410.

Joel, M.F, Pham, A.Q., Aceves, S.M. (2003) A natural gas-assisted steam electrolyzer for high-efficiency production of hydrogen. *Int J Hydrogen Energ*, **28**, 483–490.

Kar, S., Bindal, R.C., Prabhakar, S., Tewari, P.K. (2012) The application of membrane reactor technology in hydrogen production using S-I thermochemical process: A roadmap. *Int J Hydrogen Energ*, **37**, 3612–3620.

Klaysom, C., Marschall, R., Wang, L.Z., Ladewig, B.P., Lu, G.Q.M. (2010) Synthesis of composite ion-exchange membranes and their electrochemical properties for desalination applications. *J Mater Chem*, **20** (22), 4669–4674.

Koc, R., Kazantzis, N.K., Nuttall, W.J., Hua, M.Y. (2012) An economic evaluation framework for membrane reactor modules in the presence of uncertainty: The case for process safety investment and risk reduction. *J Loss Prevent Proc*, **1**, 1–10.

Kogan, A. (1998) Direct solar thermal splitting of water and on-site separation of the products—II. Experimental feasibility study. *Int J Hydrogen Energ*, **23**(2), 89–98.

Kogan, A. (2000a) Direct solar thermal splitting of water and on-site separation of the products IV. Development of porous ceramic membranes for a solar thermal water-splitting reactor. *Int J Hydrogen Energ*, **25**, 1043–1050.

Kogan, A., Spiegler, E., Wolfshtein, M. (2000b) Direct solar thermal splitting of water and on-site separation of the products. III. Improvement of reactor efficiency by steam entrainment. *Int J Hydrogen Energ*, **25**, 739–745.

Le Duigou, A., Borgard, J.M., Larousse, B. et al. (2007) HYTHEC: An EC funded search for a long term massive hydrogen production route using solar and nuclear technologies. *Int J Hydrogen Energ*, **32**, 1516–1529.

Liao, C.H., Huang, C.W., Wu, J.C.S. (2012) Novel dual-layer photoelectrode prepared by RF magnetron sputtering for photocatalytic water splitting. *Int J Hydrogen Energ*, **37**(16):11632–11639.

Lo, C.C., Huang, C.W., Liao, C.H., Wu, J.C.S. (2010) Novel twin reactor for separate evolution of hydrogen and oxygen in photocatalytic water splitting. *Int J Hydrogen Energ*, **35**, 1523–1529.

Marschall, R., Bannat, I., Feldhoff, A., Wang, L.Z., Lu, Q.G.M., Wark, M. (2009) Nanoparticles of mesoporous SO_3H-functionalized Si-MCM-41 with superior proton conductivity. *Small*, **5**(7), 854–859.

Marschall, R., Klaysom, C., Mukherji, A., Wark, M., Lu, G.Q.M., Wang, L. (2012) Composite proton-conducting polymer membranes for clean hydrogen production with solar light in a simple photoelectrochemical compartment cell. *Int J Hydrogen Energ*, **37**(5), 4012–4017.

Moeller, S., Kaucic, D., Sattler, C. (2006) Hydrogen production by solar reforming of natural gas: A comparison study of two possible process configuration. *ASME J Solar Energ Eng*, **128**, 16–23.

Molinari, R., Marino, T., Argurio, P. (2014) Photocatalytic membrane reactors for hydrogen production from water. *Int J Hydrogen Energ*, **39**(14), 7247–7261.

Mozia, S., (2010) Photocatalytic membrane reactors (PMRs) in water and wastewater treatment. A review. *Sep Purif Technol*, **73**, 71–91.

Nair, B.K.R., Harold, M.P. (2006) Hydrogen generation in a Pd membrane fuel processor: Productivity effects during methanol steam reforming. *Chem Eng Sci*, **61**, 6616–6636.

Ni, M., Leung, M.K.H., Leung, D.Y.C. (2007) Energy and energy analysis of hydrogen production by solid oxides team electrolyzer plant. *Int J Hydrogen Energ*, **32**, 4648–4660.

Ohya, H., Yatabe, H., Aihara, M., Negishi, Y., Takeuchi, T. (2002) Feasibility of hydrogen production above 2500 K by direct thermal decomposition reaction in membrane reactor using solar energy. *Int J Hydrogen Energ*, **27**, 369–376.

Onuki, K., Ioka, I., Futakawa, M., Nakajima, H., Shimizu, S., Tayama, I. (1997) Screening tests on materials of construction for the thermochemical is process. *Corros Eng*, **46**, 141–149.

Orhana, M.F., Dincer, I., Rosen, M.A., Kanoglu, M. (2012) Integrated hydrogen production options based on renewable and nuclear energy sources. *Renew Sust Energ Rev*, **16**, 6059–6082.

Patil, P., Abreu, Y.D., Botte, G.G. (2006) Electro oxidation of coal slurries on different electrode materials. *J Power Sources*, **158**, 368–377.

Rosen, M.A. (2010) Advances in hydrogen production by thermochemical water decomposition: A review. *Energy*, **35**, 1068–1076.

Saracco, G., Neomagus, H.W.J.P., Versteeg, G.F., van Swaaij, W.P.M. (1999) In: *Chemical engineering science proceedings of the 1998 15th international symposium on chemical reaction engineering*, ISCRE 15, **54**, 1997–2017.

Sasaki, Y., Kato, H., Kudo, A. (2013) $[Co(bpy)_3]^{3/2}$ and $[Co(phen)_3]^{3/2}$ electron mediators for overall water splitting under sunlight irradiation using Z-scheme photocatalyst system. *J Am Chem Soc*, **135**(14), 5441–5449.

Seehra, M.S., Bollineni, S. (2009) Nanocarbon boosts energy-efficient hydrogen production in carbon-assisted water electrolysis. *Int J Hydrogen Energ*, **34**, 6078–6084.

Seehra, M. S., Ranganathan, S., Manivannan, A. (2007) Carbon-assisted water electrolysis: An energy-efficient process to produce pure hydrogen at room temperature. *Appl Phys Lett*, **90**:044104.

Seger, B., Kamat, P.V. (2009) Fuel cell geared in reverse: Photocatalytic hydrogen production using a TiO_2/Nafion/Pt membrane assembly with no applied bias. *J Phys Chem*, **113**(43), 18946–18952.

Selli, E., Chiarello, G.L., Quartarone, E., Mustarelli, P., Rossetti, I., Forni, L. (2007) A photocatalytic water splitting device for separate hydrogen and oxygen evolution. *Chem Commun*, **47**, 5022–5024.

Sigurvinsson, J., Mansilla, C., Lovera, P., Werkoff, F. (2007) Can high temperature steam electrolysis function with geothermal heat? *Int J Hydrogen Energ*, **32**, 1174–1182.

Soumitra, S., Bindal, R.C., Prabakar, S., et al. (2011) Membrane development for applications in hydrogen production using the sulphur-iodine thermochemical route. *Int J Nucl Hydrogen Production Appl*, **2**(3), 227–236.

Tsydenov, D.E., Parmon, V.N., Vorontsov, A.V., (2012) Toward the design of asymmetric photocatalytic membranes for hydrogen production: Preparation of TiO_2-based membranes and their properties. *Int J Hydrogen Energ*, **37**(15):11046–11060.

Udagawa, J., Aguiar, P., Brandon, N.P. (2007) Hydrogen production through steam electrolysis: Model-based steady state performance of a cathode-supported intermediate temperature solid oxide electrolysis cell. *J Power Sources*, **166**, 127–136.

Utgikar, V., Thiesen, T. (2006) Life cycle assessment of high temperature electrolysis for hydrogen production via nuclear energy. *Int J Hydrogen Energ*, **31**, 939–944.

Watanuki, K., Imaeda, S., Mizukami, T., Hasuike, H., Utamura, M., Tamaura, Y. (2005) *The production of solar hybrid fuel with molten salt and membrane reactor*. Proceedings of JSES/JWEA Joint Conference, 495–498, Chino, Japan.

Wieland, I.S., Melin, I.T., Lamm, I.A. (2002) Membrane reactors for hydrogen production. *Chem Eng Sci*, **57**, 1571–1576.

Winter, C.J., Sizmann, R.L., Vant Hull, L.L. (1991) Solar Power Plants. Springer-Verlag, New York.

Xinxin, W., Kaoru, O. (2007) Thermochemical water splitting for hydrogen production utilizing nuclear heat from an HTGR. *Tsinghua Sci Technol*, **10**(2), 270–276.

Yildiz, B., Kazimi, M.S. (2006) Efficiency of hydrogen production systems using alternative nuclear energy technologies. *Int J Hydrogen Energ*, **31**, 77–92.

Yin, R.H., Zhao, Y.G., Lu, S.Y., Wang, H.M., Cao, W.M., Fan, Q.B. (2009) Electro catalytic oxidation of coal on Ti-supported metal oxides coupled with liquid catalysts for H_2 production. *Electro chim Acta*, **55**, 46–51.

Yu, B., Zhang, W.Q., Xu, J.M., Chen, J., Luo, X., Stephan, K. (2012) Preparation and electrochemical behavior of dense YSZ film for SOEC. *Int J Hydrogen Energ*, **37**, 12074–12080.

Yu, H., Ouyang, S.X., Yan, S.C., Li, Z.S., Yu, T., Zou, Z.G. (2011a) Sol—gel hydrothermal synthesis of visible-light-driven Cr-doped SrTiO3 for efficient hydrogen production. *J Mater Chem*, **21**, 11347–11351.

Yu, L.E., Hildemann, L.M., Niksa, S. (1999) Characteristics of nitrogen-containing aromatic compounds in coal tars during secondary pyrolysis. *Fuel*, **78**, 377–385.

Yu, S.C., Huanga, C.W., Liao, C.H., Wua, J.C.S., Chang, S.T., Chen, K.H. (2011b) A novel membrane reactor for separating hydrogen and oxygen in photocatalytic water splitting. *J Membr Sci*, **382**, 291–299.

Zhang, W.Q., Yu, B., Xu, J.M. (2012) Efficiency evaluation of high-temperature steam electrolytic systems couple with different nuclear reactors. *Int J Hydrogen Energ*, **37**, 12060–12068.

Zhao, W., Yao, L.P., Lin, J., Zong, Z.M. (2008) Electrolytic reduction of Nantong coal and model compounds with oxygenic functional groups in an aqueous NaCl solution. *J China Uni Mining Technol*, **18**, 112–115.

13

Membrane-Adsorption Integrated Systems/Processes

Sayed S. Madaeni[1] and Ehsan Salehi[2]

[1]*Membrane Research Center, Chemical Engineering Department, Faculty of Engineering, Razi University, Tagh Bostan, Kermanshah, Iran*

[2]*Department of Chemical Engineering, Faculty of Engineering, Arak University, Arak, Iran*

13.1 Introduction

Adsorption is a conventional chemical engineering process with a rich scientific background. Adsorption processes are applied in many industries including oil refineries, petrochemicals, and water and wastewater treatment. Adsorption is an effective separation strategy for the removal of a wide range of contaminants but it is not so favorable for the elimination of anions like boron, perchlorates, and nitrates [1,2]. Adsorption processes would be upgraded by integrating with supplementary processes to obtain hybrid processes with higher removal efficiency.

Recently, membrane technology has gained great attention as a powerful separation technique due to prominent advantages over common processes such as high removal efficiency, low energy consumption, fast kinetic, small footprint and ease of scale up. Among available membrane processes, microfiltration (MF) and ultrafiltration (UF), also called low-pressure membranes (LPMs), are favored for full-scale applications due to normal operating conditions, high productivity, and low energy consumption. On the other side, reverse osmosis (RO) and nanofiltration (NF) processes dictate higher operational costs. LPMs can efficiently eliminate many contaminants including proteins, macromolecules, natural organic matters (NOMs), dyes, and so on, from aqueous media. However, they are not capable of removing dissolved organic matter (DOM), boron, and compounds responsible for odor

Integrated Membrane Systems and Processes, First Edition. Edited by Angelo Basile and Catherine Charcosset.

and color [3, 4]. A potential solution to overcome these shortcomings is the combination of LPMs with other separation processes like adsorption. The overall removal efficiency of the hybrid process would be enhanced. The current chapter deals with the adsorption/membrane integrated systems. Some promising advantages of adsorption/membrane integrated systems are:

- Increasing separation efficiency
- Reducing process cost
- Reducing membrane fouling in some cases
- Ease of handling and control compared to individual treatments
- Lower volume of discharge
- Potential application of beneficial biosorbents
- Reusability of both membranes and adsorbents
- Fast removal kinetic
- Low energy consumption versus adsorption columns, NF and RO systems
- Low pressure drop against adsorption columns.

Generally, three different procedures for hybridization of membrane systems with adsorption processes may be found in the literature:

- Adsorption treatment before membrane filtration (pretreatment layout)
- Integrated adsorption/membrane processes (IAMPs)
- Adsorption treatment after membrane filtration (post-treatment layout).

The hybrid adsorption-membrane processes are applied to enhance the capability of the membrane systems to remove soluble contaminants, especially organic matter, polysaccharides, contaminants produced by microorganisms in biological systems, and contaminants responsible for the taste and odor. In addition, using adsorbent contactors prior to LPM system can mitigate membrane fouling as a result of a decrease in the contaminants loading in the streaming feed. This hybridization is a conventional and well-established pretreatment for the membrane systems; however, there is a strong motivation among researchers to develop the hybrid process and optimize the operating conditions.

In the integrated strategy, adsorbents (commonly fine-grade adsorbents) are directly introduced into the feed stream or into the reactor tank containing the feed solution. IAMPs may be loosely categorized as follows:

- LPM-adsorption integrated systems
- Membrane-adsorption bioreactors (MABR)
- Membrane adsorbents.

In the first class of IAMPs, fine adsorbents like powdered activated carbon (PAC) are inserted in the LPM system with a direct contact between the adsorbents and the LPMs. The second integrated system, MABR, typically works in biological and/or adsorption modes depending on the operation strategy. In such integrated systems, aeration is a necessity for the fouling mitigation, favorable mixing, and biodegradation control. Another alternative to take the mutual advantages of adsorption and membrane processes is membrane adsorbents. Adsorptive membranes are typically fabricated by incorporation of adsorption active sites into the membranes via multi-step physicochemical protocols. Such integration can result in a united separation tool, also called adsorptive membranes or membrane adsorbents.

Not only reactive functional groups but also contrived cavities in the membrane structure may be defined as adsorption sites for capturing target adsorbates. These membranes are commonly applied for the separation of macromolecules and heavy metals. Affinity membranes are conventionally applied for the purification of proteins and also as stationary phase in membrane chromatography. Novel achievements in the preparation of membrane adsorbents are 'thin membrane adsorbents' and 'imprinted membrane adsorbents'. Thin adsorptive membranes indicate fast kinetic (binding rate) and great affinity for removing trace amounts of heavy metals from aqueous phase. Besides, ion/molecular-imprinting techniques are applied to prepare adsorptive membranes with the ability of specific binding toward the target ion/molecule. This fact can enhance selectivity and specificity of the membrane adsorption.

When membrane systems are utilized in advance of the adsorption columns, the arrangement is regarded as post-treatment layout. Granular adsorbents (e.g., GAC) are commonly used in such kinds of hybridizations. The major concern is exporting fine adsorbents from the adsorption column into the finished water or the forthcoming systems. Accordingly, a supplemental separation process is usually necessary under this configuration. This arrangement has not been interestingly addressed in the literature, presumably due to full establishment and prevalent application in full scale. Among the addressed hybridization strategies, the IAMPs will be tackled with an increased focus because of attractiveness for research and preferential advantages.

13.2 Adsorption Pretreatment for Membranes

Pressure-driven membranes may be fouled by NOM, a vast range of organic compounds from low-to-high molecular weights, available in surface water and processed wastewaters [4]. Polysaccharides, proteins, amino sugars, nucleic acids, humic and organic acids, and cell debris are examples of organic matter. Generally, organic compounds consist of particulate organic matter (POM) with the average size greater than 0.45 μm and DOM with the average size lower that 0.45 μm [5]. LPMs are not potential devices to completely remove DOMs from water. Accumulation and/or adsorption of NOM on the surface or within the pores of the membranes are resulted in membrane fouling which is a menace to the membranes performance [6]. Besides that, NOM can provide a suitable medium for microbial growth and biological fouling. Accordingly, pretreatment is regarded as an important and practical tool to control fouling, especially fouling occurred by organic matter. Adsorption is an effective and well-established pretreatment for eliminating soluble nonpolar organics, natural organics, hydrophobic organics, and dissolved organics.

According to the pretreatment hybridization procedure, the adsorption system is implemented in advance of the LPMs. In this configuration, the feed is conducted to the membrane system after passing through an adsorption contactor (AC). The ACs should provide appropriate conditions for an effective contact between adsorbents and the feed. There are several alternatives available for the selection of ACs including granular activated carbon (GAC) filters, biological GAC filters, fluidized adsorption beds (FAB) and PAC contactors [2, 4]. The ACs should be equipped with the system of adsorbent dosing and discharge. Residence time, in simple words the 'age' of an adsorbent in an AC, is a crucial parameter which determines the mode of operation, that is, adsorption or biological mode. In

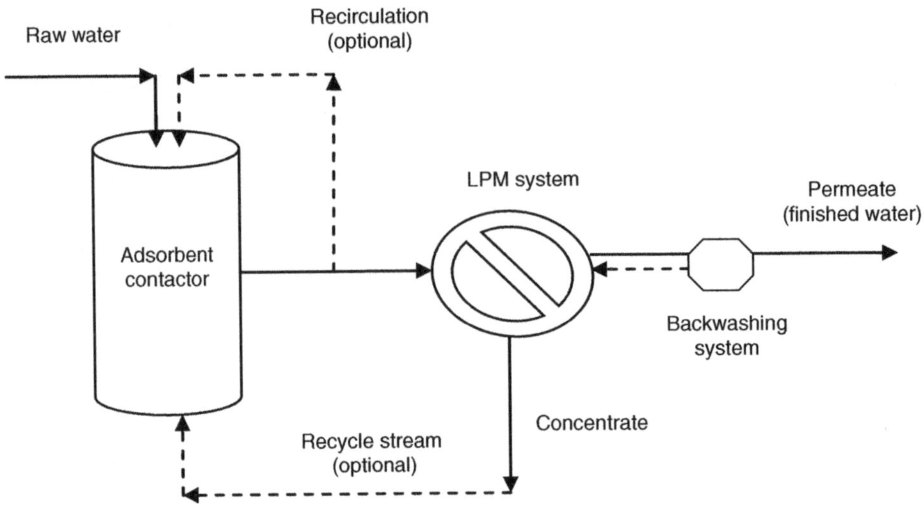

Figure 13.1 *Membrane/adsorption hybrid process with adsorption pretreatment.*

biological mode, biodegradation of contaminants by the microorganisms is the dominant removal mechanism rather than adsorption. Using pretreatment strategy makes it possible to conveniently control the adsorption system parameters including effective adsorbent age, operation mode, adsorbent dosing, and recycling activated sludge. A typical schematic of the pretreatment configuration is illustrated in Figure 13.1.

The pretreatment layout is conventionally applied for the surface water treatments. Raw water properties may be influenced by many factors such as water source, geographic region, rainfall, and season; however, the hybrid process should be designed for the worst possible conditions in which the feed contains maximum contaminants loading. Since LPMs are porous membranes, the control of membrane fouling is of greater importance to choose the best pretreatment strategy and optimize the hybrid process. Activated carbon adsorption, with GAC or PAC adsorbents, has been widely used as a pretreatment for reducing membrane fouling [4, 6, 7]. It is reported that, PAC adsorption is not capable of removing the lowest molecular weight fraction of organic matter with MW in the range 280–750 Da [8]. As a solution to the problem, adsorption is applied in combination with other supplementary processes such as coagulation or flocculation to improve the effectiveness of the pretreatment. Kweon et al. applied poly(aluminum) chloride (PACl) with concentration in the range 10–30 mg/L [9]. The coagulant dose was chosen in tune with the water turbidity. They added PAC prior to coagulant addition with concentration varied from 5 to 20 mg/L. Hollow fiber ultrafilters made of polyvinylidene fluoride (PVDF) were applied for subsequent filtration. They observed that the removal of DOMs was considerably increased with PAC addition from 38% to 60%. In another study, Shon et al. tested four different pretreatment protocols including; flocculation with ferric chloride, adsorption with PAC, flocculation followed by adsorption, and GAC biofiltration to compare their effectiveness for implementation prior to UF membranes [10]. They analyzed the surface of both clean and fouled membranes for water contact angle and obtained that hydrophilicity increased after filtration as a result of the attachment of hydrophilic organic compounds, polysaccharides, and microorganisms wastes. They observed that the hydrophilicity variation was not significant in the case of flocculation followed by adsorption pretreatment. Moreover, flux decline tabulated the lowest value for the adsorption/flocculation pretreatment scenario. Accordingly, this pretreatment was favored among the four tested protocols with the TOC removal more than 90%. It is worthy of mention that the pure flocculation pretreatment can remove nearly 80% of TOC in the best conditions. In another study, the researchers observed that pore blocking was significantly decreased after flocculation followed by adsorption pretreatment [8]. This means that fouling is better controlled using the proposed pretreatment.

The application of adsorption pretreatment is not limited to porous membranes. Gur-Reznik et al. utilized the GAC column as a pretreatment for the RO membranes applied for desalination of membrane bioreactor (MBR) effluents [6]. Adsorption experiments were conducted in both static (batch) and dynamic (continuous) modes. Experimental static adsorption curves showed a good fit to the Freundlich model indicating a relatively low adsorption capacity (0.84 mg DOC/g carbon). On the contrary, continuous adsorption in GAC columns resulted in 80–90% DOM removal. This pretreatment can provide favorable membrane permeability with an enhanced permeate quality. Dynamic adsorption provides better utilization of the external and internal active sites of adsorbents using the advantages of convection in the adsorption system, compared to slow diffusion in the static mode. In addition, additional mechanisms such as sieving and drag retardation are involved in removing DOM from the effluents [11].

Matsui et al. indicated that reducing the PAC particle size (super-powdered activated carbon, S-PAC) can enhance not only the removal of DOM from water but also the MF filtration performance when used as a pretreatment [12]. Moreover, the trans-membrane pressure (TMP) is reduced due to formation of more porous cake layer using S-PAC. PAC particles reflect both characteristics of adsorbing and flocculating agents in the super-powdered form. Interestingly, membrane fouling (both reversible and irreversible ones) was reduced causing relatively steady filtration using S-PAC adsorption pretreatment.

Some advantages of the pretreatment configuration are summarized as follows:

- The possibility of using both granular and powdered adsorbents
- The possibility of filtration in both pressurized or vacuum modes
- Indirect contact between adsorbents and membranes
- Better fouling control

Some observed disadvantages related to the pretreatment arrangement are:

- Large footprint (comparatively)
- Requirement of an intermediate separation step in the case of high adsorbent concentration in the feed
- More complicated controlling systems

13.3 Integrated Membrane-Adsorption Systems

In the integrated membrane-adsorption arrangement, fine adsorbents, like PAC, are directly introduced into the filtration tank. In the integrated configuration, due to the direct contact

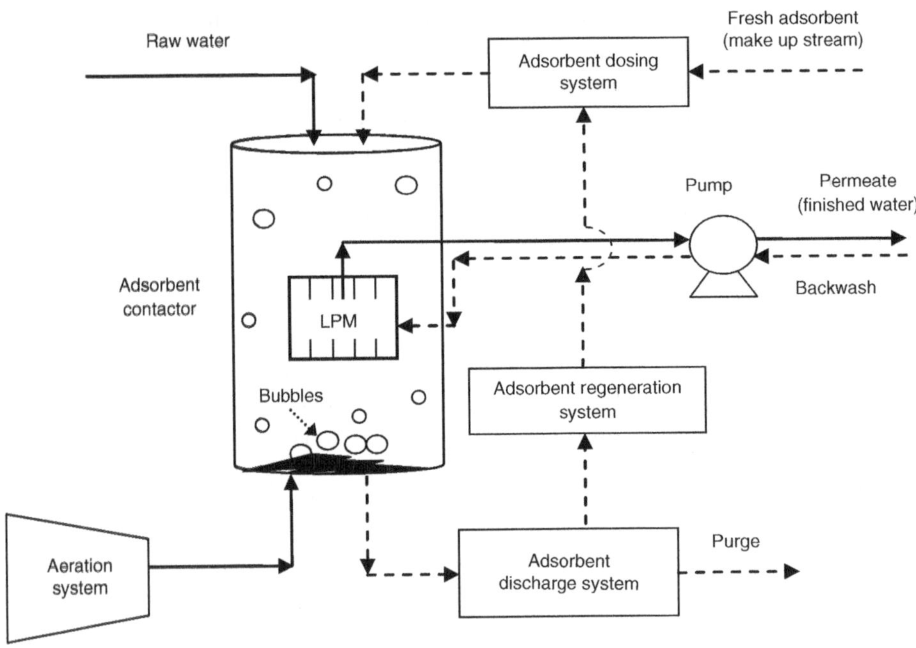

Figure 13.2 Integrated membrane/adsorption system (IMAS).

between the adsorbent and membrane, aeration is necessary to mitigate fouling, prevent adsorbent particle sedimentation, and meet better mixing conditions for adsorbent particles. Several mechanisms including concentration polarization, cake deposition, adsorption, and pore plugging may cause membrane fouling in adsorption-assisted membrane processes. The integrated system should be equipped with a system of adsorbent dosing and purging. Thus, the mode of operation, that is, adsorption or biological, would be determined depending on the adsorbent residence time (age) in the system. Both pressurized and submerged filtration modes can be selected in the integrated configuration. A schematic of the integrated membrane-adsorption system (IMAS) is depicted in Figure 13.2.

IMAS can be roughly categorized into three groups: LPM-adsorption integration, MABR, and membrane adsorbents. We will go through these integrations in the following sections.

13.3.1 LPM-Adsorption Integration

Both adsorption processes and LPMs are not favorable for elimination of low molecular weight NOM and anions like perchlorate and boron from water. Biological processes are not also cost-effective for the treatment of effluents with low anion concentrations [13]. In addition, disposal of the inserted nutrients and pathogens is another concern for the application of biological treatments. Therefore, finding an effective and dependable technology for the removal of dissolved organics and anions from water is of growing interest. The application of LPMs, especially MF and UF, in the presence of adsorptive compounds is one of the promising alternatives for water treatment, especially effective for the removal of anions and NOM. LPM-adsorption integrations take the mutual advantages of both adsorption and membrane separation techniques. The principle of LPM-adsorption integration is that the adsorbent, usually in fine or powdered grades like PAC, is introduced into the feed solution to adsorb the target contaminant [2]. Thereafter, the complexes of adsorbent contaminants are readily removed by LPM filtration. It is worthy of mention that in the integrated approach, both adsorption and filtration processes take place in a single united system.

13.3.1.1 LPM-Adsorption Operating Conditions

For anion adsorption, the solution pH should be adjusted to less than pH_{zpc} (pH at which the adsorbent surface has a zero net charge). At this pH, the net surface charge of adsorbent is positive due to protonation of active functional groups and thus, the adsorption of anions is highly favored in such condition [14]. Another important factor, which should be optimized for the membrane/adsorption combined systems, is the selection of membranes with suitable pore size or MWCO. Moreover, permeate flux is increased with increasing the membrane pore size; however, irreversible adsorptive fouling is more severe in LPMs with larger pore size. Adsorbent dose is also an important factor affecting hybrid process performance. Although adsorptive removal is enhanced with increasing the adsorbent concentration, permeate flux is declined sharply as a result of much severe fouling at higher adsorbent dosages. An optimum adsorbent dose is the concentration at which there is no significant improvement in the removal efficiency when the sorbent dosage is further increased. It is found that application of adsorbents in the form of very fine particles improves the separation efficiency of the integrated systems, that is, higher adsorption capacity, faster kinetics and enhanced rate of the separation [15]. TMP is another item which would be determined. Generally, at low TMP, flux is independent from adsorbent concentrations and remains rather steady during filtration. On the contrary, at high pressures, the cake layer resistance at the membrane surface is significant and results in a sharp flux decline [16]. An acceptable membrane performance needs an optimized backwashing procedure and selection of an adsorbent with a narrow particle size distribution and an average particle size adequately larger than the membrane pore size [17, 18]. Activated carbon derivatives, reactive and functional polymers, ion-exchange resins, metal oxides, and adsorptive biomaterials are commonly used adsorbents in the integrated adsorption-membrane systems [14–20].

13.3.1.2 Applications of LPM-Adsorption Integration

The integrated LPM-adsorption processes have been challenged for the removal of various contaminants such as DOMs, perchlorate, boron, phenol, proteins, and heavy metals in the laboratory scale. Some examples of these applications will be addressed in this section.

Higher than 90% of perchlorate was eliminated from water in the absence of other anions, using poly (diallyldimethylammonium) chloride as adsorbent and UF process, as reported by Huq et al. [19]. In the presence of other anions like sulfate and nitrate, higher concentrations of adsorbent are required to obtain the same perchlorate retention. This is due to the competition aroused in the adsorption from the mixture of adsorbates. In other words, anions compete with each other for the vacant adsorption sites. Thus, compared to adsorption from single solutions, more adsorbent doses are required to meet the anion elimination standards for a real effluent which contains a wide range of adsorbates. Xie et al. [20] studied the capability of the UF/adsorbent process using powdered chitosan (a natural polymer biosorbent) and cellulose-based UF membranes for perchlorate removal from water. The adsorption kinetic reached saturation around 3 minutes with the optimum pH adjusted at 4.3 ($<pH_{zpc} = 6.3$) for obtaining the maximum anion removal. They showed that the chitosan-perchlorate complexes did not cause irreversible fouling and the surface cake formed on the membrane can be readily removed by soaking in dilute NaOH followed by water washing.

Boron elimination from water is another challenging separation issue. Several technologies have been employed for boron removal from water such as coagulation, activated carbon adsorption, RO filtration, and ion-specific resins [21]. The application of these technologies is often limited due to some disadvantages such as production of secondary sewages, high energy consumption, high capital and operating costs, and lack of sufficient removal efficiency. Integrated membrane/adsorption processes have emerged as powerful and effective methods for boron removal. Kabay et al. [21] tested various methods including resin adsorption, RO filtration, and integrated resin-adsorbent/MF for boron removal from water. The study confirmed that the IMAS is the most effective technique for water deboronation. There are some concerns about adsorbent regeneration/recycling and waste management; however, application of ion-exchange resins and RO are faced with similar concerns. Hence, it seems that the hybrid system is the best alternative for boron elimination.

Yilmaz et al. [22] applied *N*-methyl glucamine chelating resins for boron sorption followed by filtration using hollow-fiber polypropylene MF membranes with 0.4 μm pores. Alkaline pH around 10 was favored for boron adsorption. As reported in the literature [21, 23], the pH enhancement enables ionization of boric acid and enhances its rejection by membranes. Static adsorption tests were conducted to study the sorption kinetic. A blank test (with distilled water as feed), was conducted to show the effect of adsorption on boron rejection. Boron concentration in permeate crossed a minimum at the filtration commencement, then, increased and switched to a steady amount. At the beginning of the process, adsorption was the dominant and effective removal mechanism causing maximum boron elimination by the integrated process. Subsequently, the process was mainly controlled by the rate of suspension replacement. The steady-state condition was reached when the adsorption capacity was equilibrated and the mass transfer resistances against membrane transport were fully developed. Guler et al. [24] applied fine powder boron-selective ion exchange resins (0.5 mg/L) combined with a submerged hollow-fiber module including polypropylene membranes for boron elimination from seawater. The process was interestingly controlled by permeate flux balance. When permeate flow rate was increased to 15 mL/min, the boron level in permeate started to increase and exceeded the permissible standards. When the flow rate was decreased to 10 mL/min, the contact time of feed solution with the adsorbent resins was increased. Therefore, boron rejection was enhanced and met the required standards at the flow rate of 10 mL/min. This indicates that contact time has a crucial impact on the sorption-membrane integrated systems when flux and rejection are simultaneously optimized.

Ipek et al. [25] developed a combined adsorption/membrane filtration process for the phenol removal from water. Purolite MN 200 and Purolite MN 202 hyper-crosslinked polymer adsorbents were utilized in combination to UF process. Purolite MN 200 was found to be more effective adsorbent for phenol with the optimum concentration of 3 g/L. The optimum permeate flow rate was obtained as 3 mL/min. This factor is important to control the contact time between phenol in the solution and the adsorbents.

LPM-adsorption hybrid processes are also powerful candidates for RO-desalination pretreatment. Tansakul et al. [26] applied PAC adsorption/UF integrated process for removal of DOM from seawater before conducting to RO membranes. Both dead-end and cross-flow filtration modes were tested with the latter providing better performance for the integrated process. This may be ascribed to the decrease of the concentration polarization by appearance of shear stress near the membrane surface in cross-flow mode. The DOM removal was increased from 5% without PAC to near 70% when wood-based PAC was added. Adsorption tests with different PACs showed that PAC with lower BET surface area and lower volumes of micro- and mesopores provides better condition for DOM removal. Regardless of the PAC types, PAC addition caused fouling reduction. In addition, fouling was better controlled by increasing PAC concentration in the feed. These results indicate that PAC adsorption can mitigate membrane fouling by preventing organic matters from adsorbing onto the membrane surface. It was found that for short-term PAC-adsorption/UF experiments, fouling is mostly reversible in nature.

Saravia et al. [18] examined the role of NOM on membrane performance and on removal of pharmaceuticals during the integrated PAC-adsorption/submerged membrane filtration. They indicated that NOM agglomerates PAC particles and compels them to precipitate on the surface of the membrane as a part of the foulants. Experiments in the absence of NOM showed higher removal rate of pharmaceuticals (95–100%) after 30-minute filtration with a negligible flux decline. Hence, interactions between NOM and PAC particles have an obvious effect on the fouling and flux decline. Mozia et al. [27] tested three different UF membranes, prepared from polysulfone (PSF), cellulose acetate (CA) and polyacrylonitrile (PAN), in an integrated UF/PAC-adsorption system for surface water treatment. The objective was to evaluate the effect of humic acid on membrane fouling during UF. The PAC suspended in the membrane system can adsorb low molecular weight organic compounds, which cannot be rejected by a single UF system. The UF membrane acts as a barrier against the passage of the PAC–NOM complexes. The researchers observed a significant reduction in the permeate flux during UF of HA solutions through CA and PSF membranes mainly due to the membranes surface chemistry. Accordingly, PAN membrane was favored to apply in the integrated system, regarding both NOM removal and optimum flux during PAC/UF integrated process.

Katsou et al. [28] applied a sorption-assisted UF system for the reduction of heavy metals from industrial wastewater. Zeolite, bentonite, and vermiculite low-cost minerals were used as adsorbents. The optimum pH for the heavy metal removal was obtained as 9, at which precipitation of metal hydroxides, formation of precipitates/complexes among the metal ions and the wastewater components and adsorption of metals onto the mineral sorbents cooperate with UF to treat the wastewater.

Gosh et al. [29] combined salt-induced precipitation with MF using PVDF membranes in a single unit for human IgG purification and obtained protein purity of 85%. They developed their bioseparation technique using caprylic acid as more effective precipitant to achieve higher protein purity of 95%. Two approaches were employed for hybridization: using caprylic acid together with salt in the membrane cell (integrated mode) and using caprylic acid precipitation as secondary step after the bioseparation process. An increase in TMP was reported by introducing caprylic acid into the membrane module. The results indicated that the salt–caprylic acid complex causes higher fouling tendency compared to the pure salt-induced precipitate. Bioseparation followed by caprylic acid precipitation was shown to be a more effective separation strategy providing purer IgG with inferior membrane fouling.

13.3.1.3 Advantages and Drawbacks of LPM-Adsorption integration

Introduction of fine adsorbents with their beneficial properties on membrane-adsorption integrated systems includes many advantages compared to the conventional fixed-bed and column adsorption systems. The main advantage of this integration is high removal efficiency and low operating cost compared to the classical adsorption operations. Fine adsorbent particles can be used which increases the effective interface area for adsorption. As a result, the adsorption kinetic and removal rate of the integrated process are enhanced. The major problem for application of the fine particles in adsorption columns is the high pressure drops. This problem is significantly solved in LPM–adsorption integrated processes because the thickness of the layer is much lower even when a cake of the adsorbent is built-up on the membrane surface. In addition, aeration helps to mitigate cake layer development on the surface. In integrated systems, the pressure drop is independent of particle size and the membrane dominantly controls the pressure drop. The thickness of the membrane and the cake formed on it, are much lower than the height of an adsorption column. Therefore, pressure drop and energy loss are substantially smaller for the membrane case [25]. At the vicinity of the membrane surface, the adsorbent may be rapidly saturated. Hence, the sorbent layer should be renewed regularly. Some effective methods of the layer renewal reported in the literature are: pulsatile flow and turbulence promoters, back-pulsing, and back-flushing [16]. Synergetic effects which may appear on the membrane surface can cause intensification of the process. Smaller footprints are needed in the case of integrated process compared to the other membrane/adsorption hybridization layouts.

The membrane-adsorption integrated system for the separation of soluble components from water is effective on the basis of the process yield and separation efficiency. Unfortunately, there is no straightforward connection between integrated process parameters and separation performance [16, 24]. As the majority of the process parameters suggest complicated and even opposite effects on flux and retention, process optimization is in most cases, a tailor-made method depending on the lab-scale experiments. Some drawbacks of LPM-adsorption integrated processes are summarized as follows:

- Hardship of process control to mention the integrity objectives
- Lower flux compared to other hybridization strategies
- Hardship of fouling control due to direct contact between membrane and adsorbents
- Requirement of aeration.

13.3.2 Membrane-Adsorption Bioreactors

Recently, MBRs have attracted researchers' attention worldwide due to their advantages such as low energy consumption and reliable effectiveness in water and wastewater treatment. MBR comprises of an activated sludge bioreactor and a membrane filtration system. The integration of biological degradation and membrane filtration has resulted in a powerful separation technology with a great capacity in removal of POM, biodegradable organic matter (BOM) and bacterial/pathogen community and rather acceptable potential in elimination of colloidal and DOM [30,31]. The LPM (MF or UF) can be either installed outside of the reactor (external loop) or submerged in the reactor tank. The latter, so-called SMBR, provides better conditions for reducing power consumption and membrane fouling when applied in pulsatile suction/backwash mode. The application of hollow-fiber membranes in MBR systems has gained increasing interest in recent years due to the reduction of power depletion and higher productivity. Some more advantages of MBR systems are small footprint, excellent permeate quality, high capacity for water disinfection, favorable productivity, and low sludge discharge. Accordingly, application of MBR has been advancing in replacement of conventional water and wastewater treatment technologies.

The most important limitation in the way of wide application of MBR is membrane fouling [30–32]. Fouling affects the performance of the MBR systems as a result of deposition of a cake layer on the membrane surface and/or pore blockage [30]. Fouling leads to decline of permeate flux and elevation of TMP. Cake layer formation is reported to have a major effect on MBR fouling. The mixed liquor suspended solids (MLSS) are mainly responsible for cake formation on the membranes. The mixed liquor in the bioreactor includes two main fractions including bioproducts generated as a result of the microorganisms' metabolism and colloidal/soluble compounds. Many attempts have been made to control membrane fouling in the MBR systems. Periodic suction, running under critical flux, optimizing aeration rate, selection of proper membrane type and material, adjusting solution chemistry, and addition of fouling reducers have been verified as effective methods to control fouling in MBR systems [31–33].

One of the potential solutions for fouling mitigation as well as removal efficiency enhancement is using adsorbents in the MBR system. Application of adsorbents in MBR systems results in a novel (advanced) class of MBRs called the MABR. As reported in the literature, this integration not only reduces membrane fouling but also allows taking the advantages of adsorption mechanism in combination to the ability of MBR in biodegradation. Activated carbons are well-established adsorbents for removing DOM from water for their great affinity toward medium molecular weight hydrophobic organic compounds [6]. Besides in wastewater treatment, MABRs are potential candidate in drinking water treatment owing to their capacity for removing low-molecular and non-biodegradable organic matters considering much lower hydraulic retention time (HRT) as compared to that for wastewater treatment [34]. The wastewater treated by the MABR system usually meets the stringent standards of water and wastewater treatments. Therefore, addition of adsorbents like PAC, in the MBR system can bring about couple of advantages, that is, improvement of the efficiency of organic matter removal and reduction of membrane fouling. Hydrophilic BOM can support unfavorable microbial re-growth in the MBR, resulting in a decrease in the finished water quality. Adsorbents, individually, may not be effective for the elimination of hydrophilic BOM; however, they can provide suitable substrate by developing a biofilm on their surfaces for the attachment of microorganisms and facilitate biodegradation mechanism. Moreover, the biofilm developed on the adsorbent surface can degrade organic matter more effectively and with less biomass production than suspended biomass in a conventional MBR [35]. It is also reported that the respiration rate of the microorganisms is enhanced by addition of adsorbents like PAC into the bioreactors [36]. It means that the biodegradation mechanism can be accelerated by adsorbent addition, that is, adsorption kinetic is faster than the inhibition kinetic. Fouling reduction in the MBR systems including adsorbents may be ascribed to two main factors: the first factor is the reduction of organic matter concentration in the supernatant phase and the second one is an increase in the porosity of the biofloc formed on the membrane surface. Moreover, a dynamic secondary layer with a porous structure (due to comprising adsorbent particles) is formed on the membranes. The secondary layer helps the membrane to better retain low molecular weight organics accompanied by providing adequate permeability [36, 37].

13.3.3 MABR Operating Conditions

To run MBR in the adsorption mode, first the bioreactor must be filled with sludge and acclimatized to the wastewater to support adequate microbial growth for awhile. Afterwards, a predetermined amount of adsorbent is introduced into the reactor to adsorb the organic substances. Usually, aeration is a necessity for the MBR and MABR systems. The aeration (bubbling) has three important functions in the MABR system: (1) preventing cake layer formation on the membrane surface, (2) mixing adsorbent in the reactor, and (3) supplying required oxygen for biological activities [38]. Aeration rate may vary from 8 to 20 L/min according to MBR characteristics such as size and contaminant loading. In some reports, an increase in removal efficiency with no TMP development was reported when aeration rate was increased [36]. Respirometry, is an important method for analyzing microbial growth and biological activity in bioreactor systems [30]. Greater oxygen uptake rate (OUR) means better biodegradation kinetic. The HRT is one of the most important parameters during MABR system design. Influent and permeate flow rates should be controlled to achieve a desired HRT at which adequate contact time is allowed for the proper functions of adsorption and biodegradation mechanisms. Level sensors are commonly applied to keep the reactor volume constant. In the submerged strategy, constant flow rate is obtained by suction through the membrane module. Irreversible fouling mainly caused by BOM is of greater significance in decreased HRTs [39]. Adsorbent residence time or simply adsorbent age is another crucial parameter which should be adjusted for an MABR system. It is reported that for the MABR with adsorbent age exceeding 20–30 days, the reactor mainly acts in biological than in adsorption mode and vice versa [36, 39]. Adsorption capacity increases with reducing adsorbent particle sizes due to higher accessible surface on finer particles [40]. Thus, the application of powdered (or fine grades of) adsorbents is more prevalent in the MABR. The MABR showed large dependencies on adsorbent particle size and particle size distribution. Narrow particle size distribution and an average particle size near 100 times greater than the membrane pore size helps to minimize irreversible fouling and pore blocking [41]. Adsorbent concentration is an important variable for process optimization. There should be a balance between adsorbent cost and required effectiveness in the bioreactor. As an example, PAC dose may vary from 0 to 50 g/L in MABR systems [37–39]. Minimum fouling and maximum removal efficiency should be simultaneously considered at the optimum adsorbent concentration. No considerable increase in organic removal was observed in short-term experiments when PAC concentration in the bioreactor increased from 2 to 10 g/L; however, TMP development was significantly reduced at higher PAC doses indicating a successful membrane clogging prevention [42]. Adsorbent dosing/discharge is another important factor affecting MABR operating performances. Adsorbents can be initially or continuously dosed into the bioreactor. For the initially dosed mode, the adsorbent is introduced into the tank at the process commencement. In the continuous dosing mode, the adsorbent is gradually dosed into the bioreactor during the filtration process. It is noted that in the MABR system with an initially added PAC adsorbent, an optimum daily replacement of 2.5% of the total PAC amount satisfies the requirements of normal adsorption condition and economic considerations [42]. All the MABR systems are necessary to be equipped with a backwash system, providing a short period of backwash after a longer period of filtration. Fouled membranes may be backwashed with the stored filtrate or fresh water when preferred [39]. By increasing the period between backwashing cycles, the efficiency of the membrane flux recovery is decreased due to the compactness of the cake layer formed on the membrane. Irreversible fouling becomes more highlighted in this condition causing significant flux reduction in the MABR system. Effect of pre-adsorption (running in adsorption mode before start of filtration) in the MABR was studied by researchers [42]. The TOC removal efficiency was slightly increased with lower TMP increase during filtration when running in optimized pre-adsorption conditions. Furthermore, pre-adsorption decreases the amount of adsorption sites on the PAC for organic removal in the combined stage.

13.3.4 MABR Applications

Guo et al. [38] compared the performance of the submerged MABR with the MBR system in treating a synthetic secondary wastewater and obtained interesting results. Polyethylene hollow fiber membranes with a pore size of 0.1 μm were applied in the experiments. The MABR offered somewhat higher organic removal as compared to MBR due to bilateral role of PAC, that is, adsorption and enhanced (facilitated) biodegradation by providing a proper media on its surface for the attached microorganisms. The TMP development (for long-time operation) in the MABR was lower than that in the MBR system. This may be related to the adsorption of dissolved organics on PAC, caused fouling mitigation and flux improvement. Critical flux experiments also confirmed this conclusion. Moreover, both reactors showed similar critical flux (~ 20 L/m^2 h) but, the MABR suggested lower TMP value to operate at the steady flux. Better oxygen uptake performance was observed in the MABR compared to the MBR indicating much better condition for the function of microbial substrate in the MABR system.

Tian et al. [43] observed that by addition of PAC into an MBR system, much higher removal efficiency for organic matter was obtained compared to the pure MBR which was operated in parallel. In addition, the TMP in the MABR system developed lower than that in MBR due to the impact of PAC in removing NOM and thus, mitigating membrane fouling. In another study, Tian et al. [37] integrated the MABR with coagulation to introduce an MCABR process for the treatment of surface water. Polyaluminum chloride (10 mg/L) and PAC (8 mg/L) were utilized as the coagulant and adsorbent, respectively. In this integrative process, microorganisms biodegrades BOM; PAC adsorbs organic matter with medium molecular weight and provides an excess surface for better biodegradation; coagulant flocculates large MW organics; and subsequently, the membrane together with the secondary layer separate solids (mixed sludge) from water. The high quality of the finished water is ascribed to the synergetic effects of the processes involved and not only straightforward superposition of their individual effects. Moreover, the MCABR removal efficiency cannot be recorded when the individual processes were separately rearranged. They confirmed their conclusion by comparing different combinations of UF, MBR, adsorption and coagulation processes and obtained the following order for the efficiency of the processes for DOC and UV_{254} removal:

MCABR (~70%) > MCBR (~50%) > MABR (~46%) > MBR (~7%) > UF (~11%)

It seems that the integration of the four processes is the most effective method for the water treatment.

Effects of HRT and extracellular polymeric substances (EPS), generally produced by microorganisms, on the performance of a submersed hollow fiber MF membrane in an integrated PAC–MF MBR were studied by Taimur Khan et al. [39]. Addition of PAC could enhance the biodegradation of polysaccharides and proteins up to 97% and more than 95%, respectively. Accumulation of carbohydrates on the membrane surface resulted in a significant increase in TMP. Hence, PAC is not a good adsorbent for carbohydrates. The decrease of HRT increased the backwashing frequency and irreversible fouling effects caused by BOM on the membranes. Khirani et al. [40] tested several resins to find a potential surrogate for PAC with better regeneration and economic properties for application in the MABR system. Among the tested resins, Purolite A500P offered NOM-sorption capacity and removal kinetic compared to those of PAC. Kim et al. [44] applied heated aluminum oxide particles (HAOP) as adsorbents in the integrated adsorption/membrane system. The results showed that effective adsorption of NOM by this synthetic adsorbent can reduce fouling by preventing the NOM from reaching the membrane surface. Increasing the synthetic adsorbent dosage resulted in formation of a thicker layer on the membrane surface which acted as a short and effective packed bed in removing NOM and subsequently, prolonged filtration without requirement of periodic backwashing. Lesage et al. [36] also compared the MBR and MABR systems in terms of removal efficiency, membrane fouling, and sludge production in removing a model toxic compound (dimethylphenol) present in the wastewater produced by oil refinery plants. They observed that application of PAC as adsorbent in the MBR system could reduce membrane fouling and daily sludge production, mainly due to the reduction of proteins and carbohydrates in the MBR supernatant phase. The COD removal was also slightly enhanced from 94% to 96% with the help of the adsorbent in the MBR.

13.4 Membrane Adsorbents

Adsorbent particles can be incorporated into the membrane structure by blending or surface coating methods to prepare an integrated membrane adsorbent. Membrane adsorption is an effective technique for adsorption of heavy metal ions and proteins from aqueous phase [45–47]. Membrane adsorbents show unique advantages such as low pressure drop, good removal efficiency, high productivity, rapid kinetic, low cost and energy consumption, regeneration ability, and small footprint for the adsorption of heavy metals and proteins over conventional separation processes [47,48].

Membrane adsorbents include abundant reactive functionalities, attached to the external surface or pore walls, including –COOH, $-SO_3H$, $-NH_2$, –OH, etc., or the combination of several ones by which they can bond with adsorbates through different mechanisms such as surface chelation, chemical bond formation, ion exchange complexation and H-bonding. Accordingly, adsorbates may be removed from the aqueous solution when they are in contact with the membrane adsorbent, despite the fact that the membrane pore size may be much larger than the dimensions of the adsorbates [49]. In comparison to LPMs, with the application of membrane adsorbents, the function of the membrane process is developed from pure filtration to adsorption-assisted filtration or conjugated adsorption/filtration.

Membrane adsorbents are potentially applied in either batch or dynamic adsorption modes. Batch (static) adsorption studies are necessary for the characterization of membrane adsorption process. Prevalent practical separation processes are dynamic due to its obvious advantages. Beside similar separation mechanism in both operational modes, the rate of separation in dynamic mode is much faster than that in batch condition. This is due to the fact that the convective flow in the dynamic systems can enhance and facilitate the access of solutes to the membrane active sites, compared to slow diffusion mechanisms in static mode [48, 49]. With respect to their applications, affinity membranes can be basically classified into four major classes including 'protein-adsorbent membranes,' 'metal-adsorbent membranes,' and recently developed 'ion/molecule imprinted' and 'thin-adsorptive' membranes. These four classes of membrane adsorbents will be reviewed in the following sections.

13.4.1 Protein-Adsorbent Membranes

Protein-affinity membranes are considered as a potential alternative to adsorption and chromatography columns for the separation and purification of proteins [50]. For preparation of protein-affinity membranes, reactive dendrites such as dye ligands and chelating agents are immobilized on the surface of membrane supports commonly via a three-step protocol: (i) support activation, (ii) support coating with an intermediate polymer adsorbent, and (iii) final activation of the resultant composite membrane [49, 50]. Porous membranes are prevalently used as support for their advantages such as high surface area, low mass transport resistances, and good permeability. Mass transfer resistance is significantly reduced in the dynamic adsorption mode because of the fact that adsorption is mainly conducted by the fast chelating kinetics rather than slow diffusion mechanisms. In such conditions, rapid processing of adsorption, washing, desorption and regeneration would be obtained.

Adsorption of pre-purified human serum albumin (HSA) onto Cibacron Blue F3GA (CB F3GA) incorporated nylon membranes was studied by Weissenborn et al. [51]. Dextran, hydroxyl-ethyl-cellulose and polyvinyl-alcohol were covalently coated on the activated nylon membranes to enhance the specific adsorption of the protein. Although the membrane permeability was reduced to some extent due to polymer coating resistances, covalent attachment of hydrophilic polymers onto the membrane could reduce non-specific protein adsorption. Champluvier and Kula [52] immobilized CB F3GA and some other dye ligands onto nylon 66 isotropic membranes with or without using a spacer for adsorption of albumin and lysozyme proteins. The amount of dyes immobilized was in the range of 0.56–6.65 mg/g membrane. The membranes were tested in both batch and dynamic adsorption experiments. Membrane capacity values for adsorption of the proteins were in the range of 40–120 mg/cm^2. Guo et al. [53] investigated alkaline phosphatase adsorption using a membrane affinity chromatography system in which, CB F3GA and Active Red K2BP reactive dye ligands were attached on the chemically crosslinked cellulose supports. A cartridge containing 80 sheets of adsorptive membrane was applied for dynamic adsorption tests. A protein recovery up to 60% was achieved using the affinity membrane containing active red dye.

Polyethyleneimine (PEI)-coated titania [54], poly(2-hydroxyethylmethacrylate) [55], and poly(vinyl alcohol)-coated poly(propylene) [56], cellulose and polysulfone [57] have also been used as modified membrane supports for the immobilization of CB F3GA for the separation of proteins and enzymes including albumin, HSA catalase, and *Escherichia coli* homogenate [58]. Champluvier and Kula [59] introduced an adsorptive membrane for the selective binding of enzymes, in the form of microporous membranes carrying functional groups. Suen and Tsai [60] also studied commercially available polyvinylidene Millipore membrane as the solid support for a plate-and-frame adsorptive filter. They also applied Cibacron Blue F3GA as a chelating ligand for lysozyme protein adsorption. Kassab et al. [61] stabilized CB F3GA on microporous polyamide hollow fiber membranes for HSA adsorption from plasma and reported extremely high protein adsorption. They also verified that modification of the affinity membranes with metal incorporation could significantly improve protein adsorption. Nie and Zhu [62] prepared an active microporous nylon support by coating chitosan followed by immobilizing CB F3GA on the CS-coated membranes. The researchers observed papain adsorption around 235.3 mg/g using the protein-adsorbent membrane.

Different ligands such as CB F3GA, Reactive Blue 4, Reactive Red 120, Reactive Yellow 2, Procion Brown MX 5BR, Reactive Green 19, Reactive Green 5, and Reactive Brown 10 have been used for the preparation of protein-adsorbent membranes [54–62]; however, CB F3GA is the most frequently used ligand for this purpose.

13.4.2 Metal-Adsorbent Membranes

As heavy metal contamination is a serious concern for human beings and the environment, the standards for heavy-metal-contaminated water and wastewater have become more stringent. For example, the maximum permissible content of copper, cadmium, and nickel in industrial effluents are 1.3, 1.0, and 3.0 mg/dm^{-3}, respectively [63–65]. Conventional processes for heavy metal removal from aqueous effluents are chemical precipitation, ion exchange, electrodialysis, solvent extraction, adsorption/precipitation, and membrane separation. These methods have some drawbacks such as incapability of meeting the required standards, high capital and operational cost, and generation of secondary wastes, such as the large amount of sludge in precipitation processes [65, 66]. Membrane adsorbents have attracted great attention for removing heavy metal ions from aqueous solutions. These membranes contain reactive functional sites, which can adsorb metal ions by forming surface complex or via ion exchange mechanisms [67].

Commercially available membrane adsorbents are among the highest preferences due to no requirement of any synthesis and modification; they are dependable on account of the permanently fixed properties. Nafion 117 is one of the commercially available adsorptive membranes, employed by Nasef et al. [68], for adsorption of a wide range of heavy metals including Ni(II), Co(II), Pb(II), Cu(II), and Ag(I) metal ions from aqueous solutions. The adsorption capacity of Nafion 117 for metal ion adsorption was reported to increase in the sequence of Cu(II), Ni(II), Co(II), Pb(II), and Ag(I) with a favorable kinetic around 30 minutes for all the ions. Different affinity toward different heavy metals may originate from the differences in charge density as well as hydration intensity of the ions in aqueous solution.

Application of inorganic adsorbents like zirconia in polymeric matrices can result in good metal-adsorbent membranes. PVDF/zirconia blend membrane with good capacity

(21.5 mg/g) for adsorption of arsenate from water was prepared by Zheng et al. [69]. Liu and Bai [70] prepared a porous hollow fiber membrane from chitosan (CS) and cellulose acetate (CA) for the removal of copper ions from aqueous solutions in batch adsorption mode. The membranes offered adsorption capacity around 48 mg/g with a fast adsorption rate and short equilibrium time (<20 min) for Cu(II) ions. The membranes indicated higher affinity toward copper by increasing chitosan content in the blend. Compared to film adsorbents, application of hollow fiber membrane adsorbents is rather restricted due to preparation hardships and operational difficulties. An interesting comparison between membranes and spheres of chitosan as adsorbents for Hg(II) ions has been performed by Vieira et al. [71]. Chitosan membranes exhibited higher capacity for Hg(II) adsorption and faster kinetic compared with chitosan spheres mainly for their higher surface area. In another work, the effect of membrane morphology on Cu(II) adsorption was investigated by Ghaee et al. [47]. Silica particles were utilized as pore-generator to fabricate macroporous chitosan membranes. Copper adsorption was enhanced by increasing chitosan content in the membranes. Increasing the surface roughness and membrane porosity by addition of silica to the chitosan blend resulted in greater copper adsorption. Surface area enhancement and porosity development enables more reactive groups for ion adsorption.

Chitosan is also frequently used as intermediate adsorptive layer in preparation of composite membrane adsorbers. Composite adsorptive membranes are usually applied in dynamic adsorption mode for their excellent chemical and mechanical stabilities. An activated nylon membrane was prepared by He et al. [72] for the removal of Cu^{2+} from aqueous phase. They prepared the composite membranes based on a three-stage protocol as: (a) coating a layer of chitosan on the nylon membrane, (b) stabilizing the polymer layer with crosslinking by epichlorohydrin, and (c) grafting and activation with iminodiacetic acid. The membranes were examined in dynamic adsorption experiments at various pH values, contact times, adsorption amounts, and initial metal concentrations to determine the optimum condition for membrane adsorption. A static adsorption experiment was also conducted and copper adsorption capacity of 10.794 mg/g, was obtained.

The performance of composite membrane adsorbents is usually determined based on their microporous or macroporous structure, the availability of reactive sites for metal adsorption, and physiochemical stability under harsh operating conditions [73]. Membranes in comparison to microspheres or beads are better adsorbents because they are not compressible and present low internal diffusion limitations. Pore-generators (porogens) are materials with the ability of generating membrane pores. Porogens can play couple of important roles in membrane adsorbents: internal adsorption capacity (pore capacity) enlargement and mass transport enhancement [74]. Common pore-generators for metal-adsorbent membranes are formic acid, polyethylene glycol (PEG), polyvinylpyrrolidone (PVP), silica, and inorganic salts (e.g., NaCl) [74–76]. The porogen content in the membrane adsorbent must be optimized to take the advantages of large adsorption capacity as well as proper mechanical stability. Moreover, high porogen contents results in very loose (fragile) structures with an inadequate mechanical properties. Macroporous chitosan membranes functionalized with histidine was synthesized by Beppu et al. [46] for copper ions adsorption. Compared to non-functionalized membranes, the membranes functionalized with histidine exhibited higher porosity and larger capacity for ion adsorption, that is, synergetic effect of functionalization and porosity enhancement. Highly porous CS/CA hollow fiber membrane adsorbents were prepared by Liu et al. [70] using formic acid as pore generators. They found that the membranes soaked in water as the coagulant exhibited lower surface areas than those soaked in the NaOH solution; however, the two different coagulants did not show significant effect on the porosity of the membranes. Compatibility of porogen with the polymer background is another factor affecting the porous structure of membrane adsorbents. Zeng et al. [77] compared PVP and PEG as pore generator for chitosan polymer. No pore structure was generated in the case of PVP addition because of high miscibility and strong interactions between chitosan and PVP polymers. On the contrary, a highly porous matrix was generated in the case of PEG insertion into the chitosan because of poor compatibility of the polymers and multiphase structure of the resultant blend. Accordingly, PEG is a better polymer porogen for chitosan-based membrane adsorbents.

13.4.3 Imprinted-Membrane Adsorbents

Recently, molecular imprinting technique has gained considerable attention for its high potential for increasing specific adsorption capacity of membrane adsorbents [78, 79]. This technique is used for producing imprints of template molecules or ions in polymeric membranes. Selective recognition/adsorption of adsorbates is resulted from the imprinting process. *In-situ* polymerization in the presence of the templates molecules/ions can bring about the incorporation of specific template-adsorptive sites into the polymer. In other words, the specific recognition potential is provided to specific adsorption of the host molecules by addition and subsequent extraction of template adsorbate. Extraction of template ions/molecules from the polymeric backbone leaves cavities with similar size, shape, and chemical functionality to the template. Enhanced affinity and sorption kinetic of imprinted solute on the imprinted membranes have been addressed by many researchers [78–83]. A schematic of imprinting stages is illustrated in Figure 13.3. Development of imprinted membranes can result in a powerful alternative for advanced separation and purification technologies.

Membranes adsorbents do not demonstrate sufficient selectivity in adsorption of contaminants from aqueous media. Ion-imprinting technique is a well-founded approach to develop selective membrane adsorbents. Artificial-receptor-like recognition sites, formerly contrived in the membrane structure can increase the capacity of the imprinted membrane

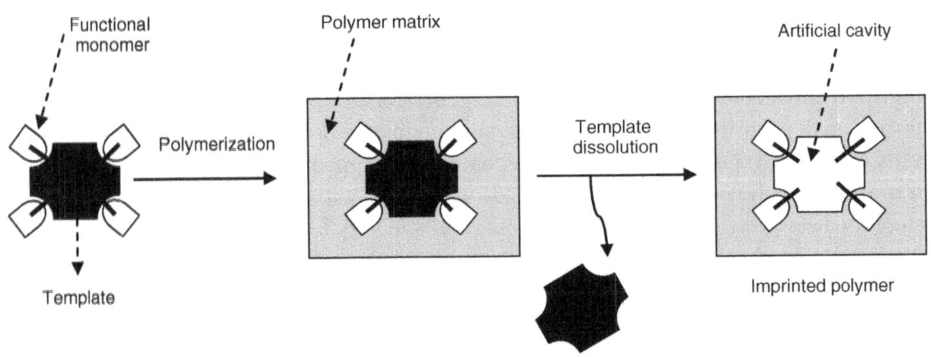

Figure 13.3 Imprinting process stages.

for a favorable specific adsorption. These fabricated reactive sites can memorize target adsorbate among other solutes available in the solution [78].

A Cr(III)-imprinted polyamine (IIP) adsorbent was prepared by An et al. [80]. PEI, a functional macromolecule, was grafted onto the surfaces of silica gel particles and formed the PEI/SiO_2 composite. Afterwards, ion-imprinting was performed using Cr(III) as template ions and epichlorohydrin as crosslinker. The adsorption capacity for the Cr(III) ion, increased almost twice as large as that of the pristine PEI/SiO_2 adsorbent. Relative selectivity of Cr^{3+} versus Zn^{2+} and Pb^{2+} were 24.63 and 59.32, respectively. Cr(III)-imprinted membrane adsorbents was prepared by Chen et al. [83]. They applied a blend of sodium alginate (SA) and polyvinyl alcohol (PVA) for the membrane polymeric background. PEG and glutaraldehyde were utilized as pore generator and crosslinker, respectively. Competitive adsorption from the binary systems of Cr(III)/Cu(II), Cr(III)/Cd(II), and the ternary system of Cr(III)/Cu(II)/Cd(II) were studied by the researchers using Cr(III)-imprinted membrane. The results suggested that the Cr(III)-imprinted membrane could have selectively adsorbed larger amounts of Cr(III) ions compared to Cu(II) and Cd(II) ions in the multi-component systems.

Su et al. [81] studied the adsorption of Ni^{2+}-imprinted ions with a novel surface imprinted adsorbent. Vatanpour et al. [82] developed a novel Ni(II)-imprinted membrane using commercial PVDF membrane as support, dithizone as chelating ligand, methacrylic acid as monomer, and ethylene glycol dimethacrylate (EGDMA) as crosslinker. Selective adsorption of Ni(II) compared to Co(II) ions was investigated in batch adsorption and dialysis permeation experiments. The results exhibited that the imprinting process supported better specification and adsorption of Ni(II) ions on the ion-imprinted membrane as compared to the non-imprinted polymer membrane. Blends of chitosan and PVA were utilized by Wang et al. [78] to fabricate a silver ion-imprinted membrane adsorbent. The results of competitive adsorption experiments indicated that Ag(I)-imprinted membrane has the ability of specificity and recognition to Ag(I). The Ag(I)-imprinted membrane offered high adsorption capacity for Ag(I) and can be used in silver enrichment as an effective adsorbent.

Sergeyeva et al. [84] synthesized a molecular-imprinted membrane via *in-situ* polymerization and employed it for solid-phase extraction of triazine herbicides. Triazine-specific membranes were prepared by the UV-initiated co-polymerization of methacrylic acid, tri(ethylene glycol) dimethacrylate, and oligourethane acrylate in the presence of the template molecule (triazine). They observed that the selective adsorption of triazine on the imprinted membrane was enhanced in solid-phase extraction process mainly, adsorption of triazine from 10^{-8} to 10^{-4} M aqueous solutions.

Imprinting technique combined to photo-degradation technology was applied by Li et al. [85] to prepare an imprinted adsorbent. They imprinted nanosized titanium dioxide in the chitosan polymer matrix. The adsorbent could adsorb organic compounds as well as heavy metal ions. The effects of chitosan concentration, TiO_2 addition, and imprinting various ions on the degradation of methyl orange and adsorption of Ni(II) ions were also examined by the researchers. Insertion of TiO_2 nanoparticles up to 40% improved methyl orange adsorption; however, further addition of TiO_2 resulted in a diverse effect. This is related to the aggregation of the nanoparticles at high concentrations causing less favorable conditions for the adsorption. To examine the selectivity of the adsorbent for different ions, ionic solutions with equal concentrations of Ni^{2+} ions and Cu^{2+} were prepared. Adsorbents imprinted by Ni^{2+} could adsorb Ni^{2+} more effectively than Cu^{2+} ions. This result confirmed higher affinity and selectivity of the Ni(II)-imprinted membrane towards nickel than copper ions. The fast adsorption kinetic of nickel ions is ascribed to the higher complex formation potential and better memory of Ni(II)-imprinted adsorbent structure for Ni^{2+} ions.

13.4.4 Thin Membrane Adsorbents

Recently, thin-film membrane adsorbents have attracted researchers' attentions [65, 77]. Mass transport limitations are greatly reduced in the thin film adsorbents and thus, the membranes can support a fast removal kinetic. Blends of chitosan and poly(vinyl alcohol) were used by Cheng et al. [65], to synthesize a thin membrane adsorbent. The membrane thickness was adjusted between 10 and 200 μm to examine the effect of membrane thickness on adsorption property. They demonstrated that adsorption is improved with the decrease of the membrane thickness. This may be connected to the mass transport limitations in thick films and lesser active sites distributed on the surface of the thicker membrane. Adsorption kinetic of copper ions onto modified thin-film membrane adsorbents made of chitosan was studied by Cestari et al. [86]. Their results suggested that higher adsorption kinetic was achievable by reducing the membrane thickness. Generally, thin membrane adsorbents suffer from lack of adequate mechanical stability, especially when they are used in dynamic processes like in membrane stack cells. Mechanical reinforcement is of great importance for the preparation of thin membrane adsorbents, since thick structure of the membranes cannot tolerate applied pressures in dynamic modes. Salehi et al. [87] applied aminated multi-walled carbon nanotubes (AMWCNTs) in chitosan/poly(vinyl alcohol) blend to elevate the thin membrane adsorbent properties. Adsorption performance (for copper adsorption) and mechanical tests indicated that both adsorption properties and mechanical stability were improved using the AMWCNTs. PEG insertion increased sorption capacity but the mechanical resistance was decreased due to the brittleness of the resultant porous structures [88]. The reinforced membranes containing optimized porogen and nanotube contents, put on proper fabric supports, were applied in dead-end adsorption/filtration cell with a desired performance, that is, satisfactory water flux and copper rejection.

13.4.5 Modeling Aspects

Thermodynamic, kinetic, and mathematical modeling are attractive aspects of membrane-adsorbent studies due to their prominent impacts on mechanism identification, process optimization, and scale up. Classical thermodynamics in conjunction with kinetic modeling are practical tools to obtain better insight into important issues such as thermo-sensitivity of adsorption, adsorption rate, equilibrium time, and the mechanisms controlling adsorption rate. Mathematical modeling is also of interest on account of its usefulness in determination of the mechanisms involved in adsorption. Moreover, a validated model decreases the volume of the experimental trails and dependent costs.

13.4.5.1 Thermodynamics

Adsorption thermodynamic is typically applied to measure the principle activation parameters that is, enthalpy, entropy, and Gibbs free energy changes. From the thermodynamic outlook, adsorption is a reaction whereupon adsorbates interact with the active sites distributed on the adsorbent surface. The principle criterion for adsorption equilibrium is that the Gibbs free energy change of the reaction be zero [89]:

$$\Delta G = \sum_{i} \nu_i \cdot \mu_i = 0 \tag{13.1}$$

where ν_i and μ_i are stoichiometric coefficient and chemical potential of the *i*th component involved in the reaction, respectively. Equation (1) is reduced to Equation (2), where equilibrium constant of the adsorption is included:

$$\Delta G^{\circ} = -RT \ln K \tag{13.2}$$

where K is the equilibrium constant of the reaction at temperature T and R is the universal gas constant. Zero-sign superscript is ascribed to the standard conditions at which the activation parameters are calculated. Different conditions may be defined as reference conditions for example, 1 bar and 0°C. Standard enthalpy change (ΔH°) and entropy change (ΔS°) can be calculated using Equation (3), the Van't Hoff equation [65, 89]:

$$\ln K = \frac{\Delta S^{\circ}}{R} - \frac{\Delta H^{\circ}}{RT} \tag{13.3}$$

Equilibrium constants can be obtained from the intercept of the linearized plots of $ln(Q_e/C_e)$ versus Q_e at different temperatures, where Q_e and C_e are equilibrium adsorption capacity and its corresponding equilibrium concentration in aqueous phase, respectively [6, 19]. Slope and intercept of the linear plot of $ln\ K$ versus $1/T$ in Equation (13.3), gives ΔH° and ΔS°, respectively. ΔG° values may be also evaluated using Equation (13.2) at various temperatures. ΔG° must be negative for a favorable and spontaneous adsorption process.

Adsorption of papain on reactive dye-ligand-incorporated membrane was thermodynamically modeled by Nie et al. [90]. Standard thermodynamic parameters for the adsorption, ΔG°, ΔH°, and ΔS° were evaluated as −12.58 kJ/mol, 17.81 kJ/mol, and 0.101 kJ/(mol K), at 298 K, respectively. The results showed an endothermic and spontaneous adsorption. Adsorption of different proteins such as lysozyme, hemoglobulins, has, and bovine serum albumin (BSA) on chitosan-based membrane adsorbents functionalized with Reactive Yellow-2 ligand was investigated by Arica et al. [91]. Negative ΔG° and positive ΔS° exhibited favorable protein adsorption on the membranes. Cestari et al. [92] fabricated and applied vanillin-modified chitosan membranes for adsorption of copper ions from water. They found that the adsorption of Cu(II) by the membrane adsorbents were greatly affected by the system temperature. The thermodynamic analysis also indicated that the adsorption of Cu(II) on the membranes is accomplished through ion-exchange and chemical bonding mechanisms based on the enthalpy and free energy change quantities.

13.4.5.2 Kinetics

Kinetics provides valuable understanding of adsorption mechanisms including mass transfer, diffusion, and surface reaction. The external mass transfer (or boundary layer diffusion), reaction on the external surface of the adsorbent, intra-particle diffusion, intra-adsorbent reaction are promising adsorption kinetics of an adsorption process [1, 89].

Ho et al. [93] indicated that adsorption rate varied inversely with the film thickness when the external mass transfer is the rate-controlling step. Cheung et al. [94] determined two distinct steps within the intra-particle diffusion plots, well-fitted to the membrane adsorption kinetic. The first region indicated a sharp adsorption rate on the external surface of the membrane. The plateau region is ascribed to the intra-particle diffusion, the rate-controlling mechanism dominated after completion of the external surface capacity.

Pseudo-first order and pseudo-second order models are the most prevalent kinetic models applied for the study of membrane adsorption rates [86–90]. The pseudo-first-order or the Lagergren model is written as follows:

$$\frac{dQ}{dt} = k_1(Q_e - Q) \tag{13.4}$$

and the pseudo-second-order model is formulated as:

$$\frac{dQ}{dt} = k_2(Q_e - Q)^2 \tag{13.5}$$

where Q and Q_e are adsorption at time t and at equilibrium, respectively. k_1 and k_2 are rate constants for the pseudo-first-order and pseudo-second-order models, respectively. It is obvious that the pseudo-second-order model gives an account of much rapid adsorption kinetic and higher affinity between adsorbent and adsorbates. Adsorption of Cu(II) ions on chitosan/poly(vinyl alcohol) membrane adsorbent was investigated by Cheng et al. [65]. They reported that the adsorption rate is better described by the pseudo-second-order model. The kinetic study also revealed a chemical adsorption through chelation/ion exchange mechanisms. External mass transfer and intra-particle diffusion were recognized as the rate-controlling steps in the light of the kinetic modeling and Fourier transform infrared analysis. Kinetic of copper adsorption on chitosan/cellulose acetate hollow fibers membrane adsorbents was investigated by Yang et al. [67]. Highly enhanced adsorption kinetic was observed with an adsorption equilibrium time in the range 20–70 minutes. He et al. [72] applied pseudo-first and pseudo-second-order kinetic models to describe Cu(II) adsorption on chitosan-coated nylon membrane adsorbents. Chen et al. [83] studied the kinetic of Cr(III) ions adsorption on Cr(III)-imprinted PVA/SA membrane adsorbents. The adsorption rate of imprinted Cr(III) ions on the imprinted membrane was larger when compared to the adsorption on the control (non-imprinted) membrane. This may be attributed to the imprinting process which increases the affinity of the target ions with the adsorption-recognition sites.

13.4.5.3 Mathematical Modeling

Mathematical modeling is applied to simulate a typical membrane transport behavior by using mass, momentum, and energy balance principles. Among various models presented in the literature, adsorption-transport models have been used to describe the transport behavior membrane adsorbents. These models apply convection, diffusion, and adsorption mechanisms in combination. A generic backbone of an adsorption-transport model is usually applied for the simulation of membrane-adsorption transport as follows [95–97]:

$$\varepsilon\frac{\partial C}{\partial t} + (1-\varepsilon)\cdot\rho_b\frac{\partial Q}{\partial t} = \varepsilon D_i\frac{\partial^2 C}{\partial x^2} - \varepsilon u_s\frac{\partial C}{\partial x} \tag{13.6}$$

where ε is the membrane bulk porosity, ρ_b is the bulk dry density of the membrane, C is the equilibrium concentration of the solute in the solution crossing through the membrane,

Q is the equilibrium adsorption of the solute, D_a is the diffusivity of the solute through the membrane, and x indicates the flow direction across the membrane.

A mathematical model comprising convection, diffusion, and adsorption described by the Langmuir equation was proposed by Guo et al. [53]. They obtained that axial dispersion may dominate over convection at low flow rates when thin membrane adsorbents were applied in dynamic mode. On the other side, majority of adsorbates passed through the membrane without binding as a result of slow adsorption kinetics at high flow rates. Nie et al. [62] modeled the adsorption of papain on the membrane adsorbents fabricated by immobilization of Cibacron blue ligands on cellulose membranes. Their analysis on a chromatography adsorption column filled with the membranes showed the significant effects of the membrane system such as pore size, pore-size distribution, and thickness on the modeling results. They found that without considering membrane characteristics, the model could not describe the separation performance properly. A novel mathematical model has been recently developed by Salehi and Madaeni [98] for describing adsorption/permeation of Ni(III) ions through a Ni(III)-imprinted membrane. The adsorption of nickel ions on the imprinted membrane was well represented by the Langmuir–Freundlich (or Sips) isotherm model. The final governing equation was obtained using the one-dimensional unsteady-state transport model coupled with the isotherm equation. The resultant model satisfactorily simulated the permeation of Ni(III) and cobalt ions through the membrane with slightly better estimations for the imprinted nickel ions.

13.4.6 Regeneration and Reuse

In practical viewpoint, regeneration capacity (regenerability) is one of the most crucial aspects of adsorption. Reusability is in direct connection with the adsorption process economy where the process cost may be enormously reduced by using an adsorbent with a favorable regeneration capacity. This factor also determines the adsorbent durability, that is, the stability of the adsorbent against chemical and mechanical deterioration in the adsorption process. Membrane adsorbents are tested in successive adsorption/regeneration cycles to measure their potential for regeneration and reuse.

Ethylenediaminetetraacetic acid (EDTA) has been utilized by Ghaee et al. [47] to desorb Cu(II) ions from the chitosan membrane adsorbent. Three successive adsorption/regeneration stages were applied to examine the membrane's regenerability and durability. The capacity of the membranes for adsorption of Cu(II) ions was slightly decreased after each regeneration step. This is ascribed to the 'hysteresis' phenomenon, that is, unavoidable sorption capacity loss due to several reasons including irreversible adsorption, decrease in the number of the active sites and adsorbent structure degradation [99]. Different eluants (extracting agents) namely, EDTA sodium salt, hydrochloric acid, and citric acid were employed to extract Cu(II) ions from the surface-activated nylon membrane adsorbents [72]. Among the tested eluting agents, Na_2EDTA showed superior performance in the extraction of the adsorbed heavy metals even at low concentration. No significant reduction was observed in the capacity of the membranes for Cu(II) ion adsorption after three adsorption/elution successive cycles. This fact indicates the potential of the composite membrane adsorbents for practical applications. Na_2EDTA and HCl eluants were also applied to extract copper ions from chitosan/cellulose acetate hollow-fiber membranes [67]. As before, Na_2EDTA showed better performance in the regeneration process compared to the acid solution. The researchers also observed that the desorption process was improved with a decrease in the eluting solution pH. Chen et al. [83] successfully regenerated Cr(III)-imprinted membranes using 0.1M HCl desorbing solution. They conducted a five-stage consecutive adsorption/regeneration experiment with a negligible adsorption capacity loss, less than 5%, after final stage.

13.5 Adsorption Post-treatment for Membranes

In post-treatment configuration, the adsorbent contactor is implemented as a subsequent to membrane filtration. A schematic of the post-treatment layout is illustrated in figure 13.4. Generally, granular-grade adsorbents (e.g., GAC) are used in post-treatment configuration due to inadequacy of the contactors filled with fine particles, that is, higher pressure drop, higher price, and probable exportation of fine particles to the finished water. The latter may be rather prevented by introducing a supplementary separation process after adsorbent contactor which unavoidably increases the hybrid process costs.

Activated carbons are commonly used as contactor-fillers applied in post-treatment configuration mainly for low price and high capacity for adsorption of a wide range of compounds [100, 101]. There are not many publications on the post-treatment configuration in the literature [102–104]. This may be due to couple of hypotheses, good establishment of the post-treatment arrangement, and/or being out of researchers' interest. BOC and DOC adsorption may be improved in the case of adsorption post-treatment due to the fact that low molecular weight NOMs are much better adsorbed by the adsorbents when their concentrations are less than 2 mg/L in the influent [104]. On the other hand, biological contamination of finished water is always a possible threat especially where the ACs work in biological mode [105]. This problem may be solved in some cases using disinfection methods like ozonation [104, 106]. Additional advantages and drawbacks of the post-treatment design are listed as follows:

Advantages:

- Hybridization of two well-established processes
- No requirement of aeration and its connected problems
- Increasing the operating life of adsorption system due to feeding with much purified influents
- Maintaining the separation efficiency in the case of membrane system failure.

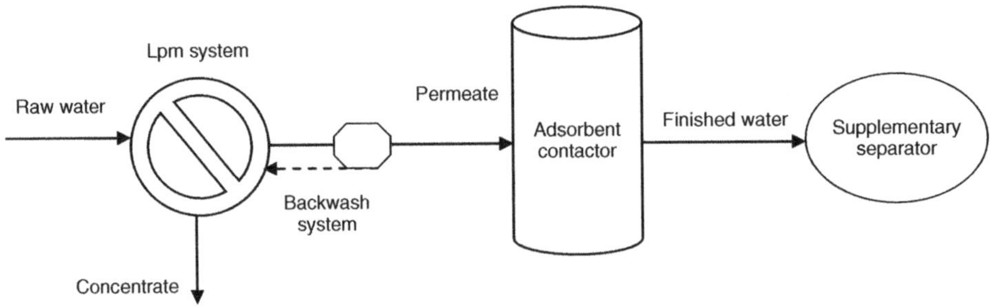

Figure 13.4 *Membrane/adsorption hybrid process with adsorption post-treatment.*

Drawbacks:

- Impossibility of the utilization of fine adsorbent particles
- Lower removal kinetic on account of using granular adsorbents
- Membrane fouling aggravation
- Larger footprint for implementation
- Risk of fine particle transfer to the finished water
- Hardship of adsorbent age control.

Finding the best alternative for the adsorption/membrane hybridization depends on many parameters such as effluent origin, type and dose of contaminants, feed flow rate, potential investment for the treatment, obligatory standard levels, and applications of the finished water. Literature review, consulting with experienced individuals and running preliminary experiments may help to choose the best alternative. Some alternatives are being industrially used and some others are under investigation and completion for the full-scale application. Typically, pre- and integrated hybridizations are superior candidates for application in drinking-water treatment where more stringent removal standards must be met. On the other hand, the applicability of IAMPs in both water and wastewater treatments is well approved by researchers; however, it is a necessity to reduce the overall costs (capital and operating costs) when these treatments are compared to available conventional processes such as ion exchangers and low-priced biosorbents.

References

1. Ruthven DW. (1984) *Principles of Adsorption and Adsorption Processes*, John Wiley & Sons, Ltd, New York.
2. Stoquart C, Servais P, Bérubé PR, *et al.* (2012) Hybrid Membrane Processes using activated carbon treatment for drinking water: A review. *J. Membr. Sci.*, **411–412**, 1–12.
3. Suzuki T, Watanabe Y, Ozawa G, *et al.* (1998) Removal of soluble organics and manganese by a hybrid MF hollow fiber membrane system. *Desalination*, **117**, 119–129.
4. Huang H, Schwab K and Jacangelo JG (2009) Pretreatment for low pressure membranes in water treatment: A review. *Environ. Sci. Technol.*, **43**, 3011–3019.
5. Imai A, Fukushima T, Matsushige K, *et al.* (2002) Characterization of dissolved organic matter in effluents from wastewater treatment plants. *Water Res.*, **36**(4), 859–870.
6. Gur-Reznik S, Katz I and Dosoretz CG (2008) Removal of dissolved organic matter by granular-activated carbon adsorption as a pretreatment to reverse osmosis of membrane bioreactor effluents. *Water Res.*, **42**, 1595–1605.
7. Jacangelo JG, DeMarco J, Owen DM, *et al.* (1995) Selected processes for removing NOM: an overview. *J. Am. Water Works Assoc.*, **87**(1), 64–77.
8. Shon HK, Kim SH, Vigneswaran S, *et al.* (2009) Physicochemical pretreatment of seawater: fouling reduction and membrane characterization. *Desalination*, **238**, 10–21.

9. Kweon JH, Hur HW, Seo GT, *et al.* (2009) Evaluation of coagulation and PAC adsorption pretreatments on membrane filtration for a surface water in Korea: A pilot study. *Desalination*, **249**, 212–216.
10. Shon HK, Vigneswaran S, Kim IS, *et al.* (2004) Effect of pretreatment on the fouling of membranes: Application in biologically treated sewage effluent. *J. Membr. Sci.*, **234**, 111–120.
11. Salehi E, Madaeni SS, Rajabi L, *et al.* (2013) Static and dynamic adsorption of copper ions on chitosan/polyvinyl alcohol thin adsorptive membranes: Combined effect of polyethylene glycol and aminated multi-walled carbon nanotubes. *Chem. Eng. J.*, **215**, 791–801.
12. Matsui Y, Hasegawa H, Ohno K, *et al.* (2009) Effects of super-powdered activated carbon pretreatment on coagulation and trans-membrane pressure buildup during microfiltration. *Water Res.*, **43**(4), 5160–5170.
13. Logan BE, Wu J and Unz RF (2001) Biological perchlorate reduction in high-salinity solutions. *Water Res.*, **35**, 3034–3038.
14. Chatterjee S and Woo SH (2009) The removal of nitrate from aqueous solutions by chitosan hydrogel beads. *J. Hazard. Mater.*, **164**, 1012–1018.
15. Bryjaka M, Wolskaa J and Kabay N (2008) Removal of boron from seawater by adsorption–membrane hybrid process: implementation and challenges. *Desalination*, **223**, 57–62.
16. Koltuniewicz AB, Witek A and Bezak K (2004) Efficiency of membrane-sorption integrated processes. *J. Membr. Sci.*, **239**, 129–141.
17. Glueckstern P and Priel M (2003) Optimization of boron removal in old and new SWFCO systems. *Desalination*, **156**, 219–228.
18. Saravia F and Frimmel FH (2008) Role of NOM in the performance of adsorption-membrane hybrid systems applied for the removal of pharmaceuticals. *Desalination*, **224**, 168–171.
19. Huq HP, Yang JS and Yang JW (2007) Removal of perchlorate from groundwater by the polyelectolyte-enhanced ultrafiltration process. *Desalination*, **204**, 335–343.
20. Xie Y, Li S, WuK, *et al.* (2011) A hybrid adsorption/ultrafiltration process for perchlorate removal. *J. Membr. Sci.*, **366**, 237–244.
21. Kabay N, Bryjak M, Schlosser S, *et al.* (2008) Adsorption-membrane filtration (AMF) hybrid process for boron removal from seawater: an overview. *Desalination*, **223**, 38–48.
22. Yilmaz İ, Kabay N, Brjyak M, *et al.* (2006) A submerged membrane–ion-exchange hybrid process for boron removal. *Desalination*, **198**, 310–315.
23. Pervov AG, Andrianov AP, Efremov RV, *et al.* (2003) A new solution for the Caspian Sea desalination: low-pressure membranes. *Desalination*, **157**, 377–384.
24. Guler E, Kabay N, Yuksel M, *et al.* (2011) Integrated solution for boron removal from seawater using RO process and sorption-membrane filtration hybrid method. *J. Membr. Sci.*, **375**, 249–257.
25. Yılmaz İ, Kabay N, Yüksel M, *et al.* (2012) Application of adsorption–ultrafiltration hybrid method for removal of phenol from water by hypercrosslinked polymer adsorbents. *Desalination*, **306**, 24–28.
26. Tansakul Ch, Laborie S and Cabassud C (2011) Adsorption combined with ultrafiltration to remove organic matter from seawater. *Water Res.*, **45**, 6362–6370.

27. Mozia S, Tomaszewska M and Morawski AW (2005) Studies on the effect of humic acids and phenol on adsorption–ultrafiltration process performance. *Water Res.*, **39**, 501–509.
28. Katsou E, Malamis S and Haralambous KJ (2011) Industrial wastewater pre-treatment for heavy metal reduction by employing a sorbent-assisted ultrafiltration system. *Chemosphere*, **82**, 557–564.
29. Ghosh R (2004) Separation of human albumin and IgG by a membrane-based integrated bioseparation technique involving simultaneous precipitation, microfiltration and membrane adsorption. *J. Membr. Sci.*, **237**, 109–117.
30. Judd S (2006) *The MBR book: Principles and Applications of Membrane Bioreactors in Water and Wastewater Treatment*, Elsevier Ltd, Oxford.
31. Yang W, Cicek N and Ilg J (2006) State-of-the-art of membrane bioreactors: Worldwide research and commercial applications in North America. *J. Membr. Sci.*, **207**, 201–211.
32. Cho BD and Fane AG (2002) Fouling transients in nominally sub-critical flux operation of a membrane bioreactor. *J. Membr. Sci.*, **209**, 391–403.
33. Pollice A, Brookes A, Jefferson B, *et al.* (2005) Sub-critical flux fouling in membrane bioreactors a review of recent literature. *Desalination*, **174**, 221–230.
34. Tian JY, Chen ZL, Liang H, *et al.* (2005) Comparison of biological activated carbon (BAC) and membrane bioreactor (MBR) for pollutants removal in drinking water treatment. *J. Water Supply Res. T.*, **60**, 1515–1523.
35. Seker S, Beyenal H and Tanyolac A (1995) The effects of biofilm density and substrate consumption rate in a differential fluidized bed biofilm reactor (DFBBR). *J. Biotechnol.*, **41**, 39–47.
36. Lesage N, Sperandio M and Cabassud C (2008) Study of a hybrid process: Adsorption on activated carbon/membrane bioreactor for the treatment of an industrial wastewater. *Chem. Eng. Pro.*, **47**, 303–307.
37. Tian JY, Chen ZL, Jun N, et al. (2010) Integrative membrane coagulation adsorption bioreactor (MCABR) for enhanced organic matter removal in drinking water treatment. *J. Membr. Sci.*, **352**, 205–212.
38. Guo W, Vigneswaran S and Ngo HH (2008) Comparison of the performance of submerged membrane bioreactor (SMBR) and submerged membrane adsorption bioreactor (SMABR). *Bioresource Technol.*, **99**, 1012–1017.
39. Khan MM, Takizawa S, Lewandowski Z, et al. (2013) Combined effects of EPS and HRT enhanced biofouling on a submerged and hybrid PAC-MF membrane bioreactor. *Water Res.*, **47**, 747–757.
40. Khirania S, Smith PJ, Manéro MH, *et al.* (eds) (2005) *Effect of periodic backwash in the submerged membrane adsorption hybrid system (SMAHS) for wastewater treatment*. Proceedings of the International Congress on Membranes and Membrane Processes (ICOM), August 21–26, 2005, Seoul, Korea.
41. Saravia F, Naab P and Frimmel FH (2006) Influence of particle size and particle size distribution on membrane-adsorption hybrid systems. *Desalination*, **200**, 446–448.
42. Vigneswaran S, Guo WS and Smith P (2007) Submerged membrane adsorption hybrid system (SMAHS): process control and optimization of operating parameters. *Desalination*, **202**, 392–399.

43. Tian JY, Liang H and Yang Y (2008) Membrane adsorption bioreactor (MABR) for treating slightly polluted surface water supplies: As compared to membrane bioreactor (MBR). *J. Membr. Sci.*, **325**, 262–270.
44. Kim J, Cai Z and Benjamin MM (2010) NOM fouling mechanisms in a hybrid adsorption/membrane system. *J. Membr. Sci.*, **349**, 35–43.
45. Vieira RS and Beppu MM (2005) Mercury ion recovery using natural and crosslinked chitosan membranes. *Adsorption*, **11**, 731–736.
46. Beppu MM, Arruda EJ, Vieira RS, *et al.* (2004) Adsorption of Cu(II) on porous chitosan membranes functionalized with histidine. *J. Membr. Sci.*, **240**, 227–235.
47. Ghaee A, Shariaty-Niassar M, Barzin J, *et al.* (2010) Effects of chitosan membrane morphology on copper ion adsorption. *Chem. Eng. J.*, **165**, 46–55.
48. Vieira RS, Guibal E, Silva EA, *et al.* (2007) Adsorption and desorption of binary mixtures of copper and mercury ions on natural and crosslinked chitosan membranes. *Adsorption*, **13**, 603–611.
49. Klein E (2000) Affinity membranes: a 10-year review. *J. Membr. Sci.*, **179**, 1–27.
50. Zeng X and Ruckenstein E (1999) Membrane chromatography: preparation and applications to protein separation, *Biotechnol. Prog.*, **15**, 1003–1019.
51. Weissenborn M, Hutter B and Singh M (1997) A study of combined filtration and adsorption on nylon–based dye–affinity membranes: separation of recombinant L–alanine dehydrogenase from crude fermentation broth. *Biotechnol. Appl. Biochem.*, **25**, 159–168.
52. Champluvier B and Kula MR (1991) Microfiltration membranes as pseudo–affinity adsorbents and comparison with gel beads. *J. Chromatogr.*, **539**, 315–325.
53. Guo W, Shang WZ, Yu Y, *et al.* (1994) Membrane affinity chromatography of alkaline phosphatase. *J. Chromatogr.*, **685**, 344–348.
54. Li Y and Spencer HG (1994) Dye–grafted poly(ethyleneimine)–coated, formed–in–place class affinity membranes for selective separation of proteins, *American Chem. Soc.*, 297–305.
55. Arica MY, Denizli A and Salih B (1997) Catalase adsorption onto Cibacron Blue F3GA and Fe(III) derivatized poly(2–hydroxyethyl methacrylate) membranes and application to a continuous system. *J. Membr. Sci.*, **129**, 65–76.
56. Schisla DK, Carr PW and Cussler EL (1995) Hollow-fiber array affinity chromatography. *Biotechnol. Prog.*, **11**, 651.
57. Suen SY, Chen RL and Tsai YD (2000) Comparison of lysozyme adsorption to immobilized Cibacron Blue F3GA using various membrane supports. *J. Liq. Chromatogr. Relat. Technol.*, **23**, 223–239.
58. Denizli A and Piskin E (2001) Dye–ligand affinity systems. *J. Biochem. Biophys. Methods*, **49**, 391–416.
59. Champluvier B and Kula MR (1992) Dye–ligand membranes as selective adsorbents for rapid purification of enzymes: a case study. *Biotechnol. Bioeng.*, **40**, 33–40.
60. Suen SY and Tsai YD (2000) Comparison of ligand density and protein adsorption on dye-affinity membranes using different spacer arms. *Sep. Sci. Technol.*, **35**, 69–87.
61. Kassab A, Yavuz H, Odabasi M, *et al.* (2000) Human serum albumin chromatography by Cibacron Blue F3GA–derived microporous polyamide hollow–fiber affinity membranes. *J. Chromatogr. B.*, **746**, 123–132.

62. Nie HL and Zhu LM (2007) Adsorption of papain with Cibacron Blue F3GA carrying chitosan-coated nylon affinity membranes. *Int. J. Biolog. Macromol.*, **40**, 261–267.
63. Pastircakova K (2004) Determination of trace metal concentrations in ashes from various biomass materials. *Energy Educ. Sci. Technol.*, **13**, 97–104.
64. Demirbas A, Pehlivan E, Gode F, *et al.* (2005) Adsorption behaviour of Cu(II), Zn(II), Ni(II), Pb(II), and Cd(II) from aqueous solution on Amberlite IR–120 synthetic resin. *J. Colloid Interface Sci.*, **282**, 20–25.
65. Cheng Z, Liu X, Han M, *et al.* (2010) Adsorption kinetic character of copper ions onto a modified chitosan transparent thin membrane from aqueous solution. *J. Hazard. Mater.*, **182**, 408–415.
66. Williams CJ, Aderhold D and Edyvean RGJ (1998) Comparison between biosorbents for the removal of metal ions from aqueous solutions. *Water Res.*, **32**, 216–224.
67. Yang L, Hsiao WW and Chen P (2002) Chitosan-cellulose composite membrane for affinity purification of biopolymers and immune adsorption. *J. Membr. Sci.*, **197**, 185–197.
68. Nasef MM and Yahaya AH (2009) Adsorption of some heavy metal ions from aqueous solutions on Nafion 117 membrane. *Desalination*, **249**, 677–681.
69. Zheng YM, Zou SW, Nadeeshani Nanayakkara KG, *et al.* (2011) Adsorptive removal of arsenic from aqueous solution by a PVDF/zirconia blend flat sheet membrane. *J. Membr. Sci.*, **374**, 1–11.
70. Liu C and Bai R (2006) Adsorptive removal of copper ions with highly porous chitosan/cellulose acetate blend hollow fiber membranes. *J. Membr. Sci.*, **284**, 313–322.
71. Vieira RS and Beppu MM (2006) Dynamic and static adsorption and desorption of Hg(II) ions on chitosan membranes and spheres. *Water. Res.*, **40**, 1726–1734.
72. He ZY, Nie HL, Branford–White C, *et al.* (2008) Removal of Cu^{2+} from aqueous solution by adsorption onto a novel activated nylon–based membrane. *Bioresource Technol.*, **99**, 7954–7958.
73. Genc O, Arpa C, Bayramoglu G, *et al.* (2002) Selective recovery of mercury by Procion Brown MX 5BR immobilized poly (hydroxyethylmethacrylate/chitosan) composite membranes. *Hydrometallurgy*, **67**, 53–62.
74. Medina-Gonzalez Y and Remigy JC (2011) Sonication-assisted preparation of pristine MWCNT–polysulfone conductive microporous membranes. *Mater. Lett.*, **65**, 229–232
75. Gumusderelioglu M and Agi P (2004) Adsorption of concanavalin A on the well–characterized macroporous chitosan and chitin membranes. *React. Funct. Polym.*, **61**, 211–220
76. Santos DES, Neto CGT, Fonseca JLC, *et al.* (2008) Chitosan macroporous asymmetric membranes—Preparation, characterization and transport of drugs. *J. Membr, Sci.*, **325**, 362–370
77. Zeng M, Fang Z and Xu Ch (2004) Effect of compatibility on the structure of the microporous membrane prepared by selective dissolution of chitosan/synthetic polymer blend membrane. *J. Membr. Sci.*, **230**, 175–181.
78. Wang X, Zhang L, Ma C, *et al.* (2009) Enrichment and separation of silver from waste solutions by metal ion imprinted membrane. *Hydrometallurgy*, **100**, 82–86.

79. Zhai Y, Liu Y, Chang X, *et al.* (2009) Metal ion–small molecule complex imprinted polymer membranes: Preparation and separation characteristics. *Reac. Funct. Polym.*, **68**, 284–291.
80. An F and Gao B (2009) Adsorption characteristics of Cr(III) ionic imprinting polyamine on silica gel surface. *Desalination*, **249**, 1390–1396.
81. Su H, Li J and Tan T (2008) Adsorption mechanism for imprinted ion (Ni^{2+}) of the surface molecular imprinting adsorbent (SMIA). *Biochem. Eng. J.*, **39**, 503–509.
82. Vatanpour V, Madaeni SS, Zinadini S, *et al.* (2011) Development of ion imprinted technique for designing nickel ion selective membrane. *J. Membr. Sci.*, **373**, 36–42.
83. Chen JH, Li GP, Liu QL, *et al.* (2010) Cr(III) ionic imprinted polyvinyl alcohol/sodium alginate (PVA/SA) porous composite membranes for selective adsorption of Cr(III) ions. *Chem. Eng. J.*, **165**, 465–473.
84. Sergeyeva TA, Brovko OO, Piletska EV, *et al.* (2007) Porous molecularly imprinted polymer membranes and polymeric particles. *Analytica Chimica Acta*, **582**, 311–319.
85. Li Q, Su H and Tan T (2008) Synthesis of ion-imprinted chitosan-TiO_2 adsorbent and its multi-functional performances. *Biochem. Eng. J.*, **38**, 212–218.
86. Cestari AR, Vieira EFS, Matos JDS, *et al.* (2005) Determination of kinetic parameters of Cu(II) interaction with chemically modified thin chitosan membranes. *J. Colloid Interf. Sci.*, **285**, 288–295.
87. Salehi E, Madaeni SS, Rajabi L, *et al.* (2012) Novel chitosan/poly(vinyl) alcohol thin adsorptive membrane modified with amino functionalized multi-walled carbon nanotubes for Cu(II) removal from water: Preparation, characterization, adsorption kinetics and thermodynamics. *Sep. Purif. Technol.*, **89**, 309–319.
88. Salehi E, Madaeni SS, Rajabi L, *et al.* (2012) Static and dynamic adsorption of copper ions on chitosan/polyvinyl alcohol thin adsorptive membranes: Combined effect of polyethylene glycol and functionalized multi–walled carbon nanotubes. *Chem. Eng. J.*, **215**, 791–801.
89. Hill CG (2003) *An Introduction to Chemical Engineering Kinetics & Reactor Design*, 5th edition, John Wiley & Sons, Ltd, New York.
90. Nie HL, Chen TX and Zhu LM (2007) Adsorption of papain on dye affinity membranes: Isotherm, kinetic, and thermodynamic analysis. *Sep. Purif. Technol.*, **57**, 121–125.
91. Arica MY, Yilmaz M, Yalçin E, *et al.* (2004) Surface properties of Reactive Yellow 2 immobilised pHEMA and HEMA/chitosan membranes: characterisation of their selectivity to different proteins. *J. Membr. Sci.*, **240**, 167–178.
92. Cestari AR, Vieira EFS and Mattos CRS (2006) Thermodynamics of the Cu(II) adsorption on thin vanillin-modified chitosan membranes. *J. Chem. Thermodynamics*, **38**, 1092–1099.
93. Ho YS, Ng JCY and McKay G (2000) Kinetics of pollutant sorption by biosorbents: review. *Sep. Purif. Methods*, **29**, 189–232.
94. Cheung WH, Szeto YS and McKay G (2007) Intra particle diffusion processes during acid dye adsorption onto chitosan. *Bioresour. Technol.*, **98**, 2897–2904.
95. Yates M and Yates S (1988) Modeling microbial fate in the subsurface environment. *Crit. Rev. Environ. Contr.*, **17**, 307–344.

96. Madaeni SS and Salehi E (2009) Adsorption–transport modeling of anions through PVD membrane in the presence of the screen phenomenon. *App. Surf. Sci.*, **255**, 3523–3529.
97. Madaeni SS and Salehi E (2009) A new adsorption–transport and porosity combined model for passage of cations through nanofiltration membrane. *J. Membr. Sci.*, **333**, 100–109.
98. Madaeni SS, Salehi E and Vatanpour V (2012) Thermodynamic investigation and mathematical modeling of ion-imprinted membrane adsorption. *J. Membr. Sci.*, **389**, 334–342.
99. Madaeni SS, Zinadini S and Vatanpour V (2012) Convective flow adsorption of nickel ions in PVDF membrane embedded with multi–walled carbon nanotubes and PAA coating. *Sep. Purif. Technol.*, **80**, 155–162.
100. Hargesheimer EE and Watson SB (1996) Drinking water treatment options for taste and odor control. *Water Res.*, **30**, 1423–1430.
101. Najm IN, Snoeyink VL, Suidan MT, *et al.* (1990) Effect of particle size and background natural organics on the adsorption efficiency of PAC. *J. Am. Water Works Assoc.*, **82**, 65–72.
102. Andersson A, Laurent P, Kihn A, *et al.* (2001) Impact of temperature on nitrification in biological activated carbon (BAC) filters used for drinking water treatment. *Water Res.*, **35**, 2923–2934.
103. Sartor M, Schlichter B, Gatjal H, *et al.* (2008) Demonstration of a new hybrid process for the decentralised drinking and service water production from surface water in Thailand. *Desalination*, **222**, 528–540.
104. Schlichter B, Mavrov V and Chmiel H (2004) Study of a hybrid process combining ozonation and membrane filtration – filtration of model solutions. *Desalination*, **156**, 257–265.
105. LeChevallier MW, Becker WC, Schorr P, *et al.* (1992) Evaluating the performance of biologically active rapid filters. *J. Am. Water Works Assoc.*, **84**, 136 146.
106. Niquette P, Hausler R and Lahaye P (2007) An innovative process for the treatment of high loaded surface waters for small communities. *Environ. Eng. Sci.*, **6**, 139–145.

Index

Note: Page numbers in bold italic refer to illustrations.

Integrated Membrane Systems and Processes, First Edition. Edited by Angelo Basile and Catherine Charcosset.

Printed and bound by CPI Group (UK) Ltd, Croydon, CR0 4YY

17/09/2026

14964302-0004